Bioethanol Fuel Production Processes. I

This book presents research on biomass pretreatments, which are a fundamental part of bioethanol fuel production to make biomass more accessible. This book also includes an introductory section on bioethanol fuels.

Bioethanol Fuel Production Processes. I: Biomass Pretreatments is the first volume in the ***Handbook of Bioethanol Fuels*** (Six-Volume Set). The primary pretreatments at the macro level are the biological chemical, hydrothermal, and mechanical pretreatments of the biomass. It also has an introductory section on the biomass pretreatments at large for bioethanol fuel production.

The major pretreatments at the micro level are the enzymatic and fungal pretreatments of the biomass as the biological pretreatments; acid, alkaline, ionic liquid, and organic solvent pretreatments of the biomass as the chemical pretreatments; steam explosion and liquid hot water pretreatments of the biomass as the hydrothermal pretreatments; and milling, ultrasonic, and microwave pretreatments of the biomass as the mechanical pretreatments.

The first volume also indicates that a wide range of pretreatments stand alone or in combination with each other to fractionate the biomass to its constituents of cellulose, lignin, and hemicellulose and improve both sugar and bioethanol fuel yield, making this bioethanol fuel more competitive in relation to crude oil- and natural gas-based fossil fuels.

This first volume is a valuable resource for stakeholders primarily in the research fields of energy and fuels, chemical engineering, environmental science and engineering, biotechnology, microbiology, chemistry, physics, mechanical engineering, agricultural sciences, food science and engineering, materials science, biochemistry, genetics, molecular biology, plant sciences, water resources, economics, business, management, transportation science and technology, ecology, public, environmental and occupational health, social sciences, toxicology, multidisciplinary sciences, and humanities among others.

Bioethanol Fuel Production Processes. I

Biomass Pretreatments

Edited by
Ozcan Konur

CRC Press
Taylor & Francis Group
Boca Raton London New York

CRC Press is an imprint of the
Taylor & Francis Group, an **informa** business

Designed cover image: © Shutterstock

First edition published 2024
by CRC Press
2385 NW Executive Center Drive, Suite 320, Boca Raton FL 33431

and by CRC Press
4 Park Square, Milton Park, Abingdon, Oxon, OX14 4RN

CRC Press is an imprint of Taylor & Francis Group, LLC

ISBN: 978-1-032-12751-4 (hbk)
ISBN: 978-1-032-12862-7 (pbk)
ISBN: 978-1-003-22653-6 (ebk)

DOI: 10.1201/9781003226536

Typeset in Times
by codeMantra

Contents

Preface

The recent supply shocks caused first by the COVID-19 pandemic and later by the Ukrainian war have shown that biofuels such as bioethanol, biohydrogen, biogas, biosyngas, and biodiesel fuels could play a vital role to maintain the energy security and indirectly food security at the global scale. These shocks have also resulted in the need for further setup of the incentive structures for the production and consumption of bioethanol fuels in blends with crude oil-based gasoline, petrodiesel, or liquefied natural gas (LNG) in gasoline and diesel engines and their direct utilization in the direct ethanol fuel cells and in the production of biohydrogen fuels for the fuel cells and valuable biochemicals from bioethanol fuels.

Thus, it is essential to assess the research on the production, evaluation, and utilization of bioethanol fuels from a wide range of biomass including the first generation starch and sugar feedstocks, wood, grass, and second generation lignocellulosic biomass including waste biomass and agricultural residues such as starch feedstock residues and sugar feedstock residues, and the third generation algal biomass.

Thus, this six-volume *Handbook of Bioethanol Fuels* assesses the research on the production, evaluation, and utilization of bioethanol fuels and presents a representative sample of this interdisciplinary research population with a sample of 110 chapters (Table 1.1).

The first two volumes provide an overview of the research on the fundamental processes for the bioethanol fuel production with a sample of 39 chapters: Pretreatments of the biomass, hydrolysis of the pretreated biomass, microbial fermentation of hydrolysates with yeasts, and separation and distillation of bioethanol fuels from the fermentation broth. They also provide an overview of the research on bioethanol fuels and production processes for bioethanol fuels (Tables 1.2 and 1.3).

The third and fourth volumes provide an overview of research on the production of bioethanol fuels from the non-waste and waste biomass, respectively, with a sample of 36 chapters. In this context, the third volume covers the production of bioethanol fuels from first generation starch feedstocks and sugar feedstocks, grass biomass, wood biomass, cellulose, biosyngas, and third generation algae (Table 1.4), while the fourth volume covers the production of second generation bioethanol fuels from residual sugar feedstocks, residual starch feedstocks, food waste, industrial waste, urban waste, forestry waste, and lignocellulosic biomass at large (Table 1.5). They also provide an overview of the research on the feedstock-based bioethanol fuels, non-waste feedstock-based bioethanol fuels, and second generation waste biomass-based bioethanol fuels (Tables 1.4 and 1.5).

Finally, the fifth and sixth volumes provide an overview of the research on the evaluation and utilization of bioethanol fuels with a sample of 37 chapters. In this context, the fifth volume covers the evaluation and utilization of bioethanol fuels in general, gasoline fuels, nanotechnology applications in bioethanol fuels, utilization of bioethanol fuels in transport engines, evaluation of bioethanol fuels, utilization of bioethanol fuels, and development and utilization of bioethanol fuel sensors (Table 1.6).

Thus, the sixth volume of this handbook provides an overview of the research on the country-based experience of bioethanol fuels at large, Chinese, the US, and European experience of bioethanol fuels, production of bioethanol fuel-based biohydrogen fuels for fuel cells, bioethanol fuel cells, and bioethanol fuel-based biochemicals with a sample of 19 chapters (Table 1.7).

Thus, the first volume of this handbook provides an overview of the research on the production, evaluation, and utilization of bioethanol fuels in general as an introduction to bioethanol fuels in the first part with three book chapters and provides an overview of the research on the pretreatments of the biomass for the sugar and bioethanol fuel production with 15 book chapters (Table 1.2). In this context, there are five key sections for the second part: Introduction to the pretreatments of the biomass, biological pretreatments of the biomass, chemical pretreatments of the biomass, hydrothermal pretreatments of the biomass, and mechanical pretreatment of the biomass.

Hence, the first volume indicates that bioethanol fuel research has intensified in recent years to become a major part of the bioenergy and biofuels research together primarily with biodiesel, biohydrogen, and biogas research as a sustainable alternative to crude oil-based gasoline and petrodiesel fuels as well as LNG.

The first volume also indicates that a wide range of pretreatments stand alone or in combination with each other fractionate the biomass to its constituents of cellulose, lignin, and hemicellulose and improve both sugar and bioethanol fuel yield, making this bioethanol fuel more competitive in relation to crude oil- and natural gas-based fuels.

The first volume also indicates that a small number of documents, authors, institutions, publication years, source titles, countries, Scopus subject categories, Scopus keywords, and research fronts have shaped the research on bioethanol fuels at large and on the pretreatments of the biomass for sugar and bioethanol fuel production.

The first volume also indicates that the level of funding for the research on both bioethanol fuels at large and on the pretreatments of the biomass for sugar and bioethanol fuel production in particular has not been sufficient with the resulting loss of momentum in the research output in recent years. Thus, there is a crucial need to improve the incentive structures for the major stakeholders such as researchers and their institutions as well source titles and academic databases to improve the volume and quality of the research output in these fields. This is a crucial need to maintain the energy security and food security indirectly at a global scale in the light of the recent supply shocks caused by the COVID-19 pandemic and the Ukrainian war.

The first volume also indicates that the contribution of social sciences and humanities to the research in these fields have been minimal, in part due to the restrictive editorial policies of the source titles in these fields towards social science- and humanities-based interdisciplinary studies. Thus, there is ample room to improve incentive structures for the inclusion of social sciences and humanities into these fields.

The first volume also indicates that China, Europe as a whole, and the USA have been major producers of the research in these research fields, and there has been heavy competition among them in terms of both volume and citation impact of the research output. The USA and Europe as a whole have had a higher citation impact in relation to China benefiting from their first-mover advantage starting their research in the 1970s. China as a late mover have had more intensive research funding initiatives in relation to the USA and Europe, improving its both research output and citation impact through the provision of the efficient incentive structures for its major stakeholders in the last two decades. In this way, China might also overtake both the USA and Europe in terms of citation impact of the research output in addition to the volume of the research output in the future.

This handbook at large and its first volume are a valuable resource for the stakeholders primarily in the research fields of energy and fuels, chemical engineering, environmental science and engineering, biotechnology, microbiology, chemistry, physics, mechanical engineering, agricultural sciences, food science and engineering, materials science, biochemistry, genetics, molecular biology, plant sciences, water resources, economics, business, management, transportation science and technology, ecology, public, environmental and occupational health, social sciences, toxicology, multidisciplinary sciences, and humanities among others.

Ozcan Konur

Acknowledgments

This handbook has been a multi-stakeholder project from its conception to its publication. CRC Press and Taylor and Francis Group have been the major stakeholders in financing and executing it. Marc Gutierrez has been the executive director of the project. A large number of teams from the Publisher have contributed immensely to the production of the handbook. Only a limited number of authors have participated in this project due to the low level of incentives, compared to journals. A small number of highly cited scholars has shaped the research on bioethanol fuels. The contribution of all these and other stakeholders to this handbook has been greatly acknowledged.

Editor

The Editor has interdisciplinary research interests and has published primarily in the areas of bioenergy and biofuels, algal bioenergy and biofuels, nanoenergy and nanofuels, nanobiomedicine, algal biomedicine, disability studies, higher education, biodiesel fuels, algal biomass, lignocellulosic biomass, scientometrics, and bioethanol fuels. He has edited a book titled *Bioenergy and Biofuels* (CRC Press, 2018), a handbook titled *Handbook of Algal Science, Technology, and Medicine* (Elsevier, 2020), and a handbook titled *Handbook of Biodiesel and Petrodiesel Fuels: Science, Technology, Health, and Environment* (CRC Press, 2021) in three volumes.

Contributors

Richa Arora
Department of Microbiology
Punjab Agricultural University
Ludhiana, India

Nor Azrini Nadiha Azmi
International Institute for Halal Research and Training (INHART)
International Islamic University Malaysia
Kuala Lumpur, Malaysia

Amal A.M. Elgharbawy
International Institute for Halal Research and Training (INHART)
and
Bioenvironmental Engineering Research Center (BERC), Department of Biotechnology Engineering, Faculty of Engineering
International Islamic University Malaysia
Kuala Lumpur, Malaysia

Anh Tuan Hoang
Institute of Engineering
HUTECH University
Ho Chi Minh City, Vietnam

Amit K. Jaiswal
School of Food Science and Environmental Health, Faculty of Science
and
Environmental Sustainability and Health Institute (ESHI)
Technological University Dublin
Dublin, Ireland

Swarna Jaiswal
School of Food Science and Environmental Health, Faculty of Science
and
Environmental Sustainability and Health Institute (ESHI)
Technological University Dublin
Dublin, Ireland

Ozcan Konur
Department of Materials Engineering
(Formerly) Ankara Yildirim Beyazit University
Ankara, Turkey

Ajay Kumar
Department of Industrial Microbiology, Jacob Institute of Biotechnology and Bioengineering
Sam Higginbottom University of Agriculture, Technology and Sciences
Allahabad, India

Sneha S. Mathew
School of Energy Materials
Mahatma Gandhi University
Kottayam, India

Muhammad Moniruzzaman
Department of Chemical Engineering
and
Centre of Research in Ionic Liquids
Universiti Teknologi PETRONAS
Seri Iskandar, Malaysia

Dileswar Pradhan
School of Food Science and Environmental Health, Faculty of Science
and
Environmental Sustainability and Health Institute (ESHI)
Technological University Dublin
Dublin, Ireland

Amit Pratush
Arboreal Bio-Innovations Pvt. Ltd.
Lucknow, India

Sonali Thakur
Department of Microbiology, School of Bioengineering and Biosciences
Lovely Professional University
Phagwara, India

Bahiru Tsegaye
Department of Food Process Engineering, College of Engineering and Technology
Wolkite University
Welkite, Ethiopia

Lei Wang
Key Laboratory of Coastal Environment and Resources of Zhejiang Province, School of Engineering
Westlake University
Zhejiang, China
and
Institute of Advanced Technology
Westlake Institute for Advanced Study
Zhejiang, China

Yi Zhang
College of Environmental and Resource Sciences
Zhejiang University
Zhejiang, China
and
Key Laboratory of Coastal Environment and Resources of Zhejiang Province, School of Engineering
Westlake University
Zhejiang, China

Part 1

Introduction to Bioethanol Fuels

1 Introduction to Bioethanol Fuels

Ozcan Konur
(Formerly) Ankara Yildirim Beyazit University

1.1 INTRODUCTION

Crude oil-based gasoline fuels (Ma et al., 2002; Newman and Kenworthy, 1989) have been widely used in the transportation sector since the 1920s. However, there have been great public concerns over the adverse environmental and human impact of these fuels (Hill et al., 2006, 2009). Hence, biomass-based bioethanol fuels (Hill et al., 2006; Konur, 2012, 2015, 2019, 2020) have increasingly been used in blending gasoline and petrodiesel fuels (Hsieh et al., 2002; Najafi et al., 2009), in fuel cells (Antolini, 2007, 2009), and in biochemical production (Angelici et al., 2013; Morschbacker, 2009) in a biorefinery context (Fernando et al., 2006; Huang et al., 2008).

However, it is necessary to pretreat the biomass (Alvira et al., 2010; Taherzadeh and Karimi, 2008) to enhance the yield of bioethanol (Hahn-Hagerdal et al., 2006; Sanchez and Cardona, 2008) prior to the bioethanol fuel production from feedstocks through hydrolysis (Sun and Cheng, 2002; Taherzadeh and Karimi, 2007) and fermentation (Lin and Tanaka, 2006; Olsson and Hahn-Hagerdal, 1996) of biomass and hydrolysates, respectively.

Research in the field of bioethanol fuels has intensified in this context in key research fronts of the production of bioethanol fuels and to a lesser extent utilization (Antolini, 2007; Hansen et al., 2005; Haryanto et al., 2005) and evaluation (Hamelinck et al., 2005; Pimentel and Patzek, 2005; Schmer et al., 2008) of bioethanol fuels. For the first research front, pretreatment (Alvira et al., 2010; Hendriks and Zeeman, 2009; Mosier et al., 2005) and hydrolysis (Alvira et al., 2010; Sun and Cheng, 2002; Zhang and Lynd, 2004) of feedstocks, fermentation of hydrolysates (Jonsson and Martin, 2016; Lin and Tanaka, 2006; Palmqvist and Hahn-Hagerdal, 2000a, b), production of bioethanol fuels (Balat, 2011; Limayem and Ricke, 2012; Lin and Tanaka, 2006), and to a lesser extent distillation of bioethanol fuels (Sano et al., 1994; Vane, 2005) are the key research areas while for the second research front, utilization of bioethanol fuels in fuel cells (Antolini, 2007, 2009; Kamarudin et al., 2013) and transport engines (Hansen et al., 2005; Niven, 2005; von Blottnitz and Curran, 2007), bioethanol-based biohydrogen fuels (Haryanto et al., 2005; Ni et al., 2007; Mattos et al., 2012), bioethanol sensors (Liu et al., 2005; Wan et al., 2004; Wang et al., 2012), and bioethanol-based biochemicals (Angelici et al., 2013; Morschbacker, 2009; Zhang and Yu, 2013) are the key research areas.

Research in this field has also intensified for feedstocks of lignocellulosic biomass at large (Hendriks and Zeeman, 2009; Mosier et al., 2005; Sun and Cheng, 2002), bioethanol fuels for their utilization (Antolini, 2007; Hansen et al., 2005; Haryanto et al., 2005), cellulose (Pinkert et al., 2009; Zhang and Lynd, 2004), starch feedstock residues such as corn stover or wheat straw (Binod et al., 2010; Talebnia et al., 2010), wood biomass such as pine or oak (Galbe and Zacchi, 2002; Zhu and Pan, 2010), and to a lesser extent industrial waste such as glycerol or vinasses (Cardona et al., 2010; Prasad et al., 2007), lignin (Bourbonnais and Paice, 1990; Kirk and Farrell, 1987), grass biomass such as switchgrass and miscanthus (Keshwani and Cheng, 2009; Pimentel and Patzek, 2005), hydrolysates (Palmqvist et al., 2000a, b), biomass at large (Lin and Tanaka, 2006; Zabed et al., 2017), sugar feedstocks such as sugarcane and sugar beet (Bai et al., 2008; Canilha et al., 2012), sugar feedstock residues such as sugarcane bagasse and straw (Cardona et al., 2010; Laser et al., 2002), algal biomass such as chlamydomonas (Ho et al., 2013; John et al., 2011), lignocellulosic wastes such as food wastes (Sanchez, 2009; Saini et al., 2015), starch feedstocks such as corn and

DOI: 10.1201/9781003226536-2

wheat (Bai et al., 2008; Bothast and Schlicher, 2005), urban wastes such as kitchen waste and used furniture (Prasad et al., 2007; Ravindran and Jaiswal, 2016), forestry wastes such as tree branches (Duff and Murray, 1996), food wastes such as potato peels (Guimaraes et al., 2010; Ravindran and Jaiswal, 2016), biosyngas (Henstra et al., 2007; Munasinghe and Khanal, 2010), and plants such as water hyacinth (Nigam, 2002; Sukumaran et al., 2009).

However, it is essential to develop efficient incentive structures (North, 1991) for the primary stakeholders to enhance research in this field (Konur, 2000, 2002a–c, 2006a, b, 2007a, b).

The recent supply shocks caused first by the COVID-19 pandemic and later by the Ukrainian war have shown that biofuels such as bioethanol, biohydrogen, biogas including biomethane, biosyngas, and biodiesel fuels could play a vital role to maintain the energy security at the global scale. These shocks have also resulted in the need for further setup of the incentive structures for the production and consumption of bioethanol fuels in blends with crude oil-based gasoline, petrodiesel, or liquefied natural gas (LNG) in gasoline and diesel engines, their direct utilization in direct ethanol fuel cells (DEFCs) and in the production of biohydrogen fuels for fuel cells, and valuable biochemicals from bioethanol fuels.

Thus, it is essential to assess the research on the production, evaluation, and utilization of bioethanol fuels and their derivatives and coproducts from a wide range of biomass including the first generation starch and sugar feedstocks, wood biomass, grass biomass, second generation lignocellulosic biomass including waste biomass and agricultural residues such as starch feedstock residues and sugar feedstock residues, and the third generation algal biomass.

Thus, this six-volume *Handbook of Bioethanol Fuels* (Konur, 2023cs) assesses the research on the production, evaluation, and utilization of bioethanol fuels and their derivatives and coproducts and presents a representative sample of this interdisciplinary research population with a collection of 110 chapters (Table 1.1).

TABLE 1.1
The Macroresearch Fronts for the Field of Bioethanol Fuels

No.	Research Fronts	No. Reviews
1.	**Bioethanol Fuel Production Processes. I: Biomass Pretreatments**	**18**
1.1.	Introduction to bioethanol fuels	3
1.2.	Biomass pretreatments	15
2	**Bioethanol Fuel Production Processes. II: Biomass Hydrolysis, Fermentation, and Bioethanol Fuel Separation**	**21**
2.1.	Introduction to production processes of bioethanol fuels	2
2.2.	Biomass hydrolysis	8
2.3.	Hydrolysate fermentation	8
2.4	Bioethanol fuel separation	3
3	**Feedstock-based Bioethanol Fuels. I. Non-Waste Feedstocks**	**19**
3.1.	Introduction to feedstock-based bioethanol fuels	2
3.2.	Introduction to non-waste feedstock-based bioethanol fuels	2
3.3.	First generation starch-based bioethanol fuels	3
3.4.	First generation sugar feedstock-based bioethanol fuels	2

(*Continued*)

TABLE 1.1 (*Continued*)
The Macroresearch Fronts for the Field of Bioethanol Fuels

No.	Research Fronts	No. Reviews
3.5.	Grass-based bioethanol fuels	2
3.6.	Wood-based bioethanol fuels	2
3.7.	Cellulose-based bioethanol fuels	2
3.8.	Third generation algal bioethanol fuels	2
3.9.	Biosyngas-based bioethanol fuels	2
4	**Feedstock-based Bioethanol Fuels. II. Waste Feedstocks**	**17**
4.1.	Introduction to second generation waste biomass-based bioethanol fuels	2
4.2.	Second generation bioethanol fuels from residual sugar feedstocks	2
4.3.	Second generation bioethanol fuels from residual starch feedstocks	2
4.4.	Second generation food waste-based bioethanol fuels	2
4.5.	Second generation industrial waste-based bioethanol fuels	2
4.6.	Second generation urban waste-based bioethanol fuels	2
4.7.	Second generation forestry waste-based bioethanol fuels	2
4.8.	Lignocellulosic biomass-based bioethanol fuels	3
5	**Evaluation and Utilization of Bioethanol Fuels. I: Evaluation of Bioethanol Fuels, Transport Engines, and Bioethanol Sensors**	**18**
5.1.	Introduction to the evaluation and utilization of bioethanol fuels	8
5.2.	Utilization of bioethanol fuels in transport engines	2
5.3.	Evaluation of bioethanol fuels	4
5.4.	Utilization of bioethanol fuels	2
5.5.	Bioethanol fuel sensors	2
6	**Evaluation and Utilization of Bioethanol Fuels. II: Biohydrogen Fuels, Fuel Cells, Biochemicals, and Country Experiences**	**17**
6.1.	Country-based experience of bioethanol fuels	9
6.2.	Bioethanol fuel-based biohydrogen fuels	3
6.3.	Bioethanol fuel cells	2
6.4.	Bioethanol fuel-based biochemical	3
	In total	**110**

The first two volumes provide an overview of the research on the fundamental processes for the bioethanol fuel production with a collection of 39 chapters (Konur, 2023ct, cu): pretreatments of the biomass, hydrolysis of the pretreated biomass, microbial fermentation of hydrolysates with yeasts, and separation and distillation of bioethanol fuels from the fermentation broth. They also provide an overview of the research on bioethanol fuels and production processes for bioethanol fuels (Tables 1.2 and 1.3).

The third and fourth volumes provide an overview of research on the production of bioethanol fuels from the non-waste and waste biomass, respectively, with a collection of 36 chapters (Konur, 2023cv, cw). In this context, the third volume covers the production of bioethanol fuels from first generation starch feedstocks and sugar feedstocks, grass biomass, wood biomass, cellulose, biosyngas, and third generation algal biomass (Table 1.4) while the fourth volume covers the production of second generation bioethanol fuels from residual sugar feedstocks, residual starch feedstocks, food waste, industrial waste, urban waste, forestry waste, and lignocellulosic biomass at large (Table 1.5). They also provide an overview of the research on the feedstock-based bioethanol fuels, non-waste feedstock-based bioethanol fuels, and second generation waste biomass-based bioethanol fuels (Tables 1.4 and 1.5).

Finally, the fifth and sixth volumes provide an overview of the research on the evaluation and utilization of bioethanol fuels with a collection of 35 chapters (Konur, 2023cx, cy). In this context, the fifth volume covers the evaluation and utilization of bioethanol fuels in general, gasoline fuels, application of nanotechnology in bioethanol fuels, utilization of bioethanol fuels in transport engines, and development and utilization of bioethanol fuel sensors (Table 1.6). Further, the sixth volume of this handbook provides an overview of the research on the country-based experience of bioethanol fuels in China, the US, and Europe, bioethanol fuel-based biohydrogen fuels for fuel cells, bioethanol fuel cells, and bioethanol fuel-based biochemicals with a collection of 19 chapters (Table 1.7).

Thus, this book chapter presents an overview of *Handbook of Bioethanol Fuels* (Konur, 2023cs) presenting a representative sample of the research in the field of bioethanol fuels.

This handbook is a valuable resource for the stakeholders engaged primarily in the research fields of energy and fuels, chemical engineering, environmental science and engineering, biotechnology, microbiology, chemistry, physics, mechanical engineering, agricultural sciences, food science and engineering, materials science, biochemistry, genetics, molecular biology, plant sciences, water resources, economics, business, management, transportation science and technology, ecology, public, environmental and occupational health, social sciences, toxicology, multidisciplinary sciences, and humanities among others.

1.2 MATERIALS AND METHODS

This book chapter presents an overview of the representative sample of research on bioethanol fuels, presented in this handbook in six volumes. Table 1.1 presents information on key research fronts for each volume while Tables 1.2–1.7 provide information on all chapters presented in this handbook in six volumes. Further, the related references are presented in the reference list. The key issues and findings from each chapter are presented and discussed briefly to highlight the research landscape in light of the population studies conducted in this field.

1.3 CONTENTS OF THE HANDBOOK

The brief information about 110 chapters on bioethanol fuels is given in Table 1.1. The primary research fronts are production, evaluation, and utilization of bioethanol fuels.

The first (Konur, 2023ct) and second volumes (Konur, 2023cu) cover the introduction to bioethanol fuels and biomass pretreatments for sugar and bioethanol fuel production with 18 and 21 chapters, respectively. The first volume has two parts: introduction to bioethanol fuels and biomass pretreatments for sugar and bioethanol fuel production with 3 and 15 chapters, respectively. The second volume has four parts on the bioethanol fuel production processes: introduction to production processes for the bioethanol fuels, hydrolysis of the pretreated biomass, hydrolysate fermentation, and bioethanol fuel separation and distillation with two, eight, eight, and three chapters, respectively.

The third (Konur, 2023cv) and fourth volumes (Konur, 2023cw) cover the bioethanol fuel production from non-waste and waste feedstocks with 19 and 17 chapters, respectively. The third volume has nine parts with two chapters each except for the first generation starch feedstock-based bioethanol fuels with three chapters: introduction to feedstock-based bioethanol fuels, introduction to non-waste feedstock-based bioethanol fuels, first generation starch-based bioethanol fuels, first generation sugar feedstock-based bioethanol fuels, grass biomass-based bioethanol fuels, wood biomass-based bioethanol fuels, cellulose-based bioethanol fuels, third generation algal bioethanol fuels, and biosyngas-based bioethanol fuels. The fourth volume has nine parts with two chapters each except for lignocellulosic biomass-based bioethanol fuels with three chapters: introduction to second generation bioethanol fuels, second generation bioethanol fuels from residual sugar feedstocks, second generation bioethanol fuels from residual starch feedstocks, second generation food waste-based bioethanol fuels, second generation industrial waste-based bioethanol fuels, second generation urban waste-based bioethanol fuels, second generation forestry waste-based bioethanol fuels, and lignocellulosic biomass-based bioethanol fuels.

Finally, fifth (Konur, 2023cx) and sixth volumes (Konur, 2023cy) cover the research on the evaluation and utilization of bioethanol fuels with 18 and 17 chapters, respectively. The fifth volume has five parts with two chapters each except for introduction to the evaluation and utilization of bioethanol fuels and evaluation of bioethanol fuels with eight and four chapters, respectively: introduction to the evaluation and utilization of bioethanol fuels, utilization of bioethanol fuels in transport engines, evaluation of bioethanol fuels, utilization of bioethanol fuels, and bioethanol fuel sensors. The final volume has four parts with nine, three, two, and three chapters, respectively: country-based experience of bioethanol fuels, bioethanol fuel-based biohydrogen fuels, bioethanol fuel cells, and bioethanol fuel-based biochemicals.

1.3.1 Volume 1. Bioethanol Fuel Production Processes. I: Biomass Pretreatments

The first volume has two major parts (Konur, 2023ct): introduction to bioethanol fuels and pretreatments of the biomass for sugar and bioethanol fuel production with 3 and 15 book chapters, respectively (Table 1.2).

1.3.1.1 Introduction to Bioethanol Fuels

This section has three chapters: introduction to bioethanol fuels, scientometric study, and review of the research on bioethanol fuels (Table 1.2).

Chapter 1 provides an overview of 110 chapters presented in this handbook under three major headings of production of bioethanol fuels with the emphasis on production processes (Volumes 1 and 2), production of bioethanol fuels with the emphasis on feedstocks (Volumes 3 and 4), and evaluation and utilization of bioethanol fuels (Volumes 5 and 6) (Konur, 2023a). Table 1.1 provides the macrodata on these 110 chapters in six volumes: the titles of the volumes and the major parts for each volume and the number of book chapters for each part and volume.

Chapter 2 provides a scientometric overview of the research on bioethanol fuels (Konur, 2023b). It finds that research in this field has intensified in key research fronts of the production of bioethanol fuels and to a lesser extent utilization and evaluation of bioethanol fuels. For the first research front, pretreatment and hydrolysis of feedstocks, fermentation of hydrolysates, production of bioethanol fuels in general, and to a lesser extent separation and distillation of bioethanol fuels are the key research areas while for the second research front, utilization of bioethanol fuels in fuel cells and transport engines, bioethanol-based biohydrogen fuels, bioethanol sensors, and bioethanol-based biochemicals are the key research areas. Research in this field has also intensified for feedstocks of lignocellulosic biomass at large, bioethanol fuels for their utilization, cellulose, starch feedstock residues, wood biomass, and to a lesser extent industrial waste, lignin, grass biomass, hydrolysates, biomass at large, sugar feedstocks, sugar feedstock residues, algal biomass, lignocellulosic wastes, starch feedstocks, urban wastes, forestry wastes, food wastes, biosyngas, and plants. This handbook has been structured in light of these findings for the primary research fronts in this field.

Chapter 3 provides a review of the influential research on bioethanol fuels complementing Chapter 2 (Konur, 2023c). It finds that influential research in this field has intensified in key research fronts of production of bioethanol fuels and to a lesser extent utilization and evaluation of bioethanol fuels. For the first research front, pretreatment and hydrolysis of feedstocks, and to a lesser extent production of bioethanol fuels in general and fermentation of the hydrolysates are the key influential research areas while for the second research front, bioethanol-based biohydrogen fuels and bioethanol sensors are the key influential research areas. Influential research in this field has also intensified for feedstocks of wood biomass, cellulose, lignin, grass biomass, bioethanol fuels, and to a lesser extent lignocellulosic biomass at large, starch feedstock residues, starch feedstocks, biomass at large, and sugar feedstock residues.

1.3.1.2 Biomass Pretreatments

This section deals with pretreatments of biomass for sugar and bioethanol production. There are four primary research fronts in this field with 15 chapters: chemical, biological, hydrothermal, and mechanical pretreatments of the biomass.

Chapter 4 provides a scientometric overview of research on biomass pretreatments (Konur, 2023d). It finds that research in this field has intensified in the research fronts of biological, chemical, and to a lesser extent mechanical and hydrothermal pretreatments of the biomass to increase the sugar and bioethanol yield in this context. On the individual terms, the most prolific pretreatments have been enzymatic hydrolysis and enzymatic, acid, ionic liquid (IL), fungal, alkaline, other chemical, hydrothermal, and to a lesser extent steam explosion, ammonia (NH_3), organic solvent, microbial, hot compressed water (HCW), milling, liquid hot water (LHW), and microwave (MW) pretreatments. Further, the most used groups of biomass have been biomass constituents, biomass in general, agricultural residues, wood biomass and a lesser extent grass biomass at the macroscale and cellulose, lignin, lignocellulosic biomass at large, biomass at large, corn stover, lignocellulose, other residues, softwood, wheat straw, switchgrass, sugarcane bagasse, and hemicellulose at the microscale.

Chapter 5 provides a review of influential research on the biomass pretreatments in general complementing Chapter 4 (Konur, 2023e). It finds that the research in the fields of biological, chemical, and to a lesser extent mechanical and hydrothermal pretreatments of the biomass has thus intensified in recent years. The most prolific pretreatments have been enzymatic hydrolysis and enzymatic, IL, fungal pretreatments, and to a lesser extent acid and alkaline pretreatments at the microscale. Further, the most used groups of biomass have been biomass constituents and to a lesser extent wood biomass, agricultural residues, and grass biomass at the macroscale while the most used individual biomass and biomass constituents have been cellulose, lignin, wood biomass, and to a lesser extent corn stover, softwood, switchgrass, wheat straw, and lignocellulosic biomass at large.

Chapter 6 provides a scientometric overview of research on the biological pretreatments of biomass (Konur, 2023f). It finds that research in this field has intensified in recent years in the research fronts of the enzymatic and fungal pretreatments of the biomass, enzymatic hydrolysis by primarily fungal enzymes, and to a lesser extent microbial pretreatments of the biomass. Further, the research has intensified in the research fronts of biomass constituents, agricultural residues, biomass in general, wood biomass, and to a lesser extent grass biomass and algal biomass at the macroscale. However, on the individual terms, the most prolific biomass have been cellulose, lignin, wood biomass, lignocellulose, and to a lesser extent lignocellulosic biomass at large, corn stover, straw, and sugarcane bagasse.

Chapter 7 provides a review of influential research on biological pretreatments of the biomass complementing Chapter 6 (Konur, 2023g). It finds that research in this field has intensified in recent years in the research fronts of the enzymatic hydrolysis of biomass by enzymes, enzymatic pretreatments of biomass, and to a lesser extent fungal pretreatments of the biomass. Further, the most prolific groups of the biomass have been biomass constituents, wood, agricultural residues, biomass at large and to a lesser extent grass biomass at the macroscale. In addition, on the individual terms, the most prolific biomass and biomass constituents have been lignin, wood biomass, cellulose, corn stover, and to a lesser extent straw, grass biomass, biomass at large, lignocellulosic biomass at large, and cellulosic biomass at large.

Chapter 8 provides a review of the lignocellulolytic enzyme-producing microbes for bioethanol production (Thakur et al., 2023). It considers cellulases, hemicellulases, and ligninases and focuses on the production mechanism of the lignocellulolytic enzyme-producing microbes and their hyper-enzyme production for efficient saccharification of lignocellulosic biomass. It finds that the most prolific microbes are aerobic bacteria and fungi as opposed to actinomycetes.

Chapter 9 provides a scientometric overview of research on chemical pretreatments of the biomass (Konur, 2023h). It finds that a number of chemical pretreatments such as acid, IL, alkaline, and to a lesser extent NH_3, organic solvent, hydrogen peroxide (H_2O_2), deep eutectic solvent (DES), wet oxidation, supercritical water (SCW), surfactant, ozone (O_3), sulfite, and supercritical carbon dioxide (CO_2) pretreatments have been widely researched in this context to increase the sugar and bioethanol yield. Further, the most prolific biomass groups used in these chemical pretreatments have been biomass constituents, agricultural residues, biomass in general, and to a lesser extent, wood biomass, grass biomass, and other biomass such as algal biomass and plants at the macroscale. Additionally, on the individual terms, the most prolific biomass used in the chemical pretreatments have been cellulose, corn stover, lignocellulosic biomass at large, and to a lesser extent wood biomass, biomass at large, grass biomass, rice straw, wheat straw, hemicellulose, sugarcane bagasse, lignocellulose, and cellobiose.

Chapter 10 provides a review of influential research on chemical pretreatments of the biomass complementing Chapter 9 (Konur, 2023i). It finds that influential research in this field has intensified in the research fronts of acid, IL, and to a lesser extent SCW, alkaline, H_2O_2, NH_3, organic solvent, O_3, and surfactant pretreatments of the biomass to increase the sugar and bioethanol yield. However, there have been three primary groups of biomass used in these pretreatment studies: biomass constituents, to a lesser extent agricultural residues, and wood biomass at the macroscale. Further, on the individual terms, the most prolific biomass have been cellulose and to a lesser extent wood biomass at large, lignin, corn stover, lignocellulose, grass biomass, rice straw, and wheat straw.

Chapter 11 provides a review of research on the utilization of ILs for pretreatment of lignocellulosic biomass for sugar and bioethanol fuel production (Elgharbawy et al., 2023). It finds that a wide range of ILs have been developed to pretreat the lignocellulosic biomass alone or in combination with other pretreatments, and nanotechnology has a lot to contribute to the efficiency of the IL pretreatments. Further, the IL pretreatment results in higher bioethanol yield, reducing the detrimental fermentation inhibitors.

Chapter 12 provides a review of emerging solvents for the biomass pretreatments for the sugar and bioethanol production (Zhang and Wang, 2023). It finds that bio-based solvents, ILs, DESs, and supercritical fluids (SCFs) have been increasingly applied in biomass pretreatment in this context. It proposes a framework incorporating the USEtox model for their sustainability assessment, which is a guidance for solvent design and development.

Chapter 13 provides a scientometric overview of research on hydrothermal pretreatments of the biomass for sugar and bioethanol production (Konur, 2023j). It finds that research in this field has intensified in recent years in the research fronts of steam explosion and LHW-based hydrothermal treatments such as LHW, HCW, and autohydrolysis pretreatments of the biomass. Further, the most prolific biomass has been agricultural residues, wood biomass, biomass constituents and to a lesser extent biomass in general and grass biomass at the macroscale. However, the most prolific individual biomass used in the hydrothermal pretreatments has been lignin, corn stover, cellulose, and to a lesser extent wheat straw, sugarcane bagasse, eucalyptus, lignocellulosic biomass at large, wood biomass at large, hemicellulose, poplar, spruce, grass biomass, pine, and corncobs.

Chapter 14 provides a review of influential research on hydrothermal pretreatments of the biomass for sugar and bioethanol production complementing Chapter 13 (Konur, 2023k). It finds that research in this field has intensified in recent years. In this context, steam explosion- and LHW-based hydrothermal treatments such as LHW, HCW, and autohydrolysis pretreatments of the biomass have been widely researched. However, the most prolific primary types of the biomass have been biomass constituents, wood biomass, agricultural residues, and to a lesser extent biomass in general at the macroscale. Further, the most prolific individual biomass has been lignin, cellulose, hemicellulose, wood biomass, corn stover, wheat straw, hardwood, and to a lesser extent softwood, sugarcane bagasse, corncobs, lignocellulosic biomass at large, and poplar.

Chapter 15 provides a scientometric overview of research on the mechanical pretreatments of the biomass for the sugar and bioethanol fuel production (Konur, 2023l). It finds that research in this field has intensified in recent years. Milling, MW, and ultrasonic pretreatments of biomass have been widely researched in this context to increase the sugar and bioethanol yield.

Chapter 16 provides a review of influential research on the milling pretreatment of biomass for the sugar and bioethanol production (Konur, 2023m). It finds that research in this field intensified in the research fronts of ball milling, disk milling, and hammer milling among others. The most prolific groups of biomass have been agricultural wastes, cellulose, wood biomass, and to a lesser extent plants at the macroscale. Further, on the individual terms, the most prolific groups of biomass have been cellulose, wood biomass at large, straw, and to a lesser extent corn stover, wood chips, sugarcane bagasse, corn, grass biomass, algal biomass, food wastes, and corncobs. It shows that milling pretreatments alone or in combination with chemical, enzymatic, and hydrothermal pretreatments have been effective in disrupting the biomass and thus enhancing the sugar and bioethanol yield.

Chapter 17 provides a review of influential research on the MW pretreatment of biomass for the sugar and bioethanol production (Hoang and Konur, 2023n). It finds that research in this field has intensified in the research fronts of straw, cellulose, sugarcane bagasse, and to a lesser extent grass biomass, wood biomass, corn stover and rice hulls. It shows that MW pretreatments alone or in combination with chemical, enzymatic, and hydrothermal pretreatments have been effective in disrupting the biomass and thus enhancing the sugar and bioethanol yield.

Chapter 18 provides a review of the research on the ultrasonic pretreatment of the lignocellulosic biomass for sugar and bioethanol fuel production (Pradhan et al., 2023). It finds that the ultrasonic pretreatment alone or in combination with alkaline, acid, IL, DES, organic solvents such as ethanol or glycerol solvents, NH_3, metal salts such as metal ion chlorides, Fenton reagent, O_3, or lime pretreatments has been effective to disintegrate the lignocellulosic biomass for the better enzymatic hydrolysis and fermentation with increased sugar and bioethanol fuel yields, respectively.

TABLE 1.2
The Microresearch Fronts for Volume 1: Bioethanol Fuel Production Processes. I: Biomass Pretreatments

No.	Research Fronts	Chapter Title	Ref.
		Part 1. Introduction to Bioethanol Fuels	
1	Overview	Introduction to bioethanol fuels	Konur (2023a)
2	Overview	Bioethanol fuels: scientometric study	Konur (2023b)
3	Overview	Bioethanol fuels: review	Konur (2023c)
		Part 2. Biomass Pretreatments for Bioethanol Production	
4	Overview	Biomass pretreatments: scientometric study	Konur (2023d)
5	Overview	Biomass pretreatments: review	Konur (2023e)
6	Biological prts.	Biological pretreatment of the biomass: scientometric study	Konur (2023f)
7	Biological prts.	Biological pretreatment of the biomass: review	Konur (2023g)
8	Biological prts.	Lignocellulolytic enzymes for biofuel production	Thakur et al. (2023)
9	Chemical prts.	Chemical pretreatment of the biomass: scientometric study	Konur (2023h)
10	Chemical prts.	Chemical pretreatment of the biomass: review	Konur (2023i)
11	IL prts.	Advances of bioethanol production using ILs	Elgharbawy et al. (2023)
12	Solvent prts.	Emerging solvents for biomass pretreatment and their sustainability	Zhang and Wang (2023)
13	Hydrothermal prts.	Hydrothermal pretreatment of the biomass: scientometric study	Konur (2023j)
14	Hydrothermal prts.	Hydrothermal pretreatment of the biomass: review	Konur (2023k)
15	Mechanical prts.	Mechanical pretreatment of the biomass: scientometric study	Konur (2023l)
16	Milling prts.	Milling pretreatment of the biomass: review	Konur (2023m)
17	Microwave prts.	Microwave pretreatment of the biomass: review	Hoang and Konur (2023n)
18	Ultrasound prts.	Ultrasound-assisted pretreatment of lignocellulosic biomass for bioethanol production	Pradhan et al. (2023)

Prts - Pretreatments.

1.3.2 Volume 2. Bioethanol Fuel Production Processes. II: Biomass Hydrolysis, Fermentation, and Bioethanol Fuel Separation

1.3.2.1 Introduction to Production Processes for Bioethanol Fuels

Chapter 19 provides a scientometric overview of research on the production processes of bioethanol fuels (Konur, 2023o). It finds that research in this field has intensified in key research fronts of pretreatment and hydrolysis of feedstocks, fermentation of hydrolysates, production of bioethanol fuels in general, and to a lesser extent evaluation, utilization, and distillation of bioethanol fuels. Research in this field has also intensified for feedstocks of lignocellulosic biomass at large, cellulose, wood biomass, starch feedstock residues, and to a lesser extent lignin, grass biomass, industrial waste, sugar feedstocks, starch feedstocks, sugar feedstock residues, algal biomass, biosyngas, urban waste, food waste, and forestry waste.

Chapter 20 provides a review of influential research on the production processes of bioethanol fuels complementing Chapter 19 (Konur, 2023p). It finds that influential research in this field has intensified in key research fronts of pretreatment and hydrolysis of feedstocks and to a lesser extent fermentation of the hydrolysates and production of bioethanol fuels in general. Research in this field has also intensified for feedstocks of wood biomass, cellulose, lignocellulosic biomass at large, and to a lesser extent starch feedstock residues and grass biomass.

1.3.2.2 Biomass Hydrolysis for Bioethanol Fuel Production

Chapter 21 provides a scientometric overview of research on the biomass hydrolysis for bioethanol fuel production (Konur, 2023q). It finds that research in this field has intensified in recent years as the hydrolysis of the agricultural residues, wood biomass, biomass constituents, and grass biomass at the macroscale biomass has been widely researched to increase the sugar and bioethanol yield in this context. Further, on the individual terms, the most prolific groups of biomass have been cellulose, corn stover, lignin, wheat straw, lignocellulose, and to a lesser extent, rice straw, sugarcane bagasse, hemicellulose, grass biomass, and food waste.

Chapter 22 provides a review of influential research on the biomass hydrolysis for bioethanol fuel production complementing Chapter 21 (Konur, 2023r). It finds that research in this field has intensified in recent years as the hydrolysis of the agricultural residues, wood biomass, biomass constituents, and to a lesser extent hydrolysates and grass biomass at the macroscale has been widely researched to increase the sugar and bioethanol yield. Further, on the individual terms, the most prolific biomass have been cellulose, corn stover, lignin, hemicellulose, and to a lesser extent lignocellulose, rice straw and wheat straw.

Chapter 23 provides a scientometric overview of research on the wood hydrolysis for the bioethanol fuel production (Konur, 2023s). It finds that wood has been one of the most studied biomass for the bioethanol production. In this context, the enzymatic, chemical, hydrothermal, and to a lesser extent mechanical hydrolysis of wood at the macrolevel have been widely researched to increase the sugar and bioethanol yield. Further, on the individual terms, the most prolific research fronts have been acid hydrolysis, steam explosion hydrolysis, and to a lesser extent autohydrolysis, solvent, IL, LHW, alkaline, and sulfite hydrolysis. However, the most prolific wood types have been hardwood, softwood, and wood in general at the macrolevel, and on the individual terms softwood in general, poplar, eucalyptus, pine, hardwood in general, bamboo, spruce, and to a lesser extent aspen, birch, and olive tree.

Chapter 24 provides a review of influential research on the wood hydrolysis for the bioethanol fuel production complementing Chapter 23 (Konur, 2023t). It finds that wood has been one of the most studied biomass for the bioethanol production. In this context, the enzymatic, chemical, hydrothermal, and to a lesser extent mechanical hydrolysis of wood at the macrolevel have been widely researched to increase the sugar and bioethanol yield. Further, on the individual terms, the most prolific types of hydrolysis have been acid, steam explosion, alkaline, sulfite, H_2O_2, LHW, milling, NH_3, hydrothermal in general, and IL hydrolysis. However, the most prolific groups of wood biomass have been hardwood and softwood at the macrolevel and on individual terms, spruce, Douglas fir, poplar, and to a lesser extent olive tree, pine, willow, aspen, bamboo, beech, birch, and eucalyptus.

Chapter 25 provides a scientometric overview of research on straw hydrolysis for the bioethanol fuel production (Konur, 2023u). It finds that wheat straw and rice straw have been among the most studied biomass for bioethanol production, and enzymatic hydrolysis of straw combined with chemical and to a lesser extent hydrothermal and mechanical pretreatments of straw has been widely researched to increase the sugar and bioethanol yield. Further, on the individual terms, the most prolific types of pretreatments and hydrolysis have been alkaline, acid, steam explosion, solvent, MW, NH_3, IL milling, H_2O_2, wet oxidation, and LHW pretreatments. However, the most prolific types of straw biomass have been wheat and rice straw and to a lesser extent barley, sorghum, soybean, and sugarcane straw.

Chapter 26 provides a review of influential research on straw hydrolysis for the bioethanol fuel production complementing Chapter 25 (Konur, 2023v). It finds that wheat straw and rice straw have been among the most studied biomass for the bioethanol production. In this context, the enzymatic hydrolysis combined with chemical, and to a lesser extent hydrothermal and mechanical pretreatments of straw has been widely researched to increase the sugar and bioethanol yield. Further, on the individual terms, the most prolific pretreatments of the straw have been acid, alkaline, IL, milling, and to a lesser extent NH_3, MW, organic solvent, surfactant, H_2O_2, HCW, LHW, O_3, steam explosion, and wet oxidation pretreatments. However, the most prolific types of straw biomass have been wheat and rice straw and to a lesser extent corn, rye, sorghum, and sugarcane straw.

Chapter 27 provides a scientometric overview of research on cellulose hydrolysis for the bioethanol fuel production (Konur, 2023w). It finds that the enzymatic, chemical, and to a lesser extent hydrothermal and mechanical hydrolysis and pretreatments of cellulose have been widely researched to increase the sugar and bioethanol yield. Further, on the individual terms, the most prolific research fronts have been acid, IL solvent, steam explosion, water, alkaline, HCW, LHW, MW, NH_3, supercritical CO_2, surfactant, milling, and ultrasound hydrolysis and pretreatments.

Chapter 28 provides a review of influential research on cellulose hydrolysis for the bioethanol fuel production complementing Chapter 27 (Konur, 2023x). In this context, research in this field has thus intensified in recent years as the enzymatic and chemical hydrolysis of cellulose has been widely researched to increase the sugar and bioethanol yield at the expense of hydrothermal and mechanical hydrolysis and pretreatments. Further, on the individual terms, the most prolific research fronts have been acid, IL, supercritical water, organic solvent, alkaline, and NH_3 hydrolysis and pretreatments.

1.3.2.3 Hydrolysate Fermentation for Bioethanol Fuel Production

Chapter 29 provides a scientometric overview of research on hydrolysate and substrate fermentation for the bioethanol fuel production (Konur, 2023y). It finds that research in this field has thus intensified in recent years as microorganism and substrate metabolic engineering, fermentation inhibitors, and to a lesser extent simultaneous saccharification and fermentation (SSF), hydrolysate detoxification, fermentation microorganisms, fermentation in general, consolidated bioprocessing (CBP), and microorganism's ethanol tolerance have been widely researched to increase the bioethanol yield in this context. However, the most prolific types of feedstocks for the fermentation processes have been hydrolysates, agricultural residues, and to a lesser extent biomass constituents, algal biomass, glycerol, biosyngas, wood biomass, grass biomass, and sugar feedstocks at the macrolevel. Further, on the individual terms, the most prolific feedstocks have been xylose, glucose, and to a lesser extent hydrolysates in general, pentose, cellulose, corn stover, hexose, wheat straw, and sugarcane bagasse.

Chapter 30 provides a review of influential research on the hydrolysate and substrate fermentation for the bioethanol fuel production complementing Chapter 29 (Konur, 2023z). It finds that research in this field has intensified in recent years as the microorganism and substrate metabolic engineering and to a lesser extent fermentation inhibitors, hydrolysate detoxification, fermentation in general, and SSF, fermentation microorganisms, other fermentation issues, and microorganism's ethanol tolerance at the macrolevel have been widely researched to increase the bioethanol yield in this context. Further, the most prolific feedstocks for the fermentation have been hydrolysates and to a lesser extent agricultural residues, wood biomass, biomass constituents, algal biomass, and grass biomass at the macrolevel. However, on the individual terms, the most prolific feedstocks have been xylose, glucose, hydrolysates in general, wheat straw, other agricultural residues, corn stover, sugarcane bagasse, and cellulose.

Chapter 31 provides a scientometric overview of research on the alternative fermentation processes for the bioethanol production (Konur, 2023aa). It finds that the research in the field of alternative fermentation processes for the bioethanol production for the separate hydrolysis and fermentation (SHF) process to improve the ethanol yield has intensified in this context in recent years. Key research fronts have been the SSF and to a lesser extent CBP, simultaneous saccharification and cofermentation (SSCF), continuous ethanol production (CEP), and continuous ethanol fermentation (CEF) of the biomass. Further, the research has focused on the agricultural residues, wood biomass, biomass in general, wastes, and to a lesser extent biomass constituents, grass biomass, hydrolysates, algal biomass, and other biomass at the macrolevel. Further, on the individual terms, the most prolific types of biomass have been corn stover, lignocellulosic biomass, wheat straw, food waste, cellulosic biomass, industrial wastes, lignocellulose, rice straw, sugarcane bagasse, corncobs, cellulose, xylose, corn, and potato.

Chapter 32 provides a review of influential research on the alternative fermentation processes for bioethanol production complementing Chapter 31 (Konur, 2023ab). It finds that research in this field to improve the ethanol yield has intensified in this context in recent years. Key research fronts have been the SSF of the biomass and to a lesser extent CBP and SSCF of the biomass. Further, the research has focused on the agricultural residues, wood biomass, other feedstocks at the macroscale, and in the individual terms, on the corn stover, wheat straw, and to a lesser extent rice straw, corncobs, sorghum bagasse, food wastes, grass biomass, industrial wastes, algal biomass, and cellulose.

Chapter 33 provides a scientometric overview of research on the metabolic engineering for the bioethanol production (Konur, 2023ac). It finds that research in this field to improve the ethanol yield has intensified in recent years. Key research fronts with regard to the microorganisms used in the metabolic engineering have been *Saccharomyces cerevisiae* and to a lesser extent *Escherichia coli*, microorganisms in general, clostridium, other microorganisms, *Pichia stipitis*, *Zymomonas mobilis*, *Bacillus subtilis,* and klebsiella. Further, the most prolific substrates have been corn, grass, and cyanobacteria. However, the most prolific feedstocks used for metabolic engineering have been hydrolysates and to a lesser extent biomass constituents and other feedstocks at the macroscale while on individual terms the feedstocks have been xylose and to a lesser extent glucose, cellulose, hydrolysates in general, pentose, lignocellulose, algal biomass, arabinose, glycerol, cellobiose, wood biomass, agricultural residues, ethane, and other biomass.

Chapter 34 provides a review of influential research on the metabolic engineering of microorganisms and substrates for the bioethanol production complementing Chapter 33 (Konur, 2023ad). It finds that research in this field has intensified in recent years. Key research fronts with regard to the microorganisms have been *Saccharomyces cerevisiae*, *Zymomonas mobilis*, other microorganisms, *Escherichia coli*, and to a lesser extent *Pichia stipitis* while the prolific substrates have been grass and cyanobacteria. Further, the most prolific feedstocks have been hydrolysates and to a lesser extent algal biomass, other feedstocks, and glycerol at the macroscale while on the individual terms, the feedstocks have been xylose, glucose, and to a lesser extent hydrolysates in general, wood, agricultural residues, grass, and biomass constituents such as cellulose.

Chapter 35 provides a scientometric overview of research on the utilization of the *Saccharomyces cerevisiae* for the bioethanol production (Konur, 2023ae). It finds that research in this field has intensified as the most studied yeast in the fermentation of feedstocks to produce bioethanol in recent years. Key research fronts with regard to feedstocks have been the hydrolysates and to a lesser extent other feedstocks, agricultural residues, wood biomass, and grass biomass at the macroscale while on the individual terms they have been xylose and to a lesser extent glucose, hydrolysates in general, arabinose, lignocellulosic biomass, food waste, cellobiose, sugarcane bagasse, cellulose, rice straw, and wheat straw. However, the most prolific thematic research fronts have been hydrolysate fermentation, metabolic engineering, biomass fermentation, ethanol stress, ethanol production in general, fermentation inhibitors, and fermentation processes at the macroscale while for the first research front they have been xylose fermentation and to a lesser extent glucose fermentation, hydrolysate fermentation in general, arabinose fermentation, and cellobiose fermentation.

Chapter 36 provides a review of influential research on the utilization of the *Saccharomyces cerevisiae* for the bioethanol production complementing Chapter 35 (Konur, 2023af). It finds that research in this field intensified as the most studied yeast in the fermentation of feedstocks to produce bioethanol in recent years. Key research fronts with regard to feedstocks have been the hydrolysates and to a lesser extent agricultural residues, wood biomass, and grass biomass at the macroscale and xylose, glucose, rice straw, corn stover, arabinose, cellobiose, wheat straw, and sorghum bagasse at the microscale. Further, the most prolific thematic research fronts have been metabolic engineering, xylose fermentation, biomass fermentation, glucose fermentation, fermentation inhibitors, and to a lesser extent ethanol stress, other hydrolysate fermentation, and fermentation processes.

1.3.2.4 Bioethanol Fuel Separation and Distillation

Chapter 37 provides a review of research on the bioethanol fuel distillation (Chen et al., 2023). It finds that in the bioethanol production process, an indispensable step is the downstream separation of the dilute product from the water-rich bioethanol fermentation broth. This separation step is one of the most energy-intensive steps in the whole bioethanol process and comprises a large percentage of the production cost. Compared with other emerging bioethanol separation technologies, distillation is a time-tested method and has been widely used in the industry.

Chapter 38 provides a review of research on the *in situ* bioethanol separation from fermentation broth (Cai et al., 2023). It considers *in situ* product recovery (ISPR) such as gas stripping, vacuum fermentation, adsorption, liquid–liquid extraction, pervaporation, perstraction, and membrane distillation. It finds that the integration of ISPR with bioethanol fermentation could significantly promote the productivity, yield, and separation efficiency of bioethanol fuels as an energy-saving downstream process. Application of multiple-stage hybrid ISPR processes to eliminate the weakness of each technique is a critical research trend while the membrane-based techniques show great potential coupled with other techniques for separation of bioethanol.

Chapter 39 provides a review of research on the production and purification of bioethanol using intensified processes (Gomez-Castro, 2023). It notes that process intensification is a tool that may enhance the performance of the bioethanol production processes as the development of advanced technologies allows using synergetic effects between phenomena, reducing the size of the required equipment, increasing the reaction yields, and reducing the energy requirements, among other advantages. It finds that through process intensification, yields can be increased and energy requirements can be reduced, which allows reducing the environmental impact associated with the generation of heat and electricity.

TABLE 1.3
The Microresearch Fronts for Volume 2: Bioethanol Fuel Production Processes. II: Biomass Hydrolysis, Fermentation, and Bioethanol Fuel Separation

No.	Research Fronts	Chapter Title	Ref.
		Part 3. Introduction to Production Processes for Bioethanol Fuels	
19	Overview	Production processes for bioethanol fuels: scientometric study	Konur (2023o)
20	Overview	Production processes for bioethanol fuels: review	Konur (2023p)
		Part 4. Biomass Hydrolysis for Bioethanol Production	
21	Overview	Biomass hydrolysis: scientometric study	Konur (2023q)
22	Overview	Biomass hydrolysis: review	Konur (2023r)
23	Wood hydrolysis	Wood hydrolysis: scientometric study	Konur (2023s)
24	Wood hydrolysis	Wood hydrolysis: review	Konur (2023t)
25	Straw hydrolysis	Straw hydrolysis: scientometric study	Konur (2023u)
26	Straw hydrolysis	Straw hydrolysis: review	Konur (2023v)
27	Cellulose hydrolysis	Cellulose hydrolysis: scientometric study	Konur (2023w)
28	Cellulose hydrolysis	Cellulose hydrolysis: review	Konur (2023x)
		Part 5. Hydrolysate Fermentation for Bioethanol Production	
29	Overview	Hydrolysate and substrate fermentation: scientometric study	Konur (2023y)
30	Overview	Hydrolysate and substrate fermentation: review	Konur (2023z)
31	Alternative fermentation processes	Alternative fermentation processes for the bioethanol production: scientometric study	Konur (2023aa)
32	Alternative fermentation processes	Alternative fermentation processes for the bioethanol production: review	Konur (2023ab)
33	Metabolic engineering	Metabolic engineering for the bioethanol production: scientometric study	Konur (2023ac)
34	Metabolic engineering	Metabolic engineering for the bioethanol production: review	Konur (2023ad)
35	Utilization of the *Saccharomyces cerevisiae*	Utilization of the *Saccharomyces cerevisiae* for the bioethanol production: scientometric study	Konur (2023ae)
36	Utilization of the *Saccharomyces cerevisiae*	Utilization of the *Saccharomyces cerevisiae* for the bioethanol production: review	Konur (2023af)
		Part 6. Bioethanol Fuel Separation and Distillation	
37	Bioethanol fuel distillation	Distillation separation of bioethanol: the current advances	Chen et al. (2023)
38	Bioethanol fuel separation	*In situ* bioethanol separation from fermentation broth	Cai et al. (2023)
39	Bioethanol fuel purification	Advances and challenges in the production and purification of bioethanol using intensified processes	Gomez-Castro (2023)

1.3.3 Volume 3. Feedstock-based Bioethanol Fuels. I. Non-Waste Feedstocks: Starch, Sugar, Grass, Wood, Cellulose, Algae, and Biosyngas-based Bioethanol Fuels

1.3.3.1 Introduction to Feedstock-based Bioethanol Fuels

Chapter 40 provides a scientometric overview of research on the feedstock-based bioethanol fuels (Konur, 2023ag). It finds that research in this field has intensified in this context in key research fronts of pretreatment and hydrolysis of feedstocks, bioethanol fuel production, fermentation of the hydrolysates, and evaluation and utilization of the bioethanol fuels. Research in this field has also intensified for feedstocks of lignocellulosic biomass, cellulose, wood, and to a lesser extent grass biomass, sugar feedstocks, starch feedstock residues, starch feedstocks, algal biomass, sugar feedstock residues, urban waste, food waste, industrial waste, and forestry waste.

Chapter 41 provides a review of influential research on feedstock-based bioethanol fuels complementing Chapter 40 (Konur, 2023ah). It finds that research in this field has intensified in key research fronts of pretreatment and hydrolysis of feedstocks and to a lesser extent evaluation of feedstock-based bioethanol fuels, fermentation of feedstock-based hydrolysates, and production and evaluation of bioethanol fuels. Research in this field has also intensified for feedstocks of wood biomass, cellulose, lignocellulosic biomass at large, grass biomass, starch feedstock residues, and sugar and starch feedstocks.

1.3.3.2 Introduction to Non-Waste Feedstock-based Bioethanol Fuels

Chapter 42 provides a scientometric overview of research on non-waste feedstock-based bioethanol fuels (Konur, 2023ai). It finds that research in this field has intensified in key research fronts of the pretreatment and hydrolysis of non-waste biomass, production of bioethanol fuels, fermentation of non-waste biomass-based hydrolysates, and to a lesser extent production and evaluation of non-waste biomass-based bioethanol fuels. Research in this field has also intensified for feedstocks of wood biomass and to a lesser extent grass biomass, algal biomass, sugar feedstocks such as sugarcane, and starch feedstocks such as corn grains. Thus, it emerges as a distinctive research field.

Chapter 43 provides a review of influential research on non-waste feedstock-based bioethanol fuels complementing Chapter 42 (Konur, 2023aj). It finds that one of the most studied feedstocks for bioethanol fuels has been non-waste biomass at large. Research in this field has intensified in key research fronts of pretreatment and hydrolysis of non-waste biomass, and to a lesser extent production of non-waste biomass-based bioethanol fuels, fermentation of non-waste biomass-based hydrolysates, and evaluation of non-waste biomass-based bioethanol fuels. Research in this field has also intensified for feedstocks of wood biomass such as poplar and pine, grass biomass such as switchgrass and miscanthus, and to a lesser extent sugar feedstocks such as sugarcane, starch feedstocks such as corn grains, and algal biomass such as chlamydomonas. Thus, it emerges as a distinctive research field.

1.3.3.3 First Generation Starch-based Bioethanol Fuels

Chapter 44 provides a scientometric overview of first generation starch feedstock-based bioethanol fuels (Konur, 2023ak). It finds that one of the most studied feedstocks for bioethanol fuels have been starch feedstocks as first generation feedstocks. Research in this field has intensified in key research fronts of pretreatment and hydrolysis of starch feedstocks, fermentation of starch feedstock-based hydrolysates, and production and evaluation of starch feedstock-based bioethanol fuels. Further, corn, and to a lesser extent cassava, starch feedstocks in general, wheat, potato, sorghum, artichoke, other starch feedstocks, rice, sago, and triticale have been studied intensively in this context.

Chapter 45 provides a review of influential research on first generation starch feedstock-based bioethanol fuels complementing Chapter 44 (Konur, 2023al). It finds that research in this field has intensified in key research fronts of evaluation of bioethanol fuels and to a lesser extent production of bioethanol fuels, fermentation of starch feedstock-based hydrolysates, pretreatments and hydrolysis of starch feedstocks, and economics of bioethanol fuels. Further, corn and to a lesser extent cassava, sorghum, potato, and wheat have been studied intensively in this context.

Chapter 46 provides a review of research on the corn pericarp-based bioethanol fuel production through steam explosion and periodic peristalsis pretreatments (Khatun et al., 2023). It finds that combinational pretreatment of steam explosion and periodic peristalsis is a new way to break the complex cell wall network with an increased bioethanol yield by 6.7% compared with untreated corn pericarp.

1.3.3.4 First Generation Sugar Feedstock-based Bioethanol Fuels

Chapter 47 provides a scientometric overview of research on first generation sugar feedstock-based bioethanol fuels (Konur, 2023am). It finds that research in this field has intensified in this context in the key research fronts of evaluation of sugar feedstock-based bioethanol fuels and to a lesser extent production of sugar feedstock-based bioethanol fuels, hydrolysate fermentation, and pretreatments and hydrolysis of sugar feedstocks. Further, sugar cane, sweet sorghum and to a lesser extent sugar beet and energy cane have been studied intensively in this context.

Chapter 48 presents a review of influential research on first generation sugar feedstock-based bioethanol fuels complementing Chapter 47 (Konur, 2023an). It finds that one of the most studied feedstocks for bioethanol fuels have been sugar feedstocks, such as sugarcane, sugar beet, sweet sorghum, and energy cane. Research in this field has intensified in key research fronts of evaluation and production of sugar feedstock-based bioethanol fuels, to a lesser extent the hydrolysate fermentation, and pretreatment and hydrolysis of sugar feedstocks.

1.3.3.5 Grass-based Bioethanol Fuels

Chapter 49 provides a scientometric overview of research on grass biomass-based bioethanol fuels (Konur, 2023ao). It finds that research in this field has intensified in this context in key research fronts of pretreatment and hydrolysis of grass biomass, to a lesser extent fermentation of grass hydrolysates, and grass-based bioethanol fuels in general. Further, switchgrass and to a lesser extent miscanthus, giant reed, bermudagrass, grass in general, ryegrass, common reed, alfalfa, napiergrass, and canarygrass have been studied intensively at the expense of other grass biomass in this context.

Chapter 50 provides a review of influential research on grass-based bioethanol fuels complementing Chapter 49 (Konur, 2023ap). It finds that research in this field has intensified in key research fronts of pretreatment and hydrolysis of grass biomass, to a lesser extent fermentation of grass hydrolysates, and production and evaluation of grass-based bioethanol fuels. Further, switchgrass and to a lesser extent miscanthus, bermudagrass, alfalfa, canarygrass, and silvergrass have been studied intensively at the expense of other grass biomass in this context.

1.3.3.6 Wood-based Bioethanol Fuels

Chapter 51 provides a scientometric overview of research on wood biomass-based bioethanol fuels (Konur, 2023aq). It finds that research in this field has intensified in key research fronts of pretreatment and hydrolysis of wood biomass, to a lesser extent fermentation of wood hydrolysates, and production and evaluation of bioethanol fuels, including the technoeconomic studies of bioethanol fuels. Further, both softwood such as pine, spruce, and Douglas fir and hardwood such as poplar, eucalyptus, and to a lesser extent aspen, birch, bamboo, and willow have been studied intensively.

Chapter 52 provides a review of influential research on wood-based bioethanol fuels complementing Chapter 51 (Konur, 2023ar). It finds that research in the field has intensified in key research fronts of pretreatment of wood biomass and to a lesser extent hydrolysis of wood biomass, fermentation of the wood hydrolysates, and the bioethanol production from wood biomass in general. Further, both softwood such as pine, spruce, and Douglas fir and hardwood such as poplar, eucalyptus, and aspen have been studied intensively. The most prolific pretreatments have been chemical and enzymatic pretreatments and to a lesser extent hydrothermal and mechanical pretreatments at the macrolevel and acid, IL, steam explosion, solvent, milling, LHW, alkaline, sulfur dioxide (SO_2), sulfite, and NH_3 pretreatments at the microlevel.

1.3.3.7 Cellulose-based Bioethanol Fuels

Chapter 53 provides a scientometric overview of research on the cellulose-based bioethanol fuels (Konur, 2023as). It finds that research in this field has intensified in key research fronts of pretreatment and hydrolysis of cellulose and a lesser extent fermentation of hydrolysates and production of bioethanol fuels. Thus, it emerges as a distinctive research field, complementing primarily the research on lignocellulosic biomass-based bioethanol fuels from agricultural residues, other wastes, wood biomass, grass biomass, and sugar and starch feedstocks.

Chapter 54 provides a review of influential research on the cellulose-based bioethanol fuels complementing Chapter 53 (Konur, 2023at). It finds that research in this field has intensified in key research fronts of pretreatment and hydrolysis of the cellulose. Thus, it emerges as a distinctive research field, complementing primarily the research on lignocellulosic biomass-based bioethanol fuels from agricultural residues, other wastes, wood biomass, grass biomass, and sugar and starch feedstocks.

1.3.3.8 Third Generation Algal Bioethanol Fuels

Chapter 55 provides a scientometric overview of research on third generation algal bioethanol fuels (Konur, 2023au). It finds that research in this field has intensified in key research fronts of the pretreatment of algal biomass and algal bioethanol production and to a lesser extent fermentation of algal hydrolysates and hydrolysis of algal biomass. Chemical and enzymatic pretreatments and to a lesser extent mechanical, microbial, and hydrothermal pretreatments have been the most prolific pretreatments at the macroscale while enzymatic hydrolysis and acid hydrolysis have been the most prolific hydrolysis. Additionally, the most prolific research fronts with regard to the fermentation and bioethanol production have been for macroalgae and to a lesser extent microalgae, cyanobacteria, and algae in general. Further, macroalgae, microalgae, and cyanobacteria have been studied intensively at the expense of diatoms, coccolithophores, and dinoflagellates in this context. The most prolific macroalgal species have been macroalgae in general, alginates, Gelidium, Ulva, Gracilaria, Saccharina, Kappaphycus, Laminaria, Sargassum, Chaetomorpha, and Undaria while the most prolific microalgal species have been microalgae in general, Chlorella, Scenedesmus, Chloroccum, Chlamydomonas, and Nannochloropsis, and the most prolific cyanobacterial species have been Synechocystis, Cyanobacteria in general, Synechococcus, and Spirulina.

Chapter 56 provides a review of influential research on third generation algal bioethanol fuels complementing Chapter 55 (Nowruzi and Konur, 2023av). It finds that research in this field has intensified in key research fronts of pretreatment and hydrolysis of algal biomass, fermentation of algal hydrolysates, and algal bioethanol production in general. The most prolific pretreatments have been chemical, enzymatic and to a lesser extent microbial and mechanical pretreatments at the macroscale. Further, the macroalgae, microalgae, and to a lesser extent cyanobacteria and algae in general have been studied intensively at the expense of the diatoms and dinoflagellates at the macrolevel in this context. However, on the individual level, the most prolific algae have been Chlamydomonas, Chlorella, Chloroccum, alginates, Laminaria, Synechocystis, Gelidium, Gracilaria, Nannochloropsis, Saccharina, Sargassum, Scenedesmus, Synechococcus, and Ulva.

1.3.3.9 Biosyngas-based Bioethanol Fuels

Chapter 57 provides a scientometric overview of research on the biosyngas-based bioethanol fuels (Konur, 2023aw). It finds that research in this field has intensified in key research fronts of microbial fermentation, catalytic reduction, electrochemical reduction of biosyngas, and to a lesser extent the production of bioethanol fuels by other processes such as photosynthetic production and photocatalytic reduction of biosyngas. Further, CO_2, biosyngas, and to a lesser extent CO have been studied intensively as feedstocks in this context.

Chapter 58 provides a review of influential research on biosyngas-based bioethanol fuels complementing Chapter 57 (Konur, 2023ax). It finds that research in this field has intensified in key research fronts of electrochemical reduction, catalytic reduction, and to a lesser extent fermentation, photosynthetic production, and photocatalytic reduction of biosyngas. Further, CO_2 and to a lesser extent biosyngas have been studied intensively in this context.

TABLE 1.4
The Microresearch Fronts for Volume 3: Feedstock-based Bioethanol Fuels. I. Non-Waste Feedstocks: Starch, Sugar, Grass, Wood, Cellulose, Algae, and Biosyngas-based Bioethanol Fuels

No.	Research Fronts	Chapter Title	Ref.
		Part 7. Introduction to Feedstock-based Bioethanol Fuels	
40	Overview	Feedstock-based bioethanol fuels: scientometric study	Konur (2023ag)
41	Overview	Feedstock-based bioethanol fuels: review	Konur (2023ah)
		Part 8. Introduction to Non-Waste Feedstock-based Bioethanol Fuels	
42	Overview	Non-waste feedstock-based bioethanol fuels: scientometric study	Konur (2023ai)
43	Overview	Non-waste feedstock-based bioethanol fuels: review	Konur (2023aj)
		Part 9. First Generation Starch-based Bioethanol Fuels	
44	Overview	First generation starch feedstock-based bioethanol fuels: scientometric study	Konur (2023ak)
45	Overview	First generation starch feedstock-based bioethanol fuels: review	Konur (2023al)
46	Overview	Corn pericarp bioethanol through steam explosion and periodic peristalsis	Khatun et al. (2023)
		Part 10. First Generation Sugar Feedstock-based Bioethanol Fuels	
47	Overview	First generation sugar feedstock-based bioethanol fuels: scientometric study	Konur (2023am)
48	Overview	First generation sugar feedstock-based bioethanol fuels: review	Konur (2023an)
		Part 11. Grass-based Bioethanol Fuels	
49	Overview	Grass-based bioethanol fuels: scientometric study	Konur (2023ao)
50	Overview	Grass-based bioethanol fuels: review	Konur (2023ap)
		Part 12. Wood-based Bioethanol Fuels	
51	Overview	Wood-based bioethanol fuels: scientometric study	Konur (2023aq)
52	Overview	Wood-based bioethanol fuels: review	Konur (2023ar)
		Part 13. Cellulose-based Bioethanol Fuels	
53	Overview	Cellulose-based bioethanol fuels: scientometric study	Konur (2023as)
54	Overview	Cellulose-based bioethanol fuels: review	Konur (2023at)
		Part 14. Third Generation Algal Bioethanol Fuels	
55	Overview	Third generation algal bioethanol fuels: scientometric study	Konur (2023au)
56	Overview	Third generation algal bioethanol fuels: review	Nowruzi and Konur (2023av)
		Part 15. Biosyngas-based Bioethanol Fuels	
57	Overview	Biosyngas-based bioethanol fuels: scientometric study	Konur (2023aw)
58	Overview	Biosyngas-based bioethanol fuels: review	Konur (2023ax)

No. reviews: number of reviews presented in this handbook.

1.3.4 Volume 4. Feedstock-based Bioethanol Fuels. II. Waste Feedstocks: Agricultural, Food, Industrial, Urban, Forestry, and Lignocellulosic Waste-based Bioethanol Fuels

1.3.4.1 Introduction to Second Generation Waste Biomass-based Bioethanol Fuels

Chapter 59 provides a scientometric overview of research on second generation waste biomass-based bioethanol fuels (Konur, 2023ay). It finds that research in this field has intensified in key research fronts of pretreatments and hydrolysis of waste biomass, and to a lesser extent fermentation of waste biomass-based hydrolysates, and production and evaluation of second generation waste biomass-based bioethanol fuels. Research in this field has also intensified for feedstocks of lignocellulosic biomass at large, residual starch feedstocks, industrial wastes, residual sugar feedstocks, and to a lesser extent urban wastes, food wastes, and forestry wastes. Thus, it emerges as a distinctive research field, complementing the primary research on first generation bioethanol fuels from the non-waste feedstocks such as starch and sugar feedstocks, grass biomass, wood biomass, algal biomass, and biosyngas among others.

Chapter 60 provides a review of influential research on second generation waste biomass-based bioethanol fuels complementing Chapter 59 (Konur, 2023az). It finds that research in this field has intensified in key research fronts of pretreatment and hydrolysis of waste biomass, to a lesser extent fermentation of waste biomass-based hydrolysates, and production and evaluation of second generation waste biomass-based bioethanol fuels. Research in this field has also intensified for feedstocks of residual starch feedstocks such as corn stover and wheat straw, lignocellulosic biomass at large, and to a lesser extent industrial wastes such as glycerol and residual sugar feedstocks such as sugarcane bagasse.

1.3.4.2 Second Generation Bioethanol Fuels from Residual Sugar Feedstocks

Chapter 61 provides a scientometric overview of research on the second generation bioethanol fuels from residual sugar feedstocks (Konur, 2023ba). It finds that research in this field has intensified in key research fronts of pretreatment and hydrolysis of residual sugar feedstocks, to a lesser extent fermentation of residual sugar feedstock-based hydrolysates, and production and evaluation of second generation bioethanol fuels from residual sugar feedstocks. The most prolific pretreatments have been chemical, enzymatic, and to a lesser extent hydrothermal, mechanical, and microbial pretreatments of biomass at the macroscale and acid, alkaline, steam explosion, LHW, and to a lesser extent IL, organic solvent, ultrasound, NH_3, surfactant, SO_2, O_3, and milling pretreatments at the microscale. Further, sugarcane bagasse and to a lesser extent sugarcane straw, sweet sorghum bagasse, sugarcane molasses, sugar beet pulp, sugarcane vinasse, sugar beet molasses, and energy cane bagasse have been studied intensively in this context.

Chapter 62 provides a review of influential research on second generation bioethanol fuels from residual sugar feedstocks complementing Chapter 61 (Konur, 2023bb). It finds that research in this field has intensified in key research fronts of pretreatment and hydrolysis of residual sugar feedstocks, fermentation of residual sugar feedstock-based hydrolysates, production of second generation bioethanol fuels from residual sugar feedstocks, and to a lesser extent evaluation of second generation bioethanol fuels from residual sugar feedstocks. The most prolific pretreatments have been chemical, acid, enzymatic, alkaline, and to a lesser extent steam explosion, mechanical, MW, LHW, organic solvent, IL, NH_3, and milling pretreatments. Further, the sugarcane bagasse and to a lesser extent sugarcane straw, sorghum bagasse, sweet sorghum straw, and energy cane bagasse have been studied intensively in this context.

1.3.4.3 Second Generation Bioethanol Fuels from Residual Starch Feedstocks

Chapter 63 provides a scientometric overview of second generation bioethanol fuels from residual starch feedstocks (Konur, 2023bc). It finds that one of the most studied feedstocks for bioethanol fuels has been the residual starch feedstocks such as corn, wheat, rice, rye straw, barley, and sorghum residues. Research in this field has intensified in key research fronts of the pretreatment and hydrolysis of residual starch feedstocks, to a lesser extent fermentation of residual starch feedstock-based hydrolysates, and production and evaluation of second generation bioethanol fuels from residual starch feedstocks. The most prolific pretreatments have been enzymatic, chemical, acid, hydrothermal, alkaline, NH_3, steam explosion, mechanical, H_2O_2, organic solvent, LHW, milling, IL, wet oxidation, SO_2, MW, HCW, surfactant, and ultrasound pretreatments. Further, corn stover and wheat straw, rice straw, and to a lesser extent other corn residues, other rice residues, barley straw, rye straw, and sorghum residues have been studied intensively in this context.

Chapter 64 provides a review of influential research on second generation bioethanol fuels from residual starch feedstocks complementing Chapter 63 (Konur, 2023bd). It finds that one of the most studied feedstocks for bioethanol fuels has been the residual starch feedstocks such as corn, wheat, rice, rye, and barley residues. Research in this field has intensified in key research fronts of the pretreatment and hydrolysis of residual starch feedstocks, to a lesser extent fermentation of the residual starch feedstock-based hydrolysates, and production and evaluation of second generation bioethanol fuels from residual starch feedstocks. The most prolific pretreatments have been chemical, acid, alkaline, hydrothermal, NH_3, LHW, and to a lesser extent steam explosion, wet oxidation, and SO_2 pretreatments. Further, corn stover, wheat straw, and to a lesser extent rice straw, rye straw, and barley straw have been studied intensively in this context.

1.3.4.4 Second Generation Food Waste-based Bioethanol Fuels

Chapter 65 provides a scientometric overview of research on second generation food waste-based bioethanol fuels (Konur, 2023be). It finds that one of the most studied feedstocks for bioethanol fuels has been food wastes such as fruit, vegetable, beverage, and dairy wastes. Research in this field has intensified in key research fronts of the pretreatment and hydrolysis of food wastes, fermentation of food waste-based hydrolysates, and production of second generation food waste-based bioethanol fuels and to a lesser extent evaluation of second generation food waste-based bioethanol fuels. Further, the most prolific food wastes have been fruit wastes, food wastes in general, vegetable wastes, dairy wastes, beverage wastes, and to a lesser extent olive wastes, and the most prolific fruit wastes have been empty fruit bunches, other fruit wastes, orange peels, and to a lesser extent apple pomaces, banana peels, cashew apple bagasse, citrus peels, and pineapple waste. Further, the most prolific dairy and beverage wastes have been cheese whey and spent coffee grounds, respectively.

Chapter 66 provides a review of influential research on second generation food waste-based bioethanol fuels complementing Chapter 65 (Konur, 2023bf). It finds that one of the most studied feedstocks for bioethanol fuels has been food wastes such as fruit, vegetable, beverage, and dairy wastes. Research in this field has intensified in key research fronts of the pretreatment and hydrolysis of food wastes, fermentation of food waste-based hydrolysates, and production of second generation food waste-based bioethanol fuels. Further, the most prolific food wastes have been fruit wastes and to a lesser extent other food wastes, and food wastes in general at the macroscale and empty fruit bunches, orange peels, citrus peels, beverage wastes (spent coffee grounds), vegetable wastes such as potato peels, cashew apple bagasse, pineapple waste, grapefruit peels, olive wastes (olive stones), fish wastes, banana peels, and dairy wastes (cheese whey) at the microscale.

1.3.4.5 Second Generation Industrial Waste-based Bioethanol Fuels

Chapter 67 provides a scientometric overview of research on second generation industrial waste-based bioethanol fuels (Konur, 2023bg). It finds that one of the most studied feedstocks for bioethanol fuels has been industrial wastes. Research in this field has intensified in key research fronts of pretreatment and hydrolysis of industrial wastes, fermentation of industrial waste-based hydrolysates, and production of second generation industrial waste-based bioethanol fuels and to a lesser extent evaluation of second generation industrial waste-based bioethanol fuels. Further, milling industrial wastes and to a lesser extent food processing industrial wastes and other industrial wastes have been studied intensively. Biodiesel industrial wastes, ethanol, brewing, plastic, textile, and leather industrial wastes have been the other industrial wastes while the primary wood and paper industrial wastes have been wood chips, sawdust, and paper sludge. Similarly, the primary food industrial wastes have been the fruit, beverage, dairy, and vegetable wastes.

Chapter 68 provides a review of influential research on second generation industrial waste-based bioethanol fuels complementing Chapter 67 (Konur, 2023bh). It finds that one of the most studied feedstocks for bioethanol fuels has been industrial wastes such as milling, food processing, wood and paper, and other industrial wastes such as biodiesel industrial wastes. Research in this field has intensified in key research fronts of pretreatment and hydrolysis of industrial wastes and to a lesser extent fermentation of industrial waste-based hydrolysates, and production and evaluation of second generation industrial waste-based bioethanol fuels. Further, milling and to a lesser extent biodiesel, food processing, wood and paper, ethanol, and brewing industrial wastes have been studied intensively in this context

1.3.4.6 Second Generation Urban Waste-based Bioethanol Fuels

Chapter 69 provides a scientometric overview of research on second generation urban waste-based bioethanol fuels (Konur, 2023bi). It finds that one of the most studied feedstocks for bioethanol fuels has been urban wastes such as food, municipal solid, paper, textile, plastic, and wood wastes. Research in this field has intensified in key research fronts of pretreatment and hydrolysis of urban wastes, fermentation of urban waste-based hydrolysates, and production and evaluation of second generation urban waste-based bioethanol fuels. Further, food and to a lesser extent paper, municipal solid, plastic, textile, and wood wastes have been studied intensively in this context. However, the most prolific food wastes have been fruit wastes, food waste in general, and to a lesser extent vegetable, dairy, kitchen, beverage, and other food wastes.

Chapter 70 provides a review of influential research on second generation urban waste-based bioethanol fuels complementing Chapter 69 (Konur, 2023bj). It finds that one of the most studied feedstocks for bioethanol fuels has been urban wastes such as food, paper, textile, plastic, and wood wastes. Research in this field has intensified in key research fronts of pretreatment and hydrolysis of urban wastes and to a lesser extent fermentation of urban waste-based hydrolysates and production of second generation urban waste-based bioethanol fuels. Further, food waste and to a lesser extent paper, municipal solid, textile, plastic, and wood wastes have been studied intensively in this context, and the most prolific food wastes have been fruit wastes such as orange peel waste and to a lesser extent food waste in general, beverage wastes such as spent coffee grounds, other food wastes, vegetable wastes such as potato peels, dairy wastes such as cheese whey, and kitchen wastes in general such as waste meals and bread.

1.3.4.7 Second Generation Forestry Waste-based Bioethanol Fuels

Chapter 71 provides a scientometric overview of research on second generation forestry waste-based bioethanol fuels (Konur, 2023bk). It finds that one of the most studied feedstocks for bioethanol fuels has been forestry wastes such as wood chips and to a lesser extent tree prunings, sawdust, bark, spent sulfite liquor, and wood wastes. Research in this field has intensified in key research fronts of pretreatment and hydrolysis of forestry wastes, and to a lesser extent fermentation of forestry waste-based hydrolysates, and production and evaluation of second generation forestry waste-based bioethanol fuels.

Chapter 72 provides a review of the influential research on second generation forestry waste-based bioethanol fuels complementing Chapter 71 (Konur, 2023bl). It finds that one of the most studied feedstocks for bioethanol fuels has been forestry wastes. Research in this field has intensified in this context in key research fronts of pretreatment and hydrolysis of forestry wastes and to a lesser extent fermentation of forestry waste-based hydrolysates, and production of second generation forestry waste-based bioethanol fuels. Further, wood chips, forest residues, and to a lesser extent sawdust, tree prunings, bark, wood waste, and spent sulfite liquor have been studied intensively in this context.

1.3.4.8 Lignocellulosic Biomass-based Bioethanol Fuels

Chapter 73 provides a scientometric overview of research on lignocellulosic biomass-based bioethanol fuels (Konur, 2023bm). It finds that one of the most studied feedstocks for bioethanol fuels has been lignocellulosic biomass at large. Research in this field has intensified in key research fronts of pretreatment of lignocellulosic biomass and to a lesser extent hydrolysis of lignocellulosic biomass, fermentation of lignocellulosic biomass-based hydrolysates, and production and evaluation of lignocellulosic biomass-based bioethanol fuels. Thus, it emerges as a distinctive research field, complementing the research on the second generation bioethanol fuels from agricultural residues as well as food, industrial, urban, and forestry wastes among others.

Chapter 74 provides a review of influential research on lignocellulosic biomass-based bioethanol fuels complementing Chapter 73 (Konur, 2023bn). It finds that one of the most studied feedstocks for bioethanol fuels has been lignocellulosic biomass at large. Research in this field has intensified in this context in key research fronts of pretreatment and hydrolysis of lignocellulosic biomass, to a lesser extent fermentation of lignocellulosic biomass-based hydrolysates, and production and evaluation of lignocellulosic biomass-based bioethanol fuels. Thus, it emerges as a distinctive research field, complementing primarily the research on second generation bioethanol fuels from agricultural residues as well as food, industrial, urban, and forestry wastes among others as well as from grass and wood.

Chapter 75 provides a review of research on the production and utilization of bioethanol fuels in a biorefinery context using agro- and forest-industrial waste-based lignocellulosic biomass (Mendieta et al., 2023). It focuses on a wide range of lignocellulosic biomass as feedstocks for bioethanol fuel production and biofuels and biochemicals as derivatives of bioethanol fuels.

TABLE 1.5
The Microresearch Fronts for Volume 4: Feedstock-based Bioethanol Fuels. II. Waste Feedstocks: Agricultural, Food, Industrial, Urban, Forestry, and Lignocellulosic Waste-based Bioethanol Fuels

No.	Research Fronts	Chapter Title	Ref.
	Part 16. Introduction to Second Generation Waste Biomass-based Bioethanol Fuels		
59	Overview	Second generation waste biomass-based bioethanol fuels: scientometric study	Konur (2023ay)
60	Overview	Second generation waste biomass-based bioethanol fuels: review	Konur (2023az)
	Part 17. Second Generation Bioethanol Fuels from Residual Sugar Feedstocks		
61	Overview	Second generation bioethanol fuels from residual sugar feedstocks: scientometric study	Konur (2023ba)
62	Overview	Second generation bioethanol fuels from residual sugar feedstocks: review	Konur (2023bb)
	Part 18. Second Generation Bioethanol Fuels from Residual Starch Feedstocks		
63	Overview	Second generation bioethanol fuels from residual starch feedstocks: scientometric study	Konur (2023bc)
64	Overview	Second generation bioethanol fuels from residual starch feedstocks: review	Konur (2023bd)

(Continued)

TABLE 1.5 (*Continued*)
The Microresearch Fronts for Volume 4: Feedstock-based Bioethanol Fuels. II. Waste Feedstocks: Agricultural, Food, Industrial, Urban, Forestry, and Lignocellulosic Waste-based Bioethanol Fuels

No.	Research Fronts	Chapter Title	Ref.
	Part 19. Second Generation Food Waste-based Bioethanol Fuels		
65	Overview	Second generation food waste-based bioethanol fuels: scientometric study	Konur (2023be)
66	Overview	Second generation food waste-based bioethanol fuels: review	Konur (2023bf)
	Part 20. Second Generation Industrial Waste-based Bioethanol Fuels		
67	Overview	Second generation industrial waste-based bioethanol fuels: scientometric study	Konur (2023bg)
68	Overview	Second generation industrial waste-based bioethanol fuels: review	Konur (2023bh)
	Part 21. Second Generation Urban Waste-based Bioethanol Fuels		
69	Overview	Second generation urban waste-based bioethanol fuels: scientometric study	Konur (2023bi)
70	Overview	Second generation urban waste-based bioethanol fuels: review	Konur (2023bj)
		Part 22. Second Generation Forestry Waste-based Bioethanol Fuels	
71	Overview	Second generation forestry waste-based bioethanol fuels: scientometric study	Konur (2023bk)
72	Overview	Second generation forestry waste-based bioethanol fuels: review	Konur (2023bl)
		Part 23. Lignocellulosic Biomass-based Bioethanol Fuels	
73	Overview	Lignocellulosic biomass-based bioethanol fuels: scientometric study	Konur (2023bm)
74	Overview	Lignocellulosic biomass-based bioethanol fuels: review	Konur (2023bn)
75	Overview	Production and uses of bioethanol in an integrated biorefinery from agro- and forest-industrial waste	Mendieta et al. (2023)

No. reviews: number of reviews presented in this handbook.

1.3.5 Volume 5. Evaluation and Utilization of Bioethanol Fuels. I: Evaluation of Bioethanol Fuels, Transport Engines, and Bioethanol Sensors

1.3.5.1 Introduction to the Evaluation and Utilization of Bioethanol Fuels

Chapter 76 provides a scientometric overview of research on the evaluation and utilization of bioethanol fuels (Konur, 2023bo). It finds that research in this field has intensified in recent years. The primary focus of the research has been the utilization of bioethanol fuels in fuel cells, evaluation of bioethanol fuels, utilization of bioethanol fuels in the transport engines, production of biohydrogen fuels from bioethanol fuels, and to a lesser extent the experiences of countries, development and utilization of bioethanol sensors, and production of biochemicals from bioethanol fuels. Additionally, research on gasoline fuels and applications of nanotechnology has also been related to this field.

Chapter 77 provides a review of influential research on the evaluation and utilization of bioethanol fuels complementing Chapter 76 (Konur, 2023bp). Research in this field has also intensified in research fronts of direct utilization of bioethanol fuels in fuel cells, evaluation of bioethanol fuels, utilization of bioethanol fuels for biohydrogen production for fuel cells, to a lesser extent bioethanol sensors, and utilization of bioethanol fuels in transportation such as gasoline engines. Additionally, research on the gasoline fuels and applications of nanotechnology in this field has also been related to this field.

Chapter 78 provides a scientometric overview of research on the evaluation and utilization of bioethanol fuels published between 2016 and 2021 (Konur, 2023bq). It finds that research in this field has intensified in recent years. The primary focus of the research has been utilization of bioethanol fuels in transportation such as gasoline engines, direct utilization of bioethanol fuels in fuel cells, bioethanol sensors, and to a lesser extent evaluation of bioethanol fuels, utilization of bioethanol fuels for biohydrogen production for fuel cells, utilization of bioethanol fuels for biochemical production, and country-based experiences of bioethanol fuels. Additionally, research on gasoline fuels and applications of nanotechnology has also been related to this field.

Chapter 79 provides a review of influential research on the evaluation and utilization of bioethanol fuels published between 2016 and 2021 complementing Chapter 78 (Konur, 2023br). It finds that research in the fields of direct utilization of bioethanol fuels in fuel cells, bioethanol sensors, utilization of bioethanol fuels in transportation such as gasoline engines, utilization of bioethanol fuels for biochemical production, evaluation of bioethanol fuels, and utilization of bioethanol fuels for biohydrogen production for fuel cells has intensified in recent years. Further, the application of nanotechnology in this field has been very intensive.

Chapter 80 provides a scientometric overview of research on gasoline fuels (Konur, 2023bs). It finds that biomass-based bioethanol fuels and other bioalcohol fuels have increasingly been used in blending gasoline fuels in recent years to reduce their adverse effects. In this context, there has been intense research in research fronts of emissions, blends with bioethanol fuels and other bioalcohols, combustion, economics, production, and to a lesser extent the environmental impact, properties, desulfurization, and health impact of gasoline fuels.

Chapter 81 provides a review of influential research on gasoline fuels and their blends with bioalcohol fuels complementing Chapter 80 (Konur, 2023bt). It finds that biomass-based bioethanol fuels and other bioalcohol fuels have increasingly been used in blending gasoline fuels in recent years to reduce their adverse effects. In this context, there has been intense research in research fronts of emissions, blends with bioethanol fuels and other bioalcohols, combustion, economics, production, and to a lesser extent the environmental impact, properties, desulfurization, and health impact of gasoline fuels.

Chapter 82 provides a scientometric overview of research on the application of nanotechnology with regard to bioethanol fuels and their derivatives (Konur, 2023bu). It finds that research in this field has particularly intensified in research fronts of sensing of bioethanol gas, oxidation of bioethanol fuels for fuel cell applications, and to a lesser extent production of bioethanol fuels, production of biohydrogen fuels from bioethanol fuels, utilization of bioethanol fuels in the transport engines, and production of biochemicals from bioethanol fuels.

Chapter 83 provides a review of influential research on the application of nanotechnology in bioethanol fuels complementing Chapter 82 (Konur, 2023bv). It finds that research in this field has particularly intensified in research fronts of sensing of bioethanol gas, oxidation of bioethanol fuels for fuel cell applications, and to a lesser extent production of bioethanol fuels from biosyngas and production of biohydrogen fuels from bioethanol fuels for fuel cells.

1.3.5.2 Utilization of Bioethanol Fuels in the Transport Engines

Chapter 84 provides a scientometric overview of research on the utilization of bioethanol fuels in transport engines (Konur, 2023bw). It finds that research has intensified in research fronts of engine emissions, fuel combustion, engine performance, fuel properties, and to a lesser extent gasoline engine and health impact of bioethanol fuels. The focus has also been on the utilization of bioethanol fuels in blending with both gasoline fuels and petrodiesel fuels in gasoline and diesel engines, respectively.

Chapter 85 provides a review of influential research on the utilization of bioethanol fuels in transport engines complementing Chapter 84 (Konur, 2023bx). It finds that biomass-based bioethanol fuels have increasingly been used in blending gasoline fuels as well as petrodiesel fuels in recent years. In this context, the research has intensified in the research fronts of engine emissions, fuel combustion, and to a lesser extent engine performance, fuel properties, gasoline engine, and health impact of bioethanol fuels. The focus has also been on the utilization of bioethanol fuels in blending with both gasoline fuels and petrodiesel fuels in gasoline and diesel engines, respectively.

1.3.5.3 Evaluation of Bioethanol Fuels

Chapter 86 provides a scientometric overview of research on the evaluation of bioethanol fuels (Konur, 2023by). It finds that there has been a significant focus on evaluative studies in bioethanol fuels. The research has progressed in research fronts of environmental issues, technoeconomics, technoassessment, and to a lesser extent policy, economics, and business. Furthermore, environmental issues included the research fronts of emissions and to a lesser extent life cycle assessment (LCA), land use, water consumption, and other issues.

Chapter 87 provides a review of influential research on the evaluation of bioethanol fuels complementing Chapter 86 (Konur, 2023bz). It finds that research in this field has progressed in research fronts of the environmental impact of bioethanol fuels, technoeconomics of bioethanol fuels, and to a lesser extent energy assessments of bioethanol fuels and policy and economic issues in bioethanol fuel production. Furthermore, the secondary research fronts have been emissions during the production and utilization of bioethanol fuels and to a lesser extent LCA of bioethanol fuels and other environmental issues.

Chapter 88 discusses the impact of corn ethanol production on land use in Brazil (Cortez, 2023). It develops a strategy for stopping deforestation of the Amazon forest and freeing land by implementing a large-scale corn ethanol program in addition to the sugarcane ethanol production in Brazil. The key issue is to integrate the corn ethanol production with the beef production based on the US example to reduce deforestation and free land for food production, reforestation, and biodiversity. This also improves the corn ethanol LCA with a negative indirect land use change (iLUC).

Chapter 89 discusses the contribution of first generation sugarcane-based bioethanol fuels to climate change mitigation (Moreira and Goldemberg, 2023). It finds that their potential has been frequently underestimated. Hybrid vehicles fed with neat bioethanol or high ethanol blends can lead to greater reduction in GHG emission compared with conventional fuels and battery. Furthermore, the land area required for its feedstock production is significantly reduced when compared with conventional cars using pure bioethanol. The traditional first generation sugar cane ethanol production route already provides lower GHG emissions compared with fossil fuels, and technical improvements can widen this gap, making it a serious competitor with electric vehicles.

1.3.5.4 Utilization of Bioethanol Fuels

Chapter 90 provides a scientometric overview of research on the utilization of bioethanol fuels (Konur, 2023ca). It finds that the primary focus of research in this area has been the direct utilization of bioethanol fuels in fuel cells, utilization of bioethanol fuels in gasoline and diesel engines, production of biohydrogen fuels from bioethanol fuels for fuel cells, and to a lesser extent development and utilization of bioethanol sensors and production of biochemicals from bioethanol fuels. Additionally, research on the gasoline fuels and applications of nanotechnology in this field has also been closely related to this field.

Chapter 91 provides a review of influential research on the utilization of bioethanol fuels complementing Chapter 90 (Konur, 2023cb). It finds that research in this field has intensified in recent years. The primary focus of the research in this area has been the direct utilization of bioethanol fuels in fuel cells, utilization of bioethanol fuels in the gasoline and diesel engines, production of biohydrogen fuels from bioethanol fuels, and to a lesser extent production of biochemicals from bioethanol fuels and development and utilization of bioethanol sensors. Additionally, research on gasoline fuels and applications of nanotechnology in this field has also been closely related to this field.

1.3.5.5 Bioethanol Fuel Sensors

Chapter 92 provides a scientometric overview of research on bioethanol fuel sensors (Konur, 2023cc). It finds that research in this field has intensified in research fronts of zinc oxide (ZnO)-based bioethanol fuel sensors, tin oxide (SnO_2)-based bioethanol fuel sensors, and other material-based bioethanol fuel sensors. Furthermore, there have been further research fronts within each of these research areas: nanomaterial-based sensors and conventional material-based sensors. However, the focus of the research has been on the development of nanomaterial-based bioethanol sensors. In the meantime, research in the nanomaterials and nanotechnology and sensors has intensified in recent years to become major research fields in the scientific research.

Chapter 93 provides a review of influential research on bioethanol fuel sensors complementing Chapter 92 (Konur, 2023cd). It finds that research in this field has intensified in research fronts of zinc oxide (ZnO)-based bioethanol fuel sensors, tin oxide (SnO_2)-based bioethanol fuel sensors, and other material-based bioethanol fuel sensors such as thin film- and oxide-based bioethanol fuel sensors. Furthermore, there have been further research fronts within each of these research areas: nanomaterial-based bioethanol sensors and conventional material-based bioethanol sensors. However, the focus of the research has been on nanomaterial-based bioethanol sensors.

TABLE 1.6
The Microresearch Fronts for Volume 5: Evaluation and Utilization of Bioethanol Fuels. I: Evaluation of Bioethanol Fuels, Transport Engines, and Bioethanol Sensors

No.	Research Fronts	Chapter Title	Ref.
		Part 24. Introduction to the Evaluation and Utilization of Bioethanol Fuels	
76	Overview	Bioethanol fuel evaluation and utilization: scientometric study.	Konur (2023bo)
77	Overview	Bioethanol fuel evaluation and utilization: review	Konur (2023bp)
78	Overview	Bioethanol fuel evaluation and utilization: hot papers, scientometric study	Konur (2023bq)
79	Overview	Bioethanol fuel evaluation and utilization: hot papers, review	Konur (2023br)
80	Gasoline fuels	Gasoline fuels: scientometric study	Konur (2023bs)
81	Gasoline fuels	Gasoline fuels: review	Konur (2023bt)
82	Nanotechnology applications	Nanotechnology applications in bioethanol fuels: scientometric study	Konur (2023bu)
83	Nanotechnology applications	Nanotechnology applications in bioethanol fuels	Konur (2023bv)
		Part 25. Utilization of Bioethanol Fuels in the Transport Engines	
84	Overview	Utilization of bioethanol fuels in the transport engines: scientometric study	Konur (2023bw)
85	Overview	Utilization of bioethanol fuels in the transport engines: review	Konur (2023bx)

(Continued)

TABLE 1.6 (*Continued*)
The Microresearch Fronts for Volume 5: Evaluation and Utilization of Bioethanol Fuels. I: Evaluation of Bioethanol Fuels, Transport Engines, and Bioethanol Sensors

No.	Research Fronts	Chapter Title	Ref.
		Part 26. Evaluation of Bioethanol Fuels	
86	Overview	Evaluation of bioethanol fuels: scientometric study	Konur (2023by)
87	Overview	Evaluation of bioethanol fuels: review	Konur (2023bz)
88	Corn-based bioethanol fuels	Impact of corn ethanol production in land use in Brazil	Cortez (2023)
89	Sugarcane-based bioethanol fuels	Ethanol fuel from sugar cane-present and future contribution to climate change mitigation	Moreira and Goldemberg (2023)
		Part 27. Utilization of Bioethanol Fuels	
90	Overview	Utilization of bioethanol fuels: scientometric study	Konur (2023ca)
91	Overview	Utilization of bioethanol fuels: review	Konur (2023cb)
		Part 28. Bioethanol Fuel Sensors	
92	Overview	Bioethanol fuel sensors: scientometric study	Konur (2023cc)
93	Overview	Bioethanol fuel sensors: review	Konur (2023cd)

No. reviews: number of reviews presented in this handbook.

1.3.6 Volume 6. Evaluation and Utilization of Bioethanol Fuels. II: Biohydrogen Fuels, Fuel Cells, Biochemicals, and Country Experiences

1.3.6.1 Country-based Experience of Bioethanol Fuels

Chapter 94 provides a scientometric overview of research on the country-based experience of bioethanol fuels (Konur, 2023ce). It finds that the primary focus of the research in this area has been the optimization of bioethanol and biomass production and development of policies to promote the production and use of bioethanol fuels worldwide. In this context, the experiences of Brazil, the USA, China, and to a lesser extent, Thailand, Europe, India, Malaysia, Sweden, Colombia, Canada, and the UK have been extensively studied. There has been a significant focus on the evaluative studies in bioethanol fuels such as technoeconomic and life cycle studies.

Chapter 95 provides a review of influential research on the country-based experience of bioethanol fuels complementing Chapter 94 (Konur, 2023cf). It finds that research in this field has intensified in recent years. The primary focus of the research has been the optimization of the bioethanol and biomass production and the development of policies to promote the production and use of bioethanol fuels worldwide. In this context, the experiences of Brazil, the USA, China, and to a lesser extent, Europe, India, Malaysia, and Colombia have been extensively studied. Further, there has been a significant focus on the evaluative studies in bioethanol fuels such as technoeconomic and life cycle studies.

Chapter 96 provides a review of research on the production of first generation bioethanol fuels from corn and sugarcane in the USA and Brazil, respectively (Senatore et al., 2023). It finds that the major producers of bioethanol fuels are the USA from corn and Brazil from sugarcane. Dry grind (DG) and to a lesser extent wet milling (WM) processes were used in the USA while Melle-Boinot (MB) and modified Melle-Boinot (MMB) processes were used in Brazil to convert sugarcane into bioethanol. The key steps used in these processes are physical, chemical, and enzymatic pretreatments of biomass, fermentation of hydrolysates by yeasts, and distillation of bioethanol fuels from the fermentation broth. It compares the productivity and production costs for these three main production processes.

Chapter 97 provides a scientometric overview of research on the Chinese experience of bioethanol fuels (Konur, 2023cg). It finds that China has been one of the most prolific countries engaged in bioethanol fuel research. The Chinese research in the field of bioethanol fuels has intensified in key research fronts of production of bioethanol fuels, utilization of bioethanol fuels, and to a lesser extent evaluation of bioethanol fuels. For the first research front, pretreatment and hydrolysis of the feedstocks, and to a lesser extent fermentation of the feedstock-based hydrolysates, production of bioethanol fuels, and distillation of bioethanol fuels are the key research areas while for the second research front, utilization of bioethanol fuels in fuel cells and transport engines, bioethanol-based biohydrogen fuels, bioethanol sensors, and to a lesser extent bioethanol-based biochemicals are the key research areas. Research in this field has also intensified for the feedstocks of bioethanol fuels for their utilization, starch feedstock residues, cellulose, wood biomass, lignocellulosic biomass at large, and to a lesser extent industrial waste, lignin, grass biomass, hydrolysates, biomass at large, sugar feedstocks, sugar feedstock residues, algal biomass, lignocellulosic wastes, starch feedstocks, urban wastes, forestry wastes, food wastes, and plants.

Chapter 98 provides a review of influential research on the Chinese experience of bioethanol fuels complementing Chapter 97 (Konur, 2023ch). It finds that China has been one of the most prolific countries engaged in bioethanol fuel research. The Chinese research in this field has intensified in key research fronts of utilization of bioethanol fuels and to a lesser extent the production of bioethanol fuels. For the first research front, direct utilization of bioethanol fuels in fuel cells and to a lesser extent utilization of bioethanol fuels in transport engines, bioethanol biosensors, and utilization of bioethanol fuel for the production of biohydrogen fuels for fuel cells have been prolific research areas. Further, for the second research front, hydrolysis and pretreatments of biomass have been the prolific research areas. Research in this field has also intensified for feedstocks of bioethanol fuels for their utilization and to a lesser extent cellulose, starch feedstock residues, wood biomass, lignocellulosic biomass at large, industrial waste, and sugar feedstock residues.

Chapter 99 provides a scientometric overview of research on the US experience of bioethanol fuels (Konur, 2023ci). It finds that the USA has been one of the most prolific countries engaged in bioethanol fuel research. The US research in this field has intensified in key research fronts of production of bioethanol fuels and to a lesser extent utilization and evaluation of bioethanol fuels. For the first research front, pretreatment and hydrolysis of feedstocks and to a lesser extent fermentation of the feedstock-based hydrolysates, production of bioethanol fuels, and distillation of bioethanol fuels have been the key research areas while for the second research front, utilization of the bioethanol fuels in fuel cells and transport engines, and the production of biohydrogen fuels for fuel cells have been the key research areas. Research in this field has also intensified for feedstocks of lignocellulosic biomass at large, cellulose, starch feedstock residues, wood biomass, bioethanol fuels, lignin, and to a lesser extent grass biomass, industrial waste, sugars, biomass at large, starch feedstocks, hydrolysates, lignocellulosic wastes, sugar feedstock residues, sugar feedstocks, and algal biomass.

Chapter 100 provides a review of influential research on the US experience of bioethanol fuels complementing Chapter 99 (Konur, 2023cj). It finds that the USA has been one of the most prolific countries engaged in the bioethanol fuel research. US research in this field has intensified in key research fronts of production of bioethanol fuels and to a lesser extent utilization and evaluation of bioethanol fuels. For the first research front, pretreatment and hydrolysis of feedstocks and to a lesser extent fermentation of hydrolysates and production of bioethanol fuels have been the key research areas while for the second research front, direct utilization of bioethanol fuels in fuel cells and bioethanol-based biohydrogen fuels have been the key research areas. Research in this field has also intensified for the feedstocks of wood biomass, starch feedstock residues, grass biomass, bioethanol fuels, cellulose, lignin, and to a lesser extent starch feedstocks, lignocellulosic biomass at large, and sugar feedstock residues.

Chapter 101 provides a scientometric overview of the European experience of bioethanol fuels (Konur, 2023ck). It finds that Europe has been one of the most prolific producers of research in bioethanol fuels, producing 28% of the global research output in this field. The European research in this field has intensified in this context in key research fronts of production of bioethanol fuels, utilization of bioethanol fuels, and to a lesser extent evaluation of bioethanol fuels. For the first research front, pretreatment and hydrolysis of feedstocks, fermentation of the feedstock-based hydrolysates, production of bioethanol fuels, and to a lesser extent distillation of bioethanol fuels are the key research areas while for the second research front, utilization of bioethanol fuels in fuel cells and transport engines, bioethanol-based biohydrogen fuels, and to a lesser extent bioethanol sensors and bioethanol-based biochemicals are the key research areas. Research in this field has also intensified for the feedstocks of lignocellulosic biomass at large, bioethanol fuels, cellulose, wood biomass, starch feedstock residues, lignin, sugars, and to a lesser extent industrial wastes, hydrolysates, biomass at large, sugar feedstock residues, urban wastes, and food wastes.

Chapter 102 provides a review of influential research on the European experience of bioethanol fuels complementing Chapter 101 (Konur, 2023cl). It finds that Europe has been one of the most prolific producers engaged in the bioethanol fuel research, producing 28% of the global research output in this field. The European research in the field of bioethanol fuels has intensified in this context in key research fronts of production of bioethanol fuels, utilization of bioethanol fuels, and to a lesser extent evaluation of bioethanol fuels. For the first research front, pretreatment of feedstocks and to a lesser extent hydrolysis of feedstocks, fermentation of hydrolysates, and production of bioethanol fuels are the key research areas while for the second research front, direct utilization of bioethanol fuels in fuel cells and bioethanol-based biohydrogen fuels are the key research areas. Research in this field has also intensified for feedstocks of wood biomass, bioethanol fuels, cellulose, lignocellulosic biomass at large, starch feedstock residues, lignin, sugars, and biomass at large.

1.3.6.2 Bioethanol Fuel-based Biohydrogen Fuels

Chapter 103 provides a scientometric overview of research on bioethanol fuel-based biohydrogen fuels (Konur, 2023cm). It finds that research in this field has progressed in research fronts of biohydrogen production by steam reforming of bioethanol fuels and to a lesser photocatalytic biohydrogen production from bioethanol fuels, by oxidative steam reforming of bioethanol fuels, by partial oxidation of bioethanol fuels, by autothermal reforming of bioethanol fuels, and by biohydrogen production from bioethanol fuels in general.

Chapter 104 provides a reviews of influential research on bioethanol fuel-based biohydrogen fuels complementing Chapter 103 (Konur, 2023cn). It finds that research in this field has progressed in the research fronts of biohydrogen production by steam reforming of bioethanol and to a lesser extent photocatalytic biohydrogen production, oxidative steam reforming of bioethanol, partial oxidation of bioethanol, autothermal reforming of bioethanol, and biohydrogen production from bioethanol in general.

Chapter 105 provides a review of research on steam reforming of bioethanol fuels for the production of biohydrogen fuels (Phung and Vu, 2023). It outlines the research on the catalysts, their supports and promoters, the reaction conditions, and the mechanisms. It covers noble metal catalysts and nickel (Ni) and cobalt (Co) catalysts for steam reforming of bioethanol fuels and discusses effects of the operating and reaction conditions, membrane-assisted ethanol steam reforming, and reaction mechanisms of ethanol steam reforming. These studies hint that the operating conditions of ethanol steam reforming can be optimized using a wide set of catalysts and catalyst supports as well as catalyst promoters to obtain a high biohydrogen yield.

1.3.6.3 Bioethanol Fuel Cells

Chapter 106 provides a scientometric overview of research on bioethanol fuel cells (Konur, 2023co). However, research on direct utilization of bioethanol fuels in DEFCs has intensified in recent years, primarily in research fronts of ethanol electrooxidation on platinum (Pt) electrocatalysts and palladium (Pd) electrocatalysts for DEFCs. The other research fronts are ethanol electrooxidation on both Pt and Pd electrocatalysts and the other electrocatalysts, other issues in the ethanol electrooxidation, and the DEFCs themselves. Further, the focus of research has been on the development of nanomaterials and conventional materials for most efficient processes for ethanol electrooxidation and DEFCs.

Chapter 107 provides a review of influential research on bioethanol fuel cells complementing Chapter 106 (Konur, 2023cp). It finds that research in this field has intensified on polymer exchange membrane fuel cells (PEMFCs). However, research on direct utilization of bioethanol fuels in DEFCs has intensified in recent years, primarily in the research fronts of ethanol electrooxidation on Pt electrocatalysts and on Pd electrocatalysts for DEFCs. The other research fronts are ethanol electrooxidation on both Pt and Pd electrocatalysts and on other electrocatalysts, other issues in the ethanol electrooxidation, and DEFCs themselves. Further, the recent research focus has been on the development of nanomaterial-based electrocatalyst systems to improve the efficiency of DEFCs.

1.3.6.4 Bioethanol Fuel-based Biochemicals

Chapter 108 provides a scientometric overview of research on bioethanol fuel-based biochemicals (Konur, 2023cq). It finds that research in this field has primarily intensified in the research fronts of bioethylenes, biobutadienes, biobutanols, bioacetaldehydes, and to a lesser extent bio-ethyl acetates, biopropylene, bio-acetic acids, bio-diethyl ethers, and bioisobutenes. Further, the focus of research has been on the development of catalysts and catalyst supports, and applications of nanotechnology have been relatively scarce.

Chapter 109 provides a review of influential research on bioethanol fuel-based biochemicals complementing Chapter 108 (Konur, 2023cr). It finds that research in this field has primarily intensified in the research fronts of biobutanols, bioethylenes, bioacetaldehydes, bio-acetic acids, biobutadienes, bio-ethyl acetates, and to a lesser extent bio-diethyl ethers, bioisobutenes, and biopropylene. Further, the focus of research has been more on the development of catalysts and catalyst supports rather than the operating conditions.

Chapter 110 provides a review of research on the conversion of bioethanol fuels into hydrocarbons including biochemicals (Caballero et al., 2023). It finds that ethanol can be selectively converted into several chain hydrocarbons ranging from C_3 to C_4 light olefins, such as propylene, n-butenes, and 1,3-butadiene, or even to C_5–C_{12} oligomers, aromatics, and paraffins that fall into gasoline and middle distillate fuel classifications. Further, enhancing catalytic selectivity, activity, and stability has been a focus of research by modifying catalytic composition by tuning the Si/Al ratio in zeolites, adding secondary dopants, or optimizing reaction conditions. As a result, a number of highly effective catalytic systems and possible reaction pathways involved in these systems have been studied.

TABLE 1.7
The Microresearch Fronts for Volume 6: Evaluation and Utilization of Bioethanol Fuels. II: Biohydrogen Fuels, Fuel Cells, Biochemicals, and Country Experiences

No.	Research Fronts	Chapter Title	Ref.
	Part 29. Country-based Experience of Bioethanol Fuels		
94	Overview	Country-based experience of bioethanol fuels: scientometric study	Konur (2023ce)
95	Overview	Country-based experience of bioethanol fuels: review	Konur (2023cf)
96	US and Brazilian experiences	Ethanol: brief review of the major production methods in the world	Senatore et al. (2023)

(Continued)

TABLE 1.7 (*Continued*)
The Microresearch Fronts for Volume 6: Evaluation and Utilization of Bioethanol Fuels. II: Biohydrogen Fuels, Fuel Cells, Biochemicals, and Country Experiences

No.	Research Fronts	Chapter Title	Ref.
97	Chinese experience	Chinese experience of bioethanol fuels: scientometric study	Konur (2023cg)
98	Chinese experience	Chinese experience of bioethanol fuels: review	Konur (2023ch)
99	The US experience	The US experience of bioethanol fuels: scientometric study	Konur (2023ci)
100	The US experience	The US experience of bioethanol fuels: review	Konur (2023cj)
101	European experience	European experience of bioethanol fuels: scientometric study	Konur (2023ck)
102	European experience	European experience of bioethanol fuels: review	Konur (2023cl)
	Part 30. Bioethanol Fuel-based Biohydrogen Fuels		
103	Overview	Bioethanol fuel-based biohydrogen fuels: scientometric study	Konur (2023cm)
104	Overview	Bioethanol fuel-based biohydrogen fuels: review	Konur (2023cn)
105	Steam reforming of bioethanol fuels	Recent advances in the steam reforming of bioethanol for the production of biohydrogen fuels	Phung and Vu (2023)
	Part 31. Bioethanol Fuel Cells		
106	Overview	Bioethanol fuel cells: scientometric study	Konur (2023co)
107	Overview	Bioethanol fuel cells: review	Konur (2023cp)
	Part 32. Bioethanol Fuel-based Biochemicals		
108	Overview	Bioethanol fuel-based biochemicals: scientometric study	Konur (2023cq)
109	Overview	Bioethanol fuel-based biochemicals: review	Konur (2023cr)
110	Overview	An overview of bioethanol conversion to hydrocarbons	Caballero et al. (2023)

No. reviews: number of reviews presented in this handbook.

1.4 DISCUSSION

The first volume of this handbook provides an overview of research on the production, evaluation, and utilization of bioethanol fuels in general as an introduction to bioethanol fuels in the first part with three book chapters and provides an overview of research on the pretreatments of biomass for sugar and bioethanol fuel production with 15 book chapters (Table 1.2). In this context, there are five key sections for the second part: introduction to the pretreatments of biomass, biological pretreatments of biomass, chemical pretreatments of biomass, hydrothermal pretreatments of biomass, and mechanical pretreatments of biomass.

The first volume also indicates that a wide range of pretreatments alone or in combination fractionate the biomass to its constituents of cellulose, lignin, and hemicellulose and improve both the sugar and bioethanol fuel yield, making bioethanol fuels more competitive in relation to crude oil- and natural gas-based fuels.

The second volume of this handbook provides an overview of research on the production processes for bioethanol fuels in general, hydrolysis of the pretreated biomass for bioethanol production, microbial fermentation of hydrolysates and substrates with yeasts for bioethanol production, and separation and distillation of bioethanol fuels from the fermentation broth, complementing the research on the biomass pretreatments presented in the first volume (Table 1.3).

The second volume of this handbook presents an overview of research on the biomass hydrolysis in general, wood hydrolysis, straw hydrolysis, and cellulose hydrolysis for bioethanol fuel production in the first section for biomass hydrolysis.

It continues to provide an overview of research on microbial hydrolysate fermentation for bioethanol production in general, alternative fermentation processes for bioethanol fuel production such as SSF and CBP compared with the SHF process, metabolic engineering of microorganisms and substrates for bioethanol fuel production, and utilization of *Saccharomyces cerevisiae* for microbial fermentation of hydrolysates for bioethanol fuel production in the second section for the hydrolysate fermentation.

It continues to provide an overview of research on bioethanol fuel separation from the fermentation broth in the last section.

The second volume also indicates that a wide range of pretreatments alone or in combination fractionate the biomass to its constituents of cellulose, lignin, and hemicellulose, and improve both the sugar and bioethanol fuel yield, making the bioethanol fuels from waste and non-waste biomass more competitive in relation to crude oil- and natural gas-based fuels.

The second volume also indicates that hydrolysis of the biomass, microbial hydrolysate fermentation, and separation and distillation of bioethanol fuels from fermentation broth together with the biomass pretreatments are the fundamental production processes for bioethanol fuel production, making bioethanol fuels more competitive in relation to crude oil- and natural gas-based fuels.

The third volume of this handbook, with 19 chapters, provides an overview of research on the production of bioethanol fuels from the non-waste feedstocks such as first generation starch and sugar feedstocks, grass biomass, wood biomass, cellulose, third generation algal feedstocks, and biosyngas (Table 1.4). In this context, there are nine key sections where the first four chapters cover the production of bioethanol fuels from feedstocks at large and non-waste feedstocks.

The third volume indicates that research on the production of bioethanol fuels from these non-waste feedstocks has intensified in recent years to become a major part of the bioenergy and biofuel research together primarily with biodiesel, biohydrogen, and biogas research as a sustainable alternative to crude oil-based gasoline and petrodiesel fuels as well as LNG.

The third volume also indicates that a wide range of pretreatments alone or in combination fractionate the biomass to its constituents of cellulose, lignin, and hemicellulose, and improve both sugar and bioethanol fuel yield, making the bioethanol fuels from these non-waste feedstocks with the exception of the biosyngas feedstock more competitive in relation to crude oil- and natural gas-based fuels.

The third volume also indicates that hydrolysis of the biomass, microbial hydrolysate fermentation, and separation and distillation of bioethanol fuels from fermentation broth together with biomass pretreatments are the fundamental production processes for bioethanol fuel production from these non-waste feedstocks with the exception of the biosyngas feedstock, making bioethanol fuels more competitive in relation to crude oil- and natural gas-based fuels.

The fourth volume of this handbook, with 17 chapters, provides an overview of research on the production of bioethanol fuels from waste feedstocks such as second generation residual sugar and starch feedstocks, food waste, industrial waste, urban waste, forestry waste, and lignocellulosic biomass at large (Table 1.5). In this context, there are nine sections where the first two chapters cover the production of bioethanol fuels from waste feedstocks at large.

Hence, the fourth volume indicates that research on the production of bioethanol fuels from these waste feedstocks has intensified in recent years to become a major part of the bioenergy and biofuel research together primarily with biodiesel, biohydrogen, and biogas research as a sustainable alternative to crude oil-based gasoline and petrodiesel fuels as well as LNG.

The fourth volume also indicates that a wide range of pretreatments alone or in combination fractionate the biomass to its constituents of cellulose, lignin, and hemicellulose and improve both the sugar and bioethanol fuel yield, making bioethanol fuels from these waste feedstocks more competitive in relation to crude oil- and natural gas-based fuels.

The fourth volume also indicates that hydrolysis of the biomass, microbial hydrolysate fermentation, and separation and distillation of bioethanol fuels from fermentation broth together with biomass pretreatments are the fundamental production processes for bioethanol fuel production from these waste feedstocks, making bioethanol fuels more competitive in relation to crude oil- and natural gas-based fuels.

The fifth volume of this handbook, with a collection of 18 chapters, provides an overview of research on the evaluation and utilization of bioethanol fuels in general, gasoline fuels, application of nanotechnology in bioethanol fuels, utilization of bioethanol fuels in transport engines, evaluation of bioethanol fuels, utilization of bioethanol fuels, and development and utilization of bioethanol fuel sensors (Table 1.6).

Hence, the fifth volume indicates that research on the evaluation and utilization of bioethanol fuels has intensified in recent years to become a major part of the bioenergy and biofuels research together primarily with biodiesel, biohydrogen, and biogas research as a sustainable alternative to crude oil-based gasoline and petrodiesel fuels as well as natural gas and syngas.

The fifth volume also indicates that the production of bioethanol fuels and their derivatives and coproducts in a biorefinery context increases the utility of bioethanol fuels and reduce their production cost in relation to the crude oil- , natural gas-, syngas-, and coal-based fuels as well as other biofuels such as biodiesel fuels.

The fifth volume also indicates that bioethanol fuels are primarily used in transport engines such as gasoline and diesel engines blending gasoline and petrodiesel fuels as well as LNG, directly in DEFCs as an alternative to crude oil- and natural gas-based fuels, the production of biohydrogen fuels for fuel cells as an alternative to their direct use in DEFCs, and the production of biochemicals as an alternative to crude oil-based chemicals. Further, bioethanol sensors are developed and utilized to sense the bioethanol fuels in the light of the increasing utilization of bioethanol fuels at a global scale.

The fifth volume also indicates that research on the evaluation of bioethanol fuels has progressed in research fronts of the environmental impact of bioethanol fuels, technoeconomics of bioethanol fuels and to a lesser extent energy assessments of bioethanol fuels, and policy and economic issues in bioethanol fuel production. Furthermore, the secondary research fronts for first research area have been emissions during the production and utilization of bioethanol fuels and to a lesser extent LCA of bioethanol fuels and other environmental issues.

The fifth volume also indicates that hydrolysis of biomass, microbial hydrolysate fermentation, and separation and distillation of bioethanol fuels from fermentation broth together with biomass pretreatments are the fundamental production processes for bioethanol fuel production from the first, second, and third generation feedstocks, making bioethanol fuels more competitive in relation to crude oil- and natural gas-based fuels.

The sixth volume indicates that research on the evaluation and utilization of bioethanol fuels has intensified in recent years to become a major part of the bioenergy and biofuels research together primarily with biodiesel, biohydrogen, and biogas research as a sustainable alternative to crude oil-based gasoline and petrodiesel fuels as well as natural gas and syngas.

The sixth volume also indicates that the production of bioethanol fuels and their derivatives and coproducts in a biorefinery context increases the utility of bioethanol fuels and reduces their production cost in relation to crude oil-, natural gas-, syngas-, and coal-based fuels as well as other biofuels such as biodiesel fuels.

The sixth volume also indicates that bioethanol fuels are primarily used directly in DEFCs as an alternative to crude oil- and natural gas-based fuels, the production of biohydrogen fuels for fuel cells as an alternative to their direct use in DEFCs, and the production of biochemicals as an alternative to crude oil-based chemicals. It also shows that the USA, Brazil, Europe, and China have been the major producers and consumers of bioethanol fuels and their derivatives, with national policies for setting up incentive structures for the production and consumption of bioethanol fuels and their derivatives and coproducts.

This handbook indicates that a small number of documents, authors, institutions, publication years, source titles, countries, Scopus subject categories, Scopus keywords, and research fronts have shaped the research on the production, evaluation, and utilization of bioethanol fuels.

This handbook also indicates that the level of funding for research on the production, evaluation and utilization of bioethanol fuels has not been sufficient with the resulting loss of momentum in the research output in recent years. Thus, there is a crucial need to improve the incentive structures for the major stakeholders such as researchers and their institutions as well source titles and academic databases to improve the volume and quality of the research output in these fields. This is a crucial need to maintain the energy security at a global scale in light of the recent supply shocks caused by the COVID-19 pandemic and the Ukrainian war.

This handbook also indicates that the contribution of social sciences and humanities to the research in these fields has been minimal, due to in part by the restrictive editorial policies of the source titles in these fields toward social science- and humanities-based interdisciplinary studies. Thus, there is ample room to improve incentive structures for the inclusion of social sciences and humanities into these fields.

This handbook also indicates that China, Europe as a whole, and the USA have been major producers of the research in these research fields, and there has been heavy competition among them in terms of both volume and citation impact of the research output. The USA and Europe as a whole have had a higher citation impact in relation to China benefiting from their first-mover advantage, starting their research in these fields in the 1970s. China as a late mover have had more intensive research funding initiatives in relation to the USA and Europe, improving both its research output and citation impact through the provision of efficient incentive structures for its major stakeholders in the last two decades. In this way, China might also overtake both the USA and Europe in terms of citation impact of the research output in addition to the volume of the research output in the future.

This handbook at large and sixth volume are a valuable resource for the stakeholders primarily in the research fields of energy and fuels, chemical engineering, environmental science and engineering, biotechnology, microbiology, chemistry, physics, mechanical engineering, agricultural sciences, food science and engineering, materials science, biochemistry, genetics, molecular biology, plant sciences, water resources, economics, business and management, transportation science and technology, ecology, public, environmental and occupational health, social sciences, toxicology, multidisciplinary sciences, and humanities among others.

1.5 CONCLUSION AND FUTURE RESEARCH

The brief information about the key research fronts covered by 110 chapters is given under 6 headings for the 6 volumes of the handbook: Bioethanol fuel production processes I and II, feedstock-based bioethanol fuels I and II, and evaluation and utilization of bioethanol fuels I and II.

The usual characteristics of production processes for the bioethanol fuel production are that pretreatments, hydrolysis of the feedstocks and fermentation of the resulting hydrolysates, and the separation and distillation of the fermentation broth are the primary processes for the bioethanol fuel production from the first, second, and third generation biomass to improve the ethanol yield.

The first volume of this handbook provides an overview of research on the production, evaluation, and utilization of bioethanol fuels in general as an introduction to bioethanol fuels in the first part with three book chapters and provides an overview of the research on the pretreatments of the biomass for the sugar and bioethanol fuel production with 15 book chapters (Table 1.2). In this context, there are five key sections for the second part: introduction to pretreatments of biomass, biological pretreatments of biomass, chemical pretreatments of biomass, hydrothermal pretreatments of biomass, and mechanical pretreatments of biomass.

The second volume of this handbook provides an overview of research on production processes for bioethanol fuels in general, hydrolysis of the pretreated biomass for bioethanol production, microbial fermentation of the hydrolysates and substrates with yeasts for bioethanol production, and separation and distillation of the bioethanol fuels from the fermentation broth, complementing the research on biomass pretreatments presented in the first volume (Table 1.3).

The third volume of this handbook, with 19 chapters, provides an overview of research on the production of bioethanol fuels from non-waste feedstocks such as first generation starch and sugar feedstocks, grass biomass, wood biomass, cellulose, third generation algal feedstocks, and biosyngas (Table 1.4). In this context, there are nine key sections where the first four chapters cover the production of bioethanol fuels from feedstocks at large and non-waste feedstocks.

The fourth volume of this handbook, with 17 chapters, provides an overview of research on the production of bioethanol fuels from the waste feedstocks such as second generation residual sugar and starch feedstocks, food waste, industrial waste, urban waste, forestry waste, and lignocellulosic biomass at large (Table 1.5). In this context, there are nine sections where the first two chapters cover the production of bioethanol fuels from waste feedstocks at large.

The fifth volume of this handbook, with 18 chapters, provides an overview of research on the evaluation and utilization of bioethanol fuels in general, gasoline fuels, application of nanotechnology in bioethanol fuels, utilization of bioethanol fuels in transport engines, evaluation of bioethanol fuels, utilization of bioethanol fuels, and development and utilization of bioethanol fuel sensors (Table 1.6).

The sixth volume indicates that bioethanol fuels are primarily used directly in DEFCs as an alternative to crude oil- and natural gas-based fuels, the production of biohydrogen fuels for fuel cells as an alternative to their direct use in DEFCs, and the production of biochemicals as an alternative to crude oil-based chemicals. It also shows that the USA, Brazil, Europe, and China have been the major producers and consumers of bioethanol fuels and their derivatives and coproducts, with national policies for setting up incentive structures for the production and consumption of bioethanol fuels and their derivatives.

This handbook indicates that a small number of documents, authors, institutions, publication years, source titles, countries, Scopus subject categories, Scopus keywords, and research fronts have shaped the research on the production, evaluation, and utilization of bioethanol fuels.

This handbook also indicates that the level of funding for the research on the production, evaluation, and utilization of bioethanol fuels has not been sufficient with the resulting loss of momentum in the research output in recent years. Thus, there is a crucial need to improve the incentive structures for the major stakeholders such as researchers and their institutions as well as source titles and academic databases to improve the volume and quality of the research output in these fields. This is a crucial need to maintain the energy security at a global scale in light of the recent supply shocks caused by the COVID-19 pandemic and the Ukrainian war.

This handbook also indicates that the contribution of the social sciences and humanities to the research in these fields has been minimal, due to in part by the restrictive editorial policies of the source titles in these fields toward social science- and humanities-based interdisciplinary studies. Thus, there is ample room to improve incentive structures for the inclusion of social sciences and humanities into these fields.

This handbook also indicates that China, Europe as a whole, and the USA have been major producers of the research in these research fields, and there has been heavy competition among them in terms of both volume and citation impact of the research output. The USA and Europe as a whole have had a higher citation impact in relation to China benefiting from their first-mover advantage, starting their research in these fields in the 1970s. China as a late mover have had more intensive research funding initiatives in relation to the USA and Europe, improving both its research output and citation impact through the provision of the efficient incentive structures for its major stakeholders in the last two decades. In this way, China might also overtake both the USA and Europe in terms of citation impact of the research output in addition to the volume of the research output in the future.

The key findings from this handbook should be read in light of the increasing public concerns about climate change, GHG emissions, and global warming as these concerns have been certainly behind the boom in the research on the bioethanol fuels as an alternative to crude oil-based gasoline and petrodiesel fuels in the last decades. The recent supply shocks caused by the COVID-19 pandemic and the Russian invasion of Ukraine also highlight the importance of the production, utilization, and evaluation of the bioethanol fuels and their derivatives and coproducts as an alternative to the crude oil-based gasoline and petrodiesel fuels.

These studies emphasize the importance of proper incentive structures for the efficient production, utilization, and evaluation of bioethanol fuels and their derivatives and coproducts in light of North's institutional framework (North, 1991). In this context, the major producers and users of bioethanol fuels such as the USA and Brazil with vast forests and farmlands have developed strong incentive structures for efficient bioethanol fuels. In light of the recent supply shocks caused primarily by the COVID-19 pandemic and Russian invasion of Ukraine, it is expected that the incentive structures such as public funding would be enhanced to increase the share of bioethanol fuels in the global fuel portfolio as a strong alternative to crude oil-based gasoline and petrodiesel fuels.

In this context, it is expected that the most prolific researchers, institutions, countries, funding bodies, and source titles in this field would have a first-mover advantage to benefit from such potential incentives. This is especially true for the US, European, and Chinese stakeholders as these countries have become the global leaders in both the production and utilization of second generation bioethanol fuels and their derivatives and coproducts.

It is expected the research would focus more on the algal, wood, and grass biomass-based bioethanol fuels as well as the agricultural residues- and waste biomass-based bioethanol fuels at the expense of the first generation starch- and sugar-based bioethanol fuels due to the large societal concerns about the food security in the future. Similarly, it is expected that the research would focus more on the pretreatments and hydrolysis of the feedstocks for the production of bioethanol fuels and utilization in cars, fuel cells, biochemicals, biohydrogen fuels, and evaluation of bioethanol fuels in the future.

This handbook is a valuable resource for the stakeholders primarily in the research fields of energy and fuels, chemical engineering, environmental science and engineering, biotechnology, microbiology, chemistry, physics, mechanical engineering, agricultural sciences, food science and engineering, materials science, biochemistry, genetics, molecular biology, plant sciences, water resources, economics, business and management, transportation science and technology, ecology, public, environmental and occupational health, social sciences, toxicology, multidisciplinary sciences, and humanities among others.

ACKNOWLEDGMENTS

The contribution of the highly cited researchers in the field of bioethanol fuels has been gratefully acknowledged.

REFERENCES

Alvira, P., E. Tomas-Pejo, M. Ballesteros and M. J. Negro. 2010. Pretreatment technologies for an efficient bioethanol production process based on enzymatic hydrolysis: A review. *Bioresource Technology* 101:4851–4861.

Angelici, C., B. M. Weckhuysen and P. C. A. Bruijnincx. 2013. Chemocatalytic conversion of ethanol into butadiene and other bulk chemicals. *ChemSusChem* 6:1595–1614.

Antolini, E. 2007. Catalysts for direct ethanol fuel cells. *Journal of Power Sources* 170:1–12.

Antolini, E. 2009. Palladium in fuel cell catalysis. *Energy and Environmental Science* 2:915–931.

Bai, F. W., W. A. Anderson and M. Moo-Young. 2008. Ethanol fermentation technologies from sugar and starch feedstocks. *Biotechnology Advances* 26:89–105.

Balat, M. 2011. Production of bioethanol from lignocellulosic materials via the biochemical pathway: A review: *Energy Conversion and Management* 52:858–875.

Binod, P., R. Sindhu and R. R. Singhania, et al. 2010. Bioethanol production from rice straw: An overview. *Bioresource Technology* 101:4767–4774.

Bothast, R. J. and M. A. Schlicher. 2005. Biotechnological processes for conversion of corn into ethanol. *Applied Microbiology and Biotechnology* 67:19–25.

Bourbonnais, R. and M. G. Paice. 1990. Oxidation of non-phenolic substrates. An expanded role for laccase in lignin biodegradation. *FEBS Letters* 267:99–102.

Caballero, V., A. W. Savoy and Y. Wang. 2023. An overview of bioethanol conversion to hydrocarbons. In *Evaluation and Utilization of Bioethanol Fuels. II: Biohydrogen Fuels, Fuel Cells, Biochemicals, and Country Experiences, Handbook of Bioethanol Fuels Volume 6,* Ed. O. Konur. Boca Raton, FL: CRC Press.

Cai, D., Z. Si, H. We, Y. Zhuang and P. Qin. 2023. *In situ* bioethanol separation from fermentation broth. In *Bioethanol Fuel Production Processes. II: Biomass Hydrolysis, Fermentation, and Bioethanol Fuel Separation. Handbook of Bioethanol Fuels Volume 2,* Ed. O. Konur. Boca Raton, FL: CRC Press.

Canilha, L., A. K. Chandel and T. S. dos Santos Milessi, et al. 2012. Bioconversion of sugarcane biomass into ethanol: An overview about composition, pretreatment methods, detoxification of hydrolysates, enzymatic saccharification, and ethanol fermentation. *Journal of Biomedicine and Biotechnology* 2012:989572.

Cardona, C.A., J. A. Quintero and I. C. Paz. 2010. Production of bioethanol from sugarcane bagasse: Status and perspectives. *Bioresource Technology* 101:4754–4766.

Chen, H., D. Cai, J. Wen, C. Su and J. Wang. 2023. Distillation separation of bioethanol: The current advances. In *Bioethanol Fuel Production Processes. II: Biomass Hydrolysis, Fermentation, and Bioethanol Fuel Separation. Handbook of Bioethanol Fuels Volume 2,* Ed. O. Konur. Boca Raton, FL: CRC Press.

Cortez, L. 2023. Impact of corn ethanol production in land use in Brazil. In *Evaluation and Utilization of Bioethanol Fuels. I: Evaluation of Bioethanol Fuels, Transport Engines, and Bioethanol Sensors. Handbook of Bioethanol Fuels Volume 5,* Ed. O. Konur. Boca Raton, FL: CRC Press.

Duff, S. J. B. and W. D. Murray. 1996. Bioconversion of forest products industry waste cellulosics to fuel ethanol: A review. *Bioresource Technology* 55:1–33.

Elgharbawy, A. A. M., N. A. N. Azmi and M. Moniruzzaman. 2023 Advances of bioethanol production using ionic liquids. In *Bioethanol Fuel Production Processes. I: Biomass Pretreatments. Handbook of Bioethanol Fuels Volume 1,* Ed. O. Konur. Boca Raton, FL: CRC Press.

Fernando, S., S. Adhikari, C. Chandrapal and M. Murali. 2006. Biorefineries: Current status, challenges, and future direction. *Energy & Fuels* 20:1727–1737.

Galbe, M. and G. Zacchi. 2002. A review of the production of ethanol from softwood. *Applied Microbiology and Biotechnology* 59:618–628.

Gomez-Castro, F. I. 2023. Advances and challenges in the production and purification of bioethanol using intensified processes. In *Bioethanol Fuel Production Processes. II: Biomass Hydrolysis, Fermentation, and Bioethanol Fuel Separation. Handbook of Bioethanol Fuels Volume 2,* Ed. O. Konur. Boca Raton, FL: CRC Press.

Guimaraes, P. M. R., J. A. Teixeira and L. Domingues. 2010. Fermentation of lactose to bio-ethanol by yeasts as part of integrated solutions for the valorisation of cheese whey. *Biotechnology Advances* 28:375–384.

Hahn-Hagerdal, B., M. Galbe, M. F. Gorwa-Grauslund, G. Liden and G. Zacchi. 2006. Bio-ethanol – The fuel of tomorrow from the residues of today. *Trends in Biotechnology* 24:549–556.

Hamelinck, C. N., G. Van Hooijdonk and A. P. C. Faaij. 2005. Ethanol from lignocellulosic biomass: Techno-economic performance in short-, middle- and long-term. *Biomass and Bioenergy* 28:384–410.

Hansen, A. C., Q. Zhang and P. W. L. Lyne. 2005. Ethanol-diesel fuel blends – A review. *Bioresource Technology* 96:277–285.

Haryanto, A., S. Fernando, N. Murali and S. Adhikari. 2005. Current status of hydrogen production techniques by steam reforming of ethanol: A review. *Energy and Fuels* 19:2098–2106.

Hendriks, A. T. W. M and G. Zeeman. 2009. Pretreatments to enhance the digestibility of lignocellulosic biomass. *Bioresource Technology* 100:10–18.

Henstra, A. M., J. Sipma, A. Rinzema and A. J. Stams. 2007. Microbiology of synthesis gas fermentation for biofuel production. *Current Opinion in Biotechnology* 18:200–206.

Hill, J., E. Nelson, D. Tilman, S. Polasky and D. Tiffany. 2006. Environmental, economic, and energetic costs and benefits of biodiesel and ethanol biofuels. *Proceedings of the National Academy of Sciences of the United States of America* 103:11206–11210.

Hill, J., S. Polasky and E. Nelson, et al. 2009. Climate change and health costs of air emissions from biofuels and gasoline. *Proceedings of the National Academy of Sciences of the United States of America* 106:2077–2082.

Ho, S. H., S. W. Huang and C. Y. Chen, et al. 2013. Bioethanol production using carbohydrate-rich microalgae biomass as feedstock. *Bioresource Technology* 135:191–198.

Hoang A H and O. Konur. 2023n. Microwave pretreatment of the biomass: Review. In Bioethanol Fuel Production Processes. *I: Biomass Pretreatments. Handbook of Bioethanol Fuels Volume 1*, Ed. O. Konur. Boca Raton, FL: CRC Press.

Hsieh, W. D., R. H. Chen, T. L. Wu and T. H. Lin. 2002. Engine performance and pollutant emission of an SI engine using ethanol-gasoline blended fuels. *Atmospheric Environment* 36:403–410.

Huang, H. J., S. Ramaswamy, U. W. Tschirner and B. V. Ramarao. 2008. A review of separation technologies in current and future biorefineries. *Separation and Purification Technology* 62:1–21.

John, R. P., G. S. Anisha, K. M. Nampoothiri and A. Pandey. 2011. Micro and macroalgal biomass: A renewable source for bioethanol. *Bioresource Technology* 102:186–193.

Jonsson, L. J. and C. Martin. 2016. Pretreatment of lignocellulose: Formation of inhibitory by-products and strategies for minimizing their effects. *Bioresource Technology* 199:103–112.

Kamarudin, M. Z. F., S. K. Kamarudin, M. S. Masdar and W. R. W. Daud. 2013. Review: Direct ethanol fuel cells. *International Journal of Hydrogen Energy* 38:9438–9453.

Keshwani, D. R. and J. J. Cheng. 2009. Switchgrass for bioethanol and other value-added applications: A review. *Bioresource Technology* 100:1515–1523.

Khatun, M. H. A., H. Z. Chen and L. Wang. 2023 Corn pericarp bioethanol through steam explosion and periodic peristalsis. In *Feedstock-based Bioethanol Fuels. I. Non-Waste Feedstocks: Starch, Sugar, Grass, Wood, Cellulose, Algae, and Biosyngas-based Bioethanol Fuels. Handbook of Bioethanol Fuels Volume 3,* Ed. O. Konur. Boca Raton, FL: CRC Press.

Kirk, T. K. and R. L. Farrell. 1987. Enzymatic "combustion": The microbial degradation of lignin. *Annual Review of Microbiology* 41:465–505.

Konur, O. 2000. Creating enforceable civil rights for disabled students in higher education: An institutional theory perspective. *Disability & Society* 15:1041–1063.

Konur, O. 2002a. Access to nursing education by disabled students: Rights and duties of nursing programs. *Nurse Education Today* 22:364–374.

Konur, O. 2002b. Assessment of disabled students in higher education: Current public policy issues. *Assessment and Evaluation in Higher Education* 27:131–52.

Konur, O. 2002c. Access to employment by disabled people in the UK: Is the Disability Discrimination Act working? *International Journal of Discrimination and the Law* 5:247–279.

Konur, O. 2006a. Participation of children with dyslexia in compulsory education: Current public policy issues. *Dyslexia* 12:51–67.

Konur, O. 2006b. Teaching disabled students in higher education. *Teaching in Higher Education* 11:351–363.

Konur, O. 2007a. A judicial outcome analysis of the Disability Discrimination Act: A windfall for the employers? *Disability & Society* 22:187–204.

Konur, O. 2007b. Computer-assisted teaching and assessment of disabled students in higher education: The interface between academic standards and disability rights. *Journal of Computer Assisted Learning* 23:207–219.

Konur, O. 2012. The evaluation of the research on the bioethanol: A scientometric approach. *Energy Education Science and Technology Part A: Energy Science and Research* 28:1051–1064.

Konur, O. 2015. Current state of research on algal bioethanol. In *Marine Bioenergy: Trends and Developments*, Eds. S. K. Kim and C. G. Lee, pp. 217–244. Boca Raton, FL: CRC Press.

Konur, O. 2019. Cyanobacterial bioenergy and biofuels science and technology: A scientometric overview. In *Cyanobacteria: From Basic Science to Applications*, Eds. A. K. Mishra, D. N. Tiwari and A. N. Rai, pp. 419–442. Amsterdam: Elsevier.

Konur, O. 2020. The scientometric analysis of the research on the bioethanol production from green macroalgae. In *Handbook of Algal Science, Technology and Medicine*, Ed. O. Konur, pp. 385–401. London: Academic Press.

Konur, O. 2023a. Introduction to bioethanol fuels. In *Bioethanol Fuel Production Processes. I: Biomass Pretreatments. Handbook of Bioethanol Fuels Volume 1,* Ed. O. Konur. Boca Raton, FL: CRC Press.

Konur, O. 2023b. Bioethanol fuels: Scientometric study. In *Bioethanol Fuel Production Processes. I: Biomass Pretreatments. Handbook of Bioethanol Fuels Volume 1,* Ed. O. Konur. Boca Raton, FL: CRC Press.

Konur, O. 2023c. Bioethanol fuels: Review. In *Bioethanol Fuel Production Processes. I: Biomass Pretreatments. Handbook of Bioethanol Fuels Volume 1,* Ed. O. Konur. Boca Raton, FL: CRC Press.

Konur, O. 2023d. Biomass pretreatments: Scientometric study. In *Bioethanol Fuel Production Processes. I: Biomass Pretreatments. Handbook of Bioethanol Fuels Volume 1,* Ed. O. Konur. Boca Raton, FL: CRC Press.

Konur, O. 2023e. Biomass pretreatments: Review. In *Bioethanol Fuel Production Processes. I: Biomass Pretreatments. Handbook of Bioethanol Fuels Volume 1,* Ed. O. Konur. Boca Raton, FL: CRC Press.

Konur, O. 2023f. Biological pretreatment of the biomass: Scientometric study. In *Bioethanol Fuel Production Processes. I: Biomass Pretreatments. Handbook of Bioethanol Fuels Volume 1,* Ed. O. Konur. Boca Raton, FL: CRC Press.

Konur, O. 2023g. Biological pretreatment of the biomass: Review. In *Bioethanol Fuel Production Processes. I: Biomass Pretreatments. Handbook of Bioethanol Fuels Volume 1,* Ed. O. Konur. Boca Raton, FL: CRC Press.

Konur, O. 2023h. Chemical pretreatment of the biomass: Scientometric study. In *Bioethanol Fuel Production Processes. I: Biomass Pretreatments. Handbook of Bioethanol Fuels Volume 1,* Ed. O. Konur. Boca Raton, FL: CRC Press.

Konur, O. 2023i. Chemical pretreatment of the biomass: Review. In *Bioethanol Fuel Production Processes. I: Biomass Pretreatments. Handbook of Bioethanol Fuels Volume 1,* Ed. O. Konur. Boca Raton, FL: CRC Press.

Konur, O. 2023ai. Non-waste feedstock-based bioethanol fuels: Scientometric study. In *Feedstock-based Bioethanol Fuels. I. Non-Waste Feedstocks: Starch, Sugar, Grass, Wood, Cellulose, Algae, and Biosyngas-based Bioethanol Fuels. Handbook of Bioethanol Fuels Volume 3,* Ed. O. Konur. Boca Raton, FL: CRC Press.

Konur, O. 2023j. Hydrothermal pretreatment of the biomass: Scientometric study. In *Bioethanol Fuel Production Processes. I: Biomass Pretreatments. Handbook of Bioethanol Fuels Volume 1,* Ed. O. Konur. Boca Raton, FL: CRC Press.

Konur, O. 2023aj. Non-waste feedstock-based bioethanol fuels: Review. In *Feedstock-based Bioethanol Fuels. I. Non-Waste Feedstocks: Starch, Sugar, Grass, Wood, Cellulose, Algae, and Biosyngas-based Bioethanol Fuels. Handbook of Bioethanol Fuels Volume 3,* Ed. O. Konur. Boca Raton, FL: CRC Press.

Konur, O. 2023bj. Second generation urban waste-based bioethanol fuels: Review. In *Feedstock-based Bioethanol Fuels. II. Waste Feedstocks: Agricultural, Food, Industrial, Urban, Forestry, and Lignocellulosic Waste-based Bioethanol Fuels. Handbook of Bioethanol Fuels Volume 4,* Ed. O. Konur. Boca Raton, FL: CRC Press.

Konur, O. 2023k. Hydrothermal pretreatment of the biomass: Review. In *Bioethanol Fuel Production Processes. I: Biomass Pretreatments. Handbook of Bioethanol Fuels Volume 1,* Ed. O. Konur. Boca Raton, FL: CRC Press.

Konur, O. 2023bk. Second generation forestry waste-based bioethanol fuels: Scientometric study. In *Feedstock-based Bioethanol Fuels. II. Waste Feedstocks: Agricultural, Food, Industrial, Urban, Forestry, and Lignocellulosic Waste-based Bioethanol Fuels. Handbook of Bioethanol Fuels Volume 4,* Ed. O. Konur. Boca Raton, FL: CRC Press.

Konur, O. 2023l. Mechanical pretreatment of the biomass: Scientometric study. In *Bioethanol Fuel Production Processes. I: Biomass Pretreatments. Handbook of Bioethanol Fuels Volume 1,* Ed. O. Konur. Boca Raton, FL: CRC Press.

Konur, O. 2023bl. Second generation forestry waste-based bioethanol fuels: Review. In *Feedstock-based Bioethanol Fuels. II. Waste Feedstocks: Agricultural, Food, Industrial, Urban, Forestry, and Lignocellulosic Waste-based Bioethanol Fuels. Handbook of Bioethanol Fuels Volume 4,* Ed. O. Konur. Boca Raton, FL: CRC Press.

Konur, O. 2023m. Milling pretreatment of the biomass: Review. In *Bioethanol Fuel Production Processes. I: Biomass Pretreatments. Handbook of Bioethanol Fuels Volume 1,* Ed. O. Konur. Boca Raton, FL: CRC Press.

Konur, O. 2023o. Production processes for bioethanol fuels: Scientometric study. In *Bioethanol Fuel Production Processes. II: Biomass Hydrolysis, Fermentation, and Bioethanol Fuel Separation. Handbook of Bioethanol Fuels Volume 2,* Ed. O. Konur. Boca Raton, FL: CRC Press.

Konur, O. 2023p. Production processes for bioethanol fuels: Review. In *Bioethanol Fuel Production Processes. II: Biomass Hydrolysis, Fermentation, and Bioethanol Fuel Separation. Handbook of Bioethanol Fuels Volume 2,* Ed. O. Konur. Boca Raton, FL: CRC Press.

Konur, O. 2023q. Biomass hydrolysis: Scientometric study. In *Bioethanol Fuel Production Processes. II: Biomass Hydrolysis, Fermentation, and Bioethanol Fuel Separation. Handbook of Bioethanol Fuels Volume 2,* Ed. O. Konur. Boca Raton, FL: CRC Press.

Konur, O. 2023r. Biomass hydrolysis: Review. In *Bioethanol Fuel Production Processes. II: Biomass Hydrolysis, Fermentation, and Bioethanol Fuel Separation. Handbook of Bioethanol Fuels Volume 2,* Ed. O. Konur. Boca Raton, FL: CRC Press.

Konur, O. 2023s. Wood hydrolysis: Scientometric study. In *Bioethanol Fuel Production Processes. II: Biomass Hydrolysis, Fermentation, and Bioethanol Fuel Separation. Handbook of Bioethanol Fuels Volume 2,* Ed. O. Konur. Boca Raton, FL: CRC Press.

Konur, O. 2023t. Wood hydrolysis: Review. In *Bioethanol Fuel Production Processes. II: Biomass Hydrolysis, Fermentation, and Bioethanol Fuel Separation. Handbook of Bioethanol Fuels Volume 2,* Ed. O. Konur. Boca Raton, FL: CRC Press.

Konur, O. 2023u. Straw hydrolysis: Scientometric study. In *Bioethanol Fuel Production Processes. II: Biomass Hydrolysis, Fermentation, and Bioethanol Fuel Separation. Handbook of Bioethanol Fuels Volume 2,* Ed. O. Konur. Boca Raton, FL: CRC Press.

Konur, O. 2023v. Straw hydrolysis: Review. In *Bioethanol Fuel Production Processes. II: Biomass Hydrolysis, Fermentation, and Bioethanol Fuel Separation. Handbook of Bioethanol Fuels Volume 2,* Ed. O. Konur. Boca Raton, FL: CRC Press.

Konur, O. 2023w. Cellulose hydrolysis: Scientometric study. In *Bioethanol Fuel Production Processes. II: Biomass Hydrolysis, Fermentation, and Bioethanol Fuel Separation. Handbook of Bioethanol Fuels Volume 2,* Ed. O. Konur. Boca Raton, FL: CRC Press.

Konur, O. 2023x. Cellulose hydrolysis: Review. In *Bioethanol Fuel Production Processes. II: Biomass Hydrolysis, Fermentation, and Bioethanol Fuel Separation. Handbook of Bioethanol Fuels Volume 2,* Ed. O. Konur. Boca Raton, FL: CRC Press.

Konur, O. 2023y. Hydrolysate and substrate fermentation: Scientometric study. In *Bioethanol Fuel Production Processes. II: Biomass Hydrolysis, Fermentation, and Bioethanol Fuel Separation. Handbook of Bioethanol Fuels Volume 2,* Ed. O. Konur. Boca Raton, FL: CRC Press.

Konur, O. 2023z. Hydrolysate and substrate fermentation: Review. In *Bioethanol Fuel Production Processes. II: Biomass Hydrolysis, Fermentation, and Bioethanol Fuel Separation. Handbook of Bioethanol Fuels Volume 2,* Ed. O. Konur. Boca Raton, FL: CRC Press.

Konur, O. 2023aa. The alternative fermentation processes for the bioethanol production: Scientometric study. In *Bioethanol Fuel Production Processes. II: Biomass Hydrolysis, Fermentation, and Bioethanol Fuel Separation. Handbook of Bioethanol Fuels Volume 2,* Ed. O. Konur. Boca Raton, FL: CRC Press.

Konur, O. 2023ab. The alternative fermentation processes for the bioethanol production: Review. In *Bioethanol Fuel Production Processes. II: Biomass Hydrolysis, Fermentation, and Bioethanol Fuel Separation. Handbook of Bioethanol Fuels Volume 2,* Ed. O. Konur. Boca Raton, FL: CRC Press.

Konur, O. 2023ac. Metabolic engineering for the bioethanol production: Scientometric study. In *Bioethanol Fuel Production Processes. II: Biomass Hydrolysis, Fermentation, and Bioethanol Fuel Separation. Handbook of Bioethanol Fuels Volume 2,* Ed. O. Konur. Boca Raton, FL: CRC Press.

Konur, O. 2023ad. Metabolic engineering for the bioethanol production: Review. In *Bioethanol Fuel Production Processes. II: Biomass Hydrolysis, Fermentation, and Bioethanol Fuel Separation. Handbook of Bioethanol Fuels Volume 2,* Ed. O. Konur. Boca Raton, FL: CRC Press.

Konur, O. 2023ae. The utilization of the *Saccharomyces cerevisiae* for the bioethanol production: Scientometric study. In *Bioethanol Fuel Production Processes. II: Biomass Hydrolysis, Fermentation, and Bioethanol Fuel Separation. Handbook of Bioethanol Fuels Volume 2,* Ed. O. Konur. Boca Raton, FL: CRC Press.

Konur, O. 2023af. The utilization of the *Saccharomyces cerevisiae* for the bioethanol production: Review. In *Bioethanol Fuel Production Processes. II: Biomass Hydrolysis, Fermentation, and Bioethanol Fuel Separation. Handbook of Bioethanol Fuels Volume 2,* Ed. O. Konur. Boca Raton, FL: CRC Press.

Konur, O. 2023ag. Feedstock-based bioethanol fuels: Scientometric study. In *Feedstock-based Bioethanol Fuels. I. Non-Waste Feedstocks: Starch, Sugar, Grass, Wood, Cellulose, Algae, and Biosyngas-based Bioethanol Fuels. Handbook of Bioethanol Fuels Volume 3,* Ed. O. Konur. Boca Raton, FL: CRC Press.

Konur, O. 2023ah. Feedstock-based bioethanol fuels: Review. In *Feedstock-based Bioethanol Fuels. I. Non-Waste Feedstocks: Starch, Sugar, Grass, Wood, Cellulose, Algae, and Biosyngas-based Bioethanol Fuels. Handbook of Bioethanol Fuels Volume 3,* Ed. O. Konur. Boca Raton, FL: CRC Press.

Konur, O. 2023ak. First generation starch feedstock-based bioethanol fuels: Scientometric study. In *Feedstock-based Bioethanol Fuels. I. Non-Waste Feedstocks: Starch, Sugar, Grass, Wood, Cellulose, Algae, and Biosyngas-based Bioethanol Fuels. Handbook of Bioethanol Fuels Volume 3,* Ed. O. Konur. Boca Raton, FL: CRC Press.

Konur, O. 2023al. First generation starch feedstock-based bioethanol fuels: Review. In *Feedstock-based Bioethanol Fuels. I. Non-Waste Feedstocks: Starch, Sugar, Grass, Wood, Cellulose, Algae, and Biosyngas-based Bioethanol Fuels. Handbook of Bioethanol Fuels Volume 3,* Ed. O. Konur. Boca Raton, FL: CRC Press.

Konur, O. 2023am. First generation sugar feedstock-based bioethanol fuels: Scientometric study. In *Feedstock-based Bioethanol Fuels. I. Non-Waste Feedstocks: Starch, Sugar, Grass, Wood, Cellulose, Algae, and Biosyngas-based Bioethanol Fuels. Handbook of Bioethanol Fuels Volume 3,* Ed. O. Konur. Boca Raton, FL: CRC Press.

Konur, O. 2023an. First generation sugar feedstock-based bioethanol fuels: Review. In *Feedstock-based Bioethanol Fuels. I. Non-Waste Feedstocks: Starch, Sugar, Grass, Wood, Cellulose, Algae, and Biosyngas-based Bioethanol Fuels. Handbook of Bioethanol Fuels Volume 3,* Ed. O. Konur. Boca Raton, FL: CRC Press.

Konur, O. 2023ao. Grass-based bioethanol fuels: Scientometric study. In *Feedstock-based Bioethanol Fuels. I. Non-Waste Feedstocks: Starch, Sugar, Grass, Wood, Cellulose, Algae, and Biosyngas-based Bioethanol Fuels. Handbook of Bioethanol Fuels Volume 3,* Ed. O. Konur. Boca Raton, FL: CRC Press.

Konur, O. 2023ap. Grass-based bioethanol fuels: Review. In *Feedstock-based Bioethanol Fuels. I. Non-Waste Feedstocks: Starch, Sugar, Grass, Wood, Cellulose, Algae, and Biosyngas-based Bioethanol Fuels. Handbook of Bioethanol Fuels Volume 3,* Ed. O. Konur. Boca Raton, FL: CRC Press.

Konur, O. 2023aq. Wood-based bioethanol fuels: Scientometric study. In *Feedstock-based Bioethanol Fuels. I. Non-Waste Feedstocks: Starch, Sugar, Grass, Wood, Cellulose, Algae, and Biosyngas-based Bioethanol Fuels. Handbook of Bioethanol Fuels Volume 3,* Ed. O. Konur. Boca Raton, FL: CRC Press.

Konur, O. 2023ar. Wood-based bioethanol fuels: Review. In *Feedstock-based Bioethanol Fuels. I. Non-Waste Feedstocks: Starch, Sugar, Grass, Wood, Cellulose, Algae, and Biosyngas-based Bioethanol Fuels Handbook of Bioethanol Fuels Volume 3,* Ed. O. Konur. Boca Raton, FL: CRC Press.

Konur, O. 2023as. Cellulose-based bioethanol fuels: Scientometric study. In *Feedstock-based Bioethanol Fuels. I. Non-Waste Feedstocks: Starch, Sugar, Grass, Wood, Cellulose, Algae, and Biosyngas-based Bioethanol Fuels. Handbook of Bioethanol Fuels Volume 3,* Ed. O. Konur. Boca Raton, FL: CRC Press.

Konur, O. 2023at. Cellulose-based bioethanol fuels: Review. In *Feedstock-based Bioethanol Fuels. I. Non-Waste Feedstocks: Starch, Sugar, Grass, Wood, Cellulose, Algae, and Biosyngas-based Bioethanol Fuels. Handbook of Bioethanol Fuels Volume 3,* Ed. O. Konur. Boca Raton, FL: CRC Press.

Konur, O. 2023au. Third generation algal bioethanol fuels: Scientometric study. In *Feedstock-based Bioethanol Fuels. I. Non-Waste Feedstocks: Starch, Sugar, Grass, Wood, Cellulose, Algae, and Biosyngas-based Bioethanol Fuels. Handbook of Bioethanol Fuels Volume 3,* Ed. O. Konur. Boca Raton, FL: CRC Press.

Konur, O. 2023aw. Biosyngas-based bioethanol fuels: Scientometric study. In *Feedstock-based Bioethanol Fuels. I. Non-Waste Feedstocks: Starch, Sugar, Grass, Wood, Cellulose, Algae, and Biosyngas-based Bioethanol Fuels. Handbook of Bioethanol Fuels Volume 3,* Ed. O. Konur. Boca Raton, FL: CRC Press.

Konur, O. 2023ax. Biosyngas-based bioethanol fuels: Review. In *Feedstock-based Bioethanol Fuels. I. Non-Waste Feedstocks: Starch, Sugar, Grass, Wood, Cellulose, Algae, and Biosyngas-based Bioethanol Fuels. Handbook of Bioethanol Fuels Volume 3,* Ed. O. Konur. Boca Raton, FL: CRC Press.

Konur, O. 2023ay. Second generation waste biomass-based bioethanol fuels: Scientometric study. In *Feedstock-based Bioethanol Fuels. II. Waste Feedstocks: Agricultural, Food, Industrial, Urban, Forestry, and Lignocellulosic Waste-based Bioethanol Fuels. Handbook of Bioethanol Fuels Volume 4,* Ed. O. Konur. Boca Raton, FL: CRC Press.

Konur, O. 2023az. Second generation waste biomass-based bioethanol fuels: Review. In *Feedstock-based Bioethanol Fuels. II. Waste Feedstocks: Agricultural, Food, Industrial, Urban, Forestry, and Lignocellulosic Waste-based Bioethanol Fuels. Handbook of Bioethanol Fuels Volume 4,* Ed. O. Konur. Boca Raton, FL: CRC Press.

Konur, O. 2023ba. Second generation bioethanol fuels from residual sugar feedstocks: Scientometric study. In *Feedstock-based Bioethanol Fuels. II. Waste Feedstocks: Agricultural, Food, Industrial, Urban, Forestry, and Lignocellulosic Waste-based Bioethanol Fuels. Handbook of Bioethanol Fuels Volume 4,* Ed. O. Konur. Boca Raton, FL: CRC Press.

Konur, O. 2023bb. Second generation bioethanol fuels from residual sugar feedstocks: Review. In *Feedstock-based Bioethanol Fuels. II. Waste Feedstocks: Agricultural, Food, Industrial, Urban, Forestry, and Lignocellulosic Waste-based Bioethanol Fuels. Handbook of Bioethanol Fuels Volume 4,* Ed. O. Konur. Boca Raton, FL: CRC Press.

Konur, O. 2023bc. Second generation bioethanol fuels from residual starch feedstocks: Scientometric study. In *Feedstock-based Bioethanol Fuels. II. Waste Feedstocks: Agricultural, Food, Industrial, Urban, Forestry, and Lignocellulosic Waste-based Bioethanol Fuels. Handbook of Bioethanol Fuels Volume 4,* Ed. O. Konur. Boca Raton, FL: CRC Press.

Konur, O. 2023bd. Second generation bioethanol fuels from residual starch feedstocks: Review. In *Feedstock-based Bioethanol Fuels. II. Waste Feedstocks: Agricultural, Food, Industrial, Urban, Forestry, and Lignocellulosic Waste-based Bioethanol Fuels. Handbook of Bioethanol Fuels Volume 4,* Ed. O. Konur. Boca Raton, FL: CRC Press.

Konur, O. 2023be. Second generation food waste-based bioethanol fuels: Scientometric study. In *Feedstock-based Bioethanol Fuels. II. Waste Feedstocks: Agricultural, Food, Industrial, Urban, Forestry, and Lignocellulosic Waste-based Bioethanol Fuels. Handbook of Bioethanol Fuels Volume 4,* Ed. O. Konur. Boca Raton, FL: CRC Press.

Konur, O. 2023bf. Second generation food waste-based bioethanol fuels: Review. In *Feedstock-based Bioethanol Fuels. II. Waste Feedstocks: Agricultural, Food, Industrial, Urban, Forestry, and Lignocellulosic Waste-based Bioethanol Fuels. Handbook of Bioethanol Fuels Volume 4,* Ed. O. Konur. Boca Raton, FL: CRC Press.

Konur, O. 2023bg. Second generation Industrial waste-based bioethanol fuels: Scientometric study. In *Feedstock-based Bioethanol Fuels. II. Waste Feedstocks: Agricultural, Food, Industrial, Urban, Forestry, and Lignocellulosic Waste-based Bioethanol Fuels. Handbook of Bioethanol Fuels Volume 4,* Ed. O. Konur. Boca Raton, FL: CRC Press.

Konur, O. 2023bh. Second generation industrial waste-based bioethanol fuels: Review. In *Feedstock-based Bioethanol Fuels. II. Waste Feedstocks: Agricultural, Food, Industrial, Urban, Forestry, and Lignocellulosic Waste-based Bioethanol Fuels. Handbook of Bioethanol Fuels Volume 4,* Ed. O. Konur. Boca Raton, FL: CRC Press.

Konur, O. 2023bi. Second generation urban waste-based bioethanol fuels: Scientometric study. In *Feedstock-based Bioethanol Fuels. II. Waste Feedstocks: Agricultural, Food, Industrial, Urban, Forestry, and Lignocellulosic Waste-based Bioethanol Fuels. Handbook of Bioethanol Fuels Volume 4,* Ed. O. Konur. Boca Raton, FL: CRC Press.

Konur, O. 2023bm. Lignocellulosic biomass-based bioethanol fuels: Scientometric study. In *Feedstock-based Bioethanol Fuels. II. Waste Feedstocks: Agricultural, Food, Industrial, Urban, Forestry, and Lignocellulosic Waste-based Bioethanol Fuels. Handbook of Bioethanol Fuels Volume 4,* Ed. O. Konur. Boca Raton, FL: CRC Press.

Konur, O. 2023bn. Lignocellulosic biomass-based bioethanol fuels: Review. In *Feedstock-based Bioethanol Fuels. II. Waste Feedstocks: Agricultural, Food, Industrial, Urban, Forestry, and Lignocellulosic Waste-based Bioethanol Fuels. Handbook of Bioethanol Fuels Volume 4,* Ed. O. Konur. Boca Raton, FL: CRC Press.

Konur, O. 2023bo. Bioethanol fuel evaluation and utilization: Scientometric study. In *Evaluation and Utilization of Bioethanol Fuels. I: Evaluation of Bioethanol Fuels, Transport Engines, and Bioethanol Sensors. Handbook of Bioethanol Fuels Volume 5,* Ed. O. Konur. Boca Raton, FL: CRC Press.

Konur, O. 2023bp. Bioethanol fuel evaluation and utilization: Review. In *Evaluation and Utilization of Bioethanol Fuels. I: Evaluation of Bioethanol Fuels, Transport Engines, and Bioethanol Sensors. Handbook of Bioethanol Fuels Volume 5,* Ed. O. Konur. Boca Raton, FL: CRC Press.

Konur, O. 2023bq. Bioethanol fuel evaluation and utilization: Hot papers, Scientometric study. In *Evaluation and Utilization of Bioethanol Fuels. I: Evaluation of Bioethanol Fuels, Transport Engines, and Bioethanol Sensors. Handbook of Bioethanol Fuels Volume 5,* Ed. O. Konur. Boca Raton, FL: CRC Press.

Konur, O. 2023br. Bioethanol fuel evaluation and utilization: Hot papers, review. In *Evaluation and Utilization of Bioethanol Fuels. I: Evaluation of Bioethanol Fuels, Transport Engines, and Bioethanol Sensors. Handbook of Bioethanol Fuels Volume 5,* Ed. O. Konur. Boca Raton, FL: CRC Press.

Konur, O. 2023bs. Gasoline fuels: Scientometric study. In *Evaluation and Utilization of Bioethanol Fuels. I: Evaluation of Bioethanol Fuels, Transport Engines, and Bioethanol Sensors. Handbook of Bioethanol Fuels Volume 5,* Ed. O. Konur. Boca Raton, FL: CRC Press.

Konur, O. 2023bt. Gasoline fuels: Review. In *Evaluation and Utilization of Bioethanol Fuels. I: Evaluation of Bioethanol Fuels, Transport Engines, and Bioethanol Sensors. Handbook of Bioethanol Fuels Volume 5,* Ed. O. Konur. Boca Raton, FL: CRC Press.

Konur, O. 2023bu. Nanotechnology applications in bioethanol fuels: Scientometric study. In *Evaluation and Utilization of Bioethanol Fuels. I: Evaluation of Bioethanol Fuels, Transport Engines, and Bioethanol Sensors. Handbook of Bioethanol Fuels Volume 5,* Ed. O. Konur. Boca Raton, FL: CRC Press.

Konur, O. 2023bv. Nanotechnology applications in bioethanol fuels: Review. In *Evaluation and Utilization of Bioethanol Fuels. I: Evaluation of Bioethanol Fuels, Transport Engines, and Bioethanol Sensors. Handbook of Bioethanol Fuels Volume 5,* Ed. O. Konur. Boca Raton, FL: CRC Press.

Konur, O. 2023bw. Utilization of bioethanol fuels in the transport engines: Scientometric study. In *Evaluation and Utilization of Bioethanol Fuels. I: Evaluation of Bioethanol Fuels, Transport Engines, and Bioethanol Sensors. Handbook of Bioethanol Fuels Volume 5,* Ed. O. Konur. Boca Raton, FL: CRC Press.

Konur, O. 2023bx. Utilization of bioethanol fuels in the transport engines: Review. In *Evaluation and Utilization of Bioethanol Fuels. I: Evaluation of Bioethanol Fuels, Transport Engines, and Bioethanol Sensors. Handbook of Bioethanol Fuels Volume 5,* Ed. O. Konur. Boca Raton, FL: CRC Press.

Konur, O. 2023by. Evaluation of bioethanol fuels: Scientometric study. In *Evaluation and Utilization of Bioethanol Fuels. I: Evaluation of Bioethanol Fuels, Transport Engines, and Bioethanol Sensors. Handbook of Bioethanol Fuels Volume 5,* Ed. O. Konur. Boca Raton, FL: CRC Press.

Konur, O. 2023bz. Evaluation of bioethanol fuels: Review. In *Evaluation and Utilization of Bioethanol Fuels. I: Evaluation of Bioethanol Fuels, Transport Engines, and Bioethanol Sensors. Handbook of Bioethanol Fuels Volume 5,* Ed. O. Konur. Boca Raton, FL: CRC Press.

Konur, O. 2023ca. Utilization of bioethanol fuels: Scientometric study. In *Evaluation and Utilization of Bioethanol Fuels. I: Evaluation of Bioethanol Fuels, Transport Engines, and Bioethanol Sensors. Handbook of Bioethanol Fuels Volume 5,* Ed. O. Konur. Boca Raton, FL: CRC Press.

Konur, O. 2023cb. Utilization of bioethanol fuels: Review. In *Evaluation and Utilization of Bioethanol Fuels. I: Evaluation of Bioethanol Fuels, Transport Engines, and Bioethanol Sensors. Handbook of Bioethanol Fuels Volume 5,* Ed. O. Konur. Boca Raton, FL: CRC Press.

Konur, O. 2023cc. Bioethanol fuel sensors: Scientometric study. In *Evaluation and Utilization of Bioethanol Fuels. I: Evaluation of Bioethanol Fuels, Transport Engines, and Bioethanol Sensors. Handbook of Bioethanol Fuels Volume 5,* Ed. O. Konur. Boca Raton, FL: CRC Press.

Konur, O. 2023cd. Bioethanol fuel sensors: Review. In *Evaluation and Utilization of Bioethanol Fuels. I: Evaluation of Bioethanol Fuels, Transport Engines, and Bioethanol Sensors. Handbook of Bioethanol Fuels Volume 5,* Ed. O. Konur. Boca Raton, FL: CRC Press.

Konur, O. 2023ce. Country-based experience of bioethanol fuels: Scientometric study. In *Evaluation and Utilization of Bioethanol Fuels. II: Biohydrogen Fuels, Fuel Cells, Biochemicals, and Country Experiences, Handbook of Bioethanol Fuels Volume 6,* Ed. O. Konur. Boca Raton, FL: CRC Press.

Konur, O. 2023cf. Country-based experience of bioethanol fuels: Review. In *Evaluation and Utilization of Bioethanol Fuels. II: Biohydrogen Fuels, Fuel Cells, Biochemicals, and Country Experiences. Handbook of Bioethanol Fuels Volume 6,* Ed. O. Konur. Boca Raton, FL: CRC Press.

Konur, O. 2023cg. Chinese experience of bioethanol fuels: Scientometric study. In *Evaluation and Utilization of Bioethanol Fuels. II: Biohydrogen Fuels, Fuel Cells, Biochemicals, and Country Experiences, Handbook of Bioethanol Fuels Volume 6,* Ed. O. Konur. Boca Raton, FL: CRC Press.

Konur, O. 2023ch. Chinese experience of bioethanol fuels: Review. In *Evaluation and Utilization of Bioethanol Fuels. II: Biohydrogen Fuels, Fuel Cells, Biochemicals, and Country Experiences. Handbook of Bioethanol Fuels Volume 6,* Ed. O. Konur. Boca Raton, FL: CRC Press.

Konur, O. 2023ci. The US experience of bioethanol fuels: Scientometric study. In *Evaluation and Utilization of Bioethanol Fuels. II: Biohydrogen Fuels, Fuel Cells, Biochemicals, and Country Experiences. Handbook of Bioethanol Fuels Volume 6,* Ed. O. Konur. Boca Raton, FL: CRC Press.

Konur, O. 2023cj. The US experience of bioethanol fuels: Review. In *Evaluation and Utilization of Bioethanol Fuels. II: Biohydrogen Fuels, Fuel Cells, Biochemicals, and Country Experiences. Handbook of Bioethanol Fuels Volume 6,* Ed. O. Konur. Boca Raton, FL: CRC Press.

Konur, O. 2023ck. European experience of bioethanol fuels: Scientometric study. In *Evaluation and Utilization of Bioethanol Fuels. II: Biohydrogen Fuels, Fuel Cells, Biochemicals, and Country Experiences. Handbook of Bioethanol Fuels Volume 6,* Ed. O. Konur. Boca Raton, FL: CRC Press.

Konur, O. 2023cl. European experience of bioethanol fuels: Review. In *Evaluation and Utilization of Bioethanol Fuels. II: Biohydrogen Fuels, Fuel Cells, Biochemicals, and Country Experiences. Handbook of Bioethanol Fuels Volume 6,* Ed. O. Konur. Boca Raton, FL: CRC Press.

Konur, O. 2023cm. Bioethanol fuel-based biohydrogen fuels: Scientometric study. In *Evaluation and Utilization of Bioethanol Fuels. II: Biohydrogen Fuels, Fuel Cells, Biochemicals, and Country Experiences. Handbook of Bioethanol Fuels Volume 6,* Ed. O. Konur. Boca Raton, FL: CRC Press.

Konur, O. 2023cn. Bioethanol fuel-based biohydrogen fuel production: Review. In *Evaluation and Utilization of Bioethanol Fuels. II: Biohydrogen Fuels, Fuel Cells, Biochemicals, and Country Experiences. Handbook of Bioethanol Fuels Volume 6,* Ed. O. Konur. Boca Raton, FL: CRC Press.

Konur, O. 2023co. Bioethanol fuel cells: Scientometric study. In *Evaluation and Utilization of Bioethanol Fuels. II: Biohydrogen Fuels, Fuel Cells, Biochemicals, and Country Experiences. Handbook of Bioethanol Fuels Volume 6,* Ed. O. Konur. Boca Raton, FL: CRC Press.

Konur, O. 2023cp. Bioethanol fuel cells: Review. In *Evaluation and Utilization of Bioethanol Fuels. II: Biohydrogen Fuels, Fuel Cells, Biochemicals, and Country Experiences. Handbook of Bioethanol Fuels Volume 6,* Ed. O. Konur. Boca Raton, FL: CRC Press.

Konur, O. 2023cq. Bioethanol fuel-based biochemicals: Scientometric study. In *Evaluation and Utilization of Bioethanol Fuels. II: Biohydrogen Fuels, Fuel Cells, Biochemicals, and Country Experiences. Handbook of Bioethanol Fuels Volume 6,* Ed. O. Konur. Boca Raton, FL: CRC Press.

Konur, O. 2023cr. Bioethanol fuel-based biochemicals: Review. In *Evaluation and Utilization of Bioethanol Fuels. II: Biohydrogen Fuels, Fuel Cells, Biochemicals, and Country Experiences, Handbook of Bioethanol Fuels Volume 6,* Ed. O. Konur. Boca Raton, FL: CRC Press.

Konur, O., Ed. 2023cs. *Handbook of Bioethanol Fuels*. Boca Raton, FL: CRC Press.

Konur, O., Ed. 2023ct. *Bioethanol Fuel Production Processes. I: Biomass Pretreatments. Handbook of Bioethanol Fuels Volume 1*. Boca Raton, FL: CRC Press.

Konur, O., Ed. 2023cu. *Bioethanol Fuel Production Processes. II: Biomass Hydrolysis, Fermentation, and Bioethanol Fuel Separation. Handbook of Bioethanol Fuels Volume 2*. Boca Raton, FL: CRC Press.

Konur, O., Ed. 2023cv. *Feedstock-based Bioethanol Fuels. I. Non-Waste Feedstocks: Starch, Sugar, Grass, Wood, Cellulose, Algae, and Biosyngas-based Bioethanol Fuels. Handbook of Bioethanol Fuels Volume 3*. Boca Raton, FL: CRC Press.

Konur, O., Ed. 2023cw. *Feedstock-based Bioethanol Fuels. II. Waste Feedstocks: Agricultural, Food, Industrial, Urban, Forestry, and Lignocellulosic Waste-based Bioethanol Fuels. Handbook of Bioethanol Fuels Volume 4*. Boca Raton, FL: CRC Press.

Konur, O., Ed. 2023cx. *Evaluation and Evaluation of Bioethanol Fuels I. Handbook of Bioethanol Fuels Volume 5*. Boca Raton, FL: CRC Press.

Konur, O., Ed. 2023cy. *Evaluation and Utilization of Bioethanol Fuels. II: Biohydrogen Fuels, Fuel Cells, Biochemicals, and Country Experiences. Handbook of Bioethanol Fuels Volume 6*. Boca Raton, FL: CRC Press.

Laser, M., D. Schulman and S. G. Allen, et al. 2002. A comparison of liquid hot water and steam pretreatments of sugar cane bagasse for bioconversion to ethanol. *Bioresource Technology* 81:33–44.

Limayem, A. and S. C. Ricke. 2012. Lignocellulosic biomass for bioethanol production: Current perspectives, potential issues and future prospects. *Progress in Energy and Combustion Science* 38:449–467.

Lin, Y. and S. Tanaka. 2006. Ethanol fermentation from biomass resources: Current state and prospects. *Applied Microbiology and Biotechnology* 69:627–642.

Liu, J., X. Wang, Q. Peng and Y. Li. 2005. Vanadium pentoxide nanobelts: Highly selective and stable ethanol sensor materials. *Advanced Materials* 17:764–767.

Ma, X., L. Sun and C. Song. 2002. A new approach to deep desulfurization of gasoline, diesel fuel and jet fuel by selective adsorption for ultra-clean fuels and for fuel cell applications. *Catalysis Today* 77:107–116.

Mattos, L.V., G. Jacobs, B. H. Davis and F. B. Noronha. 2012. Production of hydrogen from ethanol: Review of reaction mechanism and catalyst deactivation. *Chemical Reviews* 112:4094–4123.

Mendieta, C. M., J. Kruyeniski, M. E. Vallejos and M. C. Area. 2023. Production and uses of bioethanol in an integrated biorefinery from agro- and forest-industrial waste. In *Feedstock-based Bioethanol Fuels. II. Waste Feedstocks: Agricultural, Food, Industrial, Urban, Forestry, and Lignocellulosic Waste-based Bioethanol Fuels. Handbook of Bioethanol Fuels Volume 4,* Ed. O. Konur. Boca Raton, FL: CRC Press.

Moreira, J. R. and J. Goldemberg. 2023. Ethanol fuel from sugar cane-present and future contribution to climate change mitigation. In *Evaluation and Utilization of Bioethanol Fuels. I: Evaluation of Bioethanol Fuels, Transport Engines, and Bioethanol Sensors. Handbook of Bioethanol Fuels Volume 5,* Ed. O. Konur. Boca Raton, FL: CRC Press.

Morschbacker, A. 2009. Bio-ethanol based ethylene. *Polymer Reviews* 49:79–84.

Mosier, N., C. Wyman and B. Dale, et al. 2005. Features of promising technologies for pretreatment of lignocellulosic biomass. *Bioresource Technology* 96:673–686.

Munasinghe, P. C. and S. K. Khanal. 2010. Biomass-derived syngas fermentation into biofuels: Opportunities and challenges. *Bioresource Technology* 101:5013–5022.

Najafi, G., B. Ghobadian and T. Tavakoli, et al. 2009. Performance and exhaust emissions of a gasoline engine with ethanol blended gasoline fuels using artificial neural network. *Applied Energy* 86:630–639.

Newman, P. W. G. and J. R. Kenworthy. 1989. Gasoline consumption and cities: A comparison of U.S. cities with a global survey. *Journal of the American Planning Association* 55:24–37.

Ni, M., D. Y. C. Leung and M. K. H. Leung. 2007. A review on reforming bio-ethanol for hydrogen production. *International Journal of Hydrogen Energy* 32:3238–3247.

Nigam, J. N. 2002. Bioconversion of water-hyacinth *Eichhornia crassipes* hemicellulose acid hydrolysate to motor fuel ethanol by xylose-fermenting yeast. *Journal of Biotechnology* 97:107–116.

Niven, R. K. 2005. Ethanol in gasoline: Environmental impacts and sustainability review article. *Renewable and Sustainable Energy Reviews* 9:535–555.

North, D. C. 1991. Institutions. *Journal of Economic Perspectives* 5:97–112.

Nowruzi, B. and O. Konur, 2023av. Third generation algal bioethanol fuels: Review. In *Feedstock-based Bioethanol Fuels. I. Non-Waste Feedstocks: Starch, Sugar, Grass, Wood, Cellulose, Algae, and Biosyngas-based Bioethanol Fuels. Handbook of Bioethanol Fuels Volume 3,* Ed. O. Konur. Boca Raton, FL: CRC Press.

Olsson, L. and B. Hahn-Hagerdal. 1996. Fermentation of lignocellulosic hydrolysates for ethanol production. *Enzyme and Microbial Technology* 18:312–331.

Palmqvist, E. and B. Hahn-Hagerdal. 2000a. Fermentation of lignocellulosic hydrolysates. I: Inhibition and detoxification. *Bioresource Technology* 74:17–24.

Palmqvist, E. and B. Hahn-Hagerdal. 2000b. Fermentation of lignocellulosic hydrolysates. II: Inhibitors and mechanisms of inhibition. *Bioresource Technology* 74:25–33.

Phung T. K. and K. B. Vu. 2023 Recent advances in the steam reforming of bioethanol for the production of biohydrogen fuels. In *Evaluation and Utilization of Bioethanol Fuels. II: Biohydrogen Fuels, Fuel Cells, Biochemicals, and Country Experiences. Handbook of Bioethanol Fuels Volume 6,* Ed. O. Konur. Boca Raton, FL: CRC Press.

Pimentel, D. and T. W. Patzek. 2005. Ethanol production using corn, switchgrass, and wood; Biodiesel production using soybean and sunflower. *Natural Resources Research* 14:65–76.

Pinkert, A., K. N. Marsh, S. Pang and M. P. Staiger. 2009. Ionic liquids and their interaction with cellulose. *Chemical Reviews* 109:6712–6728.

Pradhan, D., B. Tsegaye, S. S. Mathew, S. Jiaswal and A. K. Jaiswal. 2023. Ultrasound-assisted pretreatment of lignocellulosic biomass for bioethanol production. In *Bioethanol Fuel Production Processes. I: Biomass Pretreatments. Handbook of Bioethanol Fuels Volume 1,* Ed. O. Konur. Boca Raton, FL: CRC Press.

Prasad, S., A. Singh and H. C. Joshi. 2007. Ethanol as an alternative fuel from agricultural, industrial and urban residues. *Resources, Conservation and Recycling* 50:1–39.

Ravindran, R., and A. K. Jaiswal. 2016. A comprehensive review on pre-treatment strategy for lignocellulosic food industry waste: Challenges and opportunities. *Bioresource Technology* 199:92–102.

Saini, J. K., R. Saini and L. Tewari. 2015. Lignocellulosic agriculture wastes as biomass feedstocks for second-generation bioethanol production: Concepts and recent developments. *3 Biotech* 5:337–353.

Sanchez, C. 2009. Lignocellulosic residues: Biodegradation and bioconversion by fungi. *Biotechnology Advances* 27:185–194.

Sanchez, O. J. and C. A. Cardona. 2008. Trends in biotechnological production of fuel ethanol from different feedstocks. *Bioresource Technology* 99:5270–5295.

Sano, T., H. Yanagishita, Y. Kiyozumi, F. Mizukami and K. Haraya. 1994. Separation of ethanol/water mixture by silicalite membrane on pervaporation. *Journal of Membrane Science* 95:221–228.

Schmer, M. R., K. P. Vogel, R. B. Mitchell and R. K. Perrin. 2008. Net energy of cellulosic ethanol from switchgrass. *Proceedings of the National Academy of Sciences of the United States of America* 105:464–469.

Senatore, A., F. Dalena and A. Basile. 2023 Ethanol: Brief review of the major production methods in the world. In *Evaluation and Utilization of Bioethanol Fuels. II: Biohydrogen Fuels, Fuel Cells, Biochemicals, and Country Experiences. Handbook of Bioethanol Fuels Volume 6,* Ed. O. Konur. Boca Raton, FL: CRC Press.

Sukumaran, R. K., R. R. Singhania, G. M. Mathew and A. Pandey. 2009. Cellulase production using biomass feed stock and its application in lignocellulose saccharification for bio-ethanol production. *Renewable Energy* 34:421–424.

Sun, Y. and J. Cheng. 2002. Hydrolysis of lignocellulosic materials for ethanol production: A review. *Bioresource Technology* 83:1–11.

Taherzadeh, M. J. and K. Karimi. 2007. Enzyme-based hydrolysis processes for ethanol from lignocellulosic materials: A review. *Bioresources* 2:707–738.

Taherzadeh, M. J. and K. Karimi. 2008. Pretreatment of lignocellulosic wastes to improve ethanol and biogas production: A review. *International Journal of Molecular Sciences* 9:1621–1651.

Talebnia, F., D. Karakashev and I. Angelidaki. 2010. Production of bioethanol from wheat straw: An overview on pretreatment, hydrolysis and fermentation. *Bioresource Technology* 101:4744–4753.

Thakur, S., R. Arora, A. Kumar and A. Pratush. 2023. Lignocellulolytic enzymes for biofuel production. In *Bioethanol Fuel Production Processes. I: Biomass Pretreatments. Handbook of Bioethanol Fuels Volume 1,* Ed. O. Konur. Boca Raton, FL: CRC Press.

Vane, L. M. 2005. A review of pervaporation for product recovery from biomass fermentation processes. *Journal of Chemical Technology and Biotechnology* 80:603–629.

von Blottnitz, H. and M. A. Curran. 2007. A review of assessments conducted on bio-ethanol as a transportation fuel from a net energy, greenhouse gas, and environmental life cycle perspective. *Journal of Cleaner Production* 15:607–619.

Wan, Q., Q. H. Li and Y. J. Chen, et al. 2004. Fabrication and ethanol sensing characteristics of ZnO nanowire gas sensors. *Applied Physics Letters* 84:3654–3656.

Wang, L., Y. Kang and X. Liu, et al. 2012. ZnO nanorod gas sensor for ethanol detection. *Sensors and Actuators, B: Chemical* 162:237–243.

Zabed, H., J. N. Sahu, A. Suely, A. N. Boyce and G. Faruq. 2017. Bioethanol production from renewable sources: Current perspectives and technological progress. *Renewable and Sustainable Energy Reviews* 71:475–501.

Zhang, M. and Y. Yu. 2013. Dehydration of ethanol to ethylene. *Industrial and Engineering Chemistry Research* 52:9505–9514.

Zhang, Y and L. Wang. 2023. Emerging solvents for biomass pretreatment and their sustainability. In *Bioethanol Fuel Production Processes. I: Biomass Pretreatments. Handbook of Bioethanol Fuels Volume 1,* Ed. O. Konur. Boca Raton, FL: CRC Press.

Zhang, Y. H. P. and L. R. Lynd. 2004. Toward an aggregated understanding of enzymatic hydrolysis of cellulose: Noncomplexed cellulase systems. *Biotechnology and Bioengineering* 88:797–824.

Zhu, J. Y. and X. J. Pan. 2010. Woody biomass pretreatment for cellulosic ethanol production: Technology and energy consumption evaluation. *Bioresource Technology* 101:4992–5002.

2 Bioethanol Fuels
Scientometric Study

Ozcan Konur
(Formerly) Ankara Yildirim Beyazit University

2.1 INTRODUCTION

The crude oil-based gasoline fuels (Ma et al., 2002; Newman and Kenworthy, 1989) have been widely used in the transportation sector since the 1920s. However, there have been great public concerns over the adverse environmental and human impact of these fuels (Hill et al., 2006, 2009). Hence, biomass-based bioethanol fuels (Hill et al., 2006; Konur, 2012e, 2015, 2020a) have increasingly been used in blending gasoline fuels (Hsieh et al., 2002; Najafi et al., 2009), in the fuel cells (Antolini, 2007, 2009), and in the biochemical production (Angelici et al., 2013; Morschbacker, 2009) in a biorefinery context (Fernando et al., 2006; Huang et al., 2008). Bioethanol fuels also play a critical role in maintaining the energy security (Kruyt et al., 2009; Winzer, 2012) in the supply shocks (Kilian, 2008, 2009) related to oil price shocks (Hamilton, 1983, 2003), coronavirus disease 2019 (COVID-19) pandemics (Fauci et al., 2020; Li et al., 2020), or wars (Hamilton, 1983; Jones, 2012) in the aftermath of the Russian invasion of Ukraine (Reeves, 2014).

However, it is necessary to pretreat the biomass (Taherzadeh and Karimi, 2008; Yang and Wyman, 2008) to enhance the yield of the bioethanol (Hahn-Hagerdal et al., 2006; Sanchez and Cardona, 2008) before the bioethanol production through the hydrolysis (Sun and Cheng, 2002; Taherzadeh and Karimi, 2007) and fermentation (Lin and Tanaka, 2006; Olsson and Hahn-Hagerdal, 1996) of the biomass and the resulting hydrolysates, respectively.

The research in the field of the bioethanol fuels has intensified in this context in the key research fronts of the production of bioethanol fuels and, to a lesser extent, utilization (Antolini, 2007; Hansen et al., 2005; Haryanto et al., 2005) and evaluation (Hamelinck et al., 2005; Pimentel and Patzek, 2005; Schmer et al., 2008) of bioethanol fuels. For the first research front, pretreatment (Alvira et al., 2010; Hendriks and Zeeman, 2009; Mosier et al., 2005) and hydrolysis (Alvira et al., 2010; Sun and Cheng, 2002; Zhang and Lynd, 2004) of the feedstocks, fermentation of the resulting hydrolysates (Jonsson and Martin, 2016; Lin and Tanaka, 2006; Palmqvist and Hahn-Hagerdal, 2000a, b), production of bioethanol fuels (Balat, 2011; Limayem and Ricke, 2012; Lin and Tanaka, 2006), and to a lesser extent separation and distillation of bioethanol fuels (Sano et al., 1994; Vane, 2005) are the key research areas, while for the second research front, utilization of bioethanol fuels in fuel cells and transport engines, bioethanol-based biohydrogen fuels, bioethanol sensors, and bioethanol-based biochemicals are the key research areas.

The research in this field has also intensified for the feedstocks of lignocellulosic biomass at large (Hendriks and Zeeman, 2009; Mosier et al., 2005; Sun and Cheng, 2002), bioethanol fuels for their utilization (Antolini, 2007; Hansen et al., 2005; Haryanto et al., 2005), cellulose (Pinkert et al., 2009; Zhang and Lynd, 2004), starch feedstock residues (Binod et al., 2010; Talebnia et al., 2010), wood biomass (Galbe and Zacchi, 2002; Zhu and Pan, 2010), and to a lesser extent industrial waste (Cardona et al., 2010; Prasad et al., 2007), lignin (Bourbonnais and Paice, 1990; Kirk and Farrell, 1987), grass (Keshwani and Cheng, 2009; Pimentel and Patzek, 2005), hydrolysates (Palmqvist et al., 2000a, b), biomass at large (Lin and Tanaka, 2006; Zabed

DOI: 10.1201/9781003226536-3

et al., 2017), sugar feedstocks (Bai et al., 2008; Canilha et al., 2012), sugar feedstock residues (Cardona et al., 2010; Laser et al., 2002), algal biomass (Ho et al., 2013; John et al., 2011), lignocellulosic wastes (Sanchez, 2009; Saini et al., 2015), starch feedstocks (Bai et al., 2008; Bothast and Schlicher, 2005), urban wastes (Prasad et al., 2007; Ravindran and Jaiswal, 2016), forestry wastes (Duff and Murray, 1996), food wastes (Guimaraes et al., 2010; Ravindran and Jaiswal, 2016), syngas (Henstra et al., 2007; Munasinghe and Khanal, 2010), and plants (Nigam, 2002; Sukumaran et al., 2009).

However, it is essential to develop efficient incentive structures (North, 1991) for the primary stakeholders to enhance the research in this field (Konur, 2000, 2002a–c, 2006a, b, 2007a, b). The scientometric analysis has been used in this context to inform the primary stakeholders about the current state of the research in a selected research field (Garfield, 1955; Konur, 2011, 2012a–i, 2015, 2018b, 2019, 2020a).

As there have been no published current scientometric studies in this field, this book chapter presents a scientometric study of the research on the bioethanol fuels. It examines the scientometric characteristics of both the sample and population data presenting scientometric characteristics of these both datasets in the order of documents, authors, publication years, institutions, funding bodies, source titles, countries, Scopus subject categories, Scopus keywords, and research fronts.

2.2 MATERIALS AND METHODS

The search for this study was carried out using the Scopus database (Burnham, 2006) in October 2022.

As a first step for the search of the relevant literature, the keywords were selected using the most-cited first 300 population papers for each research front. The selected keyword list was then optimized to obtain a representative sample of papers for the each research field. These keyword lists were then integrated to obtain the keyword list for this research field (Konur, 2023a–c).

As a second step, two sets of data were used for this study. First, a population sample of 77,433 papers was used to examine the scientometric characteristics of the population data. Second, a sample of 775 most-cited papers, corresponding to 1% of the population papers, was used to examine the scientometric characteristics of these citation classics.

The scientometric characteristics of these sample and population datasets were presented in the order of documents, authors, publication years, institutions, funding bodies, source titles, countries, Scopus subject categories, Scopus keywords, and research fronts.

Lastly, the key scientometric findings for both datasets were discussed to highlight the research landscape for the bioethanol fuels. Additionally, a number of brief conclusions were drawn and a number of relevant recommendations were made to enhance the future research landscape.

2.3 RESULTS

2.3.1 The Most Prolific Documents in the Bioethanol Fuels

The information on the types of documents for both datasets is given in Table 2.1. The articles and conference papers, published in journals, dominate both the sample (76%) and population (94%) papers with 18% deficit. Furthermore, review papers and short surveys have a 21% surplus as they are over-represented in the sample papers as they constitute 24% and 3% of the sample and population papers, respectively. Additionally, 2% of the sample papers were published as book chapters.

It is further notable that 97%, 2%, and 1% of the population papers were published in journals, book series, and books, respectively. Similarly, 99% and 1% of the sample papers were published in the journals and book series, respectively.

TABLE 2.1
Documents in the Bioethanol Fuels

Documents	Sample Dataset (%)	Population Dataset (%)	Surplus (%)
Article	72.9	90.3	−17.4
Review	21.8	2.7	19.1
Conference paper	2.6	3.9	−1.3
Short survey	2.6	0.2	2.4
Note	0.1	0.5	−0.4
Book chapter	0.0	1.9	−1.9
Letter	0.0	0.4	−0.4
Book	0.0	0.1	−0.1
Editorial	0.0	0.1	−0.1
Sample size	775	77,433	

Sample dataset: the number of papers (%) in the sample set of 775 highly cited papers.
Population dataset: the number of papers (%) in the set of the 77,433 population papers.

2.3.2 The Most Prolific Authors in the Bioethanol Fuels

The information about the 42 most prolific authors with at least 0.6% of sample papers each is given in Table 2.2. The most prolific authors are Charles E. Wyman and John N. Saddler with 2.8% and 2.1% of the sample papers, respectively, followed by Bruce E. Dale, Barbel Hahn-Hagerdal, and Lee R. Lynd with 1.9%, 1.9%, and 1.5% of the sample papers, respectively. The other prolific authors are Guido Zacchi, Michael E. Himmel, Michael R. Ladisch, and Blake A. Simmons with 1.2%–1.4% of the sample papers each.

However, the most influential author is Charles E. Wyman with 2.6% of the sample papers, followed by John N. Saddler, Barbel Hahn-Hagerdal, and Bruce E. Dale with 1.9%, 1.8%, and 1.7% surplus, respectively. The other influential authors are Lee R. Lynd, Guido Zacchi, Michael E. Himmel, and Michael R. Ladisch with 1.2%–1.4% surplus each.

The most prolific institution for the sample dataset is the Lund University with four authors, followed by National Renewable Energy Laboratory (NREL) with three authors. The other prolific institutions are National Technical University of Athens, Purdue University, University of British Columbia, and The U.S. Department of Agriculture (USDA) Forest Service with two authors each. However, the most prolific country for the sample dataset is the USA with 21 authors, followed by Sweden with six authors. The other prolific countries are France, Canada, and Greece with two to three authors each. In total, only 12 countries house these top authors.

The most prolific research front for these top authors is the pretreatments of the feedstocks with 37 authors followed by the hydrolysis of the feedstocks with 31 authors. The other prolific research fronts are the fermentation of the hydrolysates and the bioethanol production in general with 20 and 18 authors, respectively. The other minor research fronts are utilization of bioethanol fuels in the transport engines and fuel cells with two authors each.

However, there is significant gender deficit (Beaudry and Lariviere, 2016) for the sample dataset as surprisingly only five of these top researchers are female with a representation rate of 12%.

Additionally, there are other authors with the relatively low citation impact and with 0.1%–0.3% of the population papers each: Runcang Sun, Akihiko Kondo, Juan C. Parajo, Venkatesh Balan, Qiang Yong, Feng Xu, Mercedes Ballesteros, Lisbeth Olsson, Jie Bao, Carlos R. Soccol, Yong-Su Jin, Bruce S. Dien, Hasan Jameel, Ying Xu, Zhongshun Yuan, Ye Ni, Herbert Sixta, Yunqiao Pu, Eulogio Castro, Hongzhang Chen, Paul Christakopoulos, Navadol Laosiripojana,

TABLE 2.2
Most-Prolific Authors in the Bioethanol Fuels

No.	Author Name	Author Code	Sample Papers (%)	Population Papers (%)	Surplus	Institution	Country	HI	N	Res. Front
1	Wyman, Charles E.	7004396809	2.8	0.2	2.6	Univ. Calf. Riverside	USA	80	287	P, H, F, R, E
2	Saddler, John N.	7005297559 57202481615	2.1	0.2	1.9	Univ. British Columbia	Canada	99	420	P, H, F, R
3	Dale, Bruce E.	7201511969	1.9	0.2	1.7	Michigan State Univ.	USA	92	430	P, H, F, R
4	Hahn-Hagerdal, Barbel*	7005389381	1.9	0.1	1.8	Lund Univ.	Sweden	76	258	P, H, F, R
5	Lynd, Lee R.	35586183800	1.5	0.1	1.4	Dartmouth Coll.	USA	75	287	P, H, F, R
6	Zacchi, Guido	7006727748	1.4	0.2	1.2	Lund Univ.	Sweden	68	204	P, H, F, R
7	Himmel, Michael E.	7007125552	1.3	0.1	1.2	Natl. Renew. Ener. Lab.	USA	74	423	P, H
8	Ladisch, Michael R.	7005670397	1.3	0.1	1.2	Purdue Univ.	USA	60	292	P, H, F, R
9	Simmons, Blake A.	7102183263	1.2	0.2	1.0	Lawrence Berkeley Natl. Lab.	USA	76	446	P, H
10	Ragauskas, Arthur J.	7006265204	1.0	0.2	0.8	Univ. Tennessee	USA	95	766	P, H, F, R
11	Galbe, Mats	7003788758	1.0	0.1	0.9	Lund Univ.	Sweden	51	131	P, H, F, R, E
12	Holtzapple, Mark T.	7004167004	1.0	0.1	0.9	Texas A&M Univ.	USA	47	199	P
13	Gilkes, Neil R.	35493889000	1.0	0.0	1.0	Univ. British Columbia	Canada	51	87	P, H
14	Taherzadeh, Mohammad J.	6701407496	0.9	0.2	0.7	Univ. Boras	Sweden	67	426	P, H, F, R
15	Ingram, Lonnie O.	7102962097	0.9	0.1	0.8	Univ. Florida	USA	75	279	P, F, R
16	Jonsson, Leif J.	7102349315	0.9	0.1	0.8	Umea Univ.	Sweden	41	148	P, H, F, R
17	Henrissat, Bernard	7005911606	0.9	0.1	0.8	King Abdulaziz Univ.	S. Arabia	136	598	P, H
18	Van Dijken, Johannes P.	7102979857	0.9	0.1	0.8	Delft Univ. Technol.	Netherlands	69	190	P, F, R
19	Pandey, Ashok	7201771319	0.8	0.2	0.6	CSIR	India	93	920	P, H, F, R
20	Singh, Seema*	35264950300	0.8	0.1	0.7	Sandia Natl. Lab.	USA	59	186	P, H
21	Martinez, Angel T.*	55622506400	0.8	0.1	0.7	CSIC	Spain	81	345	P, H
22	Yang, Bin	7404473046	0.8	0.1	0.7	Washington State Univ.	USA	41	98	P, H
23	Gold, Michael H.	7402444296	0.8	0.1	0.7	Oregon Hlth. Sci. Univ.	USA	57	148	P, H

(*Continued*)

TABLE 2.2 (*Continued*)
Most-Prolific Authors in the Bioethanol Fuels

No.	Author Name	Author Code	Sample Papers (%)	Population Papers (%)	Surplus	Institution	Country	HI	N	Res. Front
24	Mosier, Nathan S.	6602426392	0.8	0.1	0.7	Purdue Univ.	USA	43	115	P, H, F
25	Pan, Xuejun	57203296000	0.8	0.1	0.7	Univ. Wisconsin	USA	45	118	P, H
26	Cullen, Dan	7202109135	0.8	0.0	0.8	USDA Forest Serv.	USA	46	120	P, H
27	Grigoriev, Igor V.	25027225800	0.8	0.0	0.8	Jnt. Genome Inst.	USA	116	354	P, H
28	Lamy, Claude	7007017658	0.8	0.0	0.8	Univ. Montpellier	France	72	235	C
29	Rogers, Robin D.	35474829200	0.8	0.0	0.8	Univ. Alabama	USA	118	897	P
30	Karimi, Keikhosro	10046195700	0.6	0.2	0.4	Vrije Univ.	Belgium	57	124	P, H, F, R
31	Zhu, Junyong	7405692678	0.6	0.1	0.5	USDA Forest Serv.	USA	64	311	P, H
32	Jeffries, Thomas W.	7005806269	0.6	0.1	0.5	Xylome Inc.	USA	59	156	P, F, R
33	Cotta, Michael A.	7006656876	0.6	0.1	0.5	USDA Agr. Res. Serv.	USA	53	186	P, H, F, R
34	Gorwa-Grauslund, Marie F.*	6603563787	0.6	0.1	0.5	Lund Univ.	Sweden	40	102	P, F, R
35	Rakopoulos, Constantine D.	35570765900	0.6	0.1	0.5	Natl. Tech. Univ. Athens	Greece	61	223	I
36	Zhang, Yi H. P.	34876090400	0.6	0.1	0.5	Tianjin Inst. Ind. Biotech.	China	58	178	P, H
37	Beckham, Gregg T.	16240926200	0.6	0.0	0.6	Natl. Renew. Ener. Lab.	USA	73	271	P
38	Rakopoulos, Dimitrios C	6603012578	0.6	0.0	0.6	Natl. Tech. Univ. Athens	Greece	42	95	I
39	Coutanceau, Christophe	8714035200	0.6	0.0	0.6	Univ. Poitiers	France	56	154	C
40	Coutinho, Pedro M	7006153340	0.6	0.0	0.6	Aix Marseille Univ.	France	73	145	P, H
41	Nilvebrant, Nils-Olof	57209815309	0.6	0.0	0.6	Borregaard	Norway	22	43	H, F
42	Elander, Richard T.	7004167004	0.6	0.0	0.6	Natl. Renew. Ener. Lab.	USA	31	61	P, H

Author code: the unique code given by Scopus to the authors. Sample papers: the number of papers authored in the sample dataset. Population papers: the number of papers authored in the population dataset.
HI, H-index; N, number of papers published by each author; *, female; P, pretreatment of the feedstocks; H, hydrolysis of the feedstocks; F, fermentation of the hydrolysates; R, bioethanol fuel production; E, evaluation of bioethanol fuels; C, utilization of bioethanol fuels in fuel cells; I, utilization of bioethanol fuels in the transport engines.

Gunnar Liden, Jinming Zhang, Fengwu Bai, Caoxing Huang, T. Shao, Jose A. Teixeira, Chiaki Ogino, Kyoung Heon Kim, Ignacio Ballesteros, Tomohisa Hasunuma, Liangcai Peng, Jinguang Hu, Jinguang Hu, Chenhuan Lai, Donghai Wang, Guangxing Yang, Xinshu Zhuang, Anuj K. Chandel, Jordi Llorca, Anne S. Meyer, Almir O. Neto, Wei Qi, Tianwei Tan, and Verawat Champreda.

Author code: the unique code given by Scopus to the authors. Sample papers: the number of papers authored in the sample dataset. Population papers: the number of papers authored in the population dataset. HI: H-index. N: number of papers published by each author. *: female. P: pretreatment of the feedstocks, H: hydrolysis of the feedstocks, F: fermentation of the hydrolysates. R: bioethanol fuel production. E: evaluation of bioethanol fuels. C: utilization of bioethanol fuels in fuel cells. I: utilization of bioethanol fuels in the transport engines.

2.3.3 The Most Prolific Research Output by Years in the Bioethanol Fuels

Information about papers published between 1970 and 2022 is given in Figure 2.1. This figure clearly shows that the bulk of the research papers in the population dataset were published primarily in the 2010s, the early 2020s, and the 2000s with 49%, 19%, and 14% of the population dataset, respectively. Similarly, the publication rates for the 1990s, 1980s, and 1970s were 8%, 6%, and 2%, respectively. Furthermore, the rate for the pre-1970s was 2%.

Similarly, the bulk of the research papers in the sample dataset were published in the 2000s, 2010s, and 1990s with 50% and 25%, and 12% of the sample dataset, respectively. Similarly, the publication rates for the 1980s and 1970s were 4% and 1% of the sample papers, respectively. Furthermore, the rate for the pre-1970s was 0%.

The most prolific publication years for the population dataset were 2021, 2022, and 2020 with 6.7%, 6.2%, and 6.1% of the dataset, respectively, while 73% of the population papers were published between 2008 and 2022. Similarly, 76% of the sample papers were published between 2002 and 2016, while the most prolific publication years were 2009, 2010, 2005, and 2007 with 8.3%, 7.6%, 7.1%, and 7.1% of the sample papers, respectively.

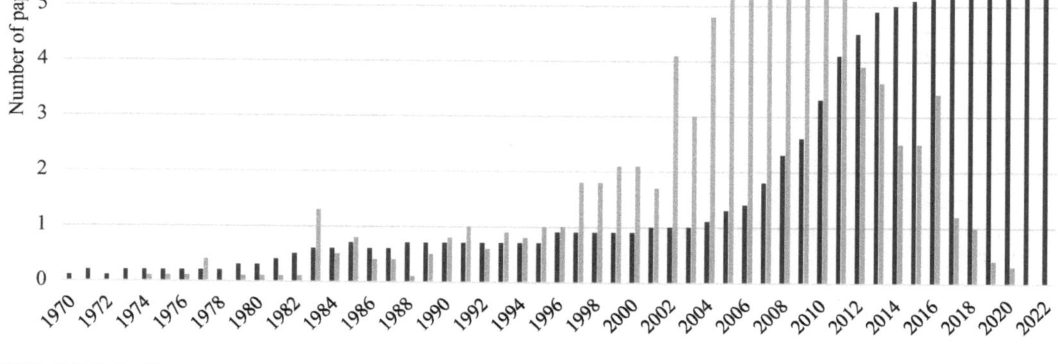

FIGURE 2.1 Research output by years regarding the bioethanol fuels.

2.3.4 The Most Prolific Institutions in the Bioethanol Fuels

Information about the 45 most prolific institutions publishing papers on the bioethanol fuels with at least 0.8% of the sample papers each is given in Table 2.3.

The most prolific institutions are the NREL with 4.5% of the sample papers, followed by the Lund University, Dartmouth College, and University of British Columbia with 3.1%–3.9% of the sample papers each. The other prolific institutions are Chinese Academy of Sciences, Michigan State University, USDA Forest Service, Purdue University, and University of Wisconsin Madison with 1.9% and 1.8% of the sample papers, respectively. Similarly, the top country for these most prolific institutions is the USA with 26 institutions. The other prolific countries are China, Denmark, France, Netherlands, and Sweden with two to three institutions each. In total, only 11 countries house these top institutions.

However, the institutions with the most impact are the NREL, Dartmouth College, Lund University, and University of British Columbia with 3.3%, 3.2%, and 2.6% of the sample papers, respectively. The other influential institutions are Michigan State University, USDA Forest Service, Purdue University, and University of Wisconsin Madison with 1.3%–2.1% surplus each.

Additionally, there are other institutions with the relatively low citation impact and with 0.4%–1.0% of the population papers each: South China University of Technology, State University of Campinas, Nanjing Forestry University, Tianjin University, Beijing Forestry University, Kyoto University, State Key Laboratory of Pulp and Paper Engineering, Russian Academy of Sciences, State University of Paulista, Federal University of Rio de Janeiro, Zhejiang University, China Agricultural University, University of Tokyo, Beijing University of Chemical Technology, Ministry of Agriculture of China, Tianjin University of Science & Technology, Shanghai Jiao Tong University, Korea University, University of Putra Malaysia, National Research Council (CNR), National Council for Scientific and Technical Research (CSIC), East China University of Science and Technology, Qilu University of Technology, Aalto University, Chalmers University of Technology, Iowa State University, Seoul National University, University of Tennessee, and Shandong University.

2.3.5 The Most Prolific Funding Bodies in the Bioethanol Fuels

Information about the 15 most prolific funding bodies funding at least 0.8% of the sample papers each is given in Table 2.4. Furthermore, only 35% and 43% of the sample and population papers each were funded, respectively.

The most prolific funding body is the U.S. Department of Energy with 4% of the sample papers. The other prolific funding bodies are the National Natural Science Foundation of China, European Commission, and National Science Foundation with 3.6%, 2.7%, and 2.3% of the sample papers, respectively. Furthermore, the other funding bodies are Natural Sciences and Engineering Research Council of Canada, Research Support Foundation of the State of Sao Paulo, and Office of Science with 1.3%–1.5% of the sample papers each.

However, the most prolific countries for these top funding bodies are the USA, China, and Brazil with six, four, and two funding bodies, respectively. In total, only five countries and the European Union (EU) house these top funding bodies.

The funding body with the most citation impact is the U.S. Department of Energy with 2.6% surplus, followed by the European Commission and National Science Foundation with 1.9% and 1.0% surplus, respectively. Furthermore, the funding bodies with the least citation impact are the National Science Foundation and Natural Sciences, National Council for Scientific and Technological Development, and National Key Research and Development Program of China with 5.9%, 1.7%, and 1.0% deficit, respectively.

The other funding bodies with the relatively low citation impact and with 0.4%–2.0% of the population papers each are the Higher Education Personnel Improvement Coordination, Fundamental Research Funds for the Central Universities, Japan Society for the Promotion of Science, National

TABLE 2.3
The Most Prolific Institutions in the Bioethanol Fuels

No.	Institutions	Country	Sample Papers (%)	Population Papers (%)	Surplus (%)
1	Natl. Renew. Ener. Lab.	USA	4.5	0.6	3.9
2	Lund Univ.	Sweden	3.9	0.7	3.2
3	Dartmouth Coll.	USA	3.5	0.2	3.3
4	Univ. British Columbia	Canada	3.1	0.5	2.6
5	Chinese Acad. Sci.	China	2.8	2.3	0.5
6	Michigan State Univ.	USA	2.6	0.5	2.1
7	USDA Forest Serv.	USA	2.6	0.5	2.1
8	Purdue Univ.	USA	2.2	0.5	1.7
9	Univ. Wisconsin Madison	USA	1.9	0.6	1.3
10	Univ. Sao Paulo	Brazil	1.8	1.7	0.1
11	Oak Ridge Natl. Lab.	USA	1.7	0.6	1.1
12	USDA Agr. Res. Serv.	USA	1.5	0.8	0.7
14	Tech. Univ. Denmark	Denmark	1.5	0.6	0.9
15	NC State Univ.	USA	1.5	0.6	0.9
16	Univ. Calif. Berkeley	USA	1.5	0.4	1.1
17	Univ. Calif. Riverside	USA	1.3	0.2	1.1
18	CNRS	France	1.2	0.6	0.6
19	Univ. Ill. U. C.	USA	1.2	0.6	0.6
20	Tsinghua Univ.	China	1.2	0.6	0.6
21	Georgia Inst. Technol.	USA	1.2	0.2	1.0
22	Jnt. Bioener. Inst.	USA	1.2	0.2	1.0
23	Pacific NW Natl. Lab.	USA	1.2	0.2	1.0
24	Sandia Natl. Lab.	USA	1.2	0.2	1.0
25	CSIC	Spain	1.0	0.5	0.5
26	Imperial Coll.	UK	1.0	0.2	0.8
27	Aix Marseille Univ.	France	1.0	0.2	0.8
28	Novozymes Biotech Inc.	USA	1.0	0.2	0.8
29	VTT Tech. Res. Ctr.	Finland	0.9	0.4	0.5
30	Univ. Minnesota	USA	0.9	0.4	0.5
31	Univ. Florida	USA	0.9	0.4	0.5
32	Texas A&M Univ.	USA	0.9	0.3	0.6
33	Lawrence Berkeley Natl. Lab.	USA	0.9	0.3	0.6
34	Delft Univ. Technol.	Netherlands	0.9	0.3	0.6
35	Auburn Univ.	USA	0.9	0.2	0.7
36	Joint Genome Inst.	USA	0.9	0.2	0.7
37	Wageningen Univ. Res.	Netherlands	0.8	0.4	0.4
38	CSIR	India	0.8	0.3	0.5
39	Univ. Copenhagen	Denmark	0.8	0.3	0.5
40	Virginia Polytech. Inst. State Univ.	USA	0.8	0.2	0.6
41	Massachusetts Inst. Technol.	USA	0.8	0.2	0.6
42	Sun Yat-Sen Univ.	China	0.8	0.2	0.6
43	Univ. Alabama	USA	0.8	0.2	0.6
44	Univ. Boras	Sweden	0.8	0.2	0.6
45	Univ. Poitiers	France	0.8	0.2	0.6

TABLE 2.4
The Most Prolific Funding Bodies in the Bioethanol Fuels

No.	Funding Bodies	Country	Sample Paper No. (%)	Population Paper No. (%)	Surplus (%)
1	U.S. Department of Energy	USA	4.0	1.4	2.6
2	National Natural Science Foundation of China	China	3.6	9.5	–5.9
3	European Commission	EU	2.7	0.8	1.9
4	National Science Foundation	USA	2.3	1.3	1.0
5	Natural Sciences and Engineering Research Council of Canada	Canada	1.5	0.9	0.6
6	Research Support Foundation of the State of Sao Paulo	Brazil	1.3	1.6	–0.3
7	Office of Science	USA	1.3	0.7	0.6
8	National Council for Scientific and Technological Development	Brazil	0.9	2.6	–1.7
9	Chinese Academy of Sciences	China	0.9	0.6	0.3
10	U.S. Department of Agriculture	USA	0.9	0.5	0.4
11	National Key Research and Development Program of China	China	0.8	1.8	–1.0
12	Ministry of Education of China	China	0.8	0.4	0.4
13	Engineering and Physical Sciences Research Council	UK	0.8	0.3	0.5
14	National Institute of General Medical Sciences	USA	0.8	0.3	0.5
15	National Renewable Energy Laboratory	USA	0.8	0.2	0.6

Research Foundation of Korea, European Regional Development Fund, China Postdoctoral Science Foundation, Priority Academic Program Development of Jiangsu Higher Education Institutions, China Scholarship Council, National Council of Science and Technology, Ministry of Education, Culture, Sports, Science and Technology, National Basic Research Program of China (973 Program), Natural Science Foundation of Jiangsu Province, Ministry of Economy and Competitiveness, Ministry of Science and Technology India, Ministry of Education, National High-Tech Research and Development Program, Seventh Framework Program, Foundation for Science and Technology, Ministry of Education, Science and Technology, Thailand Research Fund, Biological and Environmental Research, Ministry of Higher Education Malaysia, and National Institute of Food and Agriculture.

2.3.6 The Most Prolific Source Titles in the Bioethanol Fuels

Information about the 21 most prolific source titles publishing at least 1.2% of the sample papers each in the bioethanol fuels is given in Table 2.5.

The most prolific source title is the Bioresource Technology with 13.4% of the sample papers, followed by Biotechnology and Bioengineering, Applied Microbiology and Biotechnology, and Applied and Environmental Microbiology with 5.2%, 3.0%, and 2.8% of the sample papers, respectively. The other prolific titles are the Enzyme and Microbial Technology, Biomass and Bioenergy, Green Chemistry, and Renewable and Sustainable Energy Reviews with 2.1%–2.3% of the sample papers each.

However, the source title with the most impact is the Bioresource Technology with 8.2% surplus, followed by Biotechnology and Bioengineering with 3.8% surplus. The other influential titles

TABLE 2.5
The Most Prolific Source Titles in the Bioethanol Fuels

No.	Source Titles	Sample Papers (%)	Population Papers (%)	Surplus (%)
1	Bioresource Technology	13.4	5.2	8.2
2	Biotechnology and Bioengineering	5.2	1.4	3.8
3	Applied Microbiology and Biotechnology	3.0	1.2	1.8
4	Applied and Environmental Microbiology	2.8	0.9	1.9
5	Enzyme and Microbial Technology	2.3	0.7	1.6
6	Biomass and Bioenergy	2.2	1.1	1.1
7	Green Chemistry	2.2	0.5	1.7
8	Renewable and Sustainable Energy Reviews	2.1	0.2	1.9
9	Biotechnology for Biofuels	1.9	1.1	0.8
10	Proceedings of the National Academy of Sciences of the United States of America	1.7	0.1	1.6
11	Fuel	1.5	1.1	0.4
12	Biotechnology Progress	1.5	0.4	1.1
13	Science	1.5	0.1	1.4
14	International Journal of Hydrogen Energy	1.4	1.0	0.4
15	Current Opinion in Biotechnology	1.4	0.1	1.3
16	Applied Biochemistry and Biotechnology Part A	1.3	0.4	0.9
17	Journal of Biotechnology	1.3	0.4	0.9
18	Journal of Power Sources	1.3	0.4	0.9
19	Renewable Energy	1.2	0.7	0.5
20	Applied Catalysis B Environmental	1.2	0.3	0.9
21	Biotechnology Advances	1.2	0.1	1.1

are the Applied and Environmental Microbiology, Renewable and Sustainable Energy Reviews, Applied Microbiology and Biotechnology, Green Chemistry, Enzyme and Microbial Technology, and Proceedings of the National Academy of Sciences of the United States of America with 1.6%–1.9% surplus each.

The other source titles with the relatively low citation impact with 0.4%–1.4% of the population papers each are Bioresources, Industrial Crops and Products, Applied Biochemistry and Biotechnology, Biotechnology Letters, Society of Automotive Engineers (SAE) Technical Papers, Cellulose, Biomass Conversion and Biorefinery, Process Biochemistry, Industrial and Engineering Chemistry Research, American Chemical Society (ACS) Sustainable Chemistry and Engineering, Royal Society of Chemistry (RSC) Advances, Holzforschung, Journal of Agricultural and Food Chemistry, Carbohydrate Polymers, Energy and Fuels, Sensors and Actuators B Chemical, Journal of Cleaner Production, Journal of Chemical Technology and Biotechnology, Advanced Materials Research, Biochemical Engineering Journal, Waste and Biomass Valorization, Chemical Engineering Journal, Chemical Engineering Transactions, Journal of Industrial Microbiology and Biotechnology, Bioenergy Research, Bioprocess and Biosystems Engineering, PLOS One, Energy, International Journal of Biological Macromolecules, Scientific Reports, Applied Energy, Energies, Journal of Membrane Science, World Journal of Microbiology and Biotechnology, and Journal of Bioscience and Bioengineering.

2.3.7 The Most Prolific Countries in the Bioethanol Fuels

Information about the 25 most prolific countries publishing at least 1.0% of sample papers each in the bioethanol fuels is given in Table 2.6.

The most prolific country is the USA with 37% of the sample papers, followed by China, Sweden, and Canada with 11%, 7%, and 7% of the sample papers, respectively. The other prolific countries

TABLE 2.6
The Most Prolific Countries in the Bioethanol Fuels

No.	Countries	Sample Papers (%)	Population Papers (%)	Surplus (%)
1	USA	37.4	16.7	20.7
2	China	11.3	20.4	–9.1
3	Sweden	6.7	2.4	4.3
4	Canada	6.6	3.8	2.8
5	India	5.3	7.4	–2.1
6	Japan	5.2	6.4	–1.2
7	Spain	4.8	3.6	1.2
8	UK	4.6	3.5	1.1
9	France	4.3	2.8	1.5
10	Brazil	3.9	6.5	–2.6
11	Germany	3.7	3.2	0.5
12	Netherlands	3.6	1.4	2.2
13	Denmark	3.1	1.2	1.9
14	S. Korea	2.3	3.7	–1.4
15	Greece	2.1	0.7	1.4
16	Australia	1.9	1.7	0.2
17	Malaysia	1.7	2.0	–0.3
18	Finland	1.7	1.5	0.2
19	Portugal	1.4	1.2	0.2
20	Italy	1.3	2.4	–1.1
21	Taiwan	1.3	1.4	–0.1
22	Iran	1.0	1.7	–0.7
23	Turkey	1.0	1.3	–0.3
24	S. Africa	1.0	0.9	0.1
25	Norway	1.0	0.4	0.6

are India, Japan, Spain, the UK, France, Brazil, Germany, and Netherlands with 4%–5% of the sample papers each. Thirteen European countries listed in Table 2.6 as a whole produce 39% and 26% of the sample and population papers, respectively, with 14% surplus.

However, the country with the most citation impact is the USA with 21% surplus, followed by Sweden, Canada, Netherlands, and Denmark with 2%–4% surplus each. The other prolific country is Denmark with 1% surplus. Similarly, the country with the least citation impact is China with 10% deficit, followed by Brazil, India, S. Korea, Japan, and Italy with 1%–3% deficit each.

Additionally, there are other countries with relatively low citation impact and with 0.4%–1.9% of the sample papers each: Thailand, Mexico, Russia, Poland, Indonesia, Egypt, Belgium, Argentina, Austria, Pakistan, Saudi Arabia, Czech Republic, Switzerland, Colombia, New Zealand, Singapore, Vietnam, Israel, Chile, Romania, Hungary, Slovakia, and Ireland.

2.3.8 The Most Prolific Scopus Subject Categories in the Bioethanol Fuels

Information about the 11 most prolific Scopus subject categories indexing at least 3.7% of the sample papers each is given in Table 2.7.

The most prolific Scopus subject categories in the bioethanol fuels are the Chemical Engineering with 48% of the sample papers, followed by Biochemistry, Genetics and Molecular Biology, Environmental Science, Energy, and Immunology and Microbiology with 27%–28% of the sample

TABLE 2.7
The Most Prolific Scopus Subject Categories in the Bioethanol Fuels

No.	Scopus Subject Categories	Sample Papers (%)	Population Papers (%)	Surplus (%)
1	Chemical Engineering	48.3	34.7	13.6
2	Biochemistry, Genetics, and Molecular Biology	38.2	29.0	9.2
3	Environmental Science	34.1	24.5	9.6
4	Energy	32.8	21.7	11.1
5	Immunology and Microbiology	27.1	18.7	8.4
6	Chemistry	19.2	22.9	–3.7
7	Engineering	9.9	15.1	–5.2
8	Agricultural and Biological Sciences	9.8	19.8	–10.0
9	Materials Science	7.5	13.1	–5.6
10	Physics and Astronomy	4.5	6.4	–1.9
11	Multidisciplinary	3.7	1.9	1.8

papers each. It is notable that Social Sciences including Economics and Business accounts for only 0.5% and 3.0% of the sample and population studies, respectively, mostly published outside the energy and fuel journals.

However, the Scopus subject category with the most citation impact is the Chemical Engineering with 14% surplus. The other influential subject areas are Energy, Environmental Science, Biochemistry, Genetics and Molecular Biology, and Immunology and Microbiology with 8%–11% of the sample papers each. Similarly, the least influential subject category is the Agricultural and Biological Sciences with 10% deficit. The other least influential subjects are Materials Science, Engineering, and Chemistry with 4%–6% deficits each.

2.3.9 The Most Prolific Keywords in the Bioethanol Fuels

Information about the Scopus keywords used with at least 4.4% or 2.8% of the sample or population papers, respectively, is given in Table 2.8. For this purpose, keywords related to the keyword set given in the appendix of the related papers are selected from a list of the most prolific keyword set provided by Scopus database.

These keywords are grouped under five headings: feedstocks, pretreatments, fermentation, hydrolysis and hydrolysates, and products. Furthermore, the most prolific keyword related to the biomass and biomass constituents is cellulose with 41% of the sample papers, followed by lignin, biomass, and lignocellulose with 22%–36% of the sample papers each. The other prolific keywords are hemicellulose, lignocellulosic biomass, zea, and carbohydrate with 10%–13% of the sample papers each.

Furthermore, the most prolific keyword related to the pretreatments is enzymes with 17% of the sample papers, followed by cellulases and pretreatments with 15% and 11% of the sample papers, respectively. The other prolific keywords are the ionic liquids (ILs), pretreatments, solvents, temperature, and degradation with 7%–9% of the sample papers each. The most prolific keyword related to the fermentation is fermentation with 28% of the sample papers, followed by fungi and bacteria with 18% and 12% of the sample papers, respectively.

Furthermore, the most prolific keyword related to the hydrolysis and hydrolysates is hydrolysis with 33% of the sample papers, followed by sugars, enzyme activity, cellulases, glucose, enzymatic hydrolysis, and saccharification with 11%–21% of the sample papers each. Finally, the most prolific keyword related to the products is ethanol with 47% of the sample papers, followed by biofuels and bioethanol with 24% and 11% of the sample papers, respectively.

TABLE 2.8
The Most Prolific Keywords in the Bioethanol Fuels

No.	Keywords	Sample Papers (%)	Population Papers (%)	Surplus (%)
1	**Feedstocks**			
	Cellulose	41.4	17.3	24.1
	Lignin	35.9	13.8	22.1
	Biomass	32.1	12.7	19.4
	Lignocellulose	21.8	5.7	16.1
	Hemicellulose	13.0	3.2	9.8
	Lignocellulosic biomass	11.6	4.1	7.5
	Zea	10.1	3.6	6.5
	Carbohydrate	9.5	4.7	4.8
	Wood	7.7	4.3	3.4
	Corn	4.9		4.9
	Bagasse	3.1	2.8	0.3
2	**Pretreatments**			
	Enzymes	17.3	8.3	9.0
	Cellulases	15.1	5.4	9.7
	Pretreatment	11.2	4.1	7.1
	ILs	9.3	3.8	5.5
	Pretreatment	8.4	3.7	4.7
	Solvents	8.1	3.5	4.6
	Temperature	7.6	5.3	2.3
	Degradation	7.1	3.6	3.5
	Sulfuric acid	5.5	1.9	3.6
	Biodegradation	5.2	3.8	1.4
	Carbon dioxide	4.9	2.9	2.0
	Genetic engineering	4.5		4.5
	pH	3.9	5.2	−1.3
	Water	3.9	3.1	0.8
	Genetics		3.7	−3.7
3	**Fermentation**			
	Fermentation	27.9	21.5	6.4
	Fungi	18.3	9.9	8.4
	Bacteria	12.0	7.5	4.5
	Saccharomyces	9.4	6.6	2.8
	Yeast	7.6	7.0	0.6
	Acetic acid	4.9	3.2	1.7
4	**Hydrolysis and Hydrolysates**			
	Hydrolysis	33.4	15.0	18.4
	Sugar	20.8	7.1	13.7
	Enzyme activity	16.8	9.0	7.8
	Glucose	13.8	9.9	3.9
	Enzymatic hydrolysis	13.0	7.8	5.2
	Saccharification	10.6	6.2	4.4
	Xylose	9.0	4.1	4.9

(*Continued*)

TABLE 2.8 (*Continued*)
The Most Prolific Keywords in the Bioethanol Fuels

No.	Keywords	Sample Papers (%)	Population Papers (%)	Surplus (%)
5	**Products**			
	Ethanol	47.4	32.3	15.1
	Biofuels	24.3	9.9	14.4
	Bioethanol	10.6	8.4	2.2
	Oxidation	8.3	3.6	4.7
	Ethanol production	5.9	3.2	2.7
	Biofuel production	5.4	1.5	3.9
	Hydrogen	5.0	1.9	3.1
	Fuel cells	4.4		4.4
	Bioethanol production	3.0	3.1	–0.1

However, the most prolific keywords across all of the research fronts are ethanol, cellulose, lignin, hydrolysis, biomass, fermentation, biofuels, lignocellulose, sugar, fungi, enzymes, enzyme activity, and cellulases with 15%–47% of the sample papers each. The other prolific keywords are glucose, hemicellulose, enzymatic hydrolysis, bacteria, lignocellulosic biomass, pretreatment, saccharification, bioethanol, zea, and carbohydrate with 10%–14% of the sample papers each.

Similarly, the most influential keywords are cellulose, lignin, biomass, hydrolysis, lignocellulose, ethanol, biofuels, sugar, hemicellulose, and cellulases with 10%–24% surplus each. The other influential keywords are enzymes, fungi, enzyme activity, lignocellulosic biomass, pretreatment, zea, fermentation, ILs, enzymatic hydrolysis, xylose, and corn with 11.5%–9% surplus each.

2.3.10 The Most Prolific Research Fronts in the Bioethanol Fuels

Information about the research fronts for feedstocks used the sample papers in the bioethanol fuels is given in Table 2.9. As this table shows, the most prolific research front for this field is the lignocellulosic biomass at large with 23% of the sample papers, followed by bioethanol fuels and cellulose with 20% and 16% of the sample papers, respectively. The other prolific research fronts are the starch feedstock residues, wood biomass, industrial wastes, and lignin with 10%, 9%, 7%, and 6% of the sample papers, respectively.

The other prolific research fronts are the grass biomass, hydrolysates, biomass at large, sugar feedstocks, sugar feedstock residues, algal biomass, lignocellulosic wastes at large, and starch feedstocks. Furthermore, the other minor research fronts are urban wastes, forestry wastes, hemicellulose, xylan, food wastes, syngas, cellobiose, and plants with 0.3%–1.08% of the sample papers each.

Information about the thematic research fronts for the sample papers in the bioethanol fuels is given in Table 2.10. As this table shows, the most prolific research front is the production of bioethanol fuels with 74% of the sample papers, followed by the utilization and evaluation of bioethanol fuels with 20% and 6% of the sample papers, respectively.

However, the most prolific research front for the production of bioethanol fuels is the biomass pretreatment with 59% of the sample papers, followed by biomass hydrolysis, bioethanol production, and hydrolysate fermentation with 30%, 22%, and 18% of the sample papers. The other minor research front is separation and distillation of bioethanol fuels with 1% of the sample papers.

Furthermore, the most prolific research front for the utilization of bioethanol fuels is the utilization of bioethanol fuels in fuel cells, followed by utilization of bioethanol fuels in the transport engines and bioethanol-based biohydrogen fuels with 6% and 4% of the sample papers, respectively. The other research fronts are the bioethanol sensors and bioethanol fuel-based biochemicals with 2% and 0.4% of the sample papers, respectively.

TABLE 2.9
The Most Prolific Research Fronts for the Bioethanol Fuels

No.	Research Fronts	N Paper % Sample
1	Lignocellulosic biomass at large	23.0
2	Bioethanol fuels	19.6
3	Cellulose	15.1
4	Starch feedstock residues	9.7
5	Wood biomass	9.0
6	Industrial waste	6.5
7	Lignin	6.2
8	**Minor Feedstocks**	**18.3**
	Grass	3.1
	Hydrolysates	2.8
	Biomass	2.6
	Sugar feedstocks	2.2
	Sugar feedstock residues	2.2
	Algal biomass	1.9
	Lignocellulosic wastes	1.8
	Starch feedstocks	1.7
9	**Highly Minor Feedstocks**	**5.1**
	Urban wastes	1.0
	Forestry wastes	0.8
	Hemicellulose	0.8
	Xylan	0.8
	Food wastes	0.6
	Syngas	0.4
	Cellobiose	0.4
	Plants	0.3
	Sample size	775

N paper (%) sample: the number of papers in the sample of 775 papers.

2.4 DISCUSSION

2.4.1 Introduction

The crude oil-based gasoline fuels have been widely used in the transportation sector since the 1920s. However, there have been great public concerns over the adverse environmental and human impact of these fuels. Hence, biomass-based bioethanol fuels have increasingly been used in blending gasoline fuels, in the fuel cells, and in the biochemical production and biohydrogen fuel in a biorefinery context.

However, it is necessary to pretreat the biomass to enhance the yield of the bioethanol before the bioethanol production through the hydrolysis and fermentation. The research in the field of the bioethanol fuels has intensified in this context in the key research fronts of the production of bioethanol fuels and, to a lesser extent, utilization and evaluation of bioethanol fuels. For the first research

TABLE 2.10
The Most Prolific Thematic Research Fronts for the Bioethanol Fuels

No.	Research Fronts	N Paper % Sample
1	**Bioethanol Fuel Production**	**74.1**
	Biomass pretreatments	58.5
	Biomass hydrolysis	29.8
	Bioethanol production	21.5
	Hydrolysate fermentation	17.9
	Bioethanol fuel distillation	0.8
2	Biofuel Fuel Evaluation	6.2
3	**Bioethanol Fuel Utilization**	**19.7**
	Bioethanol fuels in fuel cells	6.8
	Bioethanol fuels in engines	6.1
	Bioethanol-based biohydrogen fuels	4.3
	Bioethanol sensors	2.2
	Bioethanol-based biochemicals	0.4

N paper (%) sample: the number of papers in the population sample of 775 papers.

front, pretreatment and hydrolysis of the feedstocks, fermentation of the hydrolysates, production of bioethanol fuels, and to a lesser extent separation and distillation of bioethanol fuels are the key research areas, while for the second research front, utilization of bioethanol fuels in fuel cells and transport engines, bioethanol-based biohydrogen fuels, bioethanol sensors, and bioethanol-based biochemicals are the key research areas.

The research in this field has also intensified for the feedstocks of lignocellulosic biomass at large, bioethanol fuels for their utilization, cellulose, starch feedstock residues, wood biomass, and to a lesser extent industrial waste, lignin, grass, hydrolysates, biomass, sugar feedstocks, sugar feedstock residues, algal biomass, lignocellulosic wastes, starch feedstocks, urban wastes, forestry wastes, hemicellulose, xylan, food wastes, syngas, cellobiose, and plants.

However, it is essential to develop efficient incentive structures for the primary stakeholders to enhance the research in this field. This is especially important to maintain energy security in the cases of supply shocks such as oil shocks, war-related chocks as in the case of Russian invasion of Ukraine, or COVID-19 shocks.

The scientometric analysis has been used in this context to inform the primary stakeholders about the current state of the research in a selected research field. As there has been no current scientometric study in this field, this book chapter presents a scientometric study of the research in the bioethanol fuels. It examines the scientometric characteristics of both the sample and population data presenting scientometric characteristics of these both datasets in the order of documents, authors, publication years, institutions, funding bodies, source titles, countries, Scopus subject categories, Scopus keywords, and research fronts.

As a first step for the search of the relevant literature, the keywords were selected using the most-cited first 300 population papers for each feedstock. The selected keyword list was then optimized to obtain a representative sample of papers for the each research field. These keyword lists were then integrated to obtain the keyword list for this research field (Konur, 2023a–c).

As a second step, two sets of data were used for this study. First, a population sample of 77,433 papers was used to examine the scientometric characteristics of the population data. Second, a sample of 775 most-cited papers, corresponding to 1% of the population papers, was used to examine the scientometric characteristics of these citation classics.

The scientometric characteristics of these sample and population datasets were presented in the order of documents, authors, publication years, institutions, funding bodies, source titles, countries, Scopus subject categories, Scopus keywords, and research fronts.

Lastly, the key scientometric findings for both datasets were discussed to highlight the research landscape for bioethanol fuels. Additionally, a number of brief conclusions were drawn and a number of relevant recommendations were made to enhance the future research landscape.

2.4.2 The Most Prolific Documents in the Bioethanol Fuels

Articles (together with conference papers) dominate both the sample (76%) and population (94%) papers with 18% deficit (Table 2.1). Furthermore, review papers have a surplus (21%) and the representation of the reviews in the sample papers is quite extraordinary (24%).

Scopus differs from the Web of Science database in differentiating and showing articles (73%) and conference papers (3%) published in the journals separately. However, it should be noted that these conference papers are also published in journals as articles, compared with those published only in the conference proceedings. Hence, the total number of articles and review papers in the sample dataset is 76% and 24%, respectively.

It is observed during the search process that there has been inconsistency in the classification of the documents in Scopus and in other databases such as Web of Science. This is especially relevant for the classification of papers as reviews or articles as the papers not involving a literature review may be erroneously classified as a review paper. There is also a case of review papers being classified as articles. For example, the total number of the reviews in the sample dataset was manually found as nearly 29% compared with 24% as indexed by Scopus, decreasing the number of articles and conference papers to 69% for the sample dataset. It is notable that many techno-economic and life cycle studies were often indexed as reviews by the Scopus database.

In this context, it would be helpful to provide a classification note for the published papers in the books and journals at the first instance. It would also be helpful to use the document types listed in Table 2.1 for this purpose. Book chapters may also be classified as articles or reviews as an additional classification to differentiate review chapters from the ordinary chapters as it is done by the Web of Science. It would be further helpful to additionally classify the conference papers as articles or review papers, and it is done in the Web of Science database.

2.4.3 The Most Prolific Authors in the Bioethanol Fuels

There have been 42 most prolific authors with at least 0.6% of the sample papers each as given in Table 2.2. These authors have shaped the development of the research in this field.

The most prolific authors are Charles E. Wyman, John N. Saddler, and to a lesser extent Bruce E. Dale, Barbel Hahn-Hagerdal, Lee R. Lynd, Guido Zacchi, Michael E. Himmel, Michael R. Ladisch, and Blake A. Simmons. Furthermore, the most influential authors are Charles E. Wyman, John N. Saddler, Barbel Hahn-Hagerdal, Bruce E. Dale, and to a lesser extent Lee R. Lynd, Guido Zacchi, Michael E. Himmel, and Michael R. Ladisch. It is notable that all of these top authors, except one, are male and are primarily from the USA and Europe.

It is important to note the inconsistencies in indexing of the author names in Scopus and other databases. It is especially an issue for the names with more than two components such as 'Blake Sam de Hyun Simmons'. The probable outcomes are 'Simmons, B.S.D.H.', 'de Hyun Simmons, B.S.', or 'Hyun Simmons, B.S.D.'. The first choice is the gold standard of the publishing sector as the last word in the name is taken as the last name. In most of the academic databases such as PubMed and EBSCO databases, this version is used predominantly. The second choice is a strong alternative, while the last choice is an undesired outcome as two last words are taken as the last name. It is good practice to combine the words of the last name by a hyphen: 'Hyun-Simmons, B.S.D.'. It is notable that inconsistent indexing of the author names may cause substantial inefficiencies in the search process for the papers and allocating credit to the authors as there are different author entries for each outcome in the databases.

There are also inconsistencies in the shortening Chinese names. For example, 'Hongzhang Chen' is often shortened as 'Chen, H.Z.', 'Chen, H.-Z.', and 'Chen H' as it is done in the Web of Science database as well. However, the gold standard in this case is 'Chen, H' where the last word is taken as the last name and the first word is taken as a single forename. In most of the academic databases such as PubMed and EBSCO, this first version is used predominantly. However, it makes sense to use the second version to highly differentiate Chinese names: 'Chen, H.Z.'. Therefore, there have been difficulties in locating papers for the Chinese authors. In such cases, the use of the unique author codes provided for each author by the Scopus database has been helpful.

There is also a difficulty in allowing credit for the authors, especially for the authors with common names such as 'Wang, X.' in conducting scientometric studies. These difficulties strongly influence the efficiency of the scientometric studies and allocating credit to the authors as there are the same author entries for different authors with the same name, e.g., 'Wang, X.' in the databases.

In this context, the coding of authors in Scopus database is a welcome innovation compared with the other databases such as Web of Science. In this process, Scopus allocates a unique number to each author in the database (Aman, 2018). However, there might still be substantial inefficiencies in this coding system, especially for common names. For example, some of the papers for a certain author may be allocated to another researcher with a different author code. It is possible that Scopus uses a number of software programs to differentiate the author names and the program may not be false-proof (Kim, 2018).

In this context, it does not help that author names are not given in full in some journals and books. This makes it difficult to differentiate authors with common names and makes the scientometric studies further difficult in the author domain. Therefore, the author names should be given in full in all books and journals at the first instance. There is also a cultural issue where some authors do not use their full names in their papers. Instead, they use initials for their forenames: 'Parajo, H.J.', 'Parajo, H.', or 'Parajo, J.' instead of 'Parajo, Hyun Jae'.

There are also inconsistencies in naming of the authors with more than two components by the authors themselves in journal papers and book chapters. For example, 'Parajo, A.P.C.' might be given as 'Parajo, A.' or 'Parajo, A.C.' or 'Parajo, A.P.' or 'Parajo, C' in the journals and books. This also makes the scientometric studies difficult in the author domain. Hence, contributing authors should use their name consistently in their publications.

The other critical issue regarding the author names is the inconsistencies in the spelling of the author names in the national spellings (e.g., Özğümüş and Çiğdem) rather than in the English spellings (e.g., Ozgumus and Cigdem) in Scopus database. Scopus differs from the Web of Science database and many other databases in this respect where the author names are given only in the English spellings. It is observed that national spellings of the author names do not help much in conducting scientometric studies as well in allocating credits to the authors as sometimes there are often the different author entries for the English and National spellings in the Scopus database.

The most prolific institutions for the sample dataset are Lund University, NREL, and to a lesser extent the National Technical University of Athens, Purdue University, University of British Columbia, and USDA Forest Service. Furthermore, the most prolific countries for the sample dataset are the USA and to a lesser extent France, Canada, and Greece. These findings confirm the dominance of the USA, Europe, and to a lesser extent Canada in this field. However, pretreatments and hydrolysis of the feedstocks and hydrolysates, the bioethanol production, and to a lesser extent utilization of bioethanol fuels in the transport engines and fuel cells are the key research fronts studied by these top authors.

It is also notable that there is significant gender deficit for the sample dataset as surprisingly with a representation rate of 12%. This finding is the most thought-provoking with strong public policy implications. Hence, institutions, funding bodies, and policymakers should take efficient measures to reduce the gender deficit in this field and other scientific fields with a strong gender deficit. In this context, it is also worth to note the level of representation of the researchers from the minority groups in science on the basis of race, sexuality, age, and disability, besides the gender (Blankenship, 1993; Dirth and Branscombe, 2017; Konur, 2000, 2002a–c, 2006a, b, 2007a, b).

2.4.4 The Most Prolific Research Output by Years in the Bioethanol Fuels

The research output observed between 1970 and 2022 is illustrated in Figure 2.1. This figure clearly shows that the bulk of the research papers in the population dataset were published primarily in the 2010s and to a lesser extent in the early 2020s and 2000s. Similarly, the bulk of the research papers in the sample dataset were published in the 2000s and to a lesser extent 2010s and 1990s.

These findings suggest that the most prolific sample and population papers were primarily published in the 2010s. Furthermore, a significant portion of the sample and population papers were published in the early 2020s and 2000s, respectively.

These are the thought-provoking findings as there has been significant research boom since 2008 and 2002 for the population and sample papers, respectively. In this context, the increasing public concerns about climate change (Change, 2007), greenhouse gas emissions (Carlson et al., 2017), and global warming (Kerr, 2007) have been certainly behind the boom in the research in this field since 2007. Furthermore, the recent supply shocks experienced due to the COVID-19 pandemics and the Ukrainian war might also be behind the research boom in this field since 2019.

Based on these findings, the size of the population papers likely to more than double in the current decade, provided that the public concerns about climate change, greenhouse gas emissions, and global warming, and the supply shocks are translated efficiently to the research funding in this field.

2.4.5 The Most Prolific Institutions in the Bioethanol Fuels

The 45 most prolific institutions publishing papers on the bioethanol fuels with at least 0.8% of the sample papers each given in Table 2.3 have shaped the development of the research in this field.

The most prolific institutions are the NREL, Lund University, Dartmouth College, University of British Columbia, and to a lesser extent Chinese Academy of Sciences, Michigan State University, USDA Forest Service, Purdue University, and University of Wisconsin Madison. Similarly, the top countries for these most prolific institutions are the USA and to a lesser extent China, Denmark, France, Netherlands, and Sweden. In total, 11 countries house these top institutions.

However, the institutions with the most impact are the NREL, Dartmouth College, Lund University, University of British Columbia, and to a lesser extent Michigan State University, USDA Forest Service, Purdue University, and University of Wisconsin Madison. These findings confirm the dominance of the institutions from the USA, Europe, and to a lesser extent Canada.

2.4.6 The Most Prolific Funding Bodies in the Bioethanol Fuels

The 15 most prolific funding bodies funding at least 0.8% of the sample papers each is given in Table 2.4. It is notable that only 35% and 43% of the sample and population papers were funded, respectively.

The most prolific funding bodies are the U.S. Department of Energy, National Natural Science Foundation of China, European Commission, National Science Foundation, and to a lesser extent Natural Sciences and Engineering Research Council of Canada, Research Support Foundation of the State of Sao Paulo, and Office of Science. However, the most prolific countries for these top funding bodies are the USA, China, and Brazil. In total, only five countries and the EU house these top funding bodies.

The funding bodies with the most citation impact are the U.S. Department of Energy and European Commission and National Science Foundation. Furthermore, the funding bodies with the least citation impact are the National Science Foundation and Natural Sciences and to a lesser extent National Council for Scientific and Technological Development, and National Key Research and Development Program of China.

These findings on the funding of the research in this field suggest that the level of the funding for the population papers, mostly since 2008, is moderately intensive and it has been largely instrumental in enhancing the research in this field (Ebadi and Schiffauerova, 2016) in light of North's institutional framework (North, 1991). However, the funding rate for the sample papers since 2002 is relatively low. It is also notable that the funding rate in this field is relatively modest compared with other scientific research fields. Furthermore, it is expected that this funding rate would increase in light of the recent supply shocks. Furthermore, it emerges that the USA, China, and to a lesser extent Brazil have heavily funded the research on the bioethanol fuels. It is notable that these countries have extensive research programs in bioethanol fuels and they are the major producers and consumers of bioethanol fuels.

2.4.7 The Most Prolific Source Titles in the Bioethanol Fuels

The 21 most prolific source titles publishing at least 1.2% of the sample papers each in the bioethanol fuels have shaped the development of the research in this field (Table 2.5).

The most prolific source titles are the Bioresource Technology, Biotechnology and Bioengineering, and to a lesser extent Applied Microbiology and Biotechnology, Applied and Environmental Microbiology, Enzyme and Microbial Technology, Biomass and Bioenergy, Green Chemistry, and Renewable and Sustainable Energy Reviews. However, the source titles with the most impact are the Bioresource Technology, Biotechnology and Bioengineering, and to a lesser extent Applied and Environmental Microbiology, Renewable and Sustainable Energy Reviews, Applied Microbiology and Biotechnology, Green Chemistry, Enzyme and Microbial Technology, and Proceedings of the National Academy of Sciences of the United States of America.

It is notable that these top source titles are primarily related to the bioresources, biotechnology, microbiology, and bioenergy. This finding suggests that Bioresource Technology and the other prolific journals in these fields have significantly shaped the development of the research in this field as they focus primarily on the bioethanol fuels with a high yield. In this context, the influence of two top journals is quite extraordinary.

This top journal covers more the technical issues such as waste treatment, biofuels, bioprocesses, bioproducts, and physicochemical and thermochemical processes, excluding the social science-based interdisciplinary studies with the only exceptions for the 'circular bioeconomy and energy and environmental sustainability' and 'system analysis and techno-economics of biofuels and chemicals production'. It thus excludes other social science-based interdisciplinary studies such as scientometric (Konur, 2012a–i), user (Huijts et al., 2012; Wustenhagen et al., 2007), policy (Greening et al., 2000; Hook and Tang, 2013), environmental (Popp et al., 2014; Tilman et al., 2009), and economic (Asafu-Adjaye, 2000; Stern, 1993) studies as in the case of the most journals in the field of energy and fuels excepting social science-oriented journals.

2.4.8 The Most Prolific Countries in the Bioethanol Fuels

The 25 most prolific countries publishing at least 1.0% of the sample papers each have significantly shaped the development of the research in this field (Table 2.6).

The most prolific countries are the USA, China, Sweden, Canada, and to a lesser extent India, Japan, Spain, the UK, France, Brazil, Germany, and the Netherlands. However, the countries with the most citation impact are the USA and to a lesser extent Sweden, Canada, Netherlands, Denmark, Sweden, Canada, Netherlands, and Denmark. Similarly, the countries with the least impact are

China and to a lesser extent Brazil, India, S. Korea, Japan, and Italy. Furthermore, 13 European countries listed in Table 2.6 as a whole produce 39% and 26% of the sample and population papers, respectively, with 14% surplus.

It is notable that the USA and China produce 17% and 20% of the population papers, respectively, while they produce 37% and 11% of the sample papers, respectively. These findings show that the USA has a greater citation influence with respect to China, possibly reflecting the first mover advantage (Lieberman and Montgomery, 1988) of the USA in the bioethanol fuel research intensifying the bioethanol fuel production and research in the 1970s (McMillan, 1997). Similarly, Europe as a whole has also greater citation influence in relation to China.

The close examination of these findings suggests that the USA, Europe, China, and to a lesser extent Canada, India, Japan, and Brazil are the major producers of the research in this field. It is a fact that the USA has been a major player in science (Leydesdorff and Wagner, 2009). The USA has further developed a strong research infrastructure to support its corn and grass-based bioethanol industry (Gillon, 2010).

However, China has been a rising mega star in scientific research in competition with the USA and Europe (Leydesdorff and Zhou, 2005). China is also a major player in this field as a major producer of bioethanol (Fang et al., 2010).

Next, Europe as a whole has been a persistent player in the scientific research in competition with both the USA and China (Leydesdorff, 2000). Europe has also been a persistent producer of bioethanol along with the USA and Brazil (Gnansounou, 2010).

Furthermore, Canada (Tahmooresnejad et al., 2015), Japan (Negishi et al., 2004), and India (Karpagam et al., 2011) are the other countries with substantial research activities in bioethanol fuels.

2.4.9 The Most Prolific Scopus Subject Categories in the Bioethanol Fuels

The 11 most prolific Scopus subject categories indexing at least 3.7% of the sample papers each, given in Table 2.7, have shaped the development of the research in this field.

The most prolific Scopus subject categories in the bioethanol fuels are Chemical Engineering, followed by Biochemistry, Genetics and Molecular Biology, Environmental Science, Energy, and Immunology and Microbiology. It is also notable that Social Sciences including Economics and Business have a minimal presence in both sample and population studies. This might be in part due to the exclusive editorial policies of energy and fuel journals.

However, the Scopus subject categories with the most citation impact are Chemical Engineering, Energy, Environmental Science, Biochemistry, Genetics and Molecular Biology, and Immunology and Microbiology. Similarly, the least influential subject categories are Agricultural and Biological Sciences and to a lesser extent Materials Science, Engineering, and Chemistry.

These findings are thought-provoking suggesting that the primary subject categories are related to chemical engineering, biochemistry, and to a lesser extent microbiology, environmental science, and energy as the core of the research in this field concerns with the production, evaluation, and utilization of bioethanol fuels. The other finding is that social sciences are not well represented in both the sample and population papers as in line with the most fields in bioethanol fuels. The social, environmental, and economic studies account for the field of social sciences.

As discussed briefly in Section 2.4.7, most journals in this field have exclusive editorial policies toward social science-based interdisciplinary studies. This development has been in contrast to the interdisciplinarity (Jacobs and Frickel, 2009; Nissani, 1997) of this field in light of the pressures for increasing incentives for the primary stakeholders (North, 1991). Thus, for the healthy development of this research field, the editorial policies of the journals in this field should adapt more inclusive policies toward social science- and humanities-based interdisciplinary studies.

2.4.10 The Most Prolific Keywords in the Bioethanol Fuels

A limited number of keywords have shaped the development of the research in this field as shown in Table 2.8. These keywords are grouped under five headings: feedstocks, pretreatments, fermentation, hydrolysis and hydrolysates, and products.

The most prolific keywords across all of the research fronts are ethanol, cellulose, lignin, hydrolysis, biomass, fermentation, biofuels, lignocellulose, sugar, fungi, enzymes, enzyme activity, cellulases, and to a lesser extent glucose, hemicellulose, enzymatic hydrolysis, bacteria, lignocellulosic biomass, pretreatment, saccharification, bioethanol, zea, and carbohydrate.

Similarly, the most influential keywords are cellulose, lignin, biomass, hydrolysis, lignocellulose, ethanol, biofuels, sugar, hemicellulose, cellulases, and to a lesser extent enzymes, fungi, enzyme activity, lignocellulosic biomass, pretreatment, zea, fermentation, ILs, enzymatic hydrolysis, xylose, and corn.

These findings suggest that it is necessary to determine the keyword set carefully to locate the relevant research in each of these research fronts. Additionally, the size of the sample for each keyword highlights the intensity of the research in the relevant research areas for both sample and population datasets.

These findings also highlight different spelling of some strategic keywords such as pretreatment v. pretreatment v. treatment and bioethanol v. ethanol v. bioethanol v. bioethanol fuels. However, there is tendency toward the use of the connected keywords without using a hyphen: bioethanol fuels or pretreatment, respectively. It is particularly notable that the use of treatment and ethanol instead of pretreatment and bioethanol, respectively, in the paper titles, makes the literature search less efficient and time-consuming.

2.4.11 The Most Prolific Research Fronts in the Bioethanol Fuels

Information about the research fronts for the sample papers in the bioethanol fuels is given in Table 2.9.

As this table shows, the most prolific feedstocks for this field are the lignocellulosic biomass, bioethanol fuels, cellulose, starch feedstock residues, wood biomass, industrial wastes, lignin, and to a lesser extent grass biomass, hydrolysates, biomass at large, sugar feedstocks, sugar feedstock residues, algal biomass, lignocellulosic wastes at large, starch feedstocks, urban wastes, forestry wastes, hemicellulose, xylan, food wastes, syngas, cellobiose, and plants.

Thus, the first seven research fields have substantial importance, complementing the remaining bioethanol fuel research fields. It is important to note that the research on the production of the first generation bioethanol fuels from sugar and starch feedstocks for bioethanol fuels comprises only 3.9% of the sample papers in total. These first generation bioethanol fuels are not much desirable as they undermine the food security (Makenete et al., 2008; Wu et al., 2012).

Table 2.10 shows that the most prolific thematic research fronts are the production of bioethanol fuels and to a lesser extent utilization and evaluation of bioethanol fuels. However, the most prolific research fronts for the production of bioethanol fuels are the biomass pretreatment, biomass hydrolysis, and to a lesser extent bioethanol fuel production at large, hydrolysate fermentation, and separation and distillation of bioethanol fuels. Furthermore, the most prolific research fronts for the utilization of bioethanol fuels are the utilization of bioethanol fuels in fuel cells and the transport engines, and to a lesser extent bioethanol-based biohydrogen fuels, bioethanol sensors, and bioethanol fuel-based biochemicals. In this context, the USA and Brazil have been the global leaders in the production and use of the corn- and sugarcane-based bioethanol fuels, respectively, since the 1970s in the aftermath of the global crude oil crisis in the early 1970s. Separation and distillation of the bioethanol fuels are also an important research field.

These findings are thought-provoking in seeking ways to increase feedstock-based bioethanol yield at the global scale and the utilization of bioethanol fuels. It is clear that all of these research fronts have public importance and merit substantial funding and other incentives. Furthermore, it is notable that the utilization and evaluation of bioethanol fuels have also become a core unit of the bioethanol research to make it more competitive with the crude oil-based gasoline and diesel fuels, especially for the USA, Europe, and China.

It is notable that the pretreatment and hydrolysis of the feedstocks emerge as primary research fronts for this field. These processes are required to improve the ethanol yield. However, the research fronts of the fermentation of the hydrolysates and the bioethanol production from the feedstock-based hydrolysates are also important. The research in the field of separation and distillation of bioethanol fuels is negligible.

Furthermore, the field of the evaluation of bioethanol fuels is also a neglected area. This suggests that the primary stakeholders have been primarily interested in these key processes of the bioethanol production. It is also notable that evaluation of the bioethanol fuels such as techno-economics, life cycle, economics, social, land use, labor, and environment-related studies emerges as a case study for the bioethanol fuels.

In the end, these most-cited papers in this field hint that the bioethanol fuels could be optimized using the structure, processing, and property relationships of these feedstocks in the fronts of the feedstock pretreatment and hydrolysis, hydrolysate fermentation, and the utilization of bioethanol fuels (Formela et al., 2016; Konur, 2018a, 2020b, 2021a–d; Konur and Matthews, 1989).

2.5 CONCLUSION AND FUTURE RESEARCH

The research on the bioethanol fuels has been mapped through a scientometric study of both sample (775 papers) and population (77,433 papers) datasets.

The critical issue in this study has been to obtain a representative sample of the research as in any other scientometric study. Therefore, the keyword set has been carefully devised and optimized after a number of runs in the Scopus database. It is a representative sample of the wider population studies. This keyword set was provided in the appendix of the related studies, and the relevant keywords are presented in Table 2.8. However, it should be noted that it has been very difficult to compile a representative keyword set since this research field has been connected closely with many other fields. Therefore, it has been necessary to compile a keyword list to exclude papers concerned with the other research fields.

The other issue has been the selection of a multidisciplinary database to carry out the scientometric study of the research in this field. For this purpose, the Scopus database has been selected. The journal coverage of this database has been notably wider than that of Web of Science and other multi-subject databases.

The key scientometric properties of the research in this field have been determined and discussed in this book chapter. It is evident that a limited number of documents, authors, institutions, publication years, institutions, funding bodies, source titles, countries, Scopus subject categories, Scopus keywords, and research fronts have shaped the development of the research in this field.

There is ample scope to increase the efficiency of the scientometric studies in this field in the author and document domains by developing consistent policies and practices in both domains across all the academic databases. In this respect, it seems that authors, journals, and academic databases have a lot to do. Furthermore, the significant gender deficit as in most scientific fields emerges as a public policy issue. The potential deficits on the basis of age, race, disability, and sexuality need also to be explored in this field as in other scientific fields.

The research in this field has boomed since 2008 and 2002 for the population and sample papers, respectively, possibly promoted by the public concerns on global warming, greenhouse gas emissions, and climate change. Furthermore, the recent COVID-19 pandemics and Russian invasion of Ukraine have resulted in a global supply shocks shifting the recent focus of the stakeholders from the crude oil-based fuels to biomass-based fuels such as bioethanol fuels. It is expected that there would be further incentives for the key stakeholders to carry out the research for the bioethanol fuels to increase the ethanol yield and their utilization and to make it more competitive with the crude oil-based gasoline and diesel fuels. This might be truer for the crude oil-deficient and foreign exchange-deficient countries to maintain the energy and food security at the face of the global supply shocks.

The relatively modest funding rate of 35% and 43% for the sample and population papers, respectively, suggests that funding in this field significantly enhanced the research in this field primarily since 2008, possibly more than doubling in the current decade. However, it is evident that there is ample room for more funding and other incentives to enhance the research in this field further.

The institutions from the USA and to a lesser extent China, Denmark, France, Netherlands, and Sweden have mostly shaped the research in this field. Furthermore, USA, Europe, and to a lesser extent China, Canada, India, Japan, and Brazil have been the major producers of the research in this field as the major producers and users of bioethanol fuels. It is evident that these countries have well-developed research infrastructure in bioethanol fuels and their derivatives.

It emerges that ethanol is more popular than bioethanol as a keyword with strong implications for the search strategy. In other words, the search strategy using only bioethanol keyword would not be much helpful. It is recommended that the term 'bioethanol fuels' is used instead of ethanol or bioethanol in the future. However, the Scopus keywords are grouped under five headings: biomass, pretreatments, fermentation, hydrolysis and hydrolysates, and products. These prolific keywords highlight the major fields of the research in this field for both sample and population papers.

Table 2.9 shows that the most prolific biomass used for this field are the lignocellulosic biomass at large, bioethanol fuels for their utilization, cellulose, starch feedstock residues, wood biomass, industrial wastes, lignin, and to a lesser extent grass biomass, hydrolysates, biomass at large, sugar feedstocks, sugar feedstock residues, algal biomass, lignocellulosic wastes at large, starch feedstocks, urban wastes, forestry wastes, hemicellulose, xylan, food wastes, syngas, cellobiose, and plants.

It is important to note that the research on the production of the first generation bioethanol fuels from sugar and starch feedstocks for bioethanol fuels comprises only a small part of the sample papers in total. These first generation bioethanol fuels are not much desirable as they undermine the food security.

Furthermore, Table 2.10 shows that the most prolific thematic research fronts are the production of bioethanol fuels and to a lesser extent utilization and evaluation of bioethanol fuels. However, the most prolific research fronts for the production of bioethanol fuels are the biomass pretreatment, biomass hydrolysis, and to a lesser extent bioethanol production, hydrolysate fermentation, and separation and distillation of bioethanol fuels. Furthermore, the most prolific research fronts for the utilization of bioethanol fuels are the utilization of bioethanol fuels in fuel cells and the transport engines, and to a lesser extent bioethanol-based biohydrogen fuels, bioethanol sensors, and bioethanol fuel-based biochemicals. In this context, the USA and Brazil have been the global leaders in the production and use of the corn- and sugarcane-based bioethanol fuels since the 1970s in the aftermath of the global crude oil crisis in the early 1970s.

For the production of bioethanol fuels, the first four research fronts dominate the research in this field, while the field of the distillation of the bioethanol fuels is relatively a neglected research field. Furthermore, the field of the evaluation of bioethanol fuels is also a neglected area. This suggests that the primary stakeholders have been primarily interested in these key processes of the bioethanol production. It is also notable that evaluation of the bioethanol fuels such as techno-economics, life cycle, economics, land use, labor, and environment-related studies emerges as a case study for the bioethanol fuels.

In this context, it is notable that there is ample room for the improvement of the research on social and humanitarian aspects of the research on the bioethanol fuels. As discussed briefly in Section 2.4.7, most journals in this field have exclusive editorial policies toward social science-based interdisciplinary studies. This development has been in contrast to the interdisciplinarity of this field in light of the pressures for increasing incentives for the primary stakeholders. These social science-based interdisciplinary studies include scientometric, user, policy, environmental, and economic studies in addition to the techno-economics and life cycle studies as in the case of the most journals in the field of energy and fuels excepting social science-oriented journals. Thus, for the healthy development of this research field, the journals in this field should adapt more inclusive editorial policies toward social science- and humanities-based interdisciplinary studies.

These findings are thought-provoking in seeking ways to increase the bioethanol yield and utilization at the global scale. It is clear that all of these research fronts have public importance and merit substantial funding and other incentives. Furthermore, it is notable that bioethanol fuels together with biodiesel fuels have become a core unit of the bioenergy and biofuel research to make it more competitive with the crude oil-based gasoline and diesel fuels, especially for the USA, Europe, Brazil, and China.

Thus, the scientometric analysis has a great potential to gain valuable insights into the evolution of the research in this field as in other scientific fields, especially in the aftermath of the significant global supply shocks such as COVID-19 pandemics and the Russian invasion of Ukraine.

It is recommended that further scientometric studies are carried out for the primary research fronts. It is further recommended that reviews of the most-cited papers are carried out for each primary research front to complement these scientometric studies. Next, the scientometric studies of the hot papers in these primary fields are carried out.

ACKNOWLEDGMENTS

The contribution of the highly cited researchers in the field of the bioethanol fuels has been gratefully acknowledged.

REFERENCES

Alvira, P., E. Tomas-Pejo, M. Ballesteros and M. J. Negro. 2010. Pretreatment technologies for an efficient bioethanol production process based on enzymatic hydrolysis: A review. *Bioresource Technology* 101:4851–4861.

Aman, V. 2018. Does the Scopus author ID suffice to track scientific international mobility? A case study based on Leibniz laureates. *Scientometrics* 117:705–720.

Angelici, C., B. M. Weckhuysen and P. C. A. Bruijnincx. 2013. Chemocatalytic conversion of ethanol into butadiene and other bulk chemicals. *ChemSusChem* 6:1595–1614.

Antolini, E. 2007. Catalysts for direct ethanol fuel cells. *Journal of Power Sources* 170:1–12.

Antolini, E. 2009. Palladium in fuel cell catalysis. *Energy and Environmental Science* 2:915–931.

Asafu-Adjaye, J. 2000. The relationship between energy consumption, energy prices and economic growth: Time series evidence from Asian developing countries. *Energy Economics* 22:615–625.

Bai, F. W., W. A. Anderson and M. Moo-Young. 2008. Ethanol fermentation technologies from sugar and starch feedstocks. *Biotechnology Advances* 26:89–105.

Balat, M. 2011. Production of bioethanol from lignocellulosic materials via the biochemical pathway: A review. *Energy Conversion and Management* 52:858–875.

Beaudry, C. and V. Lariviere. 2016. Which gender gap? Factors affecting researchers' scientific impact in science and medicine. *Research Policy* 45:1790–1817.

Binod, P., R. Sindhu and R. R. Singhania, et al. 2010. Bioethanol production from rice straw: An overview. *Bioresource Technology* 101:4767–4774.

Blankenship, K. M. 1993. Bringing gender and race in: US employment discrimination policy. *Gender & Society* 7:204–226.

Bothast, R. J., and M. A. Schlicher. 2005. Biotechnological processes for conversion of corn into ethanol. *Applied Microbiology and Biotechnology* 67:19–25.

Bourbonnais, R. and M. G. Paice. 1990. Oxidation of non-phenolic substrates. An expanded role for laccase in lignin biodegradation. *FEBS Letters* 267:99–102.

Burnham, J. F. 2006. Scopus database: A review. *Biomedical Digital Libraries* 3:1–8.

Canilha, L., A. K. Chandel and T. S. dos Santos Milessi, et al. 2012. Bioconversion of sugarcane biomass into ethanol: An overview about composition, pretreatment methods, detoxification of hydrolysates, enzymatic saccharification, and ethanol fermentation. *Journal of Biomedicine and Biotechnology* 2012:989572.

Cardona, C. A., J. A. Quintero and I. C. Paz. 2010. Production of bioethanol from sugarcane bagasse: Status and perspectives. *Bioresource Technology* 101:4754–4766.

Carlson, K. M., J. S. Gerber and D. Mueller, et al. 2017. Greenhouse gas emissions intensity of global croplands. *Nature Climate Change* 7:63–68.

Change, C. 2007. Climate change impacts, adaptation and vulnerability. *Science of the Total Environment* 326:95–112.

Dirth, T. P. and N. R. Branscombe. 2017. Disability models affect disability policy support through awareness of structural discrimination. *Journal of Social Issues* 73:413–442.

Duff, S. J. B. and W. D. Murray. 1996. Bioconversion of forest products industry waste cellulosics to fuel ethanol: A review. *Bioresource Technology* 55:1–33.

Ebadi, A. and A. Schiffauerova. 2016. How to boost scientific production? A statistical analysis of research funding and other influencing factors. *Scientometrics* 106:1093–1116.

Fang, X., Y. Shen, J. Zhao, X. Bao and Y. Qu. 2010. Status and prospect of lignocellulosic bioethanol production in China. *Bioresource Technology* 101:4814–4819.

Fauci, A. S., H. C. Lane and R. R. Redfield. 2020. Covid-19-navigating the uncharted. *New England Journal of Medicine* 382:1268–1269.

Fernando, S., S. Adhikari, C. Chandrapal and M. Murali. 2006. Biorefineries: Current status, challenges, and future direction. *Energy & Fuels* 20:1727–1737.

Formela, K., A. Hejna, L. Piszczyk, M. R. Saeb and X. Colom. 2016. Processing and structure-property relationships of natural rubber/wheat bran biocomposites. *Cellulose* 23:3157–3175.

Galbe, M. and G. Zacchi. 2002. A review of the production of ethanol from softwood. *Applied Microbiology and Biotechnology* 59:618–628.

Garfield, E. 1955. Citation indexes for science. *Science* 122:108–111.

Gillon, S. 2010. Fields of dreams: Negotiating an ethanol agenda in the Midwest United States. *Journal of Peasant Studies* 37:723–748.

Gnansounou, E. 2010. Production and use of lignocellulosic bioethanol in Europe: Current situation and perspectives. *Bioresource Technology* 101:4842–4850.

Greening, L. A., D. L. Greene and C. Difiglio. 2000. Energy efficiency and consumption-the rebound effect – A survey. *Energy Policy* 28:389–401.

Guimaraes, P. M. R., J. A. Teixeira and L. Domingues. 2010. Fermentation of lactose to bio-ethanol by yeasts as part of integrated solutions for the valorisation of cheese whey. *Biotechnology Advances* 28:375–384.

Hahn-Hagerdal, B., M. Galbe, M. F. Gorwa-Grauslund, G. Liden and G. Zacchi. 2006. Bio-ethanol – The fuel of tomorrow from the residues of today. *Trends in Biotechnology* 24:549–556.

Hamelinck, C. N., G. van Hooijdonk and A. P. C. Faaij. 2005. Ethanol from lignocellulosic biomass: Techno-economic performance in short-, middle- and long-term. *Biomass and Bioenergy* 28:384–410.

Hamilton, J. D. 1983. Oil and the macroeconomy since World War II. *Journal of Political Economy* 91:228–248.

Hamilton, J. D. 2003. What is an oil shock? *Journal of Econometrics* 113:363–398.

Hansen, A. C., Q. Zhang and P. W. L. Lyne. 2005. Ethanol-diesel fuel blends – A review. *Bioresource Technology* 96:277–285.

Haryanto, A., S. Fernando, N. Murali, and S. Adhikari. 2005. Current status of hydrogen production techniques by steam reforming of ethanol: A review. *Energy and Fuels* 19:2098–2106.

Hendriks, A. T. W. M. and G. Zeeman. 2009. Pretreatments to enhance the digestibility of lignocellulosic biomass. *Bioresource Technology* 100:10–18.

Henstra, A. M., J. Sipma, A. Rinzema and A. J. Stams. 2007. Microbiology of synthesis gas fermentation for biofuel production. *Current Opinion in Biotechnology* 18:200–206.

Hill, J., E. Nelson, D. Tilman, S. Polasky and D. Tiffany. 2006. Environmental, economic, and energetic costs and benefits of biodiesel and ethanol biofuels. *Proceedings of the National Academy of Sciences of the United States of America* 103:11206–11210.

Hill, J., S. Polasky and E. Nelson, et al. 2009. Climate change and health costs of air emissions from biofuels and gasoline. *Proceedings of the National Academy of Sciences of the United States of America* 106:2077–2082.

Ho, S. H., S. W. Huang and C. Y. Chen, et al. 2013. Bioethanol production using carbohydrate-rich microalgae biomass as feedstock. *Bioresource Technology* 135:191–198.

Hook, M. and X. Tang. 2013. Depletion of fossil fuels and anthropogenic climate change – A review. *Energy Policy* 52:797–809.

Hsieh, W. D., R. H. Chen, T. L. Wu and T. H. Lin. 2002. Engine performance and pollutant emission of an SI engine using ethanol-gasoline blended fuels. *Atmospheric Environment* 36:403–410.

Huang, H. J., S. Ramaswamy, U. W. Tschirner and B. V. Ramarao. 2008. A review of separation technologies in current and future biorefineries. *Separation and Purification Technology* 62:1–21.

Huijts, N. M. A., E. J. E. Molin and L. Steg. 2012. Psychological factors influencing sustainable energy technology acceptance: A review-based comprehensive framework. *Renewable and Sustainable Energy Reviews* 16:525–531.

Jacobs, J. A. and S. Frickel. 2009. Interdisciplinarity: A critical assessment. *Annual Review of Sociology* 35:43–65.

John, R. P., G. S. Anisha, K. M. Nampoothiri and A. Pandey. 2011. Micro and macroalgal biomass: A renewable source for bioethanol. *Bioresource Technology* 102:186–193.

Jones, T. C. 2012. America, oil, and war in the Middle East. *Journal of American History* 99:208–218.

Jonsson, L. J. and C. Martin. 2016. Pretreatment of lignocellulose: Formation of inhibitory by-products and strategies for minimizing their effects. *Bioresource Technology* 199:103–112.

Karpagam, R., S. Gopalakrishnan, M. Natarajan, B. R. Babu. 2011. Mapping of nanoscience and nanotechnology research in India: A scientometric analysis, 1990–2009. *Scientometrics* 89:501–522.

Kerr, R. A. 2007. Global warming is changing the world. *Science* 316:188–190.

Keshwani, D. R. and J. J. Cheng. 2009. Switchgrass for bioethanol and other value-added applications: A review. *Bioresource Technology* 100:1515–1523.

Kilian, L. 2008. Exogenous oil supply shocks: How big are they and how much do they matter for the US economy? *Review of Economics and Statistics* 90:216–240.

Kilian, L. 2009. Not all oil price shocks are alike: Disentangling demand and supply shocks in the crude oil market. *American Economic Review* 99:1053–1069.

Kim, J. 2018. Evaluating author name disambiguation for digital libraries: A case of DBLP. *Scientometrics* 116:1867–1886.

Kirk, T. K. and R. L. Farrell. 1987. Enzymatic "combustion": The microbial degradation of lignin. *Annual Review of Microbiology* 41:465–505.

Konur, O. 2000. Creating enforceable civil rights for disabled students in higher education: An institutional theory perspective. *Disability & Society* 15:1041–1063.

Konur, O. 2002a. Access to nursing education by disabled students: Rights and duties of nursing programs. *Nurse Education Today* 22:364–374.

Konur, O. 2002b. Assessment of disabled students in higher education: Current public policy issues. *Assessment and Evaluation in Higher Education* 27:131–152.

Konur, O. 2002c. Access to employment by disabled people in the UK: Is the Disability Discrimination Act working? *International Journal of Discrimination and the Law* 5:247–279.

Konur, O. 2006a. Participation of children with dyslexia in compulsory education: Current public policy issues. *Dyslexia* 12:51–67.

Konur, O. 2006b. Teaching disabled students in Higher Education. *Teaching in Higher Education* 11:351–363.

Konur, O. 2007a. A judicial outcome analysis of the Disability Discrimination Act: A windfall for the employers? *Disability & Society* 22:187–204.

Konur, O. 2007b. Computer-assisted teaching and assessment of disabled students in higher education: The interface between academic standards and disability rights. *Journal of Computer Assisted Learning* 23:207–219.

Konur, O. 2011. The scientometric evaluation of the research on the algae and bio-energy. *Applied Energy* 88:3532–3540.

Konur, O. 2012a. The evaluation of the biogas research: A scientometric approach. *Energy Education Science and Technology Part A: Energy Science and Research* 29:1277–1292.

Konur, O. 2012b. The evaluation of the educational research: A scientometric approach. *Energy Education Science and Technology Part B: Social and Educational Studies* 4:1935–1948.

Konur, O. 2012c. The evaluation of the global energy and fuels research: A scientometric approach. *Energy Education Science and Technology Part A: Energy Science and Research* 30:613–628.

Konur, O. 2012d. The evaluation of the research on the biodiesel: A scientometric approach. *Energy Education Science and Technology Part A: Energy Science and Research* 28:1003–1014.

Konur, O. 2012e. The evaluation of the research on the bioethanol: A scientometric approach. *Energy Education Science and Technology Part A: Energy Science and Research* 28:1051–1064.

Konur, O. 2012f. The evaluation of the research on the biofuels: A scientometric approach. *Energy Education Science and Technology Part A: Energy Science and Research* 28:903–916.

Konur, O. 2012g. The evaluation of the research on the biohydrogen: A scientometric approach. *Energy Education Science and Technology Part A: Energy Science and Research* 29:323–338.

Konur, O. 2012h. The evaluation of the research on the microbial fuel cells: A scientometric approach. *Energy Education Science and Technology Part A: Energy Science and Research* 29:309–322.

Konur, O. 2012i. The scientometric evaluation of the research on the production of bioenergy from biomass. *Biomass and Bioenergy* 47:504–515.

Konur, O. 2015. Current state of research on algal bioethanol. In *Marine Bioenergy: Trends and Developments*, Eds. S. K. Kim and C. G. Lee, pp. 217–244. Boca Raton, FL: CRC Press.

Konur, O. 2018b. Bioenergy and biofuels science and technology: Scientometric overview and citation classics. In *Bioenergy and Biofuels*, Ed. O. Konur, pp. 3–63. Boca Raton: CRC Press.

Konur, O. 2019. Cyanobacterial bioenergy and biofuels science and technology: A scientometric overview. In *Cyanobacteria: From Basic Science to Applications*, Eds. A. K. Mishra, D. N. Tiwari and A. N. Rai, pp. 419–442. Amsterdam: Elsevier.

Konur, O. 2020a. The scientometric analysis of the research on the bioethanol production from green macroalgae. In *Handbook of Algal Science, Technology and Medicine*, Ed. O. Konur, pp. 385–401. London: Academic Press.

Konur, O. 2023a. Production processes for bioethanol fuels: Scientometric study. In *Bioethanol Fuel Production Processes. I: Biomass Pretreatments. Handbook of Bioethanol Fuels Volume 1*, Ed. O. Konur. Boca Raton, FL: CRC Press.

Konur, O. 2023b. Feedstock-based bioethanol fuels: Scientometric study. In *Feedstock-based Bioethanol Fuels. I. Non-Waste Feedstocks: Starch, Sugar, Grass, Wood, Cellulose, Algae, and Biosyngas-based Bioethanol Fuels. Handbook of Bioethanol Fuels Volume 3*, Ed. O. Konur. Boca Raton, FL: CRC Press.

Konur, O. 2023c. Bioethanol fuel evaluation and utilization: Scientometric study. In *Evaluation and Utilization of Bioethanol Fuels. I: Evaluation of Bioethanol Fuels, Transport Engines, and Bioethanol Sensors. Handbook of Bioethanol Fuels Volume 5*, Ed. O. Konur. Boca Raton, FL: CRC Press.

Konur, O., Ed. 2018a. *Bioenergy and Biofuels*. Boca Raton, FL: CRC Press.

Konur, O., Ed. 2020b. *Handbook of Algal Science, Technology and Medicine*. London: Academic Press.

Konur, O., Ed. 2021a. *Handbook of Biodiesel and Petrodiesel Fuels: Science, Technology, Health, and Environment*. Boca Raton, FL: CRC Press.

Konur, O., Ed. 2021b. *Handbook of Biodiesel and Petrodiesel Fuels: Science, Technology, Health, and Environment. Volume 1. Biodiesel Fuels: Science, Technology, Health, and Environment*. Boca Raton, FL: CRC Press.

Konur, O., Ed. 2021c. *Handbook of Biodiesel and Petrodiesel Fuels: Science, Technology, Health, and Environment. Volume 2. Biodiesel Fuels based on the Edible and Nonedible Feedstocks, Wastes, and Algae: Science, Technology, Health, and Environment*. Boca Raton, FL: CRC Press.

Konur, O., Ed. 2021d. *Handbook of Biodiesel and Petrodiesel Fuels: Science, Technology, Health, and Environment. Volume 3. Petrodiesel Fuels: Science, Technology, Health, and Environment*. Boca Raton, FL: CRC Press.

Konur, O. and F. L. Matthews. 1989. Effect of the properties of the constituents on the fatigue performance of composites: A review. *Composites* 20:317–328.

Kruyt, B., D. P. van Vuuren, H. J. de Vries and H. Groenenberg. 2009. Indicators for energy security. *Energy Policy* 37:2166–2181.

Laser, M., D. Schulman and S. G. Allen, et al. 2002. A comparison of liquid hot water and steam pretreatments of sugar cane bagasse for bioconversion to ethanol. *Bioresource Technology* 81:33–44.

Leydesdorff, L. 2000. Is the European Union becoming a single publication system? *Scientometrics* 47:265–280.

Leydesdorff, L. and C. Wagner. 2009. Is the United States losing ground in science? A global perspective on the world science system. *Scientometrics* 78:23-36.

Leydesdorff, L. and P. Zhou. 2005. Are the contributions of China and Korea upsetting the world system of science? *Scientometrics* 63:617–630.

Li, H., S. M. Liu, X. H. Yu, S. L. Tang and C. K. Tang. 2020. Coronavirus disease 2019 (COVID-19): Current status and future perspectives. *International Journal of Antimicrobial Agents* 55:105951.

Lieberman, M. B. and D. B. Montgomery. 1988. First-mover advantages. *Strategic Management Journal* 9:41–58.

Limayem, A. and S. C. Ricke. 2012. Lignocellulosic biomass for bioethanol production: Current perspectives, potential issues and future prospects. *Progress in Energy and Combustion Science* 38:449–467.

Lin, Y. and S. Tanaka. 2006. Ethanol fermentation from biomass resources: Current state and prospects. *Applied Microbiology and Biotechnology* 69:627–642.

Ma, X., L. Sun and C. Song. 2002. A new approach to deep desulfurization of gasoline, diesel fuel and jet fuel by selective adsorption for ultra-clean fuels and for fuel cell applications. *Catalysis Today* 77:107–116.

Makenete, A. L., W. J. Lemmer and J. Kupka. 2008. The impact of biofuel production on food security: A briefing paper with a particular emphasis on maize-to-ethanol production. *International Food and Agribusiness Management Review* 11:101–110.

McMillan, J. D. 1997. Bioethanol production: Status and prospects. *Renewable Energy* 10:295–302.

Morschbacker, A. 2009. Bio-ethanol based ethylene. *Polymer Reviews* 49:79–84.

Mosier, N., C. Wyman and B. Dale, et al. 2005. Features of promising technologies for pretreatment of lignocellulosic biomass. *Bioresource Technology* 96:673–686.

Munasinghe, P. C. and S. K. Khanal. 2010. Biomass-derived syngas fermentation into biofuels: Opportunities and challenges. *Bioresource Technology* 101:5013–5022.

Najafi, G., B. Ghobadian and T. Tavakoli, et al. 2009. Performance and exhaust emissions of a gasoline engine with ethanol blended gasoline fuels using artificial neural network. *Applied Energy* 86:630–639.

Negishi, M., Y. Sun and K. Shigi. 2004. Citation database for Japanese papers: A new bibliometric tool for Japanese academic society. *Scientometrics* 60:333–351.

Newman, P. W. G. and J. R. Kenworthy. 1989. Gasoline consumption and cities: A comparison of U.S. cities with a global survey. *Journal of the American Planning Association* 55:24–37.

Nigam, J. N. 2002. Bioconversion of water-hyacinth (*Eichhornia crassipes*) hemicellulose acid hydrolysate to motor fuel ethanol by xylose-fermenting yeast. *Journal of Biotechnology* 97:107–116.

Nissani, M. 1997. Ten cheers for interdisciplinarity: The case for interdisciplinary knowledge and research. *Social Science Journal* 34:201–216.

North, D. C. 1991. Institutions. *Journal of Economic Perspectives* 5:97–112.

Olsson, L. and B. Hahn-Hagerdal. 1996. Fermentation of lignocellulosic hydrolysates for ethanol production. *Enzyme and Microbial Technology* 18:312–331.

Palmqvist, E. and B. Hahn-Hagerdal. 2000b. Fermentation of lignocellulosic hydrolysates. II: Inhibitors and mechanisms of inhibition. *Bioresource Technology* 74:25–33.

Palmqvist, E. and B. Hahn-Hagerdal. 2000a. Fermentation of lignocellulosic hydrolysates. I: Inhibition and detoxification. *Bioresource Technology* 74:17–24.

Pimentel, D. and T. W. Patzek. 2005. Ethanol production using corn, switchgrass, and wood; Biodiesel production using soybean and sunflower. *Natural Resources Research* 14:65–76.

Pinkert, A., K. N. Marsh, S. Pang and M. P. Staiger. 2009. Ionic liquids and their interaction with cellulose. *Chemical Reviews* 109:6712–6728.

Popp, J., Z. Lakner, M. Harangi-Rakos and M. Fari. 2014. The effect of bioenergy expansion: Food, energy, and environment. *Renewable and Sustainable Energy Reviews* 32:559–578.

Prasad, S., A. Singh and H. C. Joshi. 2007. Ethanol as an alternative fuel from agricultural, industrial and urban residues. *Resources, Conservation and Recycling* 50:1–39.

Ravindran, R., and A. K. Jaiswal. 2016. A comprehensive review on pre-treatment strategy for lignocellulosic food industry waste: Challenges and opportunities. *Bioresource Technology* 199:92–102.

Reeves, S. 2014. To Russia with love: How moral arguments for a humanitarian intervention in Syria opened the door for an invasion of the Ukraine. *Michigan State University International Law Review* 23:199+.

Saini, J. K., R. Saini and L. Tewari. 2015. Lignocellulosic agriculture wastes as biomass feedstocks for second-generation bioethanol production: Concepts and recent developments. *3 Biotech* 5:337–353.

Sanchez, C. 2009. Lignocellulosic residues: Biodegradation and bioconversion by fungi. *Biotechnology Advances* 27:185–194.

Sanchez, O. J. and C. A. Cardona. 2008. Trends in biotechnological production of fuel ethanol from different feedstocks. *Bioresource Technology* 99:5270–5295.

Sano, T., H. Yanagishita, Y. Kiyozumi, F. Mizukami and K. Haraya. 1994. Separation of ethanol/water mixture by silicalite membrane on pervaporation. *Journal of Membrane Science* 95:221–228.

Schmer, M. R., K. P. Vogel, R. B. Mitchell and R. K. Perrin. 2008. Net energy of cellulosic ethanol from switchgrass. *Proceedings of the National Academy of Sciences of the United States of America* 105:464–469.

Stern, D. I. 1993. Energy and economic growth in the USA: A multivariate approach. *Energy Economics* 15:137–150.

Sukumaran, R. K., R. R. Singhania, G. M. Mathew and A. Pandey. 2009. Cellulase production using biomass feed stock and its application in lignocellulose saccharification for bio-ethanol production. *Renewable Energy* 34:421–424.

Sun, Y. and J. Cheng. 2002. Hydrolysis of lignocellulosic materials for ethanol production: A review. *Bioresource Technology* 83:1–11.

Taherzadeh, M. J. and K. Karimi. 2007. Enzyme-based hydrolysis processes for ethanol from lignocellulosic materials: A review. *Bioresources* 2:707–738.

Taherzadeh, M. J. and K. Karimi. 2008. Pretreatment of lignocellulosic wastes to improve ethanol and biogas production: A review. *International Journal of Molecular Sciences* 9:1621–1651.

Tahmooresnejad, L., C. Beaudry, C and A. Schiffauerova. 2015. The role of public funding in nanotechnology scientific production: Where Canada stands in comparison to the United States. *Scientometrics* 102:753–787.

Talebnia, F., D. Karakashev and I. Angelidaki. 2010. Production of bioethanol from wheat straw: An overview on pretreatment, hydrolysis and fermentation. *Bioresource Technology* 101:4744–4753.

Tilman, D., R. Socolow and J. A. Foley, et al. 2009. Beneficial biofuels-the food, energy, and environment trilemma. *Science* 325:270–271.

Vane, L. M. 2005. A review of pervaporation for product recovery from biomass fermentation processes. *Journal of Chemical Technology and Biotechnology* 80:603–629.

Winzer, C. 2012. Conceptualizing energy security. *Energy Policy* 46:36–48.

Wu, F., D. Zhang and J. Zhang, J. 2012. Will the development of bioenergy in China create a food security problem? Modeling with fuel ethanol as an example. *Renewable Energy* 47:127–134.

Wustenhagen, R., M. Wolsink and M. J. Burer. 2007. Social acceptance of renewable energy innovation: An introduction to the concept. *Energy Policy* 35:2683–2691.

Yang, B. and C. E. Wyman. 2008. Pretreatment: The key to unlocking low-cost cellulosic ethanol. *Biofuels, Bioproducts and Biorefining* 2:26–40.

Zabed, H., J. N. Sahu, A. Suely, A. N. Boyce and G. Faruq. 2017. Bioethanol production from renewable sources: Current perspectives and technological progress. *Renewable and Sustainable Energy Reviews* 71:475–501.

Zhang, Y. H. P. and L. R. Lynd. 2004. Toward an aggregated understanding of enzymatic hydrolysis of cellulose: Noncomplexed cellulase systems. *Biotechnology and Bioengineering* 88:797–824.

Zhu, J. Y. and X. J. Pan. 2010. Woody biomass pretreatment for cellulosic ethanol production: Technology and energy consumption evaluation. *Bioresource Technology* 101:4992–5002.

3 Bioethanol Fuels
Review

Ozcan Konur
(Formerly) Ankara Yildirim Beyazit University

3.1 INTRODUCTION

Crude oil–based gasoline fuels (Ma et al., 2002; Newman and Kenworthy, 1989) have been widely used in the transportation sector since the 1920s. However, there have been great public concerns over the adverse environmental and human impact of these fuels (Hill et al., 2006, 2009). Hence, biomass-based bioethanol fuels (Hill et al., 2006; Konur, 2012, 2015, 2020) have increasingly been used in blending gasoline fuels (Hsieh et al., 2002; Najafi et al., 2009), in fuel cells (Antolini, 2007, 2009), and in biochemical production (Angelici et al., 2013; Morschbacker, 2009) in a biorefinery context (Fernando et al., 2006; Huang et al., 2008).

However, it is necessary to pretreat the biomass (Alvira et al., 2010; Taherzadeh and Karimi, 2008) to enhance the yield of bioethanol (Hahn-Hagerdal et al., 2006; Sanchez and Cardona, 2008) prior to bioethanol fuel production from the feedstocks through hydrolysis (Sun and Cheng, 2002; Taherzadeh and Karimi, 2007) and fermentation (Lin and Tanaka, 2006; Olsson and Hahn-Hagerdal, 1996) of the biomass.

Research in the field of bioethanol fuels has intensified in this context in the key research fronts of the production of bioethanol fuels, and to a lesser extent, utilization (Antolini, 2007; Hansen et al., 2005; Haryanto et al., 2005) and evaluation (Hamelinck et al., 2005; Pimentel and Patzek, 2005; Schmer et al., 2008) of bioethanol fuels. For the first research front, pretreatment (Alvira et al., 2010; Hendriks and Zeeman, 2009; Mosier et al., 2005) and hydrolysis (Alvira et al., 2010; Sun and Cheng, 2002; Zhang and Lynd, 2004) of feedstocks, fermentation of feedstock-based hydrolysates (Jonsson and Martin, 2016; Lin and Tanaka, 2006; Palmqvist and Hahn-Hagerdal, 2000a, b), production of bioethanol fuels (Balat, 2011; Limayem and Ricke, 2012; Lin and Tanaka, 2006), and to a lesser extent separation and distillation of bioethanol fuels (Sano et al., 1994; Vane, 2005) are the key research areas while for the second research front, utilization of bioethanol fuels in fuel cells and transport engines, bioethanol-based biohydrogen fuels, bioethanol sensors, and bioethanol-based biochemicals are the key research areas.

Research in this field has also intensified for feedstocks of lignocellulosic biomass at large (Hendriks and Zeeman, 2009; Mosier et al., 2005; Sun and Cheng, 2002), bioethanol fuels for their utilization (Antolini, 2007; Hansen et al., 2005; Haryanto et al., 2005), cellulose (Pinkert et al., 2009; Zhang and Lynd, 2004), starch feedstock residues (Binod et al., 2010; Talebnia et al., 2010), wood biomass (Galbe and Zacchi, 2002; Zhu and Pan, 2010), and to a lesser extent industrial waste (Cardona et al., 2010; Prasad et al., 2007), lignin (Bourbonnais and Paice, 1990; Kirk and Farrell, 1987), grass (Keshwani and Cheng, 2009; Pimentel and Patzek, 2005), hydrolysates (Palmqvist and Hahn-Hagerdal, 2000a, b), biomass at large (Lin and Tanaka, 2006; Zabed et al., 2017), sugar feedstocks (Bai et al., 2008; Canilha et al., 2012), sugar feedstock residues (Cardona et al., 2010; Laser et al., 2002), algal biomass (Ho et al., 2013; John et al., 2011), lignocellulosic wastes (Sanchez, 2009; Saini et al., 2015), starch feedstocks (Bai et al., 2008; Bothast and Schlicher, 2005), urban wastes (Prasad et al., 2007; Ravindran and Jaiswal, 2016), forestry wastes (Duff and Murray, 1996),

DOI: 10.1201/9781003226536-4

food wastes (Guimaraes et al., 2010; Ravindran and Jaiswal, 2016), syngas (Henstra et al., 2007; Munasinghe and Khanal, 2010), and plants (Nigam, 2002; Sukumaran et al., 2009).

However, it is essential to develop efficient incentive structures (North, 1991) for the primary stakeholders to enhance research in this field (Konur, 2000, 2002a–c, 2006a, b, 2007a, b). Although there have been a large number of review papers on bioethanol fuels (Hendriks and Zeeman, 2009; Mosier et al., 2005; Sun and Cheng, 2002), there has been no review of the most-cited 25 papers in this field.

Thus, this chapter presents a review of the most-cited 25 articles in the field of bioethanol fuels. Then, it discusses the key findings of these highly influential papers and comments on future research priorities in this field.

3.2 MATERIALS AND METHODS

Search for this study was carried out using the Scopus database (Burnham, 2006) in October 2022.

As a first step for the search of the relevant literature, keywords were selected using the most-cited first 300 population papers for each feedstock. The selected keyword list was then optimized to obtain a representative sample of papers for each research field. These three keyword lists were then integrated to obtain the keyword list for this research field (Konur, 2023a–c).

As a second step, a sample dataset was used for this study. The first 25 articles with at least 726 citations each were selected for the review study. Key findings from each paper were taken from the abstracts of these papers and were briefly discussed. Additionally, a number of brief conclusions were drawn and a number of relevant recommendations were made to enhance the future research landscape.

3.3 RESULTS

Brief information about the 25 most-cited papers with at least 726 citations each on bioethanol fuels is given below. The primary research fronts are pretreatments and hydrolysis of feedstocks with eight highly cited papers (HCPs) each while the other research fronts are the production, evaluation, and utilization of bioethanol fuel with one, five, and three HCPs, respectively.

3.3.1 Feedstock Pretreatments

Brief information about the 8 most-cited papers on pretreatments of feedstocks with at least 751 citations each is given below (Table 3.1).

Bourbonnais and Paice (1990) oxidized non-phenolic substrates for lignin biodegradation in a paper with 1,189 citations. They used substrates such as Remazol Blue and 2,2′-azinobis(3-ethylbenzthiazoline-6-sulphonate) (ABTS) and found that laccases *Coriolus* (*Trametes*) *versicolor* could also oxidize non-phenolic lignin model compounds. They oxidized veratryl alcohol (I) and 1-(3,4-dimethoxyphenyl)-2-(2-methoxyphenoxy)-propane-1, 3-diol (III) by laccase and mediator to give α-carbonyl derivatives. The β-1 lignin model dimer, 1-(3,4-dimethoxyphenyl)-2-phenoxy-ethane-1, 2-diol(II) was cleaved by laccase in the presence of ABTS to give veratraldehyde and benzaldehyde. In conclusion, laccase was capable of oxidizing both phenolic and non-phenolic moieties of lignin, but the latter was dependent on the co-presence of the primary laccase substrate.

Floudas et al. (2012) mapped the detailed evolution of wood-degrading enzymes in a paper with 1,098 citations. They noted that wood was highly resistant to decay, largely due to the presence of lignin, and the only organisms capable of substantial lignin decay were white-rot fungi in the agaricomycetes, which also contained non–lignin-degrading brown rot and ectomycorrhizal species. They analyzed 31 fungal genomes and found that lignin-degrading peroxidases expanded in the lineage leading to the ancestor of the agaricomycetes, which was reconstructed as a white-rot species, and then contracted in parallel lineages leading to brown rot and mycorrhizal species. Thus,

TABLE 3.1
Pretreatment of Feedstocks

No.	Papers	Biomass	Prts.	Parameters	Keywords	Lead Authors	Affil.	Cits
6	Bourbonnais and Paice (1990)	Lignin	Enzymes	Lignin biodegradation, laccases, non-phenolic substrates, phenolic substrates	Lignin, biodegradation	Bourbonnais, Robert 7003682574	McGill Univ. Canada	1,189
7	Floudas et al. (2012)	Wood, lignin	Enzymes	Wood degradation, wood-degrading enzyme evolution, genome, peroxidases	Lignin, decomposition	Grigoriev, Igor V. 25027225800	Jnt. Genome Inst. USA	1,098
11	Tien and Kirk (1983)	Wood, lignin	Enzymes	Lignin-degrading enzymes, fungi	Lignin, degrading, enzyme	Tien, Ming 7006146774	Pennsylvania State Univ. USA	976
12	Schwanninger et al. (2004)	Wood, cellulose	Milling	Vibratory ball milling, FTIR spectroscopy, temperature, particle size, oxygen, cellulose crystallinity and depolymerization	Wood, cellulose, milling	Schwanninger, Manfred 6602877236	Univ. Natr. Res. Appl. Life Sci. Austria	943
14	Martinez et al. (2008)	Lignocellulosic biomass	Enzymes	Biomass-degrading fungi, genome, enzymes	Biomass, fungus, degrading	Martinez, Diego 7202958664	Massachusetts Inst. Technol. USA	896
18	Pandey and Pitman (2003)	Wood Pine, sapwood, beech	Enzymes	Wood decay, fungi, lignin and carbohydrate content, fungi types, wood types, FTIR spectra	Wood, decay, fungi	Pandey, Krishna K. 56252805800	Inst. Wood Sci. Technol. India	843
19	Sun et al. (2009)	Wood Pine, oak	IL	IL types, dissolution rates, wood types, delignification, cellulose recovery	Wood, delignification, IL	Rogers, Robin D. 35474829200	Univ. Alabama USA	833
24	Cai and Zhang (2005)	Cellulose	IL	Cellulose dissolution, alkaline solvents, cellulose dissolution behavior, solubility, hydrogen bonding	Cellulose, dissolution	Zhang, Lina* 55917992100	Wuhan Univ. China	751

Prt, biomass pretreatments; Na, non available; Cits., number of citations received for each paper; *, female.

the origin of lignin degradation might have coincided with a sharp decrease in the rate of organic carbon burial around the end of the carboniferous period.

Tien and Kirk (1983) determined the lignin-degrading enzyme from the wood-decomposing basidiomycete *Phanerochaete chrysosporium* Burds in a paper with 976 citations. They found that the extracellular fluid of ligninolytic cultures of this basidiomycete contained an enzyme that degraded lignin substructure model compounds as well as spruce and birch lignins. It had a molecular size of 42,000 daltons and required hydrogen peroxide for activity.

Schwanninger et al. (2004) studied the effects of short-time vibratory ball milling on the shape of Fourier-transform infrared (FTIR) spectra of wood and cellulose in a paper with 943 citations. They showed that this milling had a strong influence on the shape of FTIR spectra of wood and cellulose, even if the samples were milled for only a short time. Further, neither temperature nor particle size or oxidation processes were the most important reason for the observed changes in the structure of FTIR spectra, but the mechanical treatment itself was the main influencing factor. They observed the most evident changes in the spectra of cellulose and wood at wave numbers 1,034, 1,059, 1,110, 1,162, 1,318, 1,335, and 2,902 cm^{-1} and in the OH-stretching vibration region from 3,200 to 3,500 cm^{-1}. They concluded that these changes were mainly associated with a decrease in the degree of crystallinity and/or a decrease in the degree of polymerization of cellulose.

Martinez et al. (2008) performed genome sequencing and analysis of the biomass-degrading fungus *Trichoderma reesei* in a paper with 896 citations. They assembled 89 scaffolds to generate 34 Mbp of nearly contiguous *T. reesei* genome sequence comprising 9,129 predicted gene models. They found that its genome encoded fewer cellulases and hemicellulases than any other sequenced fungus able to hydrolyze plant cell wall polysaccharides. Further, many *T. reesei* gene-encoding carbohydrate-active enzymes were distributed nonrandomly in clusters that lied between regions of synteny with other sordariomycetes. Hence, numerous gene-encoding biosynthetic pathways for secondary metabolites might promote the survival of *T. reesei* in its competitive soil habitat.

Pandey and Pitman (2003) performed FTIR studies of changes in wood chemistry following decay by brown-rot and white-rot fungi in a paper with 843 citations. They used Scots pine, sapwood, and beech decayed by *Coniophora puteana* [(Schumacher) Karsten], *C. versicolor* [(L.) Quelet], and *Phanerochaete chrysosporium* (Burdsall). They exposed wood to these fungi for different durations up to 12 weeks. They found that in wood decayed by *C. puteana,* there was a progressive increase in lignin content relative to carbohydrate evident from increases in the relative height of lignin-associated bands and a corresponding decrease in the intensities of carbohydrate bands. At higher weight losses, spectra for wood decayed by *C. puteana* had many similarities with that of Klason lignin isolated from wood. In contrast, wood decayed by *P. chrysosporium* showed selective-type decay with a reduction in peak heights associated with lignin relative to carbohydrates. Although weight losses in samples exposed to *C. versicolor* were high, simultaneous decay resulted in little change in relative intensities of lignin and carbohydrate bands, with only a slight preference for lignin.

Sun et al. (2009) dissolved and partially delignified yellow pine and red oak in ionic liquid (IL) 1-ethyl-3-methylimidazolium acetate ([C_2mim]OAc) after mild grinding in a paper with 833 citations. They showed that [C_2mim]OAc was a better solvent for wood than 1-butyl-3-methylimidazolium chloride ([C_4mim]Cl) and that the type of wood, initial wood load, particle size, etc., affected dissolution and dissolution rates. For example, red oak dissolved better and faster than yellow pine. They obtained carbohydrate-free lignin and cellulose-rich materials using proper reconstitution solvents (e.g., acetone/water 1:1 v/v), and they achieved approximately 26.1% and 34.9% of reductions in lignin content in the reconstituted cellulose-rich materials from pine and oak, respectively, in one dissolution/reconstitution cycle. Further, for pine, they recovered 59% of holocellulose from the original wood in the cellulose-rich reconstituted material while they recovered 31% and 38% of original lignin, respectively, as carbohydrate-free lignin and as carbohydrate-bonded lignin in cellulose-rich materials.

Cai and Zhang (2005) dissolved cellulose in LiOH/urea and NaOH/urea aqueous solutions in a paper with 751 citations. They evaluated the dissolution behavior and solubility of cellulose. They ound that cellulose having a viscosity–average molecular weight of 11.4×10^4 and 37.2×10^4

could be dissolved, respectively, in 7% NaOH/12% urea and 4.2% LiOH/12% urea aqueous solutions pre-cooled to −10°C within 2 min, whereas all of them could not be dissolved in a KOH/urea aqueous solution. Further, the dissolution power of solvent systems was in the order of LiOH/urea > NaOH/urea ≫ KOH/urea aqueous solution. LiOH/urea and NaOH/urea aqueous solutions as non-derivatizing solvents broke the intra- and intermolecular hydrogen bonding of cellulose and prevented the approach toward each other of cellulose molecules, leading to good dispersion of cellulose to form an actual solution.

3.3.2 Hydrolysis of Feedstocks

There are eight HCPs for hydrolysis of feedstocks with at least 726 citations (Table 3.2).

Park et al. (2010) compared four different techniques incorporating X-ray diffraction (XRD) and solid-state ^{13}C nuclear magnetic resonance (NMR) for the measurement of crystallinity index (CI) using eight different cellulose preparations. They found that the simplest method which involved measurement of just two heights in the X-ray diffractogram produced significantly higher crystallinity values than other methods. Alternative XRD and NMR methods which considered the contributions from amorphous and crystalline cellulose to the entire XRD and NMR spectra provided a more accurate measure of crystallinity of cellulose. Although celluloses having a high amorphous content were usually more easily digested by enzymes, it was unclear whether CI actually provided a clear indication of the digestibility of a cellulose sample. Cellulose accessibility should be affected by crystallinity but was also likely to be affected by several other parameters, such as lignin/hemicellulose contents and distribution, porosity, and particle size. Given the methodological dependency of cellulose CI values and the complex nature of cellulase interactions with amorphous and crystalline celluloses, they caution against trying to correlate relatively small changes in CI with changes in cellulose digestibility.

Chen and Dixon (2007) showed that lignin modification improved fermentable sugar yields for biofuel production in a paper with 1,028 citations. They found that in stems of transgenic alfalfa lines independently downregulated in each of the six lignin biosynthetic enzymes, recalcitrance to both acid pretreatment and enzymatic digestion was directly proportional to lignin content. Further, some transgenics yield nearly twice as much sugar from cell walls as wild-type plants. Thus, lignin modification could bypass the need for acid pretreatment, thereby facilitating bioprocess consolidation.

Suganuma et al. (2008) hydrolyzed cellulose by amorphous carbon-bearing SO_3H, COOH, and OH groups in a paper with 885 citations. They found that crystalline pure cellulose was not hydrolyzed by conventional strong solid Brønsted acid catalysts such as niobic acid, H-mordenite, Nafion, and Amberlyst-15, while amorphous carbon-bearing solid catalysts functioned as an efficient catalyst for the reaction. The apparent activation energy for hydrolysis of cellulose into glucose using the carbon catalyst was 110 kJ/mol, smaller than that for sulfuric acid under optimal conditions (170 kJ/mol). Further, this carbon catalyst could be readily separated from the saccharide solution after the reaction for reuse in the reaction without a loss of activity. They attributed the catalytic performance of the carbon catalyst to the ability of the material to adsorb β-1,4 glucan, which did not adsorb to other solid acids.

Li et al. (2010) compared the efficiency of dilute acid hydrolysis and dissolution in an IL with switchgrass in a paper with 860 citations. When subject to IL pretreatment, they observed that switchgrass exhibited reduced cellulose crystallinity, increased surface area, and decreased lignin content compared to dilute acid pretreatment. Further, IL pretreatment enabled a significant enhancement in the rate of enzymatic hydrolysis of the cellulose component of switchgrass, with a rate increase of 16.7-fold, and a glucan yield of 96.0% obtained in 24 h. In conclusion, IL pretreatment offered unique advantages compared to the dilute acid pretreatment process for switchgrass.

Kilpelainen et al. (2007) dissolved the wood in ILs in a paper with 832 citations. They found that 1-butyl-3-methylimidazolium chloride ([Bmim]Cl) and 1-allyl-3-methylimidazolium chloride ([Amim]Cl) had good solvating power for Norway spruce sawdust and Norway spruce and Southern

TABLE 3.2
Hydrolysis of Feedstocks

No.	Papers	Biomass	Prts.	Parameters	Keywords	Lead Authors	Affil.	Cits
2	Park et al. (2010)	Cellulose	NA	Cellulose digestibility, crystallinity index, measurement, XRD, NMR, amorphous and crystalline celluloses	Cellulose, cellulase	Johnson, David K. 24550868900	NREL USA	2,026
8	Chen and Dixon (2007)	Alfalfa	Acids, enzymes	Biomass engineering, pretreatment, hydrolysis, lignin content, sugar yield	Lignin, sugar, fermentable	Chen, Fang 57188570847	Univ. N. Texas USA	1,028
16	Suganuma et al. (2008)	Cellulose	C catalysts	Hydrolysis, solid carbon catalysts, crystalline cellulose, activation energy, catalyst recycling	Cellulose, hydrolysis	Hara, Michkazu 7403345875	Tokyo Inst. Technol. Japan	885
17	Li et al. (2010)	Switchgrass	Acids, IL, enzymes	Pretreatment types, acid hydrolysis, delignification, saccharification, sugar yields, crystallinity, surface area, lignin content, enzymatic hydrolysis	Switchgrass, pretreatment, saccharification, recalcitrance, IL	Singh, Seema* 35264950300	Sandia Natl. Lab. USA	860
20	Kilpelainen et al. (2007)	Wood Spruce, pine	IL, enzymes	IL pretreatment, enzymatic hydrolysis, IL types, wood types	Wood, IL, dissolution	Kilpelainen, Ilkka 7006830888	Univ. Helsinki Finland	832
22	Lee et al. (2009)	Wood	IL, enzymes	Enzymatic hydrolysis, IL pretreatment, delignification, cellulose crystallinity index, enzyme recycling	Wood, IL, hydrolysis	Dordick, Jonathan S. 7102545507	Rensselaer Polytech Inst. USA	816
23	Eriksson et al. (2002)	Lignocellulose Spruce	Enzymes, surfactants	Enzymatic hydrolysis, pretreatment, mechanisms, enzyme adsorption, cellulase stability, surfactant–lignin interactions	Lignocellulose, hydrolysis	Tjerneld, Folke 7006446969	Lund Univ. Sweden	754
25	Kumar et al. (2009)	Corn stover, poplar	Enzymes	Biomass hydrolysis and pretreatments, biomass crystallinity, cellulose degree of polymerization, cellulase adsorption capacity	Corn stover, poplar, pretreatment	Charles E. Wyman 7004396809	Univ. Calif. Riverside USA	726

Prt, biomass pretreatments; NA, non-available; Cits., number of citations received for each paper; *, female.

pine thermomechanical pulp fibers as these ILs provided solutions which permitted complete acetylation of the wood. Alternatively, they obtained transparent amber solutions of wood when dissolution of the same lignocellulosic samples was attempted in 1-benzyl-3-methylimidazolium chloride ([BzMim]Cl). They then digested the cellulose of the regenerated wood to glucose by a cellulase enzymatic hydrolysis treatment. Furthermore, the completely acetylated wood was readily soluble in chloroform.

Lee et al. (2009) used 1-ethyl-3-methylimidazolium acetate ([Emim]CH_3COO) to extract lignin from wood flour in a paper with 816 citations. They observed that the cellulose in the pretreated wood flour became far less crystalline without undergoing solubilization. When 40% of lignin was removed, the cellulose CI dropped below 45, resulting in >90% of cellulose in wood flour being hydrolyzed by *Trichoderma viride* cellulase. They then reused this IL, thereby resulting in a highly concentrated solution of chemically unmodified lignin.

Eriksson et al. (2002) explored the mechanism of the surfactant effect in enzymatic hydrolysis of lignocellulose in a paper with 754 citations. They screened a number of surfactants for their ability to improve the enzymatic hydrolysis of steam-pretreated spruce. They found that nonionic surfactants were the most effective. Studies of adsorption of the dominating cellulase of *T. reesei*, Cel7A (CBHI), during hydrolysis, showed that the anionic and nonionic surfactants reduced enzyme adsorption to the lignocellulose substrate. The approximate reduction of enzyme adsorption was from 90% of the adsorbed enzyme to 80% with surfactant addition. Surfactants had only a weak effect on cellulase temperature stability. They explained the improved conversion of lignocellulose with the surfactant by a reduction of the unproductive enzyme adsorption to the lignin part of the substrate. This was due to the hydrophobic interaction of the surfactant with lignin on the lignocellulose surface, which released unspecifically bound enzymes.

Kumar et al. (2009) pretreated corn stover and poplar solids using leading pretreatment technologies in a paper with 726 citations. They used ammonia fiber expansion (AFEX), ammonia-recycled percolation, controlled pH, dilute acid, flowthrough, lime, and SO_2 pretreatments. They observed that lime pretreatment removed most acetyl groups from both corn stover and poplar, while AFEX removed the least while low pH pretreatments depolymerized cellulose and enhanced biomass crystallinity much more than higher pH approaches. Further, lime-pretreated corn stover solids and flowthrough-pretreated poplar solids had the highest cellulase adsorption capacity, while dilute acid–pretreated corn stover solids and controlled pH-pretreated poplar solids had the least. Then, enzymatically extracted AFEX lignin preparations for both corn stover and poplar had the lowest cellulase adsorption capacity while SO_2-pretreated solids had the highest surface oxygen/carbon ratio for poplar, but for corn stover, the highest value was observed for dilute acid pretreatment with a Parr reactor. In conclusion, although dependent on pretreatment and substrate, along with changes in cross-linking and chemical changes, pretreatments might also decrystallize cellulose and change the ratio of crystalline cellulose polymorphs (Iα/Iβ).

3.3.3 Bioethanol Production, Evaluation, and Utilization

There are one, five, and three HCPs for the production, evaluation, and utilization of bioethanol fuels with at least 831 citations each (Table 3.3).

3.3.3.1 Bioethanol Production

Larsson et al. (1999) studied the effect of the combined severity (CS) of dilute sulfuric acid hydrolysis of spruce on sugar yield and the fermentability of the hydrolysate by *Saccharomyces cerevisiae* in a paper with 889 citations. When the CS of hydrolysis conditions increased, they observed that the yield of fermentable sugars increased to a maximum between CS 2.0 and 2.7 for mannose and 3.0 and 3.4 for glucose, above which it decreased. Further, the decrease in the yield of monosaccharides coincided with the maximum concentrations of furfural and 5-hydroxymethylfurfural (HMF). With a further increase in CS, concentrations of furfural and 5-HMF decreased while the

TABLE 3.3
Production, Evaluation, and Utilization of Bioethanol Fuels

No.	Papers	Biomass	Res. Fronts	Prts.	Yeasts	Parameters	Keywords	Lead Authors	Affil.	Cits
1	Hill et al. (2006)	Corn grains	Evaluation	Na	Na	Life cycle assessment, net energy gain, economic competitiveness, GHG emissions, food security, large-scale production	Ethanol, economic, environmental, energetic	Hill, Jason 14053737500	Univ. Minnesota	2,091
3	Wan et al. (2004)	Bioethanol	Sensors	Na	Na	Ethanol sensor fabrication and evaluation, ZnO sensors, ethanol sensitivity	Ethanol sensing, sensors	Wang, Tai-Hong 35241217600	Southern Univ. Sci. Technol. China	1,908
4	Kim and Dale (2004)	Starch feedstock residues Sugar feedstock residues	Evaluation	Na	Na	Lignocellulosic wastes, sugar feedstock residues, starch feedstock residues, bioethanol production, Asia, United States, Europe, wheat straw, rice straw, corn stover	Bioethanol, wasted, crop residues	Dale, Bruce E. 7201511969	Michigan State Univ. USA	1,447
5	Hamelinck et al. (2005)	Biomass, lignin	Evaluation	Na	Na	Bioethanol fuel production costs, techno-economic performance, available and advanced technologies, lignin, ethanol conversion efficiency	Ethanol, lignocellulosic biomass, techno-economic	Hamelinck, Carlo N 6603008025	Ecorys Netherlands	1,234
9	Pimentel and Patzek (2005)	Corn, switchgrass, wood	Evaluation	Na	Na	Bioethanol fuels, net energy gain, biodiesel fuels, biomass types	Ethanol, corn, switchgrass, wood	Pimentel, David 7005471319	Cornell Univ. USA	1,020
10	Murdoch et al. (2011)	Bioethanol	Biohydrogen	Na	Na	Photocatalytic biohydrogen production, Au/TiO_2 photocatalysts, anatase NPs, NP size	Ethanol, hydrogen	Idriss, Hicham 7006768868	Karlsruhe Inst. Technol. Germany	1,009
13	Deluga et al. (2004)	Bioethanol	Biohydrogen	Na	Na	Biohydrogen production, ethanol autothermal reforming, hydrogen selectivity, ethanol conversion rate	Ethanol, hydrogen, reforming	Schmidt, Lanny D. 13103591600	Univ. Minnesota USA	918

(Continued)

TABLE 3.3 (*Continued*)
Production, Evaluation, and Utilization of Bioethanol Fuels

No.	Papers	Biomass	Res. Fronts	Prts.	Yeasts	Parameters	Keywords	Lead Authors	Affil.	Cits
15	Larsson et al. (1999)	Softwood Spruce	Production	Acids	S. cerevisiae	Acid hydrolysis severity, fermentation, sugar and ethanol yield, fermentation inhibitors, volumetric productivity	Softwood, fermentation, hydrolysis	Hahn-Hagerdal, Barbel* 7005389381	Lund Univ. Sweden	889
21	Schmer et al. (2008)	Switchgrass	Evaluation	Na	Na	Net energy efficiency, economic feasibility, net energy yield, economic costs, biomass yield, GHG emissions	Ethanol, switchgrass, cellulosic, net energy	Vogel, Kenneth P. 7102498441	Univ. Nebraska Lincoln USA	831

Prt, biomass pretreatments; Na, non available; Cits., number of citations received for each paper; *, female.

formation of formic acid and levulinic acid increased. The yield of ethanol decreased at approximately CS 3 while the volumetric productivity decreased at lower CS. They then assayed the effect of acetic acid, formic acid, levulinic acid, furfural, and 5-HMF on fermentability in model fermentations. Ethanol yield and volumetric productivity decreased with increasing concentrations of acetic acid, formic acid, and levulinic acid. However, furfural and 5-HMF decreased the volumetric productivity but did not influence the final yield of ethanol. The decrease in the volumetric productivity was more pronounced when 5-HMF was added to fermentation, and this compound was depleted at a lower rate than furfural. Further, the inhibition observed in hydrolysates produced in higher CS could not be fully explained by the effect of furfural, 5-HMF, acetic acid, formic acid, and levulinic acid.

3.3.3.2 Bioethanol Fuel Evaluation

Hill et al. (2006) developed four criteria to evaluate corn grain-based bioethanol and soybean-based biodiesel fuels through a life cycle accounting in a paper with 2,091 citations. They found that bioethanol yielded 25% more energy than the energy invested in its production, whereas biodiesel yielded 93% more. Compared with bioethanol, biodiesel released just 1.0%, 8.3%, and 13% of agricultural nitrogen, phosphorus, and pesticide pollutants, respectively, per net energy gain. Relative to the fossil fuels they displaced, greenhouse gas (GHG) emissions were reduced by about 12% by the production and combustion of bioethanol and 41% by biodiesel. Biodiesel also released less air pollutants per net energy gain than ethanol. These advantages of biodiesel over ethanol arise from lower agricultural inputs and more efficient conversion of feedstocks to fuel. However, neither biofuel could replace much crude oil without impacting food supplies as even dedicating all U.S. corn nor soybean production to biofuels would meet only 12% of gasoline demand and 6% of diesel demand. Until recent increases in petroleum prices, high production costs made biofuels unprofitable without subsidies. Biodiesel provided sufficient environmental advantages to merit subsidy. Thus, transportation biofuels such as synfuel hydrocarbons or lignocellulosic ethanol, if produced from low-input biomass grown on agriculturally marginal land or from waste biomass, could provide much greater supplies and environmental benefits than food-based biofuels. This study compared corn grain–based bioethanol fuels with crude oil–based diesel fuels and showed that they had a net energy gain and environmental benefits and were economically competitive but not producible in large quantities without reducing food supplies. For this reason, they recommended lignocellulosic bioethanol fuels from wood, grass, wastes, and agricultural residues.

Kim and Dale (2004) estimated the global annual potential bioethanol production from the starch feedstock residues of corn, barley, oat, rice, and wheat and the sugar feedstock residues of sorghum and sugarcane in a paper with 1,447 citations. They found that there was about 1.57 petagram (10^{15} g) of dry lignocellulosic wastes in the world that could potentially produce 491 gigaliter (GL) (10^9L) year^{-1} of bioethanol, about 16 times higher than the world's ethanol production as of 2004. Thus, the potential bioethanol production could replace 353 GL of gasoline (32% of the global gasoline consumption) when bioethanol was used in E85 biofuel for a midsize passenger vehicle. Furthermore, lignin-rich fermentation residue could potentially generate both 458 terawatt-hour (10^9 KWh) of bioelectricity (about 3.6% of the world's electricity production) and 2.6 exajoule (10^{18}J) of steam. Asia was the largest potential producer of bioethanol from lignocellulosic wastes of rice straw, wheat straw, and corn stover and could produce up to 291 GL/year of bioethanol. The next highest potential region was Europe (69.2 GL of bioethanol), in which most bioethanol came from wheat straw. Corn stover was the main feedstock in North America, from which about 38.4 GL/year of bioethanol could potentially be produced. Globally, rice straw could produce 205 GL of bioethanol. The next highest potential feedstock was wheat straw, which could produce 104 GL of bioethanol.

Hamelinck et al. (2005) evaluated the techno-economic performance of bioethanol production in the short-, middle-, and long-term using hydrolysis and fermentation technologies as well as developing technologies in a paper with 1,234 citations. They found that the available technology as of 2005, which was based on dilute acid hydrolysis, had about 35% efficiency (higher heating value, HHV) from biomass to ethanol. The overall efficiency, with electricity co-produced from the non-fermentable lignin, was about 60%. However, the improvements in pretreatment and advances in biotechnology, especially through process combinations, could bring the ethanol efficiency to 48% and the overall process efficiency to 68%. They estimated current investment costs at 2.1 k€/kWHHV (at 400 MWHHV input, i.e., a nominal 2,000 tonne dry/day input). However, future technology in a five times larger plant (2 GWHHV) could have investments of 900 k€/kWHHV. Thus, a combined effect of higher hydrolysis–fermentation efficiency, lower specific capital investments, increase of scale, and cheaper biomass feedstock costs (from 3 to 2 €/GJHHV) could bring the ethanol production cost from 22 €/GJHHV in the next 5 years to 13 €/GJ over the 10- to 15-year time scale and down to 8.7 €/GJ in 20 or more years.

Pimentel and Patzek (2005) compared the net energy gain of bioethanol fuels from corn, switchgrass, and wood and biodiesel fuels from soybean and sunflower in a paper with 1,020 citations. They found that energy outputs from ethanol produced from these feedstocks were each less than the respective fossil energy inputs. The same was true for producing biodiesel; however, the energy cost for producing soybean biodiesel was only slightly negative compared with ethanol production. Ethanol production using corn grain, switchgrass, and wood required 29%, 50%, and 57% of more fossil energy than the ethanol fuel produced, respectively. Similarly, biodiesel production using soybean and sunflower required 27% and 118% of more fossil energy than the biodiesel fuel produced, respectively. The energy yield from soy oil per hectare was far lower than the ethanol yield from corn.

Schmer et al. (2008) determined the net energy yield (NEY) and GHG emissions of lignocellulosic bioethanol fuels from switchgrass in a paper with 831 citations. They managed switchgrass in field trials of 3–9 ha on marginal cropland on ten farms across a wide precipitation and temperature gradient in the mid-continental United States to determine net energy and economic costs based on known farm inputs and harvested yields. They found that the annual biomass yields of these fields averaged 5.2–11.1 Mg/ha with a resultant average estimated NEY of 60 GJ/ha·year. Switchgrass produced 540% more renewable than nonrenewable energy consumed. Switchgrass monocultures managed for high yield produced 93% more biomass yield and an equivalent estimated NEY than previous estimates from human-made prairies that received low agricultural inputs. Estimated average GHG emissions from bioethanol fuels were 94% lower than estimated GHG from gasoline. They asserted that improved genetics and agronomics might further enhance the energy sustainability and biofuel yield of switchgrass.

3.3.3.3 Bioethanol Fuel Utilization

3.3.3.3.1 Ethanol Sensors

Wan et al. (2004) synthesized ZnO nanowire ethanol gas sensors with microelectromechanical system technology and evaluated their ethanol sensing-characteristics in a paper with 1,908 citations. They found that this sensor exhibited high sensitivity and fast response to ethanol gas at a working temperature of 300°C.

3.3.3.3.2 Bioethanol Fuel-based Biohydrogen Fuels

Murdoch et al. (2011) studied the photocatalytic biohydrogen fuel production from ethanol over Au/TiO_2 nanoparticles (NPs) in a paper with 1,009 citations. They showed that AuNPs in the size range of 3–30 nm on TiO_2 were very active in hydrogen production from ethanol. Further, AuNPs of similar size on anatase NPs delivered a rate two orders of magnitude higher than that recorded for Au on rutile nanoparticles. AuNP size did not affect the photoreaction rate over the 3–12 nm range. These catalysts resulted in a high biohydrogen yield.

Deluga et al. (2004) produced biohydrogen fuels by autothermal reforming of bioethanol fuels in a paper with 918 citations. They converted ethanol and ethanol–water mixtures directly into hydrogen with nearly 100% selectivity and more than 95% conversion by catalytic partial oxidation, with a residence time on rhodium (Rh)-ceria (Ce_2O) catalysts of less than 10 ms. They performed rapid vaporization and mixing with air with an automotive fuel injector at temperatures sufficiently low and times sufficiently fast such that homogeneous reactions producing carbon, acetaldehyde, ethylene, and total combustion products could be minimized.

3.4 DISCUSSION

3.4.1 Introduction

Crude oil–based gasoline fuels have been widely used in the transportation sector since the 1920s. However, there have been great public concerns over the adverse environmental and human impact of these fuels. Hence, biomass-based bioethanol fuels have increasingly been used in blending gasoline and petrodiesel fuels, in fuel cells, and in biochemical production in a biorefinery context.

However, it is necessary to pretreat the biomass to enhance the yield of bioethanol prior to the bioethanol fuel production from the feedstocks through hydrolysis and fermentation of the biomass.

Research in the field of bioethanol fuels has intensified in this context in the key research fronts of the production of bioethanol fuels, and to a lesser extent, utilization and evaluation of bioethanol fuels. For the first research front, pretreatment and hydrolysis of feedstocks, fermentation of hydrolysates, production of bioethanol fuels in general, and to a lesser extent separation and distillation of bioethanol fuels are the key research areas while for the second research front, utilization of bioethanol fuels in fuel cells and transport engines, bioethanol-based biohydrogen fuels, bioethanol sensors, and bioethanol-based biochemicals are the key research areas. Research in this field has also intensified for feedstocks of lignocellulosic biomass at large, bioethanol fuels for their utilization, cellulose, starch feedstock residues, wood biomass, and to a lesser extent industrial waste, lignin, grass, hydrolysates, biomass, sugar feedstocks, sugar feedstock residues, algal biomass, lignocellulosic wastes, starch feedstocks, urban wastes, forestry wastes, food wastes, syngas, and plants.

However, it is essential to develop efficient incentive structures for the primary stakeholders to enhance research in this field. Although there have been a limited number of review papers for this field, there has been no review of the most-cited 25 articles in this field. Thus, this chapter presents a review of the most-cited 25 articles on bioethanol fuels. Then, it discusses the key findings of these highly influential papers and comments on future research priorities in this field.

As a first step for the search of the relevant literature, keywords were selected using the most-cited first 300 population papers for each research front. The selected keyword list was then optimized to obtain a representative sample of papers for each research field. These keyword lists were then integrated to obtain the keyword list for this research field (Konur, 2023a–c).

As a second step, a sample dataset was used for this study. The first 25 articles with at least 726 citations each were selected for the review study. Key findings from each paper were taken from the abstracts of these papers and were discussed. Additionally, a number of brief conclusions were drawn and a number of relevant recommendations were made to enhance the future research landscape.

Information about research fronts for the sample papers in bioethanol fuels is given in Table 3.4. As this table shows, the most prolific feedstock for this field is the wood biomass such as poplar and pine with 44% of HCPs, followed by cellulose, lignin, and grass biomass with 16% of these HCPs each. Other prolific feedstocks are the lignocellulosic biomass at large, starch feedstock residues, and starch feedstocks with 8% of HCPs each while biomass at large and sugar feedstock residues have 4% of HCPs each.

On the other hand, it is notable that there are no HCPs for industrial waste, hydrolysates, sugar feedstocks, algal biomass, lignocellulosic wastes, urban wastes, forestry wastes, hemicellulose, xylan, food wastes, cellobiose, syngas, and plants. Thus, the first ten feedstocks dominate the HCPs in this field where the only major waste biomass is starch feedstock residues such as corn stover or wheat straw. On the other hand, cellulose and lignin are the key constituents of the lignocellulosic

TABLE 3.4
The Most Prolific Research Fronts for Bioethanol Fuels

No.	Research Fronts	N Paper % Review	N Paper 2 % Sample	Surplus %
1	Wood biomass	44	9.0	35.0
2	Cellulose	16	15.1	0.9
3	Lignin	16	6.2	9.8
3	Grass	16	3.1	12.9
5	Bioethanol fuels	12	19.6	–7.6
6	**Minor Feedstocks**	32	39.2	–7.2.
	Lignocellulosic biomass at large	8	23.0	–15
	Starch feedstock residues	8	9.7	–1.7
	Starch feedstocks	8	1.7	6.3
	Biomass at large	4	2.6	1.4
	Sugar feedstock residues	4	2.2	1.8
7	**Missing Feedstocks**	0	20.3	–20.3
	Industrial waste	0	6.5	–6.5
	Hydrolysates	0	2.8	–2.8
	Sugar feedstocks	0	2.2	–2.2
	Algal biomass	0	1.9	–1.9
	Lignocellulosic wastes	0	1.8	–1.8
	Urban wastes	0	1.0	–1.0
	Forestry wastes	0	0.8	–0.8
	Hemicellulose	0	0.8	–0.8
	Xylan	0	0.8	–0.8
	Food wastes	0	0.6	–0.6
	Cellobiose	0	0.4	–0.4
	Syngas	0	0.4	–0.4
	Plants	0	0.3	–0.3
	Sample size	25	775	

N Paper (%) review: the number of papers in a sample of 25 reviewed papers. N paper (%) sample: the number of papers in the population sample of 775 papers.

biomass. Further bioethanol fuels are the feedstocks for the production of biohydrogen fuels and biochemical and for the utilization in transport engines and fuel cells.

It is also notable that research on food-grade sugar and starch bioethanol fuels is not substantial for all the samples studied, reflecting the adverse effect of undermining food security (Bentivoglio et al., 2016; Elobeid and Hart, 2007.

Further, the most influential research fronts are wood biomass with a 35% surplus, followed by grass, lignin, and starch feedstocks with 13%, 10%, and 6% surplus, respectively. Further, the lignocellulosic biomass at large is the least influential feedstock with 15% deficit, followed by bioethanol fuels and industrial waste with 8% and 7% deficit each. It is also notable that a significant part of papers for lignocellulosic biomass-based bioethanol fuels are reviews and short surveys.

Information about thematic research fronts for the sample papers in bioethanol fuels is given in Table 3.5. This table shows that the most prolific research fronts for this field are the production of bioethanol fuels with 68% of HCPs. Other prolific research fronts are the evaluation and utilization of bioethanol evaluation with 20% and 12% of HCPs, respectively.

Further, the most prolific research fronts for the first group are pretreatments and hydrolysis of feedstocks with 64% and 36% of HCPs, respectively. Other research fronts are bioethanol production in general, hydrolysate fermentation, and bioethanol fuel distillation with 12%, 4%, and 0% of HCPs, respectively. For the third group, the only prolific research fronts are bioethanol-based biohydrogen fuels and bioethanol sensors with 8% and 4% of HCPs, respectively, while there are no HCPs for the utilization of bioethanol fuels in fuel cells and transport engines and production of biochemicals from bioethanol fuels.

On the other hand, biofuel evaluation is the most influential research field with a 14% surplus, followed by biomass pretreatments and hydrolysis with a 6% surplus each, while hydrolysate fermentation is the least influential research front with a 14% deficit, followed by bioethanol production, bioethanol fuel utilization, and bioethanol fuels in fuel cells with 10%, 8%, and 7% deficit, respectively.

3.4.2 Feedstock Pretreatments

Brief information about the eight most-cited papers on pretreatments of feedstocks with at least 751 citations each is given below (Table 3.1). On the other hand, it is notable that as Table 3.5 shows, 64% of these HCPs are related to pretreatments of feedstocks.

These findings show that both pretreatments and hydrolysis of feedstocks are the fundamental processes for bioethanol production from biomass. These narrated studies highlight the importance of pretreatment and hydrolysis processes for the production of bioethanol fuels from biomass with a high ethanol yield. These pretreatments, primarily enzymatic and chemical pretreatments, fractionate the feedstock and enhance the enzymatic digestibility of the biomass.

Bourbonnais and Paice (1990) oxidized non-phenolic substrates for lignin biodegradation in a paper and found that laccases *T. versicolor* could also oxidize non-phenolic lignin model compounds. Further, Floudas et al. (2012) mapped the detailed evolution of wood-degrading enzymes and noted that wood was highly resistant to decay, largely due to the presence of lignin, and the only organisms capable of substantial lignin decay were white-rot fungi in the agaricomycetes.

Tien and Kirk (1983) determined the lignin-degrading enzyme from the wood-decomposing basidiomycete *P. chrysasporium* Burds and found that the extracellular fluid of ligninolytic cultures of this basidiomycete contained an enzyme that degraded lignin substructure model compounds as well as spruce and birch lignins. Further, Schwanninger et al. (2004) studied the effects of short-time vibratory ball milling on the shape of FTIR spectra of wood and cellulose and it showed that this milling had a strong influence on the shape of FTIR spectra of wood and cellulose.

Martinez et al. (2008) performed the genome sequencing and analysis of the biomass-degrading fungus *T. reesei* and found that its genome encoded fewer cellulases and hemicellulases than any other sequenced fungus being able to hydrolyze plant cell wall polysaccharides. Further, Pandey and Pitman (2003) performed FTIR studies of changes in wood chemistry following decay by brown-rot and white-rot fungi and found that in wood decayed by *C. puteana,* there was a progressive increase in lignin content relative to carbohydrates.

Sun et al. (2009) dissolved and partially delignified yellow pine and red oak in $[C_2mim]OAc$ after mild grinding and it showed that this IL was a better solvent for wood than 1 $[C_4mim]Cl$. Further, Cai and Zhang (2005) dissolved cellulose in LiOH/urea and NaOH/urea aqueous solutions and found that cellulose could be dissolved, respectively, in 7% NaOH/12% urea and 4.2% LiOH/12% urea aqueous solutions, whereas all of them could not be dissolved in a KOH/urea aqueous solution.

3.4.3 Hydrolysis of Feedstocks

There are eight HCPs for hydrolysis of feedstocks with at least 726 citations (Table 3.2). On the other hand, it is notable that as Table 3.5 shows, 36% of these HCPs are related to hydrolysis of feedstocks.

TABLE 3.5
The Most Prolific Thematic Research Fronts for Bioethanol Fuels

No.	Research Fronts	N Paper % Review	N Paper % Sample	Surplus %
1	**Bioethanol Fuel Production**	**68**	**74.1**	**–6.1**
	Biomass pretreatments	64	58.5	5.5
	Biomass hydrolysis	36	29.8	6.2
	Bioethanol production	12	21.5	–9.5
	Hydrolysate fermentation	4	17.9	–13.9
	Bioethanol fuel distillation	0	0.8	–0.8
2	Bioethanol Fuel Evaluation	20	6.2	13.8
3	**Bioethanol Fuel Utilization**	**12**	**19.7**	**–7.7**
	Bioethanol-based biohydrogen fuels	8	4.3	3.7
	Bioethanol sensors	4	2.2	1.8
	Bioethanol fuels in fuel cells	0	6.8	–6.8
	Bioethanol fuels in engines	0	6.1	–6.1
	Bioethanol-based biochemical	0	0.4	–0.4

N Paper (%) review: the number of papers in a sample of 25 reviewed papers. N paper (%) sample: the number of papers in the population sample of 775 papers.

These findings show that both pretreatments and hydrolysis of feedstocks are the fundamental processes for bioethanol production from the feedstock. These narrated studies highlight the importance of pretreatment and hydrolysis processes for the production of bioethanol fuels from the feedstock with a high ethanol yield. These pretreatments, primarily enzymatic and chemical pretreatments, fractionate the feedstock and enhance the enzymatic digestibility of the biomass.

Park et al. (2010) compared four different techniques incorporating XRD and solid-state ^{13}C NMR for the measurement of CI using eight different cellulose preparations and found that the simplest method which involved measurement of just two heights in X-ray diffractogram produced significantly higher crystallinity values than other methods. Further, Chen and Dixon (2007) showed that lignin modification improved fermentable sugar yields for biofuel production.

Suganuma et al. (2008) hydrolyzed cellulose by amorphous carbon-bearing SO_3H, COOH, and OH groups and found that crystalline pure cellulose was not hydrolyzed by conventional strong solid Brønsted acid catalysts while amorphous carbon-bearing solid catalysts functioned as an efficient catalyst for the reaction. Further, Li et al. (2010) compared the efficiency of dilute acid hydrolysis and dissolution in an IL with switchgrass and observed that switchgrass exhibited reduced cellulose crystallinity, increased surface area, and decreased lignin content compared to dilute acid pretreatment.

Kilpelainen et al. (2007) dissolved wood in ILs in a paper with 832 citations. They found that [Bmim]Cl and [Amim]Cl had good solvating power for Norway spruce sawdust and Norway spruce and Southern pine thermomechanical pulp fibers as these ILs provided solutions that permitted complete acetylation of the wood. Lee et al. (2009) used [Emim]CH_3COO to extract lignin from wood flour and observed that the cellulose in the pretreated wood flour became far less crystalline without undergoing solubilization.

Eriksson et al. (2002) explored the mechanism of the surfactant effect in enzymatic hydrolysis of lignocellulose and found that nonionic surfactants were the most effective. Further, Kumar et al. (2009) pretreated corn stover and poplar solids using leading pretreatment technologies and found that lime-pretreated corn stover solids and flowthrough-pretreated poplar solids had the highest cellulase adsorption capacity, while dilute acid-pretreated corn stover solids and controlled pH-pretreated poplar solids had the least.

3.4.4 Bioethanol Production, Evaluation, and Utilization

There are one, five, and three HCPs for the production, evaluation, and utilization of bioethanol fuels with at least 831 citations each (Table 3.3). It is also notable that as Table 3.5 shows, 12%, 8%, and 4% of these HCPs are related to the production of bioethanol fuels, bioethanol fuel-based biohydrogen fuels, and bioethanol sensors, respectively. Further, there are no HCPs for the utilization of bioethanol fuels in fuel cells and transport engines and bioethanol-based biochemical. These narrated studies highlight the importance of studies on the evaluation and utilization of bioethanol fuels.

3.4.4.1 Bioethanol Production

Larsson et al. (1999) studied the effect of combined severity (CS) of dilute sulfuric acid hydrolysis of spruce on sugar yield and the fermentability of the hydrolysate by *S. cerevisiae,* and when the CS of hydrolysis conditions increased, they observed that the yield of fermentable sugars increased to a maximum between CS 2.0 and 2.7 for mannose and 3.0 and 3.4 for glucose, above which it decreased.

3.4.4.2 Bioethanol Fuel Evaluation

Hill et al. (2006) developed four criteria to evaluate corn grain–based bioethanol and soybean-based biodiesel fuels through a life-cycle accounting and found that bioethanol yielded 25% more energy than the energy invested in its production, whereas biodiesel yielded 93% more. Further, Kim and Dale (2004) estimated the global annual potential bioethanol production from starch feedstock residues and sugar feedstock residues and found that there were about 1.57 petagram (10^{15} g) of dry lignocellulosic wastes in the world that could potentially produce 491 gigaliter (GL) (10^9 L) year^{-1} of bioethanol, about 16 times higher than the world's ethanol production as of 2004.

Hamelinck et al. (2005) evaluated the techno-economic performance of bioethanol production in the short-, middle-, and long-term using hydrolysis and fermentation technologies as well as developing technologies and found that the available technology as of 2005 had about 35% efficiency (HHV) from biomass to ethanol. Further, Pimentel and Patzek (2005) compared the net energy gain of bioethanol fuels from corn, switchgrass, and wood and biodiesel fuels from soybean and sunflower and found that energy outputs from ethanol producing these biomasses were each less than the respective fossil energy inputs. Finally, Schmer et al. (2008) determined the NEY and GHG emissions of lignocellulosic bioethanol fuels from switchgrass and found that the annual biomass yields of these fields averaged 5.2–11.1 Mg/ha with a resulting average estimated NEY of 60 GJ/ha·year.

3.4.4.3 Bioethanol Fuel Utilization

3.4.4.3.1 Ethanol Sensors

Wan et al. (2004) synthesized ZnO nanowire ethanol gas sensors with microelectromechanical system technology and evaluated their ethanol-sensing characteristics and found that this sensor exhibited high sensitivity and fast response to ethanol gas at a working temperature of 300°C.

3.4.4.3.2 Bioethanol Fuel-based Biohydrogen Fuels

Murdoch et al. (2011) studied the photocatalytic biohydrogen fuel production from ethanol over Au/TiO_2 NPs and it showed that AuNPs in the size range of 3–30 nm on TiO_2 were very active in hydrogen production from ethanol. Further, Deluga et al. (2004) produced biohydrogen fuels by autothermal reforming of bioethanol fuels and converted ethanol and ethanol–water mixtures directly into hydrogen with nearly 100% selectivity and more than 95% conversion by catalytic partial oxidation.

3.5 CONCLUSION AND FUTURE RESEARCH

Brief information about key research fronts covered by the 25 most-cited papers with at least 726 citations each is given under three primary headings: production, evaluation, and utilization of bioethanol fuels.

The usual characteristics of these HCPs for the production of bioethanol fuels are that pretreatments and hydrolysis of feedstocks and fermentation of resultant hydrolysates are the primary processes for bioethanol fuel production from various feedstocks to improve ethanol yield.

The key findings on these research fronts should be read in light of the increasing public concerns about climate change, GHG emissions, and global warming as these concerns have been certainly behind the boom in the research on bioethanol fuels as an alternative to crude oil–based gasoline and diesel fuels in the last decades. It is also a sustainable alternative to food crop–based bioethanol fuels such as corn grain–based bioethanol fuels. Recent supply shocks caused by COVID-19 pandemic and the Russian invasion of Ukraine also highlight the importance of the production and utilization of bioethanol fuels as an alternative to crude oil–based gasoline and diesel fuels.

As Table 3.4 shows, the most prolific research fronts for this field are wood biomass, cellulose, lignin, and grass biomass. Other prolific feedstocks are the lignocellulosic biomass at large, starch feedstock residues, starch feedstocks, and biomass and sugar feedstock residues. On the other hand, it is notable that there are no HCPs for industrial wastes, hydrolysates, sugar feedstocks, algal biomass, lignocellulosic wastes, urban wastes, forestry wastes, hemicellulose, xylan, food wastes, cellobiose, syngas, and plants.

Thus, the first ten feedstocks dominate the HCPs in this field where the only major waste biomass is starch feedstock residue such as corn stover or wheat straw. On the other hand, cellulose and lignin are the key constituents of the lignocellulosic biomass. Further, bioethanol fuels are feedstocks for the production of biohydrogen fuels and biochemicals and for their utilization in transport engines and fuel cells. It is also notable that research on sugar and starch bioethanol fuels is not substantial for all the samples studied, reflecting the adverse effect of undermining food security.

Further, the most influential research fronts are biomass at large and to a lesser degree grass, lignin, and starch feedstocks while the lignocellulosic biomass at large, bioethanol fuels, and industrial waste are the least influential feedstocks.

As Table 3.5 shows, the most prolific thematic research fronts for this field are the production of bioethanol fuels, and to a lesser degree evaluation and utilization of bioethanol evaluation. Further, the most prolific research fronts for the first group are pretreatments, hydrolysis of feedstocks, and to a lesser degree bioethanol production at large, hydrolysate fermentation, and bioethanol fuel distillation. For the third group, the only prolific research fronts are bioethanol-based biohydrogen fuels and bioethanol sensors while there are no HCPs for the utilization of bioethanol fuels in fuel cells and transport engines and production of biochemicals from bioethanol fuels.

On the other hand, bioethanol fuel evaluation is the most influent research field followed by biomass pretreatments and hydrolysis while hydrolysate fermentation is the least influential research front followed by bioethanol production, bioethanol fuel utilization, and bioethanol fuels in fuel cells.

These studies emphasize the importance of proper incentive structures for the efficient production of bioethanol fuels in light of North's institutional framework (North, 1991). In this context, the major producers and users of bioethanol fuels such as the United States and Brazil with vast forests and farmlands have developed strong incentive structures for efficient bioethanol fuels.

In light of recent supply shocks caused primarily by COVID-19 pandemic and the Russian invasion of Ukraine, it is expected that incentive structures such as public funding would be enhanced to increase the share of bioethanol fuels in the global fuel portfolio as a strong alternative to crude oil–based gasoline and diesel fuels. In this context, it is expected that the most prolific researchers, institutions, countries, funding bodies, and journals in this field would have a first-mover advantage to benefit from such potential incentives. This is especially true for the US stakeholders as the United States has become the global leader in both the production and utilization of second generation bioethanol fuels from the feedstocks.

It is expected that research in the future would focus more on the algal, wood, and grass biomass-based bioethanol as well as agricultural residues and waste biomass-based bioethanol fuels at the expense of starch- and sugar-based bioethanol fuels due to the large societal concerns about the food security in the future. Similarly, it is expected that research would focus more on pretreatments and hydrolysis of feedstocks for the production of bioethanol fuels and utilization and evaluation of bioethanol fuels in the future.

It is recommended that further review studies are performed for the primary research fronts of bioethanol fuels.

ACKNOWLEDGMENTS

The contribution of the highly cited researchers in the field of bioethanol fuels has been gratefully acknowledged.

REFERENCES

Alvira, P., E. Tomas-Pejo, M. Ballesteros and M. J. Negro. 2010. Pretreatment technologies for an efficient bioethanol production process based on enzymatic hydrolysis: A review. *Bioresource Technology* 101:4851–4861.

Angelici, C., B. M. Weckhuysen and P. C. A. Bruijnincx. 2013. Chemocatalytic conversion of ethanol into butadiene and other bulk chemicals. *ChemSusChem* 6:1595–1614.

Antolini, E. 2007. Catalysts for direct ethanol fuel cells. *Journal of Power Sources* 170:1–12.

Antolini, E. 2009. Palladium in fuel cell catalysis. *Energy and Environmental Science* 2:915–931.

Bai, F. W., W. A. Anderson and M. Moo-Young. 2008. Ethanol fermentation technologies from sugar and starch feedstocks. *Biotechnology Advances* 26:89–105.

Balat, M. 2011. Production of bioethanol from lignocellulosic materials via the biochemical pathway: A review. *Energy Conversion and Management* 52:858–875.

Bentivoglio, D., A. Finco and M. R. P. Bacchi. 2016. Interdependencies between biofuel, fuel and food prices: The case of the Brazilian ethanol market. *Energies* 9:464.

Binod, P., R. Sindhu and R. R. Singhania, et al. 2010. Bioethanol production from rice straw: An overview. *Bioresource Technology* 101:4767–4774.

Bothast, R. J., and M. A. Schlicher. 2005. Biotechnological processes for conversion of corn into ethanol. *Applied Microbiology and Biotechnology* 67:19–25.

Bourbonnais, R. and M. G. Paice. 1990. Oxidation of non-phenolic substrates. An expanded role for laccase in lignin biodegradation. *FEBS Letters* 267:99–102.

Burnham, J. F. 2006. Scopus database: A review. *Biomedical Digital Libraries* 3:1–8.

Cai, J. and L. Zhang. 2005. Rapid dissolution of cellulose in LiOH/urea and NaOH/urea aqueous solutions. *Macromolecular Bioscience* 5:539–548.

Canilha, L., A. K. Chandel and T. S. dos Santos Milessi, et al. 2012. Bioconversion of sugarcane biomass into ethanol: An overview about composition, pretreatment methods, detoxification of hydrolysates, enzymatic saccharification, and ethanol fermentation. *Journal of Biomedicine and Biotechnology* 2012:989572.

Cardona, C. A., J. A. Quintero and I. C. Paz. 2010. Production of bioethanol from sugarcane bagasse: Status and perspectives. *Bioresource Technology* 101:4754–4766.

Chen, F. and R. A. Dixon. 2007. Lignin modification improves fermentable sugar yields for biofuel production. *Nature Biotechnology* 25:759–761.

Deluga, G. A., J. R. Salge, L. D. Schmidt and X. E. Verykios. 2004. Renewable hydrogen from ethanol by autothermal reforming. *Science* 303:993–997.

Duff, S. J. B. and W. D. Murray. 1996. Bioconversion of forest products industry waste cellulosics to fuel ethanol: A review. *Bioresource Technology* 55:1–33.

Elobeid, A. and C. Hart. 2007. Ethanol expansion in the food versus fuel debate: How will developing countries fare? *Journal of Agricultural & Food Industrial Organization* 5:7.

Eriksson, T., J. Borjesson and F. Tjerneld. 2002. Mechanism of surfactant effect in enzymatic hydrolysis of lignocellulose. *Enzyme and Microbial Technology* 31:353–364.

Fernando, S., S. Adhikari, C. Chandrapal and M. Murali. 2006. Biorefineries: Current status, challenges, and future direction. *Energy & Fuels* 20:1727–1737.

Floudas, D., M. Binder and R. Riley, et al. 2012. The paleozoic origin of enzymatic lignin decomposition reconstructed from 31 fungal genomes. *Science* 336:1715–1719.

Galbe, M. and G. Zacchi. 2002. A review of the production of ethanol from softwood. *Applied Microbiology and Biotechnology* 59:618–628.

Guimaraes, P. M. R., J. A. Teixeira and L. Domingues. 2010. Fermentation of lactose to bio-ethanol by yeasts as part of integrated solutions for the valorisation of cheese whey. *Biotechnology Advances* 28:375–384.

Hahn-Hagerdal, B., M. Galbe, M. F. Gorwa-Grauslund, G. Liden and G. Zacchi. 2006. Bio-ethanol – The fuel of tomorrow from the residues of today. *Trends in Biotechnology* 24:549–556.

Hamelinck, C. N., G. van Hooijdonk and A. P. C. Faaij. 2005. Ethanol from lignocellulosic biomass: Techno-economic performance in short-, middle- and long-term. *Biomass and Bioenergy* 28:384–410.

Hansen, A. C., Q. Zhang and P. W. L. Lyne. 2005. Ethanol-diesel fuel blends – A review. *Bioresource Technology* 96:277–285.

Haryanto, A., S. Fernando, N. Murali and S. Adhikari. 2005. Current status of hydrogen production techniques by steam reforming of ethanol: A review. *Energy and Fuels* 19:2098–2106.

Hendriks, A. T. W. M. and G. Zeeman. 2009. Pretreatments to enhance the digestibility of lignocellulosic biomass. *Bioresource Technology* 100:10–18.

Henstra, A. M., J. Sipma, A. Rinzema and A. J. Stams. 2007. Microbiology of synthesis gas fermentation for biofuel production. *Current Opinion in Biotechnology* 18:200–206.

Hill, J., E. Nelson, D. Tilman, S. Polasky and D. Tiffany. 2006. Environmental, economic, and energetic costs and benefits of biodiesel and ethanol biofuels. *Proceedings of the National Academy of Sciences of the United States of America* 103:11206–11210.

Hill, J., S. Polasky and E. Nelson, et al. 2009. Climate change and health costs of air emissions from biofuels and gasoline. *Proceedings of the National Academy of Sciences of the United States of America* 106:2077–2082.

Ho, S. H., S. W. Huang and C. Y. Chen, et al. 2013. Bioethanol production using carbohydrate-rich microalgae biomass as feedstock. *Bioresource Technology* 135:191–198.

Hsieh, W. D., R. H. Chen, T. L. Wu and T. H. Lin. 2002. Engine performance and pollutant emission of an SI engine using ethanol-gasoline blended fuels. *Atmospheric Environment* 36:403–410.

Huang, H. J., S. Ramaswamy, U. W. Tschirner and B. V. Ramarao. 2008. A review of separation technologies in current and future biorefineries. *Separation and Purification Technology* 62:1–21.

John, R. P., G. S. Anisha, K. M. Nampoothiri and A. Pandey. 2011. Micro and macroalgal biomass: A renewable source for bioethanol. *Bioresource Technology* 102:186–193.

Jonsson, L. J. and C. Martin. 2016. Pretreatment of lignocellulose: Formation of inhibitory by-products and strategies for minimizing their effects. *Bioresource Technology* 199:103–112.

Keshwani, D. R. and J. J. Cheng. 2009. Switchgrass for bioethanol and other value-added applications: A review. *Bioresource Technology* 100:1515–1523.

Kilpelainen, I., X. Xie and A. King, et al. 2007. Dissolution of wood in ionic liquids. *Journal of Agricultural and Food Chemistry* 55:9142–9148.

Kim, S. and B. E. Dale. 2004. Global potential bioethanol production from wasted crops and crop residues. *Biomass and Bioenergy* 26:361–375.

Kirk, T. K. and R. L. Farrell. 1987. Enzymatic "combustion": The microbial degradation of lignin. *Annual Review of Microbiology* 41:465–505.

Konur, O. 2000. Creating enforceable civil rights for disabled students in higher education: An institutional theory perspective. *Disability & Society* 15:1041–1063.

Konur, O. 2002a. Access to nursing education by disabled students: Rights and duties of nursing programs. *Nurse Education Today* 22:364–374.

Konur, O. 2002b. Assessment of disabled students in higher education: Current public policy issues. *Assessment and Evaluation in Higher Education* 27:131–152.

Konur, O. 2002c. Access to employment by disabled people in the UK: Is the Disability Discrimination Act working? *International Journal of Discrimination and the Law* 5:247–279.

Konur, O. 2006a. Participation of children with dyslexia in compulsory education: Current public policy issues. *Dyslexia* 12:51–67.

Konur, O. 2006b. Teaching disabled students in Higher Education. *Teaching in Higher Education* 11:351–363.

Konur, O. 2007a. A judicial outcome analysis of the Disability Discrimination Act: A windfall for the employers? *Disability & Society* 22:187–204.

Konur, O. 2007b. Computer-assisted teaching and assessment of disabled students in higher education: The interface between academic standards and disability rights. *Journal of Computer Assisted Learning* 23:207–219.

Konur, O. 2012. The evaluation of the research on the bioethanol: A scientometric approach. *Energy Education Science and Technology Part A: Energy Science and Research* 28:1051–1064.

Konur, O. 2015. Current state of research on algal bioethanol. In *Marine Bioenergy: Trends and Developments*, Eds. S. K. Kim and C. G. Lee, pp. 217–244. Boca Raton, FL: CRC Press.

Konur, O. 2020. The scientometric analysis of the research on the bioethanol production from green macroalgae. In *Handbook of Algal Science, Technology and Medicine*, Ed. O. Konur, pp. 385–401. London: Academic Press.

Konur, O. 2023a. Production processes for bioethanol fuels: Scientometric study. In *Bioethanol Fuel Production Processes. I: Biomass Pretreatments. Handbook of Bioethanol Fuels Volume 1*, Ed. O. Konur. Boca Raton, FL: CRC Press.

Konur, O. 2023b. Feedstock-based bioethanol fuels: Scientometric study. In *Feedstock-based Bioethanol Fuels. I. Non-Waste Feedstocks: Starch, Sugar, Grass, Wood, Cellulose, Algae, and Biosyngas-based Bioethanol Fuels. Handbook of Bioethanol Fuels Volume 3*, Ed. O. Konur. Boca Raton, FL: CRC Press.

Konur, O. 2023c. Bioethanol fuel evaluation and utilization: Scientometric study. In *Evaluation and Utilization of Bioethanol Fuels. I: Evaluation of Bioethanol Fuels, Transport Engines, and Bioethanol Sensors. Handbook of Bioethanol Fuels Volume 5*, Ed. O. Konur. Boca Raton, FL: CRC Press.

Kumar, R., G. Mago, V. Balan and C. E. Wyman. 2009. Physical and chemical characterizations of corn stover and poplar solids resulting from leading pretreatment technologies. *Bioresource Technology* 100:3948–3962.

Larsson, S., E. Palmqvist and B. Hahn-Hagerdal, et al. 1999. The generation of fermentation inhibitors during dilute acid hydrolysis of softwood. *Enzyme and Microbial Technology* 24:151–159.

Laser, M., D. Schulman and S. G. Allen, et al. 2002. A comparison of liquid hot water and steam pretreatments of sugar cane bagasse for bioconversion to ethanol. *Bioresource Technology* 81:33–44.

Lee, S. H., T. V. Doherty, R. J. Linhardt and J. S. Dordick. 2009. Ionic liquid-mediated selective extraction of lignin from wood leading to enhanced enzymatic cellulose hydrolysis. *Biotechnology and Bioengineering* 102:1368–1376.

Li, C., B. Knierim and C. Manisseri, et al. 2010. Comparison of dilute acid and ionic liquid pretreatment of switchgrass: Biomass recalcitrance, delignification and enzymatic saccharification. *Bioresource Technology* 101:4900–4906.

Limayem, A. and S. C. Ricke. 2012. Lignocellulosic biomass for bioethanol production: Current perspectives, potential issues and future prospects. *Progress in Energy and Combustion Science* 38:449–467.

Lin, Y. and S. Tanaka. 2006. Ethanol fermentation from biomass resources: Current state and prospects. *Applied Microbiology and Biotechnology* 69:627–642.

Ma, X., L. Sun and C. Song. 2002. A new approach to deep desulfurization of gasoline, diesel fuel and jet fuel by selective adsorption for ultra-clean fuels and for fuel cell applications. *Catalysis Today* 77:107–116.

Martinez, D., R. M. Berka and B. Henrissat, et al. 2008. Genome sequencing and analysis of the biomass-degrading fungus *Trichoderma reesei* (syn. *Hypocrea jecorina*). *Nature Biotechnology* 26:553–560.

Morschbacker, A. 2009. Bio-ethanol based ethylene. *Polymer Reviews* 49:79–84.

Mosier, N., C. Wyman and B. Dale, et al. 2005. Features of promising technologies for pretreatment of lignocellulosic biomass. *Bioresource Technology* 96:673–686.

Munasinghe, P. C. and S. K. Khanal. 2010. Biomass-derived syngas fermentation into biofuels: Opportunities and challenges. *Bioresource Technology* 101:5013–5022.

Murdoch, M., G. I. N. Waterhouse and M. A. Nadeem, et al. 2011. The effect of gold loading and particle size on photocatalytic hydrogen production from ethanol over Au/TiO_2 nanoparticles. *Nature Chemistry* 3:489–492.

Najafi, G., B. Ghobadian and T. Tavakoli, et al. 2009. Performance and exhaust emissions of a gasoline engine with ethanol blended gasoline fuels using artificial neural network. *Applied Energy* 86:630–639.

Newman, P. W. G. and J. R. Kenworthy. 1989. Gasoline consumption and cities: A comparison of U.S. cities with a global survey. *Journal of the American Planning Association* 55:24–37.

Nigam, J. N. 2002. Bioconversion of water-hyacinth (*Eichhornia crassipes*) hemicellulose acid hydrolysate to motor fuel ethanol by xylose-fermenting yeast. *Journal of Biotechnology* 97:107–116.

North, D. C. 1991. Institutions. *Journal of Economic Perspectives* 5:97–112.

Olsson, L. and B. Hahn-Hagerdal. 1996. Fermentation of lignocellulosic hydrolysates for ethanol production. *Enzyme and Microbial Technology* 18:312–331.

Palmqvist, E. and B. Hahn-Hagerdal. 2000a. Fermentation of lignocellulosic hydrolysates. I: Inhibition and detoxification. *Bioresource Technology* 74:17–24.

Palmqvist, E. and B. Hahn-Hagerdal. 2000b. Fermentation of lignocellulosic hydrolysates. II: Inhibitors and mechanisms of inhibition. *Bioresource Technology* 74:25–33.

Pandey, K. K. and A. J. Pitman. 2003. FTIR studies of the changes in wood chemistry following decay by brown-rot and white-rot fungi. *International Biodeterioration and Biodegradation* 52:151–160.

Park, S., J. O. Baker, M. E. Himmel, P. A. Parilla and D. K. Johnson. 2010. Cellulose crystallinity index: Measurement techniques and their impact on interpreting cellulase performance. *Biotechnology for Biofuels* 3:10.

Pimentel, D. and T. W. Patzek. 2005. Ethanol production using corn, switchgrass, and wood; Biodiesel production using soybean and sunflower. *Natural Resources Research* 14:65–76.

Pinkert, A., K. N. Marsh, S. Pang and M. P. Staiger. 2009. Ionic liquids and their interaction with cellulose. *Chemical Reviews* 109:6712–6728.

Prasad, S., A. Singh and H. C. Joshi. 2007. Ethanol as an alternative fuel from agricultural, industrial and urban residues. *Resources, Conservation and Recycling* 50:1–39.

Ravindran, R., and A. K. Jaiswal. 2016. A comprehensive review on pre-treatment strategy for lignocellulosic food industry waste: Challenges and opportunities. *Bioresource Technology* 199:92–102.

Saini, J. K., R. Saini and L. Tewari. 2015. Lignocellulosic agriculture wastes as biomass feedstocks for second-generation bioethanol production: Concepts and recent developments. *3 Biotech* 5:337–353.

Sanchez, C. 2009. Lignocellulosic residues: Biodegradation and bioconversion by fungi. *Biotechnology Advances* 27:185–194.

Sanchez, O. J. and C. A. Cardona. 2008. Trends in biotechnological production of fuel ethanol from different feedstocks. *Bioresource Technology* 99:5270–5295.

Sano, T., H. Yanagishita, Y. Kiyozumi, F. Mizukami and K. Haraya. 1994. Separation of ethanol/water mixture by silicalite membrane on pervaporation. *Journal of Membrane Science* 95:221–228.

Schmer, M. R., K. P. Vogel, R. B. Mitchell and R. K. Perrin. 2008. Net energy of cellulosic ethanol from switchgrass. *Proceedings of the National Academy of Sciences of the United States of America* 105:464–469.

Schwanninger, M., J. C. Rodrigues, H. Pereira, and B. Hinterstoisser. 2004. Effects of short-time vibratory ball milling on the shape of FT-IR spectra of wood and cellulose. *Vibrational Spectroscopy* 36:23–40.

Suganuma, S., K. Nakajima and M. Kitano, et al. 2008. Hydrolysis of cellulose by amorphous carbon bearing SO_3H, COOH, and OH groups. *Journal of the American Chemical Society* 130:12787–12793.

Sukumaran, R. K., R. R. Singhania, G. M. Mathew and A. Pandey. 2009. Cellulase production using biomass feed stock and its application in lignocellulose saccharification for bio-ethanol production. *Renewable Energy* 34:421–424.

Sun, N., M. Rahman and Y. Qin, et al. 2009. Complete dissolution and partial delignification of wood in the ionic liquid 1-ethyl-3-methylimidazolium acetate. *Green Chemistry* 11:646–655.

Sun, Y. and J. Cheng. 2002. Hydrolysis of lignocellulosic materials for ethanol production: A review. *Bioresource Technology* 83:1–11.

Taherzadeh, M. J. and K. Karimi. 2007. Enzyme-based hydrolysis processes for ethanol from lignocellulosic materials: A review. *Bioresources* 2:707–738.

Taherzadeh, M. J. and K. Karimi. 2008. Pretreatment of lignocellulosic wastes to improve ethanol and biogas production: A review. *International Journal of Molecular Sciences* 9:1621–1651.

Talebnia, F., D. Karakashev and I. Angelidaki. 2010. Production of bioethanol from wheat straw: An overview on pretreatment, hydrolysis and fermentation. *Bioresource Technology* 101:4744–4753.

Tien, M. and T. K. Kirk. 1983. Lignin-degrading enzyme from the hymenomycete *Phanerochaete chrysasporium* burds. *Science* 221:661–663.

Vane, L. M. 2005. A review of pervaporation for product recovery from biomass fermentation processes. *Journal of Chemical Technology and Biotechnology* 80:603–629.

Wan, Q., Q. H. Li and Y. J. Chen, et al. 2004. Fabrication and ethanol sensing characteristics of ZnO nanowire gas sensors. *Applied Physics Letters* 84:3654–3656.

Zabed, H., J. N. Sahu, A. Suely, A. N. Boyce and G. Faruq. 2017. Bioethanol production from renewable sources: Current perspectives and technological progress. *Renewable and Sustainable Energy Reviews* 71: 475–501.

Zhang, Y. H. P. and L. R. Lynd. 2004. Toward an aggregated understanding of enzymatic hydrolysis of cellulose: Noncomplexed cellulase systems. *Biotechnology and Bioengineering* 88:797–824.

Zhu, J. Y. and X. J. Pan. 2010. Woody biomass pretreatment for cellulosic ethanol production: Technology and energy consumption evaluation. *Bioresource Technology* 101:4992–5002.

Part 2

Biomass Pretreatments for Bioethanol Production

4 Biomass Pretreatments
Scientometric Study

Ozcan Konur
(Formerly) Ankara Yildirim Beyazit University

4.1 INTRODUCTION

The crude oil-based gasoline fuels (Ma et al., 2002; Newman and Kenworthy, 1989) have been widely used in the transportation sector since the 1920s. However, there have been great public concerns over the adverse environmental and human impact of these fuels (Hill et al., 2006, 2009). Hence, biomass-based bioethanol fuels (Hill et al., 2006; Konur, 2012e, 2015, 2019, 2020a; Sun and Cheng, 2002) have increasingly been used in blending gasoline fuels (Hsieh et al., 2002; Najafi et al., 2009), in the fuel cells (Antolini, 2007, 2009), and in the biochemical production (Angelici et al., 2013; Morschbacker, 2009) in a biorefinery context (Fernando et al., 2006; Huang et al., 2008). Bioethanol fuels also play a critical role in maintaining the energy security (Kruyt et al., 2009; Winzer, 2012) in the supply shocks (Kilian, 2008, 2009) related to oil price shocks (Hamilton, 2003, 2009), COVID-19 pandemics (Fauci et al., 2020; Li et al., 2020) or wars (Hamilton, 2003; Jones, 2012) in the aftermath of the Russian invasion of Ukraine starting in February 2022 (Reeves, 2014) as well as the COVID-19 pandemics.

The research in the fields of biomass pretreatments (Hendriks and Zeeman, 2009; Mosier et al., 2005) has also intensified in recent years. Biological (Himmel et al., 2007; Zhang and Lynd, 2004), chemical (Brandt et al., 2013; Zhu et al., 2006), mechanical (Hu and Wen, 2008; Mani et al., 2004), and hydrothermal (Chandra et al., 2007; Pu et al., 2013) pretreatments of the biomass have been widely researched to increase the sugar and bioethanol yield in this context in recent years.

However, it is essential to develop efficient incentive structures (North, 1991) for the primary stakeholders to enhance the research in this field (Konur, 2000, 2002a–c, 2006a, b, 2007a, b). The scientometric analysis has been used in this context to inform the primary stakeholders about the current state of the research in a selected research field (Garfield, 1955; Konur, 2011, 2012a–i, 2015, 2018b, 2019, 2020a).

As there have been no scientometric studies on the biomass pretreatments as of May 2022, this book chapter presents a scientometric study of the research in biomass pretreatments. It examines the scientometric characteristics of both the sample and population data presenting scientometric characteristics of both datasets in the order of documents, authors, publication years, institutions, funding bodies, source titles, countries, Scopus subject categories, keywords, and research fronts.

4.2 MATERIALS AND METHODS

The search for this study was carried out using Scopus database (Burnham, 2006) in May 2022.

As a first step for the search of the relevant literature, the keywords were selected using the first most-cited 200 papers for each of the primary research fronts. These keyword lists were then integrated to obtain a single keyword list for this research field. The selected keyword list was optimized to obtain a representative sample of papers for the searched research field. This keyword list focusing on the biological, chemical, mechanical, and hydrothermal pretreatments for 20,251 papers was used for this book chapter. Additionally, this keyword list was extended to include the

DOI: 10.1201/9781003226536-6

papers for the biomass pretreatments at the macroscale without focusing on single type of pretreatments, mostly reviews. This extended keyword list for 23,407 papers was provided in Appendix for future replication studies.

As a second step, two sets of data were used for this study. First, a population sample of around 20,251 papers was used to examine the scientometric characteristics of the population data. Secondly, a sample of 203 most-cited papers, corresponding to nearly 1% of the population papers, was used to examine the scientometric characteristics of these citation classics.

The scientometric characteristics of both sample and population datasets were presented in the order of documents, authors, publication years, institutions, funding bodies, source titles, countries, Scopus subject categories, keywords, and research fronts.

Lastly, the key scientometric findings for both datasets were discussed to highlight the research landscape for the biomass pretreatments. Additionally, a number of brief conclusions were drawn, and a number of relevant recommendations were made to enhance the future research landscape.

4.3 RESULTS

4.3.1 The Most Prolific Documents in the Biomass Pretreatments

The information on the types of documents for both datasets is given in Table 4.1. The articles and conference papers together with notes dominate both the sample (75%) and population (95%) papers as they are under-represented in the sample papers by 20%. Further, review papers and short surveys have a surplus as they are over-represented in the sample papers by 23% as they constitute 25% and 3% of the sample and population papers, respectively.

It is further notable that 97% of the population papers were published in journals while 2% and 1% of them were published in book series and books, respectively. On the contrary, 98% of the sample papers were published in the journals.

4.3.2 The Most Prolific Authors in the Biomass Pretreatments

The information about the most prolific 18 authors with at least 2.0% of sample papers each is given in Table 4.2.

TABLE 4.1
Documents in the Biomass Pretreatments

Documents	Sample Dataset (%)	Population Dataset (%)	Surplus (%)
Article	73.4	91.8	−18.4
Review	22.7	2.9	19.8
Conference paper	2.0	3.2	−1.2
Short survey	2.0	0.1	1.9
Book chapter	0.0	1.3	−1.3
Note	0.0	0.3	−0.3
Letter	0.0	0.2	−0.2
Book	0.0	0.1	−0.1
Editorial	0.0	0.1	−0.1
Sample size	203	20,251	

Sample dataset: the number of papers (%) in the set of 203 **HCPs**. Population dataset: the number of papers (%) in the set of the 20,251 population papers.

TABLE 4.2
Most Prolific Authors in the Biomass Pretreatments

No.	Author Name	Author Code	Sample Papers %	Population Papers %	Surplus	Institution	Country	HI	N	Res. Front
1	Saddler, John N.	7005297559	4.9	0.7	4.2	Univ. British Columbia	Canada	96	403	Biol.
2	Henrissat, Bernard	7005911606	3.4	0.1	3.3	Aix Marseille Univ.	France	141	591	Biol.
3	Simmons, Blake A.	7102183263	3.0	0.5	2.5	Lawrence Livermore Natl. Lab.	USA	74	434	Chem.
4	Wyman, Charles E	7004396809	3.0	0.4	2.6	Univ. Calif. Riverside	USA	80	286	Chem., biol., hydroth.
5	Grigoriev, Igor V.	25027225800	2.5	0.0	2.5	Lawrence Livermore Natl. Lab.	USA	114	339	Biol.
6	Cullen, Dan	7202109135	2.5	0.1	2.4	U.S: Forest Serv.	USA	46	119	Biol.
7	Dale, Bruce E.	7201511969	2.5	0.3	2.2	Michigan State Univ.	USA	90	446	Chem., biol.
8	Himmel, Michael E.	7007125552	2.5	0.2	2.3	Natl. Renew. Ener. Lab.	USA	73	421	Chem., biol.
9	Coutinho, Pedro M.	7006153340	2.5	0.0	2.5	Aix Marseille Univ.	France	71	145	Biol.
10	Martinez, Angel T.	55622506400	2.0	0.1	1.9	CSIC	Spain	79	343	Biol.
11	Brandt, Agnieszka*	35785816800	2.0	0.1	1.9	Imperial Coll.	UK	22	39	Chem.
12	Singh, Seema*	35264950300	2.0	0.3	1.7	Lawrence Livermore Natl. Lab.	USA	56	175	Chem.
13	Baker, Scott E.	35232609200	2.0	0.1	1.9	Pacific NW Natl. lab.	USA	48	145	Bio.
14	Martinez, Diego	7202958664	2.0	0.1	1.9	Massachusetts Inst. Technol.	USA	28	35	Biol.
15	Pandey, Ashok	7201771319	2.0	0.2	1.8	CSIR	India	88	817	Biol.
16	Holtzapple, Mark T.	7004167004	2.0	0.2	1.8	Texas A&M Univ.	USA	46	195	Chem., biol.
17	Welton, Tom	7003503272	2.0	0.1	1.9	Imperial coll.	UK	73	186	Chem.
18	Larrondo, Luis F.	6603826074	2.0	0.1	1.9	Pontificial Cath. Univ. Chile	Chile	28	75	Biol.

Author code: the unique code given by Scopus to the authors. Sample papers: the number of papers authored in the sample dataset. Population papers: the number of papers authored in the population dataset.
HI, H-index; N, number of papers published by each author; *, female; Chem., chemical pretreatments; Biol., biological pretreatments; Hydroth., hydrothermal pretreatments.

The most prolific author is John N. Saddler of University of British Columbia of Canada with 4.9% of the sample papers working primarily on the biological pretreatment of the biomass. Next, Bernard Henrissat, Blake A. Simmons, and Charles E. Wyman publish 3% of the sample papers each while Igor V. Grigoriev, Dan Cullen, Bruce E. Dale, Michael E. Himmel, and Pedro M. Coutinho publish 2.5% of the sample papers each.

The most influential author is John N. Saddler with 4.2% surplus, followed by Bernard Henrissat with 3.3% surplus. The other influential authors are Charles E. Wyman, Igor V. Grigoriev, Blake A. Simmons, and Pedro M. Coutinho with 2.5% surplus each.

The most prolific institution for the sample dataset is the Joint Bioenergy Institute of the Lawrence Livermore National Laboratory of the USA with three authors while the Aix Marseille University of France and Imperial College of the UK house two authors each. On the other hand, the most prolific country for the sample dataset is the USA with ten authors while France and the UK house two authors each.

The most prolific research front is the biological pretreatment of the biomass with 13 authors, closely followed by the chemical pretreatments with 8 authors. Additionally, one author studies primarily hydrothermal pretreatment of the biomass while there is no author studying the mechanical pretreatment of the biomass.

On the other hand, there is significant gender deficit (Beaudry and Lariviere, 2016) for the sample dataset as surprisingly only two of these top researchers are female.

Additionally, there are other authors with the relatively low citation impact and with 0.2%–0.7% of the population papers each: Runcang Sun, Feng Xu, Arthur J. Ragauskas, Qiang Yong, Guido Zacchi, Venkatesh Balan, Mercedes Ballesteros, Mats Galbe, Junyong Zhu, Zhenhong Yuan, Hasan Jameel, Juan J. Parajo, Yunqiao Pu, Wei Qi, Hongzhang Chen, Yongcan Ji, Herbert Sixta, Ignacio Ballesteros, Tae H. Kim, Guang Yang, Xinshu Zhuang, Verawat Champreda, Andre Ferraz, Xueqing Qiu, Qiang Yu, and Jinming Zhang.

4.3.3 The Most Prolific Research Output by Years in Biomass Pretreatments

Information about papers published between 1970 and 2022 is given in Figure 4.1. This figure clearly shows that the bulk of the research papers in the population dataset were published primarily in the 2010s with 54% of the population dataset. The publication rates for the 2020s, 2000s, 1990s, 1980s, and 1970s were 15%, 13%, 8%, 5%, and 2% respectively. Additionally, 1% of the population papers were published between 1892 and 1969.

Similarly, the bulk of the research papers in the sample dataset were published in the 2000s and 2010s with 58% and 24% of the sample dataset, respectively. The publication rates for the 1990s, 1980s, and 1970s were 13%, 4%, and 2% of the sample papers, respectively.

The most prolific publication year for the population dataset was 2021 with 8% while 72% of the population papers were published between 2012 and 2022. Similarly, 70% of the sample papers were published between 2001 and 2011 while the most prolific publication year was 2008 with 10.3% of the sample papers. The other prolific years were 2006 and 2010 with 9% of the sample papers each.

4.3.4 The Most Prolific Institutions in the Biomass Pretreatments

Information about the most prolific 21 institutions publishing papers on the biomass pretreatments with at least 2.5% of the sample papers each is given in Table 4.3.

The most prolific institution is the University of British Columbia with 5.9% of the sample papers, closely followed by the USDA Forest Service with 5.4% of the sample papers. The other prolific institutions are Dartmouth College, National Renewable Energy Laboratory (NREL), and Novozymes with 3.4%–4.4% of the sample papers each.

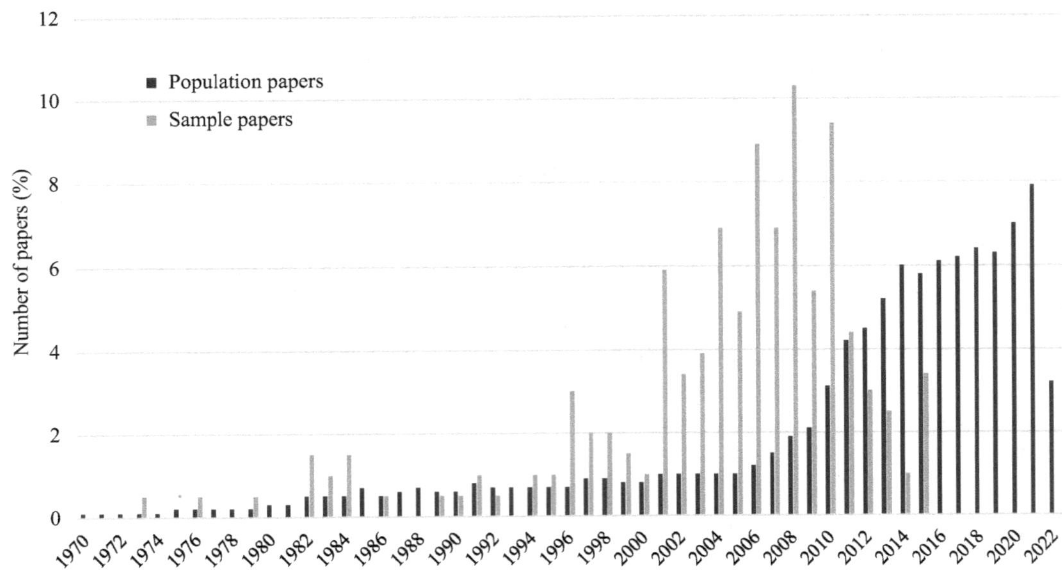

FIGURE 4.1 The research output by years regarding the biomass pretreatments.

TABLE 4.3
The Most Prolific Institutions in Biomass Pretreatments

No.	Institutions	Country	Sample Papers (%)	Population Papers (%)	Surplus (%)
1	Univ. British Columbia	Canada	5.9	1.1	4.8
2	USDA Forest Serv.	USA	5.4	1.2	4.2
3	Dartmouth Coll.	USA	4.4	0.3	4.1
4	Natl. Renew. Ener. Lab.	USA	3.9	1.0	2.9
5	Novozymes	USA	3.4	0.2	3.2
6	Univ. Wisconsin-Madison	USA	3.0	0.8	2.2
7	Michigan State Univ.	USA	3.0	0.8	2.2
8	Joint Bioener. Inst.	USA	3.0	0.5	2.5
9	Georgia Inst. Technol.	USA	3.0	0.4	2.6
10	Aix Marseille Univ.	France	3.0	0.4	2.6
11	Pacific NW Natl. Lab.	USA	3.0	0.2	2.8
12	Joint Genome Inst.	USA	3.0	0.2	2.8
13	North Carolina State Univ.	USA	2.5	1.1	1.4
14	Oak Ridge Natl. Lab.	USA	2.5	0.9	1.6
15	Lund Univ.	Sweden	2.5	0.8	1.7
16	CNRS	France	2.5	0.7	1.8
17	CSIC	Spain	2.5	0.7	1.8
18	USDA Agr. Res. Serv.	USA	2.5	0.7	1.8
19	VTT Tech. Res. Ctr.	Finland	2.5	0.7	1.8
20	Univ. Helsinki	Finland	2.5	0.6	1.9
21	Sandia Natl. Lab.	USA	2.5	0.5	2.0

The top country for these most prolific institutions is the USA with 14 institutions while France and Finland house 2 institutions each. In total, only six countries house these top institutions.

On the other hand, the institution with the most impact is the University of British Columbia with 4.8% surplus, closely followed by the USDA Forest Service and Dartmouth College with 4.2% and 4.1% surplus, respectively. The other prolific institutions are Novozymes, NREL, Joint Genome Institute, Pacific Northwest National Laboratory, Aix Marseille University, and Georgia Institute of Technology with 2.6%–3.2% surplus each. Similarly, the institution with the least impact is the North Carolina State University with 1.4% surplus.

Additionally, there are other institutions with the relatively low citation impact and with 0.6%–2.9% of the population papers each: Chinese Academy of Sciences, South China University of Technology, Nanjing Forestry University, Beijing Forestry University, University of Sao Paulo, Kyoto University, State University of Campinas, Qilu University of Technology, China Agricultural University, Aalto University, Russian Academy of Sciences, Technical University of Denmark, University of Tokyo, Tianjin University of Science & Technology, and University of Tennessee, Knoxville.

4.3.5 The Most Prolific Funding Bodies in the Biomass Pretreatments

Information about the most prolific eight funding bodies funding at least 1.5% of the sample papers each is given in Table 4.4. Only 27% and 48% of the sample and population papers were funded.

The most prolific funding body is the U.S. Department of Energy with 5.4% of the sample papers, closely followed by the European Commission with 3.9% of the sample papers. On the other hand, the most prolific country for these top funding bodies is the USA with six funding bodies while in total, two countries and the EU house these top funding bodies.

The funding body with the most citation impact is the European Commission with 2.0% surplus, closely followed by the U.S. Department of Energy with 1.9% surplus. The other prolific funding bodies are the National Nuclear Security Agency and NREL with 1.5% surplus each. Similarly, the funding body with the least citation impact is the Ministry of Education, Culture, Sports, Science and Technology of Japan and National Science Foundation of the USA with 0% surplus each.

The other funding bodies with the relatively low citation impact and with 0.7%–11.0% of the population papers each are National Natural Science Foundation of China, Ministry of Science and Technology of China, National Council for Scientific and Technological Development of

TABLE 4.4
The Most Prolific Funding Bodies in Biomass Pretreatments

No.	Funding Bodies	Country	Sample Paper No. (%)	Population Paper No. (%)	Surplus (%)
1	US Dept. Ener.	USA	5.4	3.5	1.9
2	Eur. Commis.	EU	3.9	1.9	2.0
3	US Dept. Agric.	USA	2.0	0.9	1.1
4	Natl. Renew. Ener. Lab.	USA	2.0	0.5	1.5
5	Natl. Nucl. Sec. Agcy.	USA	2.0	0.5	1.5
6	Natl. Sci. Found.	USA	1.5	1.5	0.0
7	Minist. Educ. Cult. Spor. Sci. Technol.	Japan	1.5	1.5	0.0
8	Natl. Inst. Med. Sci.	USA	1.5	0.2	1.3

Brazil, National Key Research and Development Program of China, Higher Education Personnel Improvement Coordination of Brazil, Ministry of Education of China, Fundamental Research Funds for the Central Universities, Research Support Foundation of the State of Sao Paulo, Japan Society for the Promotion of Science, European Regional Development Fund, Ministry of Science, Technology and Innovation of Brazil, Natural Sciences and Engineering Research Council of Canada, Ministry of Finance of Japan, National Research Foundation of Korea, National Basic Research Program of China (973 Program), Chinese Academy of Sciences, Priority Academic Program Development of Jiangsu Higher Education Institutions, and China Scholarship Council.

It is highly notable that China produces 1% and 11% of the sample and population papers with 10% deficit.

4.3.6 The Most Prolific Source Titles in the Biomass Pretreatments

Information about the most prolific 11 source titles publishing at least 2.0% of the sample papers each in biomass pretreatments is given in Table 4.5.

The most prolific source title is Bioresource Technology with over 14% of the sample papers, followed by Biotechnology and Bioengineering and Green Chemistry with 10% and 8% of the sample papers, respectively. Enzyme and Microbial Technology and Biomass and Bioenergy are the other prolific journals with 3% of the sample papers each.

On the other hand, the source title with the most impact is the Biotechnology and Bioengineering with 9% surplus, closely followed by Green Chemistry and Bioresource Technology with 7% and 5% surplus, respectively. Similarly, the source titles with the least impact are the Biotechnology for Biofuels and Biomass and Bioenergy with 1% surplus each.

The other source titles with the relatively low citation impact with 0.8%–2.5% of the population paper each are Bioresources, Industrial Crops and Products, Applied Biochemistry and Biotechnology, Cellulose, Biomass Conversion and Biorefinery, Holzforschung, Carbohydrate Polymers, ACS Sustainable Chemistry and Engineering, Industrial and Engineering Chemistry Research, Applied and Environmental Microbiology, International Biodeterioration and Biodegradation, Waste and Biomass Valorization, RSC Advances, Process Biochemistry, Advanced Materials Research, and Journal of Agricultural and Food Chemistry.

TABLE 4.5
The Most Prolific Source Titles in Biomass Pretreatments

No.	Source Titles	Sample Papers (%)	Population Papers (%)	Surplus (%)
1	Bioresource Technology	14.3	9.5	4.8
2	Biotechnology and Bioengineering	10.3	1.5	8.8
3	Green Chemistry	7.9	1.1	6.8
4	Enzyme and Microbial Technology	3.4	1.1	2.3
5	Biomass and Bioenergy	3.0	1.8	1.2
7	Biotechnology for Biofuels	2.5	1.9	0.6
6	Applied Microbiology and Biotechnology	2.5	1.0	1.5
8	Proceedings of the National Academy of Sciences of the United States of America	2.5	0.2	2.3
9	Journal of Biotechnology	2.0	0.4	1.6
10	Biotechnology Advances	2.0	0.1	1.9
11	Science	2.0	0.1	1.9

4.3.7 The Most Prolific Countries in the Biomass Pretreatments

Information about the most prolific 14 countries publishing at least 3% of sample papers each in biomass pretreatments is given in Table 4.6.

The most prolific country is the USA with 47% of the sample papers. Canada, China, the UK, Japan, Sweden, and Spain are the other prolific countries with 6%–11% of the sample papers each.

On the other hand, the country with the most citation impact is the USA with 29% surplus while Canada, the UK, and Sweden are the other influential countries with 4%–7% surplus each. Similarly, the country with the least citation impact is China with 15% deficit while India has 3% deficit.

Additionally, there are other countries with relatively low citation impact and with 0.6%–5.3% of the sample papers each: Brazil, South Korea, Malaysia, Italy, Thailand, Iran, Russian Federation, Portugal, Mexico, Turkey, Indonesia, Taiwan, Poland, Egypt, South Africa, Austria, Pakistan, Switzerland, Greece, Belgium, Chile, Czech Republic, and Norway.

4.3.8 The Most Prolific Scopus Subject Categories in the Biomass Pretreatments

Information about the most prolific 11 Scopus subject categories indexing at least 3% of the sample papers each is given in Table 4.7.

The most prolific Scopus subject category in the biomass pretreatments is Chemical Engineering with 52% of sample papers, closely followed by Biochemistry, Genetics and Molecular Biology with 47% of the sample papers. The other prolific subject categories are Environmental Science, Immunology and Microbiology, and Energy with 26%–36% of the sample papers each. It is notable that Social Sciences including Economics and Business account only 2.2% of the population studies.

On the other hand, the Scopus subject category with the most citation impact is the Biochemistry, Genetics and Molecular Biology with 20% surplus, closely followed by Immunology and Microbiology and Chemical Engineering with 17% and 14% surplus, respectively.

Similarly, the Scopus subject category with the least citation impact is Agricultural and Biological Sciences with 12% deficit, closely followed by Chemistry, Materials Science, and Engineering with 5%–9% deficit each.

TABLE 4.6
The Most Prolific Countries in the Biomass Pretreatments

No.	Countries	Sample Papers (%)	Population Papers (%)	Surplus (%)
1	USA	46.8	17.6	29.2
2	Canada	11.3	4.2	7.1
3	China	7.9	23.0	−15.1
4	UK	7.9	3.3	4.6
5	Japan	6.9	6.8	0.1
6	Sweden	6.9	3.2	3.7
7	Spain	6.4	4.1	2.3
8	Germany	4.9	3.0	1.9
9	Finland	4.9	2.6	2.3
10	India	4.4	7.0	−2.6
11	France	4.4	2.8	1.6
12	Denmark	3.9	1.3	2.6
13	The Netherlands	3.4	1.3	2.1
14	Australia	3.0	1.8	1.2

TABLE 4.7
The Most Prolific Scopus Subject Categories in the Biomass Pretreatments

No.	Scopus Subject Categories	Sample Papers (%)	Population Papers (%)	Surplus (%)
1	Chemical Engineering	52.2	38.4	13.8
2	Biochemistry, Genetics and Molecular Biology	47.3	27.3	20.0
3	Environmental Science	35.5	32.1	3.4
4	Immunology and Microbiology	35.5	18.5	17.0
5	Energy	26.1	26.5	−0.4
6	Chemistry	11.3	19.9	−8.6
7	Agricultural and Biological Sciences	8.9	20.4	−11.5
8	Materials Science	7.4	15.5	−8.1
9	Engineering	6.4	11.1	−4.7
10	Multidisciplinary	4.4	1.8	2.6
11	Physics and Astronomy	3.0	3.0	0.0

4.3.9 The Most Prolific Keywords in the Biomass Pretreatments

Information about the keywords used in at least 6.4% or 3.3% of the sample or population papers, respectively, is given in Table 4.8. For this purpose, keywords related to the keyword set given in Appendix are selected from a list of the most prolific keyword set provided by Scopus database.

These keywords are grouped under the four headings: biomass, pretreatments, other processes, and products of the pretreatments.

There are 13 keywords selected related to the biomass and biomass constituents: The prolific keywords are cellulose, lignin, biomass, lignocellulose, *Zea mays* (corn, maize), hemicellulose, wood, and lignocellulosic biomass with 12%–65% of the sample papers each.

The prolific keywords related to the biomass pretreatments are cellulases, enzymes, fungi, enzyme activity, enzymatic hydrolysis, ionic liquids (ILs), *Trichoderma reesei*, basidiomycota, solvents, and temperature with 11%–35% of the sample papers, each while those related to the other processes are fermentation, pretreatment, saccharification, degradation, biodegradation, dissolution, and pretreatments with 10%–16% of the sample papers each.

The prolific keywords related to the products of the pretreatments are sugar, ethanol, biofuel, glucose, polysaccharides, and alcohols with 10%–22% of the sample papers each. It is notable that only 5.4% of the indexed papers employ bioethanol keyword.

Further, the most prolific keywords are cellulose, lignin, hydrolysis, cellulases, biomass, lignocellulose, *Zea mays* (corn, maize), *Trichoderma reesei*, enzymes, fungi, enzyme activity, Basidiomycota, sugar, and hemicellulose with 10%–21% surplus each.

4.3.10 The Most Prolific Research Fronts in Biomass Pretreatments

Information about the research fronts for the sample papers in biomass pretreatments is given in Table 4.9. As Table 4.9 shows, there are two primary research fronts for this field: biological and chemical pretreatments with 74% and 61% of the 203 highly cited papers (HCPs), respectively. The other research fronts are hydrothermal and mechanical pretreatments with 9% and 4% of the HCPs, respectively.

The prolific research fronts in the first research front are enzymatic hydrolysis, enzymatic pretreatments, and fungal pretreatments with 14%–32% of the sample papers each. The prolific research fronts in the second front are acid pretreatment, IL pretreatment, and alkali pretreatment with 9%–19% of the HCPs. Further, steam explosion and milling pretreatments are the most prolific research fronts for the hydrothermal and mechanical pretreatments, respectively.

TABLE 4.8
The Most Prolific Keywords in Biomass Pretreatments

No.	Keywords	Sample Papers (%)	Population Papers (%)	Surplus (%)
1	**Biomass**			
	Cellulose	64.5	33.9	30.6
	Lignin	57.1	31.1	26.0
	Biomass	39.9	21.0	18.9
	Lignocellulose	25.6	10.3	15.3
	Zea mays (corn, maize)	22.2	7.4	14.8
	Hemicellulose	17.2	6.7	10.5
	Wood	15.3	9.1	6.2
	Lignocellulosic biomass	12.3	7.5	4.8
	Xylan	7.9	3.7	4.2
	Corn stover	7.9	2.4	5.5
	Carbohydrates	7.4	7.6	–0.2
	Straw	6.9	5.3	1.6
	Bagasse	3.4	5.3	−1.9
2	**Pretreatment**			
	Cellulases	35.0	14.5	20.5
	Enzymes	30.5	16.2	14.3
	Fungi	30.0	16.6	13.4
	Enzyme activity	29.1	16.5	12.6
	Enzymatic hydrolysis	21.2	18.7	2.5
	Ionic liquids	17.2	8.2	9.0
	Trichoderma reesei	16.7	2.2	14.5
	Basidiomycota	14.8	2.2	12.6
	Solvents	13.3	6.7	6.6
	Temperature	11.3	7.0	4.3
	Bacteria	9.9	11.0	−1.1
	Enzymolysis	9.4	4.5	4.9
	Sulfuric acids	9.4	3.9	5.5
	Hypocrea jecorina	9.4	2.0	7.4
	Phanerochaete	6.9	0.0	6.9
	Ammonia	6.4	2.0	4.4
	Enzyme kinetics	6.4	0.0	6.4
	Water	5.9	5.0	0.9
	Enzymology	4.4	3.7	0.7
	Ph	3.9	6.1	–2.2
	Alkalinity	3.9	3.3	0.6
	Enzymatic saccharification		4.2	–4.2
	Sodium hydroxide		3.8	–3.8
3	**Other Processes**			
	Fermentation	15.8	12.5	3.3
	Pretreatment	14.8	9.6	5.2
	Saccharification	13.3	12.1	1.2
	Degradation	11.8	7.6	4.2
	Biodegradation	11.3	8.3	3.0

(Continued)

TABLE 4.8 (*Continued*)
The Most Prolific Keywords in Biomass Pretreatments

No.	Keywords	Sample Papers (%)	Population Papers (%)	Surplus (%)
	Dissolution	10.3	3.6	6.7
	Pretreatment	9.9	11.3	−1.4
	Bioconversion	8.9	2.8	6.1
	Delignification	7.4	6.2	1.2
	Oxidation	6.9	3.0	3.9
	Biotransformation	6.9	2.5	4.4
	Hydrolysis	52.2	27.9	24.3
	Extraction		6.2	−6.2
4	**Products**			
	Sugar	22.2	11.2	11
	Ethanol	20.2	12.4	7.8
	Biofuel	18.7	9.9	8.8
	Glucose	18.2	13.0	5.2
	Polysaccharides	14.3	6.6	7.7
	Alcohol	9.9	5.0	4.9
	Xylose	9.4	4.4	5.0
	Bioethanol	5.4	6.1	−0.7

TABLE 4.9
The Most Prolific Thematic Research Fronts for the Biomass Pretreatments

No.	Research Fronts	N Paper % Sample
1	**Biological Pretreatments**	**73.9**
	Enzymatic hydrolysis	31.5
	Enzymatic pretreatments	26.1
	Fungal pretreatments	13.8
	Microbial pretreatments	2.5
2	**Chemical Pretreatments**	**60.6**
	Acid pretreatments	19.2
	Ionic liquid pretreatments	15.8
	Alkaline pretreatments	9.4
	Other chemical pretreatments	9.4
	Ammonia pretreatment	3.4
	Organic solvent pretreatments	3.4
3	**Hydrothermal Pretreatments**	**9.4**
	Steam explosion pretreatments	5.4
	HCW pretreatments	2.5
	LHW pretreatments	1.0
	Autohydrolysis	0.5
4	**Mechanical Pretreatments**	**3.5**
	Milling pretreatment	2.5
	Microwave pretreatment	1.0
	Ultrasonic pretreatment	0.0

N paper (%) sample: the number of papers in the population sample of 203 papers.

TABLE 4.10
The Most Prolific Research Fronts for the Biomass for the Biomass Pretreatments

No.	Research Fronts	N Paper % Sample
1	**Biomass Constituents**	**56.7**
	Cellulose	27.1
	Lignin	18.2
	Lignocellulose	5.9
	Other constituents	3.0
	Hemicellulose	2.5
2	**Wood**	**14.3**
	Wood	9.9
	Softwood	4.4
3	**Grass**	**5.4**
	Switchgrass	3.4
	Other grass	2.0
4	**Agricultural Residues**	**21.7**
	Corn stover	9.4
	Other residues	4.9
	Wheat straw	4.4
	Sugarcane bagasse	3.0
5	**Biomass in General**	**24.6**
	Lignocellulosic biomass	14.3
	Biomass at large	10.3

N paper (%) sample: the number of papers in the population sample of 203 papers.

Information about the research fronts for the sample papers in biomass pretreatments with regard to the biomass used in these pretreatments is given in Table 4.10. As Table 4.10 shows, there are four primary research fronts for this field: biomass constituents, biomass, in general, agricultural residues, and wood with 57, 25, 22, and % of the 203 HCPs, respectively. The other research front is grass with 5% of the sample papers. There is no paper for the algae.

The most prolific biomass constituents are cellulose and lignin with 27% and 18% of the sample papers, respectively, while lignocellulosic biomass and biomass, in general, account for 14% and 10% of the sample papers, respectively.

The most prolific agricultural residue is corn stover with 5% of the sample papers, switchgrass is the most prolific grass with 3% of the sample papers, and wood, in general, and softwood account for 10% and 4% of the sample papers, respectively.

4.4 DISCUSSION

4.4.1 Introduction

The crude oil-based gasoline fuels have been widely used in the transportation sector since the 1920s. However, there have been great public concerns over the adverse environmental and human impact of these fuels. Hence, biomass-based bioethanol fuels have increasingly been used in blending gasoline fuels in the fuel cells, and in the biochemical production in a biorefinery context.

The research in the field of biomass pretreatments such as biological, chemical, mechanical, and hydrothermal pretreatments to increase the sugar and bioethanol yield has also intensified in recent years.

However, it is essential to develop efficient incentive structures for the primary stakeholders to enhance the research in this field. This is especially important to maintain energy security in the cases of supply shocks such as oil shocks or war-related shocks as in the case of Russian invasion of Ukraine starting in February 2022.

The scientometric analysis has been used in this context to inform the primary stakeholders about the current state of the research in a selected research field. As there have been no scientometric studies on the biomass pretreatments, this book chapter presents a scientometric study of the research in the biomass pretreatments. It examines the scientometric characteristics of both the sample and population data presenting scientometric characteristics of both datasets in the order of documents, authors, publication years, institutions, funding bodies, source titles, countries, Scopus subject categories, keywords, and research fronts.

As a first step for the search of the relevant literature, the keywords were selected using the first most-cited 200 papers for each of the biological, chemical, mechanical, and hydrothermal pretreatments. The selected keyword lists were optimized to obtain a representative sample of papers for the searched research field. These keyword lists were integrated to obtain a single keyword set for the biomass pretreatments and used for this study. Further, this keyword list extended the papers relating to the biomass pretreatments at the macrolevel without focusing on any type of the biomass pretreatments. A copy of this extended keyword list was provided in Appendix for future replication studies. Further, a selected list of keywords is presented in Table 4.7.

As a second step, two sets of data were used for this study. First, a population sample of over 20,251 papers was used to examine the scientometric characteristics of the population data. Secondly, a sample of 203 most-cited papers, corresponding to 1% of the population dataset, was used to examine the scientometric characteristics of these citation classics.

The scientometric characteristics of both sample and population datasets were presented in the order of documents, authors, publication years, institutions, funding bodies, source titles, countries, Scopus subject categories, keywords, and research fronts.

Lastly, the key scientometric findings for both datasets were discussed to highlight the research landscape for the biomass pretreatments. Additionally, a number of brief conclusions were drawn, and a number of relevant recommendations were made to enhance the future research landscape.

4.4.2 The Most Prolific Documents in the Biomass Pretreatments

Articles (together with conference papers) dominate both the sample (75%) and population (95%) papers (Table 4.1). Further, review papers and articles have a surplus (23%) and deficit (20%), respectively. The representation of the reviews and short surveys in the sample papers is extraordinarily high (25%). This representation is higher for the extended sample dataset with over 29%.

Scopus differs from the Web of Science database in differentiating and showing articles (74%) and conference papers (2%) published in the journals separately. However, it should be noted that these conference papers are also published in journals as articles, compared to those published only in the conference proceedings. Similarly, Scopus differs from Web of Science database in introducing short surveys (2%). Hence, a total number of articles and review papers in the sample dataset are 75% and 25%, respectively.

It is observed during the search process that there has been inconsistency in the classification of the documents in Scopus as well as in other databases such as Web of Science. This is especially relevant for the classification of papers as reviews or articles as the papers not involving a literature review may be erroneously classified as a review paper. There is also a case of review papers being classified as articles. For example, the total number of the reviews in the sample dataset was manually found as over 30% compared to 25% as indexed by Scopus, reducing the number of articles and conference papers to 70% for the sample dataset.

In this context, it would be helpful to provide a classification note for the published papers in the books and journals at the first instance. It would also be helpful to use the document types listed in Table 4.1 for this purpose. Book chapters may also be classified as articles or reviews as an additional classification to differentiate review chapters from the experimental chapters as it is done by the Web of Science. It would be further helpful to additionally classify the conference papers as articles or review papers as well as it is done in the Web of Science database.

4.4.3 The Most Prolific Authors in the Biomass Pretreatments

There have been most prolific 18 authors with at least 3.1% of the sample papers each as given in Table 4.2. These authors have shaped the development of the research in this field.

The most prolific authors are John N. Saddler, Bernard Henrissat, Blake A. Simmons, Charles E. Wyman, Igor V. Grigoriev, Dan Cullen, Bruce E. Dale, Michael E. Himmel, and Pedro M. Coutinho working primarily on biological and chemical pretreatments of the biomass.

It is important to note the inconsistencies in indexing of the author names in Scopus and other databases. It is especially an issue for the names with more than two components such as 'Blake Sam de Hyun Saddler'. The probable outcomes are 'Saddler, B.S.D.H.', 'de Hyun Saddler, B.S.', or 'Hyun Saddler, B.S.D.'. The first choice is the gold standard of the publishing sector as the last word in the name is taken as the last name. In most of the academic databases such as PUBMED and EBSCO, this version is used predominantly. The second choice is a strong alternative while the last choice is an undesired outcome as two last words are taken as the last name. It is good practice to combine the words of the last name by a hyphen: 'Hyun-Saddler, B.S.D.'. It is notable that inconsistent indexing of the author names may cause substantial inefficiencies in the search process for the papers as well as allocating credit to the authors as there are different author entries for each outcome in the databases.

There are also inconsistencies in the shortening Chinese names. For example, 'Yuoyong Wang' is often shortened as 'Wang, Y.', 'Wang, Y.-Y.', and 'Wang Y.Y' as it is done in the Web of Science database as well. However, the gold standard in this case is 'Zhang Y' where the last word is taken as the last name and the first word is taken as a single forename. In most of the academic databases such as PUBMED and EBSCO, this first version is used predominantly. Therefore, there have been difficulties to locate papers for the Chinese authors. In such cases, the use of the unique author codes provided for each author by the Scopus database has been helpful.

There is also a difficulty in allowing credit for the authors especially for the authors with common names such as 'Zhang, X' or 'Huang, X' or 'Zhu, X' in conducting scientometric studies. These difficulties strongly influence the efficiency of the scientometric studies as well as allocating credit to the authors as there are the same author entries for different authors with the same name, for example, 'Zhang X' in the databases.

In this context, the coding of authors in Scopus database is a welcome innovation compared to the other databases such as Web of Science. In this process, Scopus allocates a unique number to each author in the database (Aman, 2018). However, there might still be substantial inefficiencies in this coding system especially for common names. For example, some of the papers for a certain author may be allocated to another researcher with a different author code. It is possible that Scopus uses a number of software programs to differentiate the author names and the program may not be false-proof (D'Angelo and van Eck, 2020).

In this context, it does not help that author names are not given in full in some journals and books. This makes difficult to differentiate authors with common names and makes the scientometric studies further difficult in the author domain. Therefore, the author names should be given in all books and journals at the first instance. There is also a cultural issue where some authors do not use their full names in their papers. Instead, they use initials for their forenames: 'Saddler, H.J.' or just 'Saddler' instead of 'Saddler, Hyun Jae'.

There are also inconsistencies in naming of the authors with more than two components by the authors themselves in journal papers and book chapters. For example, 'Saddler, APC', 'Welton, C.E.', and 'Larrondo, S.J.' might be given as 'Saddler, A', 'Welton, C', or 'Larrondo, S' in the journals and books, respectively. This also makes the scientometric studies difficult in the author domain. Hence, contributing authors should use their name consistently in their publications.

The other critical issue regarding the author names is the inconsistencies in the spelling of the author names in the national spellings (e.g., Gonçalves, Özoğul) rather than in the English spellings (e.g., Goncalves, Ozogul) in Scopus database. Scopus differs from the Web of Science database and many other databases in this respect where the author names are given only in the English spellings. It is observed that national spellings of the author names do not help much in conducting scientometric studies as well as in allocating credits to the authors as sometimes there are the different author entries for the English and National spellings in the Scopus database.

The most prolific institutions for the sample dataset are the Joint Bioenergy Institute of the Lawrence Livermore National Laboratory of the USA, Aix Marseille University of France, and Imperial College of the UK. The most prolific countries for the sample dataset are the USA, the UK, and France. These findings confirm the dominance of the USA and, to a lesser extent, of the Europe in this field.

The most prolific research fronts are the biological pretreatments of the biomass, followed by the chemical pretreatments. Additionally, one author studies primarily hydrothermal pretreatment of the biomass while there is no author studying the mechanical pretreatment of the biomass.

It is also notable that there is significant gender deficit for the sample dataset as surprisingly only two of these top researchers are female. This finding is the most thought-provoking with strong public policy implications. Hence, institutions, funding bodies, and policymakers should take efficient measures to reduce the gender deficit in this field as well as other scientific fields with strong gender deficit. In this context, it is worth to note the level of representation of the researchers from the minority groups in science on the basis of race, sexuality, age, and disability, besides the gender (Blankenship, 1993; Dirth and Branscombe, 2017; Konur, 2000, 2002a–c, 2006a, b, 2007a, b).

4.4.4 The Most Prolific Research Output by Years in the Biomass Pretreatments

The research output observed between 1970 and 2022 is illustrated in Figure 4.1. This figure clearly shows that the bulk of the research papers in the population dataset were published primarily in the 2010s with 54% of the population dataset. Similarly, the bulk of the research papers in the sample dataset were published in the last two decades with 58% and 24% of the sample dataset, respectively.

These findings suggest that the most prolific sample and population papers were primarily published in the last two decades. These are the thought-provoking findings as there has been no significant research in this field in the pre-2000s, but there has been significant research boom in the last two decades. In this context, the increasing public concerns about climate change (Change, 2007), greenhouse gas emissions (Carlson et al., 2017), and global warming (Kerr, 2007) have been certainly behind the boom in the research in this field in the last two decades.

Based on these findings, the size of the population papers likely to more than double in the current decade provided that the public concerns about climate change, greenhouse gas emissions, and global warming are translated efficiently to the research funding in this field.

4.4.5 The Most Prolific Institutions in the Biomass Pretreatments

The most prolific 21 institutions publishing papers on the biomass pretreatments with at least 2.5% of the sample papers each given in Table 4.3 have shaped the development of the research in this field.

The most prolific institutions are University of British Columbia, USDA Forest Service, Dartmouth College, NREL, and Novozymes. The top countries for these most prolific institutions are USA, France, and Finland. In total, only six countries house these top institutions.

On the other hand, the institutions with the most and least impact are University of British Columbia and North Carolina State University, respectively. These findings confirm the dominance of the US, Canadian, and European institutions in this research field.

4.4.6 The Most Prolific Funding Bodies in the Biomass Pretreatments

The most prolific eight funding bodies funding at least 1.5% of the sample papers each are given in Table 4.4. It is notable that only 27% and 48% of the sample and population papers were funded.

The most prolific funding bodies are the U.S. Department of Energy and the European Commission. The most prolific country for these top funding bodies is the USA. In total, two countries and the EU house these top funding bodies. The funding bodies with the most citation impact are European Commission and the one with the least impact are Ministry of Education, Culture, Sports, Science and Technology of Japan, and National Science Foundation of the USA.

These findings on the funding of the research in this field suggest that the level of the funding, mostly in the last two decades, is not highly intensive, but it has been largely instrumental in enhancing the research in this field (Ebadi and Schiffauerova, 2016) in light of North's institutional framework (North, 1991). However, considering the relatively low levels of funding especially for the sample papers, there is ample room to enhance funding in this field.

4.4.7 The Most Prolific Source Titles in Biomass Pretreatments

The most prolific 11 source titles publishing at least 2.0% of the sample papers each in biomass pretreatments have shaped the development of the research in this field (Table 4.5).

The most prolific source titles are Bioresource Technology, Biotechnology and Bioengineering, Green Chemistry, Enzyme and Microbial Technology, and Biomass and Bioenergy.

It is notable that these top source titles are primarily related to the biomass, bioresources, biotechnology, and bioenergy. This finding suggests that the journals in this field have significantly shaped the development of the research in this field as they focus on the hydrothermal pretreatment aspects of bioethanol fuels.

4.4.8 The Most Prolific Countries in the Biomass Pretreatments

The most prolific 11 countries publishing at least 3.1% of the sample papers each have significantly shaped the development of the research in this field (Table 4.6).

The most prolific countries are the USA, Canada, China, the UK, Japan, Sweden, and Spain. On the other hand, the countries with the most and least citation impact are the USA and China, respectively. Additionally, it is notable that China and India are under-represented in the sample papers while the USA, Canada, the UK, Denmark, and Spain are over-represented.

The close examination of these findings suggests that the USA, Europe, China, and Canada are the major producers of the research in this field. It is a fact that the USA has been a major player in science (Leydesdorff and Wagner, 2009; Leydesdorff et al., 2014). The USA has further developed a strong research infrastructure to support its corn- and grass-based bioethanol industry (Vadas et al., 2008).

However, China has been a rising star in scientific research in competition with the USA and Europe (Leydesdorff and Zhou, 2005). China is also a major player in this field as a major producer of bioethanol (Li and Chan-Halbrendt, 2009).

Next, Europe as a whole has been a persistent player in the scientific research in competition with both the USA and China (Leydesdorff, 2000). Europe has also been a persistent producer of bioethanol along with the USA and Brazil (Gnansounou, 2010).

Additionally, Brazil has also been a persistent player in scientific research at a moderate level (Glanzel et al., 2006). Brazil has also developed a strong research infrastructure to support its biomass-based bioethanol industry (Soccol et al., 2010). However, it is notable that Brazil has no significant presence for this research field.

4.4.9 The Most Prolific Scopus Subject Categories in Biomass Pretreatments

The most prolific 11 Scopus subject categories indexing at least 3% of the sample papers each, respectively, given in Table 4.7 have shaped the development of the research in this field.

The most prolific Scopus subject categories in the biomass pretreatments are Chemical Engineering, Biochemistry, Genetics and Molecular Biology, Environmental Science, Immunology and Microbiology, and Energy.

These findings are thought-provoking suggesting that the primary subject categories are Chemical Engineering and Biochemistry, Genetics and Molecular Biology. The other key finding is that social sciences are not well represented in both the sample and population papers, as in the most fields in bioethanol fuels. These findings are not surprising as the key research fronts in this field relate to the development and application of biomass pretreatments.

4.4.10 The Most Prolific Keywords in Biomass Pretreatments

A limited number of keywords have shaped the development of the research in this field as shown in Table 4.8 and Appendix.

These keywords are grouped under the four headings: biomass and biomass constituents, pretreatments, other processes, and products of the pretreatments.

The keywords related to the biomass and biomass constituents are cellulose, lignin, biomass, lignocellulose, *Zea mays* (corn, maize), hemicellulose, wood, and lignocellulosic biomass while those related to the pretreatments are cellulases, enzymes, fungi, enzyme activity, enzymatic hydrolysis, ILs, *Trichoderma reesei*, basidiomycota, solvents, and temperature. Those related to the other processes are fermentation, pretreatment, saccharification, degradation, biodegradation, dissolution, and pretreatments while those related to the products are sugar, ethanol, biofuel, glucose, polysaccharides, and alcohols.

These findings suggest that it is necessary to determine the keyword set carefully to locate the relevant research in each of these research fronts. Additionally, the size of the samples for each keyword highlights the intensity of the research in the relevant research areas. The relevant keywords are presented in Table 4.7 as well as in Appendix.

4.4.11 The Most Prolific Research Fronts in Biomass Pretreatments

As Table 4.9 shows, there are two primary research fronts for this field: biological and chemical pretreatments while the other research fronts are hydrothermal and mechanical pretreatments.

The prolific research fronts in the first research front are enzymatic hydrolysis, enzymatic pretreatments, and fungal pretreatments while the prolific research fronts in the second front are acid pretreatment, IL pretreatment, and alkali pretreatment. Further, steam explosion and milling pretreatments are the most prolific research fronts for the hydrothermal and mechanical pretreatments, respectively.

Information about the research fronts for the sample papers in biomass pretreatments with regard to the biomass or biomass constituents used in these pretreatments is given in Table 4.10. As Table 4.10 shows, there are four primary research fronts for this field: biomass constituents, biomass, in general, agricultural residues, and wood. The other research front is grass while there is no paper for the algae.

The most prolific biomass constituents are cellulose and lignin while lignocellulosic biomass and biomass are the other prolific fronts. Further, the most prolific agricultural residue is corn stover while switchgrass is the most prolific grass, and wood, in general, and softwood are the prolific fronts for the wood.

These findings are thought-provoking in seeking ways to increase bioethanol yield through the biomass pretreatments at the global scale. It is clear that all of these research fronts with regard to both pretreatment types and biomass types have public importance and merit substantial funding and other incentives.

In the end, these most-cited papers in this field hint that the efficiency of bioethanol fuels and their derivatives could be optimized using the structure, processing, and property relationships of these pretreatments (Formela et al., 2016; Konur, 2018a, 2020b, 2021a–d; Konur and Matthews, 1989).

4.5 CONCLUSION AND FUTURE RESEARCH

The research on the biomass pretreatments has been mapped through a scientometric study of both sample (203 papers) and population (20,251 papers) datasets. However, the extended population dataset contains 23,407 papers, including additionally the papers for the biomass pretreatments at the macroscale, mostly the reviews.

The critical issue in this study has been to obtain a representative sample of the research as in any other scientometric study. Therefore, the keyword set has been carefully devised and optimized after a number of runs in the Scopus database. It is a representative sample of the wider population studies. This keyword set was provided in Appendix, and the relevant keywords are presented in Table 4.8.

The other issue has been the selection of a multidisciplinary database to carry out the scientometric study of the research in this field. For this purpose, Scopus database has been selected. The journal coverage of this database has been notably wider than that of Web of Science and other multi-subject databases.

The key scientometric properties of the research in this field have been determined and discussed in this book chapter. It is evident that a limited number of documents, authors, institutions, publication periods, institutions, funding bodies, source titles, countries, Scopus subject categories, keywords, and research fronts as well as HCPs have shaped the development of the research in this field.

There is ample scope to increase the efficiency of the scientometric studies in this field in the author and document domains by developing consistent policies and practices in both domains across all the academic databases. In this respect, it seems that authors, journals, and academic databases have a lot to do. Furthermore, the significant gender deficit as in most scientific fields emerges as a public policy issue. The potential deficits on the basis of age, race, disability, and sexuality need also to be explored in this field as in other scientific fields.

The research in this field has boomed in the last two decades possibly promoted by the public concerns on global warming, greenhouse gas emissions, and climate change. The institutions from the USA, Canada, and Europe have mostly shaped the research in this field.

The relatively well funding rate of 48% for the population papers suggests that funding in this field significantly enhanced the research in this field primarily in the last two decades, possibly more than doubling in the current decade. However, it is evident that there is ample room for more funding and other incentives to enhance the research in this field further as only 27% of the sample papers declared any funding.

The most prolific journals have been mostly indexed by the subject categories of Chemical Engineering and Biochemistry, Genetics and Molecular Biology as the focus of the sample papers has been on the development and utilization of the biomass pretreatments to increase the sugar and bioethanol yield at large.

The USA, China, Canada, and Europe have been the major producers of the research in this field as the major producers and users of bioethanol fuels from different types of biomass such as corn, sugarcane, and grass as well as other types of biomass. It is evident that these countries have well-developed research infrastructure in bioethanol fuels and their derivatives.

The primary Scopus subject categories have been Chemical Engineering and Biochemistry, Genetics and Molecular Biology as the focus of the research has been on the development and utilization of the biomass pretreatments to increase the sugar and bioethanol yield at large. Further, social sciences are not well represented in both the sample and population papers, as in the most fields in bioethanol fuels. These findings are not surprising.

It emerges that ethanol is more popular than bioethanol as a keyword with strong implications for the search strategy. In other words, the search strategy using only bioethanol keyword would not be much helpful.

These keywords are grouped under the four headings: biomass and biomass constituents, pretreatments, other processes, and products of the pretreatments. These groups of keywords highlight the potential primary research fronts for these fields.

There are two primary research fronts for this field: biological and chemical pretreatments. The other research fronts are hydrothermal and mechanical pretreatments. The prolific research fronts for the first research front are enzymatic hydrolysis, enzymatic pretreatments, and fungal pretreatments while those in the second front are acid pretreatment, IL pretreatment, and alkali pretreatments. Further, steam explosion and milling pretreatments are the most prolific research fronts for the hydrothermal and mechanical pretreatments, respectively.

Similarly, there are four primary research fronts for the biomass or biomass constituents used for the biomass pretreatments: agricultural residues, wood, biomass constituents, and biomass, in general. Further, the most prolific biomass constituents are cellulose and lignin while lignocellulosic biomass and biomass at large are the prolific biomass. The most prolific agricultural residue is corn stover while switchgrass is the most prolific grass, and wood, in general, and softwood are the prolific fronts for the wood.

These findings are thought-provoking. The focus of these most-cited 203 papers as well as 20,251 population papers is the development and utilization of pretreatments to increase the sugar and bioethanol yield. These studies highlight strong structure–processing–property relationships for pretreatments for bioethanol fuels and their derivatives.

Thus, the scientometric analysis has a great potential to gain valuable insights into the evolution of the research in this field as in other scientific fields especially in the aftermath of the significant global supply shocks such as the Russian invasion of Ukraine starting in February 2022.

It is recommended that further scientometric studies are carried out for the primary types of the pretreatments and the biomass used. It is further recommended that reviews of the most-cited papers are carried out for each research front to complement these scientometric studies. Next, the scientometric studies of the hot papers in these primary fields are carried out.

ACKNOWLEDGMENTS

The contribution of the highly cited researchers in the field of the biomass pretreatments has been gratefully acknowledged.

APPENDIX: THE KEYWORD SET FOR BIOMASS PRETREATMENTS

(((TITLE (pretreat* OR 'pre-treat*' OR 'biological treat*' OR 'chemical treat*' OR 'physicochemical treat*') OR TITLE (microwave* OR 'micro wave*') OR TITLE (ultrasound OR ultrason* OR sonicat* OR sono* OR 'irradiation pretreat*') OR SRCTITLE (ultrasound OR ultrason*) OR TITLE (milling OR grinding OR 'extrusion pretreat*' OR 'mechanical pretreat*' OR 'popping pretreat*' OR densification OR 'mechanical pre-treat*' OR 'physical pretreat*' OR 'mechanical activation'

OR ‘washing pre-treat*’ OR ‘washing pretreat*’ OR ‘thermo-mechanical pre-treat*’ OR ‘thermo-mechanical pretreat*’ OR ‘thermomechanical pretreat*’) OR TITLE (‘steam explosion’ OR ‘steam pretreat*’ OR ‘steaming pretreat*’ OR ‘steam treat*’ OR ‘steam explod*’ OR ‘liquid hot water’ OR ‘hot water’ OR ‘hot compressed water’ OR ‘hot compressed liquid water’ OR ‘compressed hot water’ OR ‘hydrothermal pretreat*’ OR ‘hydrothermal pre-treat*’ OR ‘low steam’ OR ‘steam assisted’ OR autohydrolysis OR ‘wet torrefaction’ OR ‘hydrothermally pretreat*’ OR autoclaving OR ‘pressurized hot water’ OR ‘steam fractionat*’ OR aquasolve OR lhw OR ‘steam hydrolysis’ OR hydrothermolysis OR ‘pressure cooking’)) OR ((TITLE (‘biological pretreat*’ OR ‘biological treat*’ OR ‘microbial pretreat*’ OR ‘fungal pretreat*’ OR enzym* OR celluclast* OR novozym* OR fung* OR laccase* OR cellulosome* OR *enzyme OR bacter* OR hydrolase* OR cellobiohydrolase* OR glucosidase* OR xylanase* OR esterase* OR cellulase* OR trichoderma OR rhodosporidium OR peroxidase* OR ligninase* OR trametes OR pleurotus OR myceliophthora OR mucor OR rhizopus OR aspergillus OR esterase* OR microb* OR postia OR phanerochaete* OR oxidase* OR oudemansiella OR ceriporiopsis OR bifidobacterium OR lactobacillus OR candida OR echinodontium OR yeast* OR agaricomycetes OR hemicellulase* OR glucanase* OR endoglucanase* OR biosurfactant* OR pseudomonas OR mannanase* OR mannosidase* OR galactosidase* OR swollenin OR basidiomyc* OR pectinase* OR thielavia OR pycnoporus OR alcalase* OR ligninase* OR monooxygenase* OR neurospora OR hypocrea OR coniophora OR coriolus OR isozyme*) OR TITLE (milled OR milling OR grinding OR ‘mill pretreatment*’ OR grinded) OR TITLE (‘chemical treat*’ OR ‘chemical pretreat*’ OR acid* OR h2so4 OR ‘h 2so 4’ OR ‘hydrogen peroxide*’ OR h2o2 OR ‘h 2o 2’ OR ‘alkaline peroxide*’ OR ‘chemo-mechanical pretreat*’ OR ‘thermo-chemical pretreat*’ OR ‘thermo-chemical pre-treat*’ OR ‘CO2 pretreat*’ OR ‘carbon dioxide pretreat*’ OR ‘thermochemical pretreat*’ OR ‘leaching pretreat*’ OR ‘H2O pretreat*’ OR ‘wet explosion’ OR hcl OR salt* OR sulfite* OR sporl OR ph OR lime* OR ‘sulphur dioxide’ OR ‘sulfur dioxide’ OR so2 OR ‘so 2’ OR alkali* OR *alkali OR naoh* OR lioh OR ‘sodium hydroxide*’ OR koh OR caoh* OR ‘ca(oh)*’ OR ‘calcium hydroxide’ OR ‘sodium methoxide’ OR base OR ‘ionic liquid*’ OR solvent* OR *solvent OR *methylimidazolium OR ‘methyl imidazolium’ OR dissolv* OR *organosolv OR *organosolve OR ammonia OR nh3 OR ozonation OR ozonolysis OR ozone OR ‘wet oxidation’ OR ‘air oxidation’ OR flowthrough OR ‘supercritical co2’ OR ‘carbon dioxide explosion’ OR ‘co2 explosion’ OR ‘supercritical carbon dioxide’ OR ‘supercritical co2’ OR methylmorpholine OR nmmo* OR bmim* OR surfactant* OR afex* OR ammonolysis OR ‘recycled percolation’ OR ‘physicochemical pretreat*’ OR ‘physicochemical treat*’ OR ‘physico-chemical pretreat*’ OR ‘supercritical water’ OR ‘subcritical water’ OR fecl3 OR fenton OR ‘liquor pretreat*’ OR ‘aqueous pretreat*’)) AND TITLE (pretreat* OR ‘pre-treat*’ OR decay* OR *hydrolysis OR saccharification OR hydrolytic OR digestibili* OR disruption OR fractionation OR decomposition OR dissolution OR delignification OR depolymerization OR sugar OR sugars OR *oligosaccharides OR glucose OR deconstruction OR xylose OR disintegration OR *conversion OR recalcitrance OR *degradation OR *degradability OR *degrading OR extract* OR bioconversion OR ligninolysis OR ethanol OR bioethanol OR ‘size reduction’))) AND NOT (TITLE (*gas OR *sorption OR anaerobic OR sludge* OR pyrolysis OR lipid* OR wastewater* OR ‘thermal pretreat*’ OR gasif* OR *hydrogen OR removal OR torref* OR *methane OR composite* OR biomethan* OR biochars OR recovery OR *sorbents OR butanol OR coal OR ‘waste water*’ OR ‘thermal pre-treat*’ OR ‘bio-oil*’ OR nano* OR *composites OR hydrogen* OR benzene OR ‘fuel cell*’ OR synthesis OR forest* OR inhibit* OR soil* OR *diesel OR bioelectricity OR aromatic* OR antioxidant* OR antioxidat* OR xylem* OR levulinic OR adhesive* OR phenolic OR polyol* OR dye* OR protein* OR seed* OR purex OR ‘degradation products’ OR mariculture OR unex OR drug* OR lactic OR *furfural OR phar* OR ‘advanced mat*’ OR ‘oil prod*’ OR ‘activated carb*’ OR detox* OR furan* OR acer OR keton* OR ‘essential oil*’ OR adsorptive OR ‘value added’ OR dna OR carotene OR pcr OR concrete OR fucoidan* OR hydrolysate* OR hexitol* OR leaf) OR SRCTITLE (dairy OR phar* OR medi* OR cereal* OR nutrit* OR feed* OR brew* OR animal* OR zoo* OR lipid* OR soil* OR composite* OR atmosph* OR poult* OR postharv* OR botan* OR cellular OR nano* OR cement*) OR SUBJAREA (medi OR phar OR heal OR nurs

OR neur OR eart OR vete OR dent))) AND (TITLE (*mannan OR waste OR wastes OR *grass OR biomass* OR lignocellul* OR bagasse* OR cellulos* OR *wood OR woody OR *woods OR *algae OR *algal OR microalga OR olive OR chlamydomonas OR straw OR straws OR stover* OR eucalyptus OR miscanthus OR pine OR chlorella OR lignin OR cellulolytic OR hemicellulos* OR xylan OR porphyridium OR nannochloropsis OR poplar* OR cedar OR pulp* OR 'corn cob*' OR spruce OR beech OR oak OR pinus OR residue* OR willow* OR cypress OR sawdust OR prosopis OR 'rice hull*' OR '*bean hull*' OR husk* OR birch OR cynara OR '*cane tops' OR 'cotton stalk*' OR 'corn stalk*' OR alfalfa OR 'maize stem*' OR cornstalk* OR hyacinth OR salix OR corncob* OR 'fruit bunch*' OR populus OR sage OR '*weed stem*' OR bamboo* OR birch OR '*flower stalk*' OR 'barley hull*' OR 'wheat bran' OR pseudostem* OR bluestem* OR reed OR arundo OR cellobiose OR sawmill OR ulva OR seaweed* OR shoots OR rhizome OR agave OR coffee OR molass* OR triticale OR arborera OR sacharum OR bioresource* OR eucheuma OR laminaria OR bode OR taro OR szarvasi OR eichhornia OR sago OR kenaf OR scrap* OR fir OR maple OR chestnut)) AND (LIMIT-TO (SRCTYPE, 'j') OR LIMIT-TO (SRCTYPE, 'k') OR LIMIT-TO (SRCTYPE, 'b')) AND (LIMIT-TO (DOCTYPE, 'ar') OR LIMIT-TO (DOCTYPE, 'cp') OR LIMIT-TO (DOCTYPE, 're') OR LIMIT-TO (DOCTYPE, 'ch') OR LIMIT-TO (DOCTYPE, 'no') OR LIMIT-TO (DOCTYPE, 'le') OR LIMIT-TO (DOCTYPE, 'sh') OR LIMIT-TO (DOCTYPE, 'bk') OR LIMIT-TO (DOCTYPE, 'ed') OR LIMIT-TO (DOCTYPE, 'cr')) AND (LIMIT-TO (LANGUAGE, 'English')).

REFERENCES

Aman, V. 2018. Does the Scopus author ID suffice to track scientific international mobility? A case study based on Leibniz laureates. *Scientometrics* 117:705–720.

Angelici, C., B. M. Weckhuysen and P. C. A. Bruijnincx. 2013. Chemocatalytic conversion of ethanol into butadiene and other bulk chemicals. *ChemSusChem* 6:1595–1614.

Antolini, E. 2007. Catalysts for direct ethanol fuel cells. *Journal of Power Sources* 170:1–12.

Antolini, E. 2009. Palladium in fuel cell catalysis. *Energy and Environmental Science* 2:915–931.

Beaudry, C. and V. Lariviere. 2016. Which gender gap? Factors affecting researchers' scientific impact in science and medicine. *Research Policy* 45:1790–1817.

Blankenship, K. M. 1993. Bringing gender and race in: US employment discrimination policy. *Gender & Society* 7:204–226.

Brandt, A., J. Grasvik, J. P. Hallett and T. Welton. 2013. Deconstruction of lignocellulosic biomass with ionic liquids. *Green Chemistry* 15:550–583.

Burnham, J. F. 2006. Scopus database: A review. *Biomedical Digital Libraries* 3:1–8.

Carlson, K. M., J. S. Gerber and D. Mueller, et al. 2017. Greenhouse gas emissions intensity of global croplands. *Nature Climate Change* 7:63–68.

Chandra, R. P., R. Bura and W. E. Mabee, et al. 2007. Substrate pretreatment: The key to effective enzymatic hydrolysis of lignocellulosics? *Advances in Biochemical Engineering/Biotechnology* 108:67–93.

Change, C. 2007. Climate change impacts, adaptation and vulnerability. *Science of the Total Environment* 326:95–112.

D'Angelo, C. A. and N. J. van Eck. 2020. Collecting large-scale publication data at the level of individual researchers: A practical proposal for author name disambiguation. *Scientometrics* 123:883–907.

Dirth, T. P. and N. R. Branscombe. 2017. Disability models affect disability policy support through awareness of structural discrimination. *Journal of Social Issues* 73:413–442.

Ebadi, A. and A. Schiffauerova. 2016. How to boost scientific production? A statistical analysis of research funding and other influencing factors. *Scientometrics* 106:1093–1116.

Fauci, A. S., H. C. Lane and R. R. Redfield. 2020. Covid-19-navigating the uncharted. *New England Journal of Medicine* 382:1268–1269.

Fernando, S., S. Adhikari, C. Chandrapal and M. Murali. 2006. Biorefineries: Current status, challenges, and future direction. *Energy & Fuels* 20:1727–1737.

Formela, K., A. Hejna, L. Piszczyk, M. R. Saeb and X. Colom. 2016. Processing and structure-property relationships of natural rubber/wheat bran biocomposites. *Cellulose* 23:3157–3175.

Garfield, E. 1955. Citation indexes for science. *Science* 122:108–111.

Glanzel, W., J. Leta and B. Thijs. 2006. Science in Brazil. Part 1: A macro-level comparative study. *Scientometrics* 67:67–86.

Gnansounou, E. 2010. Production and use of lignocellulosic bioethanol in Europe: Current situation and perspectives. *Bioresource Technology* 101:4842–4850.

Hamilton, J. D. 2003. What is an oil shock? *Journal of Econometrics* 113:363–398.

Hamilton, J. D. 2009. Causes and consequences of the oil shock of 2007–08. *Brookings Papers on Economic Activity* 2009:215–261.

Hendriks, A. T. W. M. and G. Zeeman. 2009. Pretreatments to enhance the digestibility of lignocellulosic biomass. *Bioresource Technology* 100:10–18.

Hill, J., E. Nelson, D. Tilman, S. Polasky and D. Tiffany. 2006. Environmental, economic, and energetic costs and benefits of biodiesel and ethanol biofuels. *Proceedings of the National Academy of Sciences of the United States of America* 103:11206–11210.

Hill, J., S. Polasky and E. Nelson, et al. 2009. Climate change and health costs of air emissions from biofuels and gasoline. *Proceedings of the National Academy of Sciences of the United States of America* 106:2077–2082.

Himmel, M. E., S. Y. Ding and D. K. Johnson, et al. 2007. Biomass recalcitrance: Engineering plants and enzymes for biofuels production. *Science* 315:804–807.

Hsieh, W. D., R. H. Chen, T. L. Wu and T. H. Lin. 2002. Engine performance and pollutant emission of an SI engine using ethanol-gasoline blended fuels. *Atmospheric Environment* 36:403–410.

Hu, Z. and Z. Wen. 2008. Enhancing enzymatic digestibility of switchgrass by microwave-assisted alkali pretreatment. *Biochemical Engineering Journal* 38:369–378.

Huang, H. J., S. Ramaswamy, U. W. Tschirner and B. V. Ramarao. 2008. A review of separation technologies in current and future biorefineries. *Separation and Purification Technology* 62:1–21.

Jones, T. C. 2012. America, oil, and war in the Middle East. *Journal of American History* 99:208–218.

Kerr, R. A. 2007. Global warming is changing the world. *Science* 316:188–190.

Kilian, L. 2008. Exogenous oil supply shocks: How big are they and how much do they matter for the US economy? *Review of Economics and Statistics* 90:216–240.

Kilian, L. 2009. Not all oil price shocks are alike: Disentangling demand and supply shocks in the crude oil market. *American Economic Review* 99:1053–1069.

Konur, O. 2000. Creating enforceable civil rights for disabled students in higher education: An institutional theory perspective. *Disability & Society* 15:1041–1063.

Konur, O. 2002a. Access to nursing education by disabled students: Rights and duties of nursing programs. *Nurse Education Today* 22:364–374.

Konur, O. 2002b. Assessment of disabled students in higher education: Current public policy issues. *Assessment and Evaluation in Higher Education* 27:131–152.

Konur, O. 2002c. Access to employment by disabled people in the UK: Is the Disability Discrimination Act working? *International Journal of Discrimination and the Law* 5:247–279.

Konur, O. 2006a. Participation of children with dyslexia in compulsory education: Current public policy issues. *Dyslexia* 12:51–67.

Konur, O. 2006b. Teaching disabled students in Higher Education. *Teaching in Higher Education* 11:351–363.

Konur, O. 2007a. A judicial outcome analysis of the Disability Discrimination Act: A windfall for the employers? *Disability & Society* 22:187–204.

Konur, O. 2007b. Computer-assisted teaching and assessment of disabled students in higher education: The interface between academic standards and disability rights. *Journal of Computer Assisted Learning* 23:207–219.

Konur, O. 2011. The scientometric evaluation of the research on the algae and bio-energy. *Applied Energy* 88:3532–3540.

Konur, O. 2012a. Prof. Dr. Ayhan Demirbas' scientometric biography. *Energy Education Science and Technology Part A: Energy Science and Research* 28:727–738.

Konur, O. 2012b. The evaluation of the biogas research: A scientometric approach. *Energy Education Science and Technology Part A: Energy Science and Research* 29:1277–1292.

Konur, O. 2012c. The evaluation of the global energy and fuels research: A scientometric approach. *Energy Education Science and Technology Part A: Energy Science and Research* 30:613–628.

Konur, O. 2012d. The evaluation of the research on the biodiesel: A scientometric approach. *Energy Education Science and Technology Part A: Energy Science and Research* 28:1003–1014.

Konur, O. 2012e. The evaluation of the research on the bioethanol: A scientometric approach. *Energy Education Science and Technology Part A: Energy Science and Research* 28:1051–1064.

Konur, O. 2012f. The evaluation of the research on the biofuels: A scientometric approach. *Energy Education Science and Technology Part A: Energy Science and Research* 28:903–916.

Konur, O. 2012g. The evaluation of the research on the biohydrogen: A scientometric approach. *Energy Education Science and Technology Part A: Energy Science and Research* 29:323–338.

Konur, O. 2012h. The evaluation of the research on the microbial fuel cells: A scientometric approach. *Energy Education Science and Technology Part A: Energy Science and Research* 29:309–322.

Konur, O. 2012i. The scientometric evaluation of the research on the production of bioenergy from biomass. *Biomass and Bioenergy* 47:504–515.

Konur, O. 2015. Current state of research on algal bioethanol. In *Marine Bioenergy: Trends and Developments*, Eds. S. K. Kim and C. G. Lee, pp. 217–244. Boca Raton, FL: CRC Press.

Konur, O. 2018b. Bioenergy and biofuels science and technology: Scientometric overview and citation classics. In *Bioenergy and Biofuels*, Ed. O. Konur, pp. 3–63. Boca Raton: CRC Press.

Konur, O. 2019. Cyanobacterial bioenergy and biofuels science and technology: A scientometric overview. In *Cyanobacteria: From Basic Science to Applications*, Eds. A. K. Mishra, D. N. Tiwari and A. N. Rai, pp. 419–442. Amsterdam: Elsevier.

Konur, O. 2020a. The scientometric analysis of the research on the bioethanol production from green macroalgae. In *Handbook of Algal Science, Technology and Medicine*, Ed. O. Konur, pp. 385–401. London: Academic Press.

Konur, O., Ed. 2018a. *Bioenergy and Biofuels*. Boca Raton, FL: CRC Press.

Konur, O., Ed. 2020b. *Handbook of Algal Science, Technology and Medicine*. London: Academic Press.

Konur, O., Ed. 2021a. *Handbook of Biodiesel and Petrodiesel Fuels: Science, Technology, Health, and Environment*. Boca Raton, FL: CRC Press.

Konur, O., Ed. 2021b. *Handbook of Biodiesel and Petrodiesel Fuels: Science, Technology, Health, and Environment. Volume 1. Biodiesel Fuels: Science, Technology, Health, and Environment*. Boca Raton, FL: CRC Press.

Konur, O., Ed. 2021c. *Handbook of Biodiesel and Petrodiesel Fuels: Science, Technology, Health, and Environment. Volume 2. Biodiesel Fuels Based on the Edible and Nonedible Feedstocks, Wastes, and Algae: Science, Technology, Health, and Environment*. Boca Raton, FL: CRC Press.

Konur, O., Ed. 2021d. *Handbook of Biodiesel and Petrodiesel Fuels: Science, Technology, Health, and Environment. Volume 3. Petrodiesel Fuels: Science, Technology, Health, and Environment*. Boca Raton, FL: CRC Press.

Konur, O. and F. L. Matthews. 1989. Effect of the properties of the constituents on the fatigue performance of composites: A review. *Composites* 20:317–328.

Kruyt, B., D. P. van Vuuren, H. J. de Vries and H. Groenenberg. 2009. Indicators for energy security. *Energy Policy* 37:2166–2181.

Leydesdorff, L. 2000. Is the European Union becoming a single publication system? *Scientometrics* 47:265–280.

Leydesdorff, L. and C. Wagner. 2009. Is the United States losing ground in science? A global perspective on the world science system. *Scientometrics* 78:23–36.

Leydesdorff, L., C. S. Wagner and L. Bornmann. 2014. The European Union, China, and the United States in the top-1% and top-10% layers of most-frequently cited publications: Competition and collaborations. *Journal of Informetrics* 8:606–617.

Leydesdorff, L. and P. Zhou. 2005. Are the contributions of China and Korea upsetting the world system of science? *Scientometrics* 63:617–630.

Li, H., S. M. Liu, X. H. Yu, S. L. Tang and C. K. Tang. 2020. Coronavirus disease 2019 (COVID-19): Current status and future perspectives. *International Journal of Antimicrobial Agents* 55:105951.

Li, S. Z. and C. Chan-Halbrendt. 2009. Ethanol production in (the) People's Republic of China: Potential and technologies. *Applied Energy* 86:S162–S169.

Ma, X., L. Sun and C. Song. 2002. A new approach to deep desulfurization of gasoline, diesel fuel and jet fuel by selective adsorption for ultra-clean fuels and for fuel cell applications. *Catalysis Today* 77:107–116.

Mani, S., L. G. Tabil and S. Sokhansanj. 2004. Grinding performance and physical properties of wheat and barley straws, corn stover and switchgrass. *Biomass and Bioenergy* 27:339–352.

Morschbacker, A. 2009. Bio-ethanol based ethylene. *Polymer Reviews* 49:79–84.

Mosier, N., C. Wyman and B. Dale, et al. 2005. Features of promising technologies for pretreatment of lignocellulosic biomass. *Bioresource Technology* 96:673–686.

Najafi, G., B. Ghobadian and T. Tavakoli, et al. 2009. Performance and exhaust emissions of a gasoline engine with ethanol blended gasoline fuels using artificial neural network. *Applied Energy* 86:630–639.

Newman, P. W. G. and J. R. Kenworthy. 1989. Gasoline consumption and cities: A comparison of U.S. cities with a global survey. *Journal of the American Planning Association* 55:24–37.

North, D. C. 1991. Institutions. *Journal of Economic Perspectives* 5:97–112.

Pu, Y., F. Hu, F. Huang, B. H. Davison and A. J. Ragauskas. 2013. Assessing the molecular structure basis for biomass recalcitrance during dilute acid and hydrothermal pretreatments. *Biotechnology for Biofuels* 6:15.

Reeves, S. 2014. To Russia with love: How moral arguments for a humanitarian intervention in Syria opened the door for an invasion of the Ukraine. *Michigan State University International Law Review* 23:199+.

Soccol, C. R., L. P. de Souza Vandenberghe and A. B. P. Medeiros, et al. 2010. Bioethanol from lignocelluloses: Status and perspectives in Brazil. *Bioresource Technology* 101:4820–4825.

Sun, Y. and J. Cheng. 2002. Hydrolysis of lignocellulosic materials for ethanol production: A review. *Bioresource Technology* 83:1–11.

Vadas, P. A., K. H. Barnett and D. J. Undersander 2008. Economics and energy of ethanol production from alfalfa, corn, and switchgrass in the Upper Midwest, USA. *Bioenergy Research* 1:44–55.

Winzer, C. 2012. Conceptualizing energy security. *Energy Policy* 46:36–48.

Zhang, Y. H. P. and L. R. Lynd. 2004. Toward an aggregated understanding of enzymatic hydrolysis of cellulose: Noncomplexed cellulase systems. *Biotechnology and Bioengineering* 88:797–824.

Zhu, S., Y. Wu and Q. Chen. 2006. Dissolution of cellulose with ionic liquids and its application: A mini-review. *Green Chemistry* 8:325–327.

5 Biomass Pretreatments
Review

Ozcan Konur
(Formerly) Ankara Yildirim Beyazit University

5.1 INTRODUCTION

The crude oil-based gasoline fuels (Ma et al., 2002; Newman and Kenworthy, 1989) have been widely used in the transportation sector since the 1920s. However, there have been great public concerns over the adverse environmental and human impact of these fuels (Hill et al., 2006, 2009). Hence, biomass-based bioethanol fuels (Hill et al., 2006; Konur, 2012, 2015, 2019, 2020; Sun and Cheng, 2002) have increasingly been used in blending gasoline fuels (Hsieh et al., 2002; Najafi et al., 2009), in the fuel cells (Antolini, 2007, 2009), and in the biochemical production (Angelici et al., 2013; Morschbacker, 2009) in a biorefinery context (Fernando et al., 2006; Huang et al., 2008).

However, it is necessary to pretreat the biomass (Alvira et al., 2010; Taherzadeh and Karimi, 2008) to enhance the yield of the bioethanol (Hahn-Hagerdal et al., 2006; Sanchez and Cardona, 2008) prior to the bioethanol production through the hydrolysis (Alvira et al., 2010; Sun and Cheng, 2002; Taherzadeh and Karimi, 2007) and fermentation (Lin and Tanaka, 2006; Olsson and Hahn-Hagerdal, 1996) of the biomass.

The research in the fields of biomass pretreatments (Hendriks and Zeeman, 2009; Mosier et al., 2005) has thus intensified in recent years. Biological (Himmel et al., 2007; Zhang and Lynd, 2004), chemical (Brandt et al., 2013; Zhu et al., 2006), mechanical (Hu and Wen, 2008; Mani et al., 2004), and hydrothermal (Chandra et al., 2007; Pu et al., 2013) pretreatments of the biomass have been widely researched to increase the sugar and bioethanol yield in this context in recent years.

However, it is essential to develop efficient incentive structures (North, 1991) for the primary stakeholders to enhance the research in this field (Konur, 2000, 2002a–c, 2006a, b, 2007a, b). Although there has been a number of review papers on the biomass pretreatments (Brandt et al., 2013; Chandra et al., 2007; Hendriks and Zeeman, 2009; Himmel et al., 2007; Mosier et al., 2005; Zhang and Lynd, 2004; Zhu et al., 2006), there has been no updated review of the most-cited 25 articles in this field.

Thus, this book chapter presents a review of the most-cited 25 articles in the field of the biomass pretreatments. Then, it discusses the key findings of these highly influential papers and comments on the future research priorities in this field.

5.2 MATERIALS AND METHODS

The search for this study was carried out using Scopus database (Burnham, 2006) in May 2022.

As a first step for the search of the relevant literature, the keywords were selected using all of the population papers. The selected keyword list was then optimized to obtain a representative sample of papers for the searched research field. This final keyword set was provided in the Appendix of Konur (2023) for future replication studies.

As a second step, a sample dataset was used for this study. The first 25 articles with at least 461 citations each were selected for the review study. Key findings from each paper were taken from the abstracts of these papers and were discussed. Additionally, a number of brief conclusions were drawn, and a number of relevant recommendations were made to enhance the future research landscape.

DOI: 10.1201/9781003226536-7

5.3 RESULTS

The brief information about 25 most-cited papers with at least 461 citations each on the biomass pretreatments is given below. The primary research fronts are biological and chemical pretreatments of the biomass with eight papers each. Further, seven papers are related to the biological pretreatments combined with other pretreatments. Next, there are two papers related to hydrothermal and mechanical pretreatments.

5.3.1 The Biological Pretreatment of the Biomass

There are eight HCPs for the research front of the sole biological pretreatments (Table 5.1).

TABLE 5.1
The Biological Pretreatments of the Biomass

No.	Papers	Biomass	Enzymes	Parameters	Keywords	Lead Author	Affil.	Cits
1	Bourbonnais and Paice (1990)	Lignin	*Trametes versicolor (Coriolus versicolor)* laccases	Lignin biodegradation mechanisms	Laccase, lignin, biodegradation	Bourbonnais, Robert 7003682574	McGill Univ. Canada	1,171
2	Floudas et al. (2012)	Lignin	Fungal peroxidases	Lignin biodegradation, genomic analysis	Lignin, decomposition, fungal, enzymatic	Grigoriev, Igor V. 25027225800	Joint Genome Inst. USA	1,050
3	Tien and Kirk (1983)	Lignin	*Phanerochaete chrysosporium* enzymes	Lignin biodegradation	Lignin, degrading, enzyme, Phanerochaete	Tien, Ming 7006146774	USDA Forest Serv. USA	963
4	Martinez et al. (2008)	Biomass	*Trichoderma reesei (Hypocrea jecorina) enzymes*	Genomic analysis, biomass degradation	Biomass, degrading, fungus, Trichoderma, Hypocrea	Martinez, Diego 7202958664	Massachusetts Inst. Technol. USA	878
5	Pandey and Pitman (2003)	Wood, pine, beech	*Coniophora puteana, Phanerochaete chrysosporium, Coriolus versicolor*	Wood degradation, spectroscopic study	Wood, decay, fungi	Pandey, Krishna K. 56252805800	Inst. Wood Sci. Technol. India	790
6	Martinez et al. (2004)	Wood	*Phanerochaete chrysosporium* enzymes	Genomic analysis, wood degradation	Lignocellulose, fungus, degrading	Cullen, Dan. 7202109135	USDA Forest Serv. USA	668
7	Quinlan et al. (2011)	Cellulose	Glycoside hydrolase	Glycoside hydrolase structure, cellulose degradation	Cellulose, degradation, metalloenzyme, biomass	Johansen, Katja S.* 36473579400	Copenhagen Univ. Denmark	635
8	Glenn and Gold (1985)	Lignin	*Phanerochaete chrysosporium* peroxidase	Peroxide purification and characterization, lignin degradation	Lignin, peroxidase, degrading, basidiomycete, Phanerochaete	Gold, Michael H. 7402444296	Oregon Hlth. Sci. Univ.	576

Prt, biological pretreatments; Na, non-available; Cits., number of citations received for each paper; *, female.

Bourbonnais and Paice (1990) studied the lignin biodegradation by laccases of *Trametes versicolor (Coriolus versicolor)* in a paper with 1,171 citations. They observed that veratryl alcohol (I) and 1-(3,4-dimethoxyphenyl)-2-(2-methoxyphenoxy)-propane-1, 3-diol (III) were oxidized by laccase and mediator, 2,2′-azinobis(3-ethylbenzthiazoline-6-sulphonate) (ABTS), to give the α-carbonyl derivatives. The β-1 lignin model dimer, 1-(3,4-dimethoxyphenyl)-2-phenoxy-ethane-1, 2-diol(II), was cleaved by laccase in the presence of the mediator to give veratraldehyde and benzaldehyde. They asserted that laccase was capable of oxidizing both phenolic and non-phenolic moieties of lignin but that the latter was dependent on the co-presence of primary laccase substrates.

Floudas et al. (2012) studied the paleozoic origin of enzymatic lignin decomposition based on the analysis of 31 fungal genomes in a paper with 1,050 citations. They carried out a comparative analysis of 31 fungal genomes and found that lignin-degrading peroxidases expanded in the lineage leading to the ancestor of the Agaricomycetes, which was reconstructed as a white rot species, and then contracted in parallel lineages leading to brown rot and mycorrhizal species. Further, the origin of lignin degradation might have coincided with the sharp decrease in the rate of organic carbon burial around the end of the Carboniferous period.

Tien and Kirk (1983) studied the lignin-degrading enzyme from the *Phanerochaete chrysosporium* in a paper with 963 citations. They observed that this *P. chrysosporium* contained an enzyme that degraded lignin substructure model compounds as well as spruce and birch lignins. It had a molecular size of 42,000 Da and required hydrogen peroxide for activity.

Martinez et al. (2008) provided a roadmap for constructing enhanced *Trichoderma reesei* strains as a source for cellulases and hemicellulases for the biomass degradation in a paper with 878 citations. They assembled 89 scaffolds to generate 34 Mbp of nearly contiguous *T. reesei* genome sequence comprising 9,129 predicted gene models. They observed that its genome encoded fewer cellulases and hemicellulases compared to any other sequenced fungus. Many *T. reesei* genes encoding carbohydrate-active enzymes were distributed non-randomly in clusters that stood between regions of synteny with other Sordariomycetes. There was little mechanistic insight into its extraordinary capacity for protein secretion.

Pandey and Pitman (2003) studied pine (*Pinus sylvestris* L.) and beech wood (*Fagus sylvatica* L.) degradation by *Coniophora puteana*, *Coriolus versicolor* (*Trichoderma versicolor*), and *Phanerochaete chrysosporium* by a Fourier transform infrared method in a paper with 790 citations. In wood decayed by *C. puteana*, they observed that there was a progressive increase in lignin content relative to carbohydrate evident from increases in the relative height of lignin-associated bands and a corresponding decrease in the intensities of carbohydrate bands. At higher weight losses, spectra for wood decayed by *C. puteana* had many similarities with those of Klason lignin isolated from wood. In contrast, wood decayed by *P. chrysosporium* showed selective-type decay with a reduction in peak heights associated with lignin relative to carbohydrates. Although weight losses in samples exposed to *C. versicolor* were high (45.5% and 39.8% for beech and Scots pine, respectively, after 12 weeks), simultaneous decay resulted in little change in the relative intensities of the lignin and carbohydrate bands, with only a slight preference for lignin.

Martinez et al. (2004) carried out a genomic analysis of wood-degrading *Phanerochaete chrysosporium* as a source for cellulases and hemicellulases for the biomass degradation in a paper with 668 citations. They sequenced the 30-million base-pair genome of *P. chrysosporium* strain RP78 using a whole-genome shotgun approach. The *P. chrysosporium* genome revealed an impressive array of genes encoding secreted oxidases, peroxidases, and hydrolytic enzymes that cooperated in wood decay. This genome provided a high-quality draft sequence of a basidiomycete.

Quinlan et al. (2011) studied the oxidative degradation of cellulose by a glycoside hydrolase (CAZy) GH61 enzyme focusing on cellulose-enhancing factors in a paper with 635 citations. They showed that GH61 enzymes were a unique family of copper-dependent oxidases. Further, copper was needed for GH61 maximal activity and that the formation of cellodextrin and oxidized cellodextrin products by GH61 was enhanced in the presence of small-molecule redox-active cofactors such as ascorbate and gallate. The active site of GH61 contained a type II copper and, uniquely, a methylated histidine in the copper's coordination sphere.

Glenn and Gold (1985) purified and characterized an extracellular Mn(II)-dependent peroxidase from the lignin-degrading *Phanerochaete chrysosporium* in a paper with 576 citations. They observed the presence of a heme prosthetic group and the enzyme contained one molecule of heme as iron protoporphyrin IX. The absorption maximum of the native enzyme shifted in the reduced enzyme and in the reduced-CO complex. Both CN and N_3 readily bounded to the native enzyme, indicating an available coordination site, and that the heme iron was high spin. The absorption spectrum of the H_2O_2 enzyme complex was similar to that of horseradish peroxidase compound II. *P. chrysosporium* peroxidase activity was dependent on Mn(II). The enzyme was also stimulated to varying degrees by α-hydroxy acids and protein while the peroxidase was capable of oxidizing NADH. Several of the substrates were oxidized by this Mn(II)-dependent peroxidase at considerably faster rates than those catalyzed by horseradish peroxidase. The enzyme rapidly oxidized Mn(II) to Mn(III). Based on the inhibition of the oxidation of the substrate diammonium 2,2-azino-bis(3-ethyl-6-benzothiazolinesulfonate) (ABTS) by Na-pyrophosphate, they suggested that Mn(III) played a role in the enzyme mechanism.

5.3.2 The Chemical Pretreatment of the Biomass

There are eight HCPs for the research front of chemical pretreatments: four, one, and three HCPs for ionic liquid (IL), acid, and other chemical pretreatments, respectively (Table 5.2).

TABLE 5.2
The Chemical Pretreatments of the Biomass

No.	Papers	Biomass	Prt.	Parameters	Keywords	Lead Author	Affil.	Cits
1	Roman and Winter (2004)	Cellulose	Acids	Acid concentration, acid-to-cellulose ratio, sulfate group content	Hydrolysis, cellulose, degradation	Winter, William T. 7102526472	McGill Univ. Canada	966
2	Kilpelainen et al. (2007)	Wood	ILs	IL types, dissolution severity, solubility	Wood, ILs, dissolution	Kilpelainen, Ilkka 7006830888	Univ. Helsinki Finland	807
3	Sun et al. (2009)	Wood, lignin	ILs	IL types, dissolution severity, solubility, wood type	Dissolution, delignification, wood, ILs	Rogers, Robin D. 35474829200	Univ. Alabama USA	807
4	Cai and Zhang (2005)	Cellulose	Alkali	Alkali type, dissolution severity, solubility	NaOH, dissolution, cellulose	Zhang, Lina* 55917992100	Wuhan Univ. China	715
5	Fukaya et al. (2008)	Cellulose	ILs	IL type, dissolution severity, solubility	Cellulose, dissolution, ILs	Ohno, Hiroyuki 7403244652	Tokyo Univ. Japan	658
6	Remsing et al. (2006)	Cellulose	ILs	Dissolution mechanisms	Cellulose, dissolution, ILs	Rogers, Robin D. 35474829200	Univ. Alabama USA	645
7	Sasaki et al. (2000)	Cellulose	Supercritical water	Dissolution severity, temperature	Dissolution, hydrolysis, cellulose, supercritical water	Arai, Kunio 7403965625	Tohoku Univ. Japan	599
8	Sasaki et al. (1998)	Cellulose	Supercritical water	Dissolution severity, temperature	Hydrolysis, cellulose, supercritical water	Arai, Kunio 7403965625	Tohoku Univ. Japan	577

Prt, biological pretreatments; Na, non-available; Cits., number of citations received for each paper; *, female.

5.3.2.1 The Sole Ionic Liquid Pretreatment of the Biomass

Kilpelainen et al. (2007) studied the dissolution of spruce and pine wood in imidazolium-based ILs under gentle conditions in a paper with 817 citations. They observed that a variety of ILs could only partially dissolve wood chips, whereas ILs such as 1-butyl-3-methylimidazolium chloride ([Bmim]Cl) and 1-allyl-3-methylimidazolium chloride ([Amim]Cl) had good solvating power for Norway spruce sawdust and Norway spruce and Southern pine thermomechanical pulp fibers. Despite the fact that the obtained solutions were not fully clear, these ILs provided solutions which permitted the complete acetylation of the wood. Alternatively, transparent amber solutions of wood could be obtained when the dissolution of the same lignocellulosic samples was attempted in 1-benzyl-3-methylimidazolium chloride. This was based on a designed augmented interaction of the aromatic character of the cation of the IL with the lignin in the wood. After dissolution, wood could be regenerated as an amorphous mixture of its original components. The cellulose of the regenerated wood could be efficiently digested to glucose by a cellulase enzymatic hydrolysis treatment.

Sun et al. (2009) studied the complete dissolution and partial delignification of oak and pine wood in the 1-ethyl-3-methylimidazolium acetate ([C_2mim]OAc) in a paper with 808 times. Both woods could be completely dissolved in this IL after mild grinding. They obtained complete dissolution by heating the sample in an oil bath, although wood dissolution could be accelerated by microwave pulses or ultrasound irradiation. They showed that [C_2mim]OAc was a better solvent for wood than 1-butyl-3-methylimidazolium chloride ([C_4mim]Cl) and that variables such as type of wood, initial wood load, particle size, etc., affected dissolution and dissolution rates. For example, red oak dissolved better and faster than Southern yellow pine. Carbohydrate-free lignin and cellulose-rich materials could be obtained by using the proper reconstitution solvents, and approximately 26.1% and 34.9% reductions of lignin content in the reconstituted cellulose-rich materials from pine and oak, respectively, were achieved in one dissolution/reconstitution cycle. For pine, 59% of the holocellulose in the original wood could be recovered in the cellulose-rich reconstituted material, whereas 31% and 38% of the original lignin were recovered, respectively, as carbohydrate-free lignin and as carbohydrate-bonded lignin in the cellulose-rich material.

Fukaya et al. (2008) studied the cellulose dissolution with polar ILs under mild conditions in a paper with 658 citations. They obtained a series of alkylimidazolium salts containing dimethyl phosphate, methyl methylphosphonate, or methyl phosphonate prepared by a facile, one-pot procedure as room-temperature ILs, which had the potential to solubilize cellulose under mild conditions. They observed that especially N-ethyl-N′-methylimidazolium methylphosphonate enabled the preparation of 10 wt% cellulose solution by keeping it at 45°C for 30 min with stirring and rendered soluble 2–4 wt% cellulose without pretreatments and heating.

Remsing et al. (2006) studied the mechanism of cellulose dissolution in the IL 1-n-butyl-3-methylimidazolium chloride ([C_4mim]Cl) in a paper with 645 citations. ^{13}C and $^{35/37}$Cl NMR relaxation measurements on several model systems demonstrated that the solvation of cellulose by the [C_4mim]Cl involved hydrogen bonding between the carbohydrate hydroxyl protons and the IL chloride ions in a 1:1 stoichiometry.

5.3.2.2 The Sole Acid Pretreatment of the Biomass

Roman and Winter (2004) studied the impact of sulfate groups from sulfuric acid (H_2SO_4) hydrolysis on the thermal degradation behavior of bacterial cellulose in a paper with 968 citations. They hydrolyzed bacterial cellulose with sulfuric acid under different hydrolysis conditions and determined the number of sulfate groups in the crystals by potentiometric titration. They observed that the sulfate group content increased with acid concentration, acid-to-cellulose ratio, and hydrolysis time. Even at low levels, the sulfate groups caused a significant decrease in degradation temperatures and an increase in char fraction confirming that the sulfate groups acted as flame retardants. There was thermal separation of the degradation reactions by the sulfate groups into low- and high-temperature processes. The activation energies were lower at larger amounts of sulfate groups showing a catalytic effect on the degradation reactions. They recommended that for high thermostability in the crystals, low acid concentrations, small acid-to-cellulose ratios, and short hydrolysis times should be used.

5.3.2.3 The Other Sole Chemical Pretreatments of the Biomass

Cai and Zhang (2005) studied the rapid dissolution of cellulose in LiOH/urea and NaOH/urea aqueous solutions in a paper with 718 citations. They evaluated the dissolution behavior and solubility of cellulose. They found that cellulose having viscosity-average molecular weight (M_η) of 11.4×10^4 and 37.2×10^4 could be dissolved, respectively, in 7% NaOH/12% urea and 4.2% LiOH/12% urea aqueous solutions pre-cooled to −10°C within 2 min, whereas all of them could not be dissolved in KOH/urea aqueous solution. The dissolution power of the solvent systems was in the order of LiOH/urea >NaOH/urea ≫ KOH/urea aqueous solution. Further, LiOH/urea and NaOH/urea aqueous solutions as non-derivatizing solvents broke the intra- and inter-molecular hydrogen bonding of cellulose and prevented the approach toward each other of the cellulose molecules, leading to the good dispersion of cellulose to form an actual solution.

Sasaki et al. (2000) studied the dissolution and hydrolysis of cellulose in subcritical and supercritical water in a paper with 599 citations. They performed decomposition experiments of microcrystalline cellulose in subcritical and supercritical water (25 MPa, 320°C–400°C, and 0.05–10.0 s). At 400°C, they obtained hydrolysis products mainly, while in 320°C–350°C water, aqueous decomposition products of glucose were the main products. Below 350°C, the cellulose decomposition rate was slower than the glucose and cellobiose decomposition rates, while above 350°C, the cellulose hydrolysis rate drastically increased and became higher than the glucose and cellobiose decomposition rates. Below 280°C, cellulose particles became gradually smaller with increasing reaction time but, at high temperatures (300°C–320°C), cellulose particles disappeared with increasing transparency and much more rapidly than expected from the lower temperature results. They asserted that cellulose hydrolysis at high temperature took place with dissolution in water. This was probably because of the cleavage of intra- and intermolecular hydrogen linkages in the cellulose crystal. Thus, a homogeneous atmosphere was formed in supercritical water, and this resulted in the drastic increase of the cellulose decomposition rate above 350°C.

Sasaki et al. (1998) studied cellulose hydrolysis in subcritical and supercritical water in a paper with 577 citations. They proposed a new method to hydrolyze cellulose rapidly in supercritical water to recover glucose, fructose, and oligomers such as cellobiose. They performed cellulose decomposition experiments conducted with a flow type reactor in the range of temperature from 290°C to 400°C at 25 MPa. They observed that hydrolysis product yields (around 75%) in supercritical water were much higher than those in subcritical water. At a low-temperature region, the glucose or oligomer conversion rate was much faster than the hydrolysis rate of cellulose. Thus, even if the hydrolysis products, such as glucose or oligomers, were formed, their further decomposition rapidly took place and thus high yields of hydrolysis products could not be obtained. However, around the critical point, the hydrolysis rate jumped to more than an order of magnitude higher level and became faster than the glucose or oligomer decomposition rate. This was the reason why they obtained a high yield of hydrolysis products in supercritical water.

5.3.3 The Biological and Other Pretreatments of the Biomass

There are seven HCPs for the research front of the biological and other pretreatments (Table 5.3).

Li et al. (2010) compared dilute acid and IL pretreatment of switchgrass in terms of delignification, enzymatic saccharification efficiency, and saccharide yields in a paper with 832 citations. For the IL pretreatment, they observed that switchgrass exhibited reduced cellulose crystallinity, increased surface area, and decreased lignin content compared to dilute acid

TABLE 5.3
The Biological and Other Pretreatments of the Biomass

No.	Papers	Biomass	Enzymes	Other Prts.	Parameters	Keywords	Lead Author	Affil.	Cits
1	Li et al. (2010)	Switchgrass	Enzymes	IL, acid	Enzymatic hydrolysis	Switchgrass, biomass, recalcitrance, delignification, enzymatic, saccharification	Singh, Seema* 35264950300	Joint Bioenergy Inst. USA	832
2	Lee et al. (2009)	Lignin	*Trichoderma viride* cellulases	IL	Enzymatic hydrolysis	Lignin, extraction, wood, cellulose, enzymatic, hydrolysis	Doherty, Thomas V. 57213283673	Rensselaer Polytech. Inst. USA	795
3	Eriksson et al. (2002)	Spruce	*Trichoderma reesei* cellulases	Surfactants	Enzymatic hydrolysis, hydrolysis mechanisms	Enzymatic, lignocellulose, hydrolysis	Tjerneld, Folke 7006446969	Lund Univ. Sweden	740
4	Silverstein et al. (2007)	Cotton stalks	Celluclast, Novozym	Acids, alkali, ozone, hydrogen peroxide	Pretreatment type, concentration treatment temperature, time	Chemical pretreatment, saccharification, cotton stalks	Sharma-Shivappa, Ratna R. 16231216900	NC State Univ. USA	652
5	Saha et al. (2005)	Wheat straw	β-glucosidase, xylanase, esterase	Acids	Enzymatic hydrolysis, sugar and ethanol yield	Acid, pretreatment, straw, enzymatic, saccharification	Saha, Badal C. 7202946302	USDA Agr. Res. Serv.	645
6	Lloyd and Wyman (2005)	Corn stover	Enzyme	Acids	Enzymatic hydrolysis, xylose and glucose yield	Corn stover, pretreatment, enzymatic, hydrolysis	Wyman, Charles E. 7004396809	Dartmouth Coll. USA	616
7	Yang and Wyman (2004)	Corn stover cellulose	Enzyme	Water, acids	Enzymatic hydrolysis, xylan and lignin degradation	Xylan, lignin, enzymatic, digestibility, corn stover, cellulose	Wyman, Charles E. 7004396809	Dartmouth Coll. USA	556

Prt, biological pretreatments; Other prts., other pretreatments; Na, non-available; Cits., number of citations received for each paper; *, female.

pretreatment. Further, IL pretreatment enabled a significant enhancement in the rate of enzyme hydrolysis of the cellulose component of switchgrass, with a rate increase of 16.7-fold, and a glucan yield of 96.0% obtained in 24 h.

Lee et al. (2009) studied the enzymatic hydrolysis of lignin pretreated by the ILs in a paper with 795 citations. They used the IL, 1-ethyl-3 methylimidazolium acetate ([Emim]CH_3COO), as a pretreatment solvent to extract lignin from wood flour. They observed that the cellulose in the pretreated wood flour became far-less crystalline without undergoing solubilization while 40% of the lignin was removed, the cellulose crystallinity index dropped below 45, resulting in more than 90% of the cellulose in wood flour to be hydrolyzed by *Trichoderma viride* cellulase.

Eriksson et al. (2002) studied the enzymatic hydrolysis of the steam- and surfactant-pretreated spruce with cellulase of *Trichoderma reesei* in a paper with 740 citations. They found that non-ionic surfactants were the most effective. The anionic and non-ionic surfactants reduced enzyme adsorption to the spruce substrate. The approximate reduction of enzyme adsorption was from 90% adsorbed enzyme to 80% with surfactant addition. Surfactants had only a weak effect on cellulase temperature stability. They explained the improved conversion of lignocellulose with surfactant by the reduction of the unproductive enzyme adsorption to the lignin part of the substrate. This was due to hydrophobic interaction of surfactant with lignin on the lignocellulose surface, which released unspecifically bound enzyme.

Silverstein et al. (2007) compared sulfuric acid (H_2SO_4), sodium hydroxide (NaOH), hydrogen peroxide (H_2O_2), and ozone (O_3) pretreatments for improving saccharification of cotton stalks in a paper with 652 citations. They observed that solids from H_2SO_4, NaOH, and H_2O_2 pretreatments (at 2%, 60 min, 121°C/15 psi) showed significant lignin degradation and/or high sugar availability and hence were hydrolyzed by Celluclast 1.5 L and Novozym 188 at 50°C. H_2SO_4 pretreatment resulted in the highest xylan reduction (95.23% for 2% acid, 90 min, 121°C/15 psi) but the lowest cellulose-to-glucose conversion during hydrolysis (23.85%). NaOH pretreatment resulted in the highest level of delignification (65.63% for 2% NaOH, 90 min, 121°C/15 psi) and cellulose conversion (60.8%). H_2O_2 pretreatment resulted in significantly lower delignification (maximum of 29.51% for 2%, 30 min, 121°C/15 psi) and cellulose conversion (49.8%) than NaOH pretreatment, but had a higher cellulose conversion than H_2SO_4 pretreatment. Ozone did not cause any significant changes in lignin, xylan, or glucan contents over time.

Saha et al. (2005) studied the enzymatic hydrolysis of the acid-pretreated wheat straw for bioethanol production in a paper with 645 citations. They observed that the maximum yield of monomeric sugars from wheat straw (7.83%, w/v, dry solid-DS) by dilute H_2SO_4 (0.75%, v/v) pretreatment and enzymatic hydrolysis (45°C, pH 5.0, 72 h) using cellulase, β-glucosidase, xylanase, and esterase was 565 mg/g. Under this condition, there were no measurable quantities of furfural and hydroxymethylfurfural. The yield of ethanol (per liter) from acid-pretreated enzyme saccharified wheat straw (78.3 g) hydrolysate by recombinant *Escherichia coli* strain FBR5 was 19 g with a yield of 0.24 g/g DS.

Lloyd and Wyman (2005) studied the enzymatic hydrolysis of acid-pretreated corn stover in a paper with 616 citations. They observed that up to 15% of the total potential sugar in the substrate could be released as glucose during pretreatment and between 15% and 90+% of the xylose remaining in the solid residue could be recovered in subsequent enzymatic hydrolysis, depending on the enzyme loading. Glucose yields increased from as high as 56% of total maximum potential glucose plus xylose for just enzymatic digestion to 60% when glucose released in pretreatment was included. Xylose yields similarly increased from as high as 34% of total potential sugars for pretreatment alone to between 35% and 37% when credit was taken for xylose released in digestion. Yields were much lower if no acid was used. Overall, up to about 92.5% of the total sugars originally available in the corn stover used could be recovered for coupled dilute acid pretreatment and enzymatic hydrolysis. Enhanced hemicellulase activity could further improve xylose yields, particularly for low cellulase loadings.

Yang and Wyman (2004) studied the enzymatic hydrolysis of corn stover cellulose by batch and flowthrough as well as water and acid pretreatment in a paper with 556 citations. They observed that increasing flow with just water enhanced the xylan dissolution rate, more than doubled total lignin removal, and increased cellulose digestibility. Furthermore, adding dilute sulfuric acid increased the rate of xylan removal for both batch and flowthrough systems. Further, adding acid also increased the lignin removal rate with flow, but less lignin was left in solution when acid was added in batch. Although the enzymatic hydrolysis of pretreated cellulose was related to xylan removal, the digestibility was much better for flowthrough compared with batch systems, for the same degree of xylan removal. Cellulose digestibility for flowthrough reactors was related to lignin removal as well.

TABLE 5.4
The Other Pretreatments of the Biomass

No.	Papers	Biomass	Prt.	Parameters	Keywords	Lead Author	Affil.	Cits
1	Schwanninger et al. (2004)	Wood, cellulose	VBM	Milling, temperature, particle size, oxygen	Milling, wood, cellulose	Schwanninger, Manfred 6602877236	Univ. Natr. Sci. App. Sci. Austria	883
2	Li et al. (2007)	Wood, lignin	Steam explosion	Lignin degradation	Lignin, wood, steam explosion	Li, Jiebing 7410069979	Royal Inst. Technol. Sweden	641

Prt, other pretreatments; Na, non-available; Cits., number of citations received for each paper.

5.3.4 The Other Pretreatments of the Biomass

There are two HCPs for the research fronts of the other pretreatments: mechanical and hydrothermal pretreatments (Table 5.4).

Schwanninger et al. (2004) studied the effect of short-time vibratory ball milling (VBM) on the chemical and structural changes of wood and cellulose in a paper with 869 citations. They observed that the milling itself was the main influencing factor and these changes were mainly associated with a decrease in the degree of crystallinity and/or a decrease in the degree of polymerization of the cellulose.

Li et al. (2007) studied lignin depolymerization and repolymerization and its critical role for delignification of aspen wood by steam explosion in a paper with 640 citations. They observed the competition between lignin depolymerization and repolymerization and identified the conditions required for these two types of reaction. Addition of 2-naphthol inhibited the repolymerization reaction strongly, resulting in a highly improved delignification by subsequent solvent extraction and an extracted lignin of uniform structure.

5.4 DISCUSSION

5.4.1 Introduction

The crude oil-based gasoline fuels have been widely used in the transportation sector since the 1920s. However, there have been great public concerns over the adverse environmental and human impact of these fuels. Hence, biomass-based bioethanol fuels have increasingly been used in blending gasoline and petrodiesel fuels, in the fuel cells, and in the biochemical production in a biorefinery context.

The research in the field of biological, chemical, mechanical, and hydrothermal pretreatments to increase the sugar and bioethanol yield has thus intensified in recent years.

However, it is essential to develop efficient incentive structures for the primary stakeholders to enhance the research in this field. Although there have been a number of review papers for this field, there has been no review of the most-cited 25 articles in this field.

Thus, this book chapter presents a review of the most-cited 25 articles in this field. Then, it discusses the key findings of these highly influential papers and comments on the future research priorities in this field.

As a first step for the search of the relevant literature, the keywords were selected using the whole set of papers. The selected keyword list was optimized to obtain a representative sample of papers for the searched research field. This keyword list was provided in Appendix for future replication studies.

As a second step, a sample dataset was used for this study. The first 25 articles with at least 461 citations each were selected for the review study. Key findings from each paper were taken from the abstracts of these papers and were discussed. Additionally, a number of brief conclusions were drawn and a number of relevant recommendations were made to enhance the future research landscape.

There are two primary research fronts for these most-cited 25 papers: biological and chemical pretreatments with 60% and 32% of the 25 HCPs, respectively (Table 5.5). Additionally, hydrothermal and mechanical pretreatments account for 4% of the reviewed papers each.

Further, biological, chemical, and hydrothermal pretreatments are over-represented in the reviewed papers compared to the 203 HCPs with 14%, 29%, and 5% surplus, respectively. On the other hand, enzymatic hydrolysis, enzymatic pretreatment, IL pretreatment, and fungal pretreatment are the most prolific individual pretreatments with 28%, 20%, 16%, and 12% of the reviewed papers, respectively.

TABLE 5.5
The Most Prolific Research Fronts for the Biomass Pretreatments

No.	Research Fronts	N Paper % Review	N Paper % Sample	Surplus
1	**Biological Pretreatments**	**60.0**	**73.9**	**−13.9**
	Enzymatic hydrolysis	28.0	31.5	−3.5
	Enzymatic pretreatments	20.0	26.1	−6.1
	Fungal pretreatments	12.0	13.8	−1.8
	Microbial pretreatment	0.0	2.5	−2.5
2	**Chemical Pretreatments**	**32.0**	**60.6**	**−28.6**
	Acid pretreatment	4.0	19.2	−15.2
	IL pretreatment	16.0	15.8	0.2
	Alkali pretreatment	4.0	9.4	−5.4
	Other chemical pretreatments	8.0	9.4	−1.4
	Ammonia pretreatment	0.0	3.4	−3.4
	Organic solvent pretreatment	0.0	3.4	−3.4
3	**Hydrothermal Pretreatments**	**4.0**	**9.4**	**−5.4**
	Steam explosion pretreatment	4.0	5.4	−1.4
	HCW pretreatment	0.0	2.5	−2.5
	LHW pretreatment	0.0	1.0	−1.0
	Autohydrolysis	0.0	0.5	−0.5
4	**Mechanical Pretreatments**	**4.0**	**3.5**	**0.5**
	Milling pretreatment	4.0	2.5	1.5
	Microwave pretreatment	0.0	1.0	−1.0
	Ultrasonic pretreatment	0	0.0	0.0

N paper (%) sample: the number of papers in the population sample of 203 papers. N paper (%) review: the number of papers in the sample of **25 most-cited** papers.

HCW, hot-compressed water; LHW, liquid hot water; IL, ionic liquid.

TABLE 5.6
The Most Prolific Research Fronts for the Biomass Used for the Biomass Pretreatments

No.	Research Fronts	N Paper % Review	N Paper % Sample	Surplus %
1	**Biomass Constituents**	**64.0**	**56.7**	**7.3**
	Cellulose	36.0	27.1	8.9
	Lignin	28.0	18.2	9.8
	Lignocellulose	0.0	5.9	–5.9
	Other constituents	0.0	3.0	–3.0
	Hemicellulose	0.0	2.5	–2.5
2	**Wood**	**28.0**	**14.3**	**13.7**
	Wood in general	24.0	9.9	14.1
	Softwood	4.0	4.4	–0.4
3	**Grass**	**4.0**	**5.4**	**–1.4**
	Switchgrass	4.0	3.4	0.6
	Other grass	0.0	2.0	–2.0
4	**Agricultural Residues**	**16.0**	**21.7**	**–5.7**
	Corn stover	8.0	9.4	–1.4
	Other residues	4.0	4.9	–0.9
	Wheat straw	4.0	4.4	–0.4
	Sugarcane bagasse	0.0	3.0	–3.0
5	**Biomass in General**	**4.0**	**24.6**	**–20.6**
	Lignocellulosic biomass	4.0	14.3	–10.3
	Biomass	0.0	10.3	–10.3

N paper (%) sample: the number of papers in the population sample of 203 papers. N paper (%) review: the number of papers in the sample of 25 most-cited papers.

Table 5.6 provides information on these biomass used in these 25 HCPs. There are three primary types of the biomass: biomass constituents, wood, and agricultural residues with 64%, 28%, and 16% of the 25 HCPs, respectively. Further, the other front is biomass in general with 4% of the reviewed papers.

Cellulose and lignin are the primary biomass constituents with 36% and 28% of the 25 HCPs, respectively. Further, wood in general and corn stover are the other individual prolific biomass with 24% and 8% of the reviewed papers, respectively.

It is notable the wood, in general, lignin, cellulose, and biomass constituents are over-represented with 7%–14% surplus each in the reviewed papers while biomass, in general, lignocellulosic biomass, lignocellulose, and agricultural residues are under-represented with 6%–21% deficit each.

5.4.2 The Biological Pretreatment of the Biomass

There are eight HCPs for the research front of the sole biological pretreatments. Bourbonnais and Paice (1990) studied the lignin biodegradation by laccases of *T versicolor* and observed that laccase was capable of oxidizing both phenolic and non-phenolic moieties of lignin, but that the latter was dependent on the co-presence of primary laccase substrates. On the other hand, Floudas et al. (2012) studied the paleozoic origin of enzymatic lignin decomposition based on the analysis of 31 fungal genomes and found that lignin-degrading peroxidases expanded in the lineage leading to the ancestor of the Agaricomycetes.

Tien and Kirk (1983) studied the lignin-degrading enzyme from the *P. chrysosporium* and observed that this *P. chrysosporium* contained an enzyme that degraded lignin substructure model compounds as well as spruce and birch lignins. On the other hand, Martinez et al. (2008) provided a roadmap for constructing enhanced *T. reesei* strains as a source for cellulases and hemicellulases for the biomass degradation.

Pandey and Pitman (2003) studied pine and beech wood degradation by *C. puteana*, *C. versicolor* (*T. versicolor*), and *P. chrysosporium,* and observed that in wood decayed by *C. puteana*, there was a progressive increase in lignin content relative to carbohydrate evident from increases in the relative height of lignin-associated bands and a corresponding decrease in the intensities of carbohydrate bands. On the other hand, Martinez et al. (2004) carried out a genomic analysis of wood-degrading *P. chrysosporium* as a source for cellulases and hemicellulases for the biomass degradation.

Quinlan et al. (2011) studied the oxidative degradation of cellulose by a glycoside hydrolase (CAZy) GH61 enzyme focusing on cellulose-enhancing factors and showed that GH61 enzymes were a unique family of copper-dependent oxidases. On the other hand, Glenn and Gold (1985) purified and characterized an extracellular Mn(II)-dependent peroxidase from the lignin-degrading *P. chrysosporium* and observed that Mn(III) played a role in the enzyme mechanism.

Overall, these HCPs clearly show that these enzymatic pretreatments degrade the biomass to improve the hydrolysis of the biomass and fermentation of the resulting hydrolysates with an improved sugar and bioethanol yield.

5.4.3 The Chemical Pretreatment of the Biomass

There are eight HCPs for the research front of chemical pretreatments: four, one, and three HCPs for IL, acid, and other chemical pretreatments, respectively.

5.4.3.1 The Sole Ionic Liquid Pretreatment of the Biomass

Kilpelainen et al. (2007) studied the dissolution of spruce and pine wood in imidazolium-based ILs under gentle conditions and observed that a variety of ILs could only partially dissolve wood chips, whereas ILs such as 1-butyl-3-methylimidazolium chloride and 1-allyl-3-methylimidazolium chloride had good solvating power for Norway spruce sawdust and Norway spruce and Southern pine thermomechanical pulp fibers. On the other hand, Sun et al. (2009) studied the complete dissolution and partial delignification of oak and pine wood in the 1-ethyl-3-methylimidazolium acetate ([C_2mim]OAc) and observed that both woods could be completely dissolved in this IL after mild grinding.

Fukaya et al. (2008) studied the cellulose dissolution with polar ILs under mild conditions and observed that especially N-ethyl-N′-methylimidazolium methylphosphonate enabled the preparation of 10 wt% cellulose solution without pretreatments and heating. On the other hand, Remsing et al. (2006) studied the mechanism of cellulose dissolution in the IL 1-n-butyl-3-methylimidazolium chloride ([C_4mim]Cl) and observed that the solvation of cellulose by the [C_4mim]Cl involved hydrogen bonding between the carbohydrate hydroxyl protons and the IL chloride ions in a 1:1 stoichiometry.

5.4.3.2 The Sole Acid Pretreatment of the Biomass

Roman and Winter (2004) studied the impact of sulfate groups from H_2SO_4 hydrolysis on the thermal degradation behavior of bacterial cellulose and observed that the sulfate group content increased with acid concentration, acid-to-cellulose ratio, and hydrolysis time.

5.4.3.3 The Other Sole Chemical Pretreatments of the Biomass

Cai and Zhang (2005) studied the rapid dissolution of cellulose in LiOH/urea and NaOH/urea aqueous solutions and observed that the dissolution power of the solvent systems was in the order of LiOH/urea >NaOH/urea ≫ KOH/urea aqueous solution.

Sasaki et al. (2000) studied the dissolution and hydrolysis of cellulose in subcritical and supercritical water and observed that cellulose hydrolysis at high temperature took place with dissolution in water. This was probably because of the cleavage of intra- and intermolecular hydrogen linkages in the cellulose crystal. On the other hand, Sasaki et al. (1998) studied cellulose hydrolysis in subcritical and supercritical water and proposed a new method to hydrolyze cellulose rapidly in supercritical water to recover glucose, fructose, and oligomers.

Overall, these HCPs clearly show that these chemical pretreatments degrade the biomass to improve the hydrolysis of the biomass and fermentation of the resulting hydrolysates with an improved sugar and bioethanol yield.

5.4.4 The Biological and Other Pretreatments of the Biomass

Li et al. (2010) compared dilute acid and IL pretreatment of switchgrass in terms of delignification, enzymatic saccharification efficiency, and saccharide yields, and for the IL pretreatment, they observed that switchgrass exhibited reduced cellulose crystallinity, increased surface area, and decreased lignin content compared to dilute acid pretreatment. On the other hand, Lee et al. (2009) studied the enzymatic hydrolysis of lignin pretreated by the ILs and observed that the cellulose in the pretreated wood flour became far-less crystalline without undergoing solubilization while 40% of the lignin was removed, the cellulose crystallinity index dropped below 45, resulting in more than 90% of the cellulose in wood flour to be hydrolyzed by *T. viride* cellulase.

Eriksson et al. (2002) studied the enzymatic hydrolysis of the steam- and surfactant-pretreated spruce with cellulase of *T. reesei* and found that non-ionic surfactants were the most effective. On the other hand, Silverstein et al. (2007) compared H_2SO_4, NaOH, H_2O_2, and O_3 pretreatments for improving saccharification of cotton stalks and observed that solids from H_2SO_4, NaOH, and H_2O_2 pretreatments showed significant lignin degradation and/or high sugar availability and hence were hydrolyzed by Celluclast 1.5 L and Novozym 188.

Saha et al. (2005) studied the enzymatic hydrolysis of the acid-pretreated wheat straw for bioethanol production and observed that the maximum yield of monomeric sugars from wheat straw by dilute H_2SO_4 pretreatment and enzymatic hydrolysis using cellulase, β-glucosidase, xylanase, and esterase was 565 mg/g. On the other hand, Lloyd and Wyman (2005) studied the enzymatic hydrolysis of acid-pretreated corn stover and observed that up to 15% of the total potential sugar in the substrate could be released as glucose during pretreatment and between 15% and 90+% of the xylose remaining in the solid residue could be recovered in subsequent enzymatic hydrolysis, depending on the enzyme loading.

Yang and Wyman (2004) studied the enzymatic hydrolysis of corn stover cellulose by batch and flowthrough as well as water and acid pretreatment and observed that increasing flow with just water enhanced the xylan dissolution rate, more than doubled total lignin removal, and increased cellulose digestibility.

Overall, these HCPs clearly show that these enzymatic pretreatments in combination with chemical and other pretreatments degrade the biomass to improve the hydrolysis of the biomass and fermentation of the resulting hydrolysates with an improved sugar and bioethanol yield.

5.4.5 The Other Pretreatments of the Biomass

There are two HCPs for the research fronts of the other pretreatments: mechanical and hydrothermal pretreatments.

Schwanninger et al. (2004) studied the effect of short-time VBM on the chemical and structural changes of wood and cellulose and observed that the milling itself was the main influencing factor and these changes were mainly associated with a decrease in the degree of crystallinity and/or a decrease in the degree of polymerization of the cellulose. On the other hand, Li et al. (2007) studied lignin depolymerization and repolymerization and its critical role for delignification of aspen wood by steam explosion pretreatment, observed the competition between lignin depolymerization and repolymerization, and identified the conditions required for these two types of reaction.

5.5 CONCLUSION AND FUTURE RESEARCH

The brief information about the key research fronts covered by the 25 most-cited papers with at least 461 citations each is given under two primary headings: biological and chemical pretreatments of the biomass. The other two minor research fronts are mechanical and hydrothermal pretreatments of the biomass.

The usual characteristics of these highly cited papers are that these pretreatments are often used in combination with other pretreatments for the hydrolysis of the biomass as seen in Table 5.3. In this way, these biomass pretreatments are effective in disrupting the biomass constituents such as cellulose, lignin, and hemicellulose of these feedstocks resulting in improved sugar and bioethanol yield.

The key findings on these research fronts should be read in light of the increasing public concerns about climate change, GHG emissions, and global warming as these concerns have been certainly behind the boom in the research on the bioethanol production as an alternative to crude oil-based gasoline and diesel fuels in the last decades.

As Table 5.5 shows, there are two primary research fronts for these most-cited 25 papers: biological and chemical pretreatments. Further, Table 5.6 provides information on the biomass used in these 25 HCPs. There are three primary types of the biomass: biomass constituents, wood, and agricultural residues. Further, lignin, cellulose, and wood in general are the primary biomass types while corn stover is the key agricultural residue.

These studies emphasize the importance of proper incentive structures for the efficient development and application of biomass pretreatments to enhance sugar and bioethanol yield of the biomass after the hydrolysis of the biomass and the following fermentation of the resulting hydrolysates in light of North's institutional framework (North, 1991). In this context, the major producers and users of bioethanol fuels such as the USA, Canada, China, Japan, and Europe had developed strong incentive structures for the effective development and application of pretreatments for bioethanol and sugar production.

It is recommended that such review studies are performed for the primary research fronts of biomass pretreatments as well as the biomass constituents, wood, grass, and agricultural residues.

ACKNOWLEDGMENTS

The contribution of the highly cited researchers in the field of the biomass pretreatments has been gratefully acknowledged.

REFERENCES

Alvira, P., E. Tomas-Pejo, M. Ballesteros and M. J. Negro. 2010. Pretreatment technologies for an efficient bioethanol production process based on enzymatic hydrolysis: A review. *Bioresource Technology* 101:4851–4861.

Angelici, C., B. M. Weckhuysen and P. C. A. Bruijnincx. 2013. Chemocatalytic conversion of ethanol into butadiene and other bulk chemicals. *ChemSusChem* 6:1595–1614.

Antolini, E. 2007. Catalysts for direct ethanol fuel cells. *Journal of Power Sources* 170:1–12.

Antolini, E. 2009. Palladium in fuel cell catalysis. *Energy and Environmental Science* 2:915–931.

Bourbonnais, R. and M. G. Paice. 1990. Oxidation of non-phenolic substrates. An expanded role for laccase in lignin biodegradation. *FEBS Letters* 267:99–102.

Brandt, A., J. Grasvik, J. P. Hallett and T. Welton. 2013. Deconstruction of lignocellulosic biomass with ionic liquids. *Green Chemistry* 15:550–583.

Burnham, J. F. 2006. Scopus database: A review. *Biomedical Digital Libraries* 3:1–8.

Cai, J. and L. Zhang. 2005. Rapid dissolution of cellulose in LiOH/urea and NaOH/urea aqueous solutions. *Macromolecular Bioscience* 5:539–548.

Chandra, R. P., R. Bura and W. E. Mabee, et al. 2007. Substrate pretreatment: The key to effective enzymatic hydrolysis of lignocellulosics? *Advances in Biochemical Engineering/Biotechnology* 108:67–93.

Eriksson, T., J. Borjesson and F. Tjerneld. 2002. Mechanism of surfactant effect in enzymatic hydrolysis of lignocellulose. *Enzyme and Microbial Technology* 31:353–364.

Fernando, S., S. Adhikari, C. Chandrapal and M. Murali. 2006. Biorefineries: Current status, challenges, and future direction. *Energy & Fuels* 20:1727–1737.

Floudas, D., M. Binder and R. Riley, et al. 2012. The paleozoic origin of enzymatic lignin decomposition reconstructed from 31 fungal genomes. *Science* 336:1715–1719.

Fukaya, Y., K. Hayashi, M. Wada and H. Ohno. 2008. Cellulose dissolution with polar ionic liquids under mild conditions: Required factors for anions. *Green Chemistry* 10:44–46.

Glenn, J. K. and M. H. Gold. 1985. Purification and characterization of an extracellular Mn(II)-dependent peroxidase from the lignin-degrading basidiomycete, *Phanerochaete chrysosporium. Archives of Biochemistry and Biophysics* 242:329–341.

Hahn-Hagerdal, B., M. Galbe, M. F. Gorwa-Grauslund, G. Liden and G. Zacchi. 2006. Bio-ethanol – The fuel of tomorrow from the residues of today. *Trends in Biotechnology* 24:549–556.

Hendriks, A. T. W. M. and G. Zeeman. 2009. Pretreatments to enhance the digestibility of lignocellulosic biomass. *Bioresource Technology* 100:10–18.

Hill, J., E. Nelson, D. Tilman, S. Polasky and D. Tiffany. 2006. Environmental, economic, and energetic costs and benefits of biodiesel and ethanol biofuels. *Proceedings of the National Academy of Sciences of the United States of America* 103:11206–11210.

Hill, J., S. Polasky and E. Nelson, et al. 2009. Climate change and health costs of air emissions from biofuels and gasoline. *Proceedings of the National Academy of Sciences of the United States of America* 106:2077–2082.

Himmel, M. E., S. Y. Ding and D. K. Johnson, et al. 2007. Biomass recalcitrance: Engineering plants and enzymes for biofuels production. *Science* 315:804–807.

Hsieh, W. D., R. H. Chen, T. L. Wu and T. H. Lin. 2002. Engine performance and pollutant emission of an SI engine using ethanol-gasoline blended fuels. *Atmospheric Environment* 36:403–410.

Hu, Z. and Z. Wen. 2008. Enhancing enzymatic digestibility of switchgrass by microwave-assisted alkali pretreatment. *Biochemical Engineering Journal* 38:369–378.

Huang, H. J., S. Ramaswamy, U. W. Tschirner and B. V. Ramarao. 2008. A review of separation technologies in current and future biorefineries. *Separation and Purification Technology* 62:1–21.

Kilpelainen, I., X. Xie and A. King, et al. 2007. Dissolution of wood in ionic liquids. *Journal of Agricultural and Food Chemistry* 55:9142–9148.

Konur, O. 2000. Creating enforceable civil rights for disabled students in higher education: An institutional theory perspective. *Disability & Society* 15:1041–1063.

Konur, O. 2002a. Access to nursing education by disabled students: Rights and duties of nursing programs. *Nurse Education Today* 22:364–374.

Konur, O. 2002b. Assessment of disabled students in higher education: Current public policy issues. *Assessment and Evaluation in Higher Education* 27:131–52.

Konur, O. 2002c. Access to employment by disabled people in the UK: Is the Disability Discrimination Act working? *International Journal of Discrimination and the Law* 5:247–279.

Konur, O. 2006a. Participation of children with dyslexia in compulsory education: Current public policy issues. *Dyslexia* 12:51–67.

Konur, O. 2006b. Teaching disabled students in Higher Education. *Teaching in Higher Education* 11:351–363.

Konur, O. 2007a. A judicial outcome analysis of the Disability Discrimination Act: A windfall for the employers? *Disability & Society* 22:187–204.

Konur, O. 2007b. Computer-assisted teaching and assessment of disabled students in higher education: The interface between academic standards and disability rights. *Journal of Computer Assisted Learning* 23:207–219.

Konur, O. 2012. The evaluation of the research on the bioethanol: A scientometric approach. *Energy Education Science and Technology Part A: Energy Science and Research* 28:1051–1064.

Konur, O. 2015. Current state of research on algal bioethanol. In *Marine Bioenergy: Trends and Developments*, Eds. S. K. Kim and C. G. Lee, pp. 217–244. Boca Raton, FL: CRC Press.

Konur, O. 2019. Cyanobacterial bioenergy and biofuels science and technology: A scientometric overview. In *Cyanobacteria: From Basic Science to Applications*, Eds. A. K. Mishra, D. N. Tiwari and A. N. Rai, pp. 419–442. Amsterdam: Elsevier.

Konur, O. 2020. The scientometric analysis of the research on the bioethanol production from green macroalgae. In *Handbook of Algal Science, Technology and Medicine*, Ed. O. Konur, pp. 385–401. London: Academic Press.

Konur, O. 2023. Biomass pretreatments: Scientometric study. In *Bioethanol Fuel Production Processes. I: Biomass Pretreatments. Handbook of Bioethanol Fuels Volume 1*, Ed. O. Konur. Boca Raton, FL: CRC Press.

Lee, S. H., T. V. Doherty, R. J. Linhardt and J. S. Dordick. 2009. Ionic liquid-mediated selective extraction of lignin from wood leading to enhanced enzymatic cellulose hydrolysis. *Biotechnology and Bioengineering* 102:1368–1376.

Li, C., B. Knierim and C. Manisseri, et al. 2010. Comparison of dilute acid and ionic liquid pretreatment of switchgrass: Biomass recalcitrance, delignification and enzymatic saccharification. *Bioresource Technology* 101:4900–4906.

Li, J., G. Henriksson and G. Gellerstedt. 2007. Lignin depolymerization/repolymerization and its critical role for delignification of aspen wood by steam explosion. *Bioresource Technology* 98:3061–3068.

Lin, Y. and S. Tanaka. 2006. Ethanol fermentation from biomass resources: Current state and prospects. *Applied Microbiology and Biotechnology* 69:627–642.

Lloyd, T. A. and C. E. Wyman. 2005. Combined sugar yields for dilute sulfuric acid pretreatment of corn stover followed by enzymatic hydrolysis of the remaining solids. *Bioresource Technology* 96:1967–1977.

Ma, X., L. Sun and C. Song. 2002. A new approach to deep desulfurization of gasoline, diesel fuel and jet fuel by selective adsorption for ultra-clean fuels and for fuel cell applications. *Catalysis Today* 77:107–116.

Mani, S., L. G. Tabil and S. Sokhansanj. 2004. Grinding performance and physical properties of wheat and barley straws, corn stover and switchgrass. *Biomass and Bioenergy* 27:339–352.

Martinez, D., L. F. Larrondo and N. Putnam, et al. 2004. Genome sequence of the lignocellulose degrading fungus *Phanerochaete chrysosporium* strain RP78. *Nature Biotechnology* 22:695–700.

Martinez, D., R. M. Berka and B. Henrissat, et al. 2008. Genome sequencing and analysis of the biomass-degrading fungus *Trichoderma reesei* (syn. *Hypocrea jecorina*). *Nature Biotechnology* 26:553–560.

Morschbacker, A. 2009. Bio-ethanol based ethylene. *Polymer Reviews* 49:79–84.

Mosier, N., C. Wyman and B. Dale, et al. 2005. Features of promising technologies for pretreatment of lignocellulosic biomass. *Bioresource Technology* 96:673–686.

Najafi, G., B. Ghobadian and T. Tavakoli, et al. 2009. Performance and exhaust emissions of a gasoline engine with ethanol blended gasoline fuels using artificial neural network. *Applied Energy* 86:630–639.

Newman, P. W. G. and J. R. Kenworthy. 1989. Gasoline consumption and cities: A comparison of U.S. cities with a global survey. *Journal of the American Planning Association* 55:24–37.

North, D. C. 1991. Institutions. *Journal of Economic Perspectives* 5:97–112.

Olsson, L. and B. Hahn-Hagerdal. 1996. Fermentation of lignocellulosic hydrolysates for ethanol production. *Enzyme and Microbial Technology* 18:312–331.

Pandey, K. K. and A. J. Pitman. 2003. FTIR studies of the changes in wood chemistry following decay by brown-rot and white-rot fungi. *International Biodeterioration and Biodegradation* 52:151–160.

Pu, Y., F. Hu, F. Huang, B. H. Davison and A. J. Ragauskas. 2013. Assessing the molecular structure basis for biomass recalcitrance during dilute acid and hydrothermal pretreatments. *Biotechnology for Biofuels* 6:15.

Quinlan, R. J., M. D. Sweeney and L. lo Leggio, et al. 2011. Insights into the oxidative degradation of cellulose by a copper metalloenzyme that exploits biomass components. *Proceedings of the National Academy of Sciences of the United States of America* 108:15079–15084.

Remsing, R. C., R. P. Swatloski, R. D. Rogers and G. Moyna. 2006. Mechanism of cellulose dissolution in the ionic liquid 1-n-butyl-3- methylimidazolium chloride: A^{13}C and$^{35/37}$Cl NMR relaxation study on model systems. *Chemical Communications* 2006:1271–1273.

Roman, M. and W. T. Winter. 2004. Effect of sulfate groups from sulfuric acid hydrolysis on the thermal degradation behavior of bacterial cellulose. *Biomacromolecules* 5:1671–1677.

Saha, B. C., L. B. Iten, M. A. Cotta and Y. V. Wu. 2005. Dilute acid pretreatment, enzymatic saccharification and fermentation of wheat straw to ethanol. *Process Biochemistry* 40:3693–3700.

Sanchez, O. J. and C. A. Cardona. 2008. Trends in biotechnological production of fuel ethanol from different feedstocks. *Bioresource Technology* 99:5270–5295.

Sasaki, M., B. Kabyemela and R. Malaluan, et al. 1998. Cellulose hydrolysis in subcritical and supercritical water. *Journal of Supercritical Fluids* 13:261–268.

Sasaki, M., Z. Fang, Y. Fukushima, T. Adschiri and K. Arai. 2000. Dissolution and hydrolysis of cellulose in subcritical and supercritical water. *Industrial and Engineering Chemistry Research* 39:2883–2890.

Schwanninger, M., J. C. Rodrigues, H. Pereira and B. Hinterstoisser. 2004. Effects of short-time vibratory ball milling on the shape of FT-IR spectra of wood and cellulose. *Vibrational Spectroscopy* 36:23–40.

Silverstein, R. A., Y. Chen, R. R. Sharma-Shivappa, M. D. Boyette and J. Osborne. 2007. A comparison of chemical pretreatment methods for improving saccharification of cotton stalks. *Bioresource Technology* 98:3000–3011.

Sun, N., M. Rahman and Y. Qin, et al. 2009. Complete dissolution and partial delignification of wood in the ionic liquid 1-ethyl-3-methylimidazolium acetate. *Green Chemistry* 11:646–665.

Sun, Y. and J. Cheng. 2002. Hydrolysis of lignocellulosic materials for ethanol production: A review. *Bioresource Technology* 83:1–11.

Taherzadeh, M. J. and K. Karimi. 2007. Enzyme-based hydrolysis processes for ethanol from lignocellulosic materials: A review. *Bioresources* 2:707–738.

Taherzadeh, M. J. and K. Karimi. 2008. Pretreatment of lignocellulosic wastes to improve ethanol and biogas production: A review. *International Journal of Molecular Sciences* 9:1621–1651.

Tien, M. and T. K. Kirk. 1983. Lignin-degrading enzyme from the hymenomycete *Phanerochaete chrysosporium* Burds. *Science* 221:661–663.

Yang, B. and C. E. Wyman. 2004. Effect of xylan and lignin removal by batch and flowthrough pretreatment on the enzymatic digestibility of corn stover cellulose. *Biotechnology and Bioengineering* 86:88–98.

Zhang, Y. H. P. and L. R. Lynd. 2004. Toward an aggregated understanding of enzymatic hydrolysis of cellulose: Noncomplexed cellulase systems. *Biotechnology and Bioengineering* 88:797–824.

Zhu, S., Y. Wu and Q. Chen. 2006. Dissolution of cellulose with ionic liquids and its application: A mini-review. *Green Chemistry* 8:325–327.

Biological Biomass Pretreatments for Bioethanol Production

6 Biological Pretreatment of the Biomass

Scientometric Study

Ozcan Konur
(Formerly) Ankara Yildirim Beyazit University

6.1 INTRODUCTION

The crude oil-based gasoline fuels (Ma et al., 2002; Newman and Kenworthy, 1989) have been widely used in the transportation sector since the 1920s. However, there have been great public concerns over the adverse environmental and human impact of these fuels (Hill et al., 2006, 2009). Hence, biomass-based bioethanol fuels (Hill et al., 2006; Konur, 2012e, 2015, 2020a; Sun and Cheng, 2002) have increasingly been used in blending gasoline fuels (Hsieh et al., 2002; Najafi et al., 2009), in the fuel cells (Antolini, 2007, 2009), and in the biochemical production (Angelici et al., 2013; Morschbacker, 2009) in a biorefinery context (Fernando et al., 2006; Huang et al., 2008).

However, it is necessary to pretreat the biomass (Alvira et al., 2010; Taherzadeh and Karimi, 2008) to enhance the yield of the bioethanol (Hahn-Hagerdal et al., 2006; Sanchez and Cardona, 2008) prior to the bioethanol production through the hydrolysis (Alvira et al., 2010; Sun and Cheng, 2002; Taherzadeh and Karimi, 2007) and fermentation (Lin and Tanaka, 2006; Olsson and Hahn-Hagerdal, 1996) of the biomass and hydrolysates, respectively.

The research in the fields of biological pretreatments of the biomass has also intensified in recent years in the research fronts of the enzymatic pretreatments (Sindhu et al., 2016; van Dyk and Pletsche, 2012), fungal pretreatments (Dashtban et al., 2009; Sanchez, 2009), enzymatic hydrolysis by enzymes (Taherzadeh and Karimi, 2007; van Dyk and Pletsche, 2012), and microbial pretreatments (Brown and Chang, 2014; Bugg et al., 2011).

However, it is essential to develop efficient incentive structures (North, 1991) for the primary stakeholders to enhance the research in this field (Konur, 2000, 2002a–c, 2006a, b, 2007a, b). The scientometric analysis has been used in this context to inform the primary stakeholders about the current state of the research in a selected research field (Garfield, 1955; Konur, 2011, 2012a–i, 2015, 2018b, 2019, 2020a).

Although there has been a number of review papers on the biological pretreatment of the biomass (Brown et al., 2014; Bugg et al., 2011; Dashtban et al., 2009; Sanchez, 2009; Sindhu et al., 2016; Taherzadeh and Karimi, 2007; van Dyk and Pletsche, 2012), there has been no review of the most-cited 25 articles in this field.

Thus, this book chapter presents a scientometric study of the research in biological pretreatments of the biomass. It examines the scientometric characteristics of both the sample and population data presenting scientometric characteristics of both these datasets in the order of documents, authors, publication years, institutions, funding bodies, source titles, countries, Scopus subject categories, keywords, and research fronts.

DOI: 10.1201/9781003226536-9

6.2 MATERIALS AND METHODS

The search for this study was carried out using Scopus database (Burnham, 2006) in April 2022.

As a first step for the search of the relevant literature, the keywords were selected using the first most-cited 200 papers. The selected keyword list was optimized to obtain a representative sample of papers for the searched research field. This keyword list was provided in the Appendix for future replication studies.

As a second step, two sets of data were used for this study. First, a population sample of over 13,500 papers was used to examine the scientometric characteristics of the population data. Secondly, a sample of 135 most-cited papers, corresponding to 1% of the population papers, was used to examine the scientometric characteristics of these citation classics.

The scientometric characteristics of both these sample and population datasets were presented in the order of documents, authors, publication years, institutions, funding bodies, source titles, countries, Scopus subject categories, keywords, and research fronts.

Lastly, the key scientometric findings for both datasets were discussed to highlight the research landscape for the biological pretreatments of the biomass. Additionally, a number of brief conclusions were drawn, and a number of relevant recommendations were made to enhance the future research landscape.

6.3 RESULTS

6.3.1 The Most Prolific Documents in the Biological Pretreatments of the Biomass

The information on the types of documents for both datasets is given in Table 6.1. The articles and conference papers dominate both the sample (73%) and population papers (95%) as they are under-represented in the sample papers by 23%. Further, review papers and short surveys have a substantial presence in the sample papers (27%) and have a surplus (24%) as they are over-represented in the sample papers.

It is further notable that 97% of the population papers were published in journals, while 2% and 1% of them were published in book series and books, respectively. On the contrary, all of the sample papers were published in the journals.

TABLE 6.1
Documents in the Biological Pretreatments of the Biomass

Documents	Sample Dataset (%)	Population Dataset (%)	Surplus (%)
Article	70.4	92.4	–22.0
Review	24.4	2.9	21.5
Short survey	3.0	0.2	2.8
Conference paper	2.2	2.5	–0.3
Book chapter	0.0	1.3	–1.3
Letter	0.0	0.3	–0.3
Note	0.0	0.3	–0.3
Book	0.0	0.1	–0.1
Editorial	0.0	0.0	0.0
Sample size	135	13,506	

Sample dataset: the number of papers (%) in the set of 135 highly cited papers.
Population dataset: the number of papers (%) in the set of the 13,506 population papers.

6.3.2 The Most Prolific Authors in the Biological Pretreatments of the Biomass

The information about the most-prolific 13 authors with at least 3% of sample papers each is given in Table 6.2.

The most prolific author is Jack N. Saddler of British Columbia University of Canada with 5.9% of the sample papers working primarily on the enzymatic hydrolysis of the biomass by the use of enzymes. Charles E. Wyman, Bernard Henrissat, Dan Cullen, Igor V. Grigoriev, and Pedro M. Coutinho follow this top author with 3.7%–5.2% of the sample papers each, working primarily on the fungal pretreatments of the biomass.

In total, 13 institutions house these prolific authors. The most prolific country for the sample dataset is the USA with seven authors, while in total, seven countries house these authors.

TABLE 6.2
Most-Prolific Authors in the Biological Pretreatments of the Biomass

No.	Author Name	Author Code	Sample Papers %	Population Papers %	Institution	Country	HI	N	Res. Front
1	Saddler, Jack N.	7005297559	5.9	0.9	Univ. British Columbia	Canada	96	407	Enzy. Hydrol.
2	Wyman, Charles E.	7004396809	5.2	0.3	Univ. Calif. Riverside	USA	80	289	Enzy. Hydrol.
3	Henrissat, Bernard	7005911606	5.2	0.2	King Abdulaziz Univ.	S. Arabia	141	589	Fungi
4	Cullen, Dan	7004396809	4.4	0.2	USDA Forest Serv.	USA	46	119	Fungi
5	Grigoriev, Igor V.	25027225800	3.7	0.1	Univ. Calif. Berkeley	USA	113	337	Fungi
6	Coutinho, Pedro M.	7006153340	3.7	0.1	Aix-Marseille Univ.	France	71	145	Fungi
7	Pandey, Ashok	7201771319	3.0	0.2	CSIR	India	88	817	Enzy. Hydrol.
8	Holtzapple, Mark T.	7004167004	3.0	0.1	Texas A&M Univ.	USA	46	195	Enzy. Hydrol.
9	Baker, Scott E.	35232609200	3.0	0.1	Pacific NW Natl. lab.	USA	48	145	Fungi
10	Larrondo, Luis F.	6603826074	3.0	0.1	Pontificial Cath. Univ. Chile	Chile	28	75	Fungi
11	Martinez, Angel T.	55622506400	3.0	0.4	CSIC	Spain	79	337	Fungi
12	Martinez, Diego	7202958664	3.0	0.1	Massachusetts Inst. Technol.	USA	28	35	Fungi
13	Yang, Bin	7404473046	3.0	0.1	Washington State Univ.	USA	39	93	Enzy. Hydrol.

Author code: the unique code given by Scopus to the authors. Sample papers: the number of papers authored in the sample dataset. Population papers: the number of papers authored in the population dataset.
HI, H-index; N, number of papers published by each author; *, female; Enzy. Hydrol., enzymatic hydrolysis.

The most-prolific research front is the fungal pretreatment of the biomass with eight authors, followed by the enzymatic hydrolysis of the biomass by enzymes with five authors. It is notable that there are no authors related to both enzymatic and microbial pretreatments of the biomass.

On the other hand, there is significant gender deficit (Beaudry and Lariviere, 2016) for the sample dataset as surprisingly all of these top researchers are male.

Additionally, there are a large number authors with the relatively low citation impact and with 0.3%–0.4% of the population papers each: Qiang Yong, Bruce E. Dale, Runcang Sun, Hasan Jameel, Yongcan Jin, Arkady P. Sinitsyn, Venkatesh Balan, Jinguang Hu, Blake A. Simmons, Andre Ferraz, Arthur J. Ragauskas, Guido Zacchi, Xiao Zhang, Junyong Zhu, Caoxing Huang, Chenhuan Lai, Michael E. Himmell, and Feng Xu.

6.3.3 The Most Prolific Research Output by Years in the Biological Pretreatments of the Biomass

Information about papers published between 1970 and 2022 is given in Figure 6.1. This figure clearly shows that the bulk of the research papers in the population dataset were published primarily in the 2010s with 51% of the population dataset. The publication rates for the 2020s, 2000s, 1990s, 1980s, and 1970s were 14%, 14%, 9%, 6%, and 2% respectively. Additionally, 1% of the population papers were published between 1892 and 1969.

Similarly, the bulk of the research papers in the sample dataset were published in the 2010s and 2000s with 28% and 50% of the sample dataset, respectively. The publication rates for the 1990s, 1980s, and 1970s were 12%, 4%, and 2% of the sample papers, respectively.

The most prolific publication year for the population dataset was 2021 with 7.3%, while 65% of the population papers were published between 2011 and 2022. Similarly, 67% of the sample papers were published between 2002 and 2012, while the most prolific publication years were 2007 and 2009 with 10% of the sample papers each.

6.3.4 The Most Prolific Institutions in the Biological Pretreatments of the Biomass

Information about the most prolific eight institutions publishing at least 3.7% of the sample papers each is given in Table 6.3.

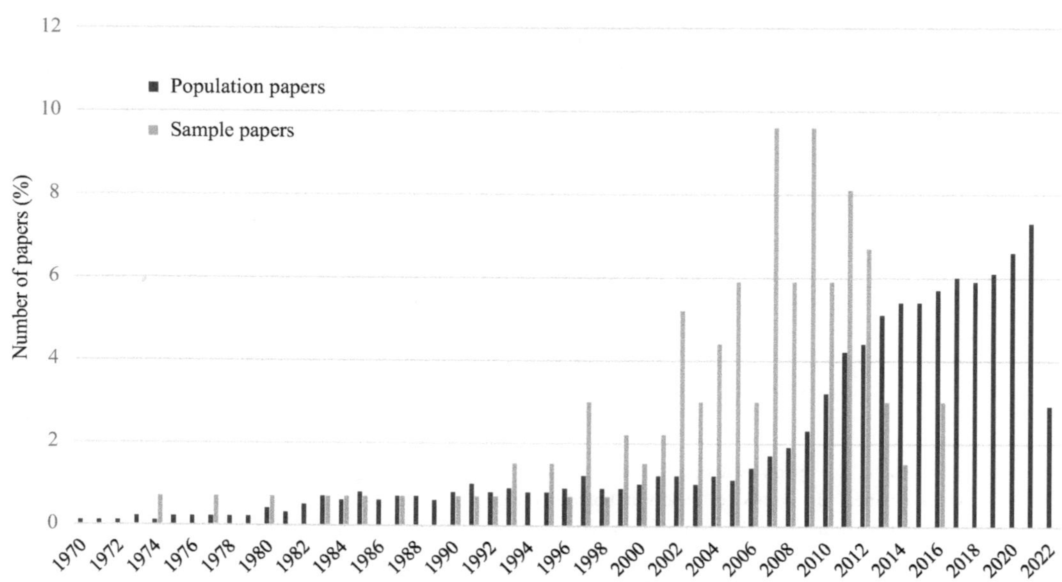

FIGURE 6.1 The research output by years regarding the biological pretreatments of the biomass.

TABLE 6.3
The Most-Prolific Institutions in Biological Pretreatments of the Biomass

No.	Institutions	Country	Sample Papers (%)	Population Papers (%)	Surplus (%)
1	USDA Forest Serv.	USA	7.4	1.4	6.0
2	Univ. British Columbia	Canada	6.7	1.3	5.4
3	Natl. Renew. Ener. Lab.	USA	6.7	0.9	5.8
4	Novozymes Inc.	USA	6.7	0.3	6.4
5	Dartmouth Coll.	USA	5.2	0.3	4.9
6	VTT Tech. Res. Ctr.	Finland	4.4	0.7	3.7
7	Aix-Marseille Univ.	France	4.4	0.5	3.9
8	Joint Genome Inst.	USA	4.4	0.2	4.2
9	CSIC	Spain	3.7	1.0	2.7
10	CNRS	France	3.7	0.8	2.9
11	Pacific NW Natl. Lab.	USA	3.7	0.3	3.4

The most prolific institution is the Forest Service of the US Department of Agriculture (USDA) with 7.4% of the sample papers. The University of British Columbia, National Renewable Energy Laboratory (NREL), and Novozymes Inc. of the USA follow them with 6.7% of the sample papers each.

The top country for these most prolific institutions is the USA with six institutions, while France have two institutions. In total, five countries house these top institutions.

On the other hand, the institutions with the most impact are the Novozymes Inc., USDA Forest Service, and NREL with 6% surplus each. Similarly, the institutions with the least impact are the Superior Council of Scientific Investigations (CSIC) of Spain and Scientific Research National Center (CNRS) of France with 3% surplus each.

Additionally, there are a large number of institutions with the relatively low citation impact and with 0.7%–2.8% of the population papers each: Chinese Academy of Sciences, South China University of Technology, Nanjing Forestry University, University of Sao Paulo, Beijing Forestry University, North Carolina State University, Kyoto University, Lund University, Michigan State University, China Agricultural University, Russian Academy of Sciences, University of Wisconsin-Madison, USDA Agricultural Research Service, and University of Helsinki.

6.3.5 The Most Prolific Funding Bodies in the Biological Pretreatments of the Biomass

Information about the most prolific seven funding bodies funding at least 2.2% of the sample papers each is given in Table 6.4. Only 33% and 47% of the sample and population papers, respectively, were funded.

The most prolific funding body is the US Department of Energy (US DOE) with 5.9% of the sample papers, followed by the European Commission with 4.4% of the sample papers.

The most prolific country for these top funding bodies is the USA with six funding bodies, while the Europe has only one funding body. In total, the USA and the EU house these top funding bodies.

The funding body with the most citation impact is the US DOE with 4% surplus. Similarly, the funding body with the least citation impact is Office of Science of the USA with 1.2% surplus.

The other funding bodies with the relatively low citation impact and with 0.8%–10.4% of the population papers each are National Natural Science Foundation of China, National Council for Scientific and Technological Development, National Key Research and Development Program of China, Higher Education Personnel Improvement Coordination, National Science Foundation,

TABLE 6.4
The Most-Prolific Funding Bodies in Biological Pretreatments of the Biomass

No.	Funding Bodies	Country	Sample Paper No. (%)	Population Paper No. (%)	Surplus (%)
1	US Dept. Ener.	USA	5.9	1.9	4.0
2	Eur. Commis.	EU	4.4	2.1	2.3
3	Off. Sci.	USA	2.2	1.0	1.2
4	US Dept. Agr.	USA	2.2	0.9	1.3
5	Natl. Inst. Gen. Med Sci.	USA	2.2	0.5	1.7
6	NREL	USA	2.2	0.4	1.8
7	Natl. Nucl. Sec. Admr.	USA	2.2	0.4	1.8

Ministry of Education of the China, Research Support Foundation of the State of Sao Paulo, Ministry of Education, Culture, Sports, Science and Technology of Japan, Fundamental Research Funds for the Central Universities, Japan Society for the Promotion of Science, Ministry of Science, Technology and Innovation, European Regional Development Fund, Priority Academic Program Development of Jiangsu Higher Education Institutions, Natural Sciences and Engineering Research Council of Canada, U.S. Department of Agriculture, Ministry of Finance of Japan, National Research Foundation of Korea, Chinese Academy of Sciences, Natural Science Foundation of Jiangsu Province, and Seventh Framework Program. Among them, it is notable that National Natural Science Foundation of China funds 0.8% and 10.4% of the sample and population papers, respectively, with 10% deficit.

6.3.6 The Most Prolific Source Titles in the Biological Pretreatments of the Biomass

Information about the most prolific 11 source titles publishing at least 2.2% of the sample papers each in biological pretreatments of the biomass is given in Table 6.5.

The most prolific source title is Bioresource Technology with 19% of the sample papers, followed by Biotechnology and Bioengineering with 10% of the sample papers.

On the other hand, the source title with the most impact is the Bioresource Technology with 9% surplus, closely followed by Biotechnology and Bioengineering with 8% surplus. The other prolific journals are Enzyme and Microbial Technology and Science with 4% surplus each. Similarly, the source titles with the least impact are Biotechnology for Biofuels and Applied and Environmental Microbiology with 1% surplus each.

The other source titles with the relatively low citation impact with 0.7%–2.2% of the population paper each are Applied Biochemistry and Biotechnology, Bioresources, Industrial Crops and Products, International Biodeterioration and Biodegradation, Biomass and Bioenergy, Process Biochemistry, Biotechnology Letters, Holzforschung (Wood Research), Cellulose, Carbohydrate Polymers, Journal of Agricultural and Food Chemistry, Biomass Conversion and Biorefinery, and Journal of Chemical Technology and Biotechnology.

6.3.7 The Most Prolific Countries in the Biological Pretreatments of the Biomass

Information about the most prolific 17 countries publishing at least 2.2% of sample papers each in biological pretreatments of the biomass is given in Table 6.6.

The most prolific country is the USA with 52% of the sample papers, followed by Canada with 14% of the sample papers. The other prolific countries are Spain, Sweden, Finland, the UK, and France with 6%–8% of the sample papers each. On the other hand, eight European countries listed in Table 6.6 produce 47% and 22% of the sample and population papers, with a resulting 25% surplus.

TABLE 6.5
The Most-Prolific Source Titles in Biological Pretreatments of the Biomass

No.	Source Titles	Sample Papers (%)	Population Papers (%)	Surplus (%)
1	Bioresource Technology	18.5	9.7	8.8
2	Biotechnology and Bioengineering	10.4	2.0	8.4
3	Enzyme and Microbial Technology	5.2	1.6	3.6
4	Applied Microbiology and Biotechnology	3.7	1.6	2.1
5	Science	3.7	0.1	3.6
6	Biotechnology for Biofuels	3.0	2.2	0.8
7	Applied And Environmental Microbiology	3.0	1.6	1.4
8	Journal of Biotechnology	3.0	0.7	2.3
9	Biotechnology Advances	3.0	0.1	2.9
10	Applied Biochemistry and Biotechnology Part A Enzyme Engineering And Biotechnology	2.2	0.5	1.7
11	Proceedings of the National Academy of Sciences of the United States of America	2.2	0.3	1.9

Further, the country with the most citation impact is the USA with 34% surplus, followed by Canada with 10% surplus. The other prolific countries are Finland, Spain, and Sweden with 4%–5% surplus each. Similarly, the country with the least citation impact is China with 17% deficit. India, Brazil, Japan, and S. Korea are the other countries with 1%–4% deficit each.

Additionally, there are other countries with relatively low citation impact and with 0.7%–2.0% of the population papers each: Italy, Malaysia, Thailand, Iran, Mexico, Russian Federation, Egypt, Taiwan, Poland, Portugal, South Africa, Indonesia, Austria, Turkey, Pakistan, Switzerland, and Israel.

6.3.8 The Most Prolific Scopus Subject Categories in the Biological Pretreatments of the Biomass

Information about the most prolific eight Scopus subject categories indexing at least 5% of the sample papers each is given in Table 6.7.

The most prolific Scopus subject category in the biological pretreatments of the biomass is Biochemistry, Genetics, and Molecular Biology with 64% of sample papers, followed by Chemical Engineering and Immunology and Microbiology with 56% and 47% of the sample papers, respectively. The other prolific subject categories are Environmental Science and Energy with 33% and 27% of the sample papers, respectively.

On the other hand, the Scopus subject category with the most citation impact is the Biochemistry, Genetics, and Molecular Biology with 26% surplus, followed by Chemical Engineering and Immunology and Microbiology with 20% surplus each. Similarly, the Scopus subject category with the least citation impact is Agricultural and Biological Sciences with 14% deficit. Engineering and Environmental Science are the other categories with 2% and 1% deficit, respectively.

TABLE 6.6
The Most-Prolific Countries in the Biological Pretreatments of the Biomass

	Countries	Sample Papers (%)	Population Papers (%)	Surplus (%)
1	USA	51.9	17.7	34.2
2	Canada	14.1	4.1	10.0
3	Spain	8.1	3.7	4.4
4	Sweden	6.7	3.0	3.7
5	Finland	6.7	2.0	4.7
6	UK	5.9	3.3	2.6
7	France	5.9	2.9	3.0
8	China	4.4	22.2	−17.8
9	India	4.4	8.6	−4.2
10	Japan	4.4	6.7	−2.3
11	Germany	4.4	3.7	0.7
12	Denmark	4.4	1.6	2.8
13	Netherlands	4.4	1.4	3.0
14	Brazil	3.0	5.5	−2.5
15	Chile	3.0	0.6	2.4
16	S. Korea	2.2	3.2	−1.0
17	Australia	2.2	1.6	0.6

TABLE 6.7
The Most-Prolific Scopus Subject Categories in the Biological Pretreatments of the Biomass

No.	Scopus Subject Categories	Sample Papers (%)	Population Papers (%)	Surplus (%)
1	Biochemistry, Genetics, and Molecular Biology	64.4	38.6	25.8
2	Chemical Engineering	55.6	35.5	20.1
3	Immunology and Microbiology	47.4	27.9	19.5
4	Environmental Science	32.6	33.9	−1.3
5	Energy	27.4	22.6	4.8
6	Agricultural and Biological Sciences	8.9	22.9	−14.0
7	Multidisciplinary	5.9	2.6	3.3
8	Engineering	5.2	7.5	−2.3

6.3.9 The Most Prolific Keywords in the Biological Pretreatments of the Biomass

Information about the keywords used in at least 7.4% or 5.5% of the sample or population papers, respectively, is given in Table 6.8. For this purpose, keywords related to the keyword set given in the Appendix are selected from a list of the most prolific keyword set provided by Scopus database.

These keywords are grouped under the five headings: biomass, biological pretreatments, other pretreatments, other processes, and products of the biological pretreatments.

There are 12 keywords selected related to the biomass and biomass constituents: The prolific keywords are cellulose, lignin, biomass, and lignocellulose with 30%–73% of the sample papers each. The prolific keywords related to the biological pretreatments of the biomass are cellulases, enzymes, fungi, and enzyme activity with 42%–50% of the sample papers each.

TABLE 6.8
The Most-Prolific Keywords in Biological Pretreatments of the Biomass

No.	Keywords	Sample Papers (%)	Population Papers (%)	Surplus (%)
1	**Biomass**			
	Cellulose	72.6	32.3	40.3
	Lignin	65.2	28.8	36.4
	Biomass	43.0	19.7	23.3
	Lignocellulose	30.4	11.9	18.5
	Hemicellulose	18.5	6.2	12.3
	Wood	17.8	8.1	9.7
	Zea mays	13.3	5.1	8.2
	Lignocellulosic biomass	12.6	6.6	6.0
	Substrate	9.6	8.2	1.4
	Corn stover	8.1	2.3	5.8
	Straw	7.4	1.6	5.8
	Sugarcane bagasse	3.7	7.0	–3.3
2	**Biological Pretreatments**			
	Cellulase	50.4	21.2	29.2
	Enzymes	46.7	24.1	22.6
	Fungi	45.2	27.1	18.1
	Enzyme activity	42.2	23.9	18.3
	Enzymatic hydrolysis	28.1	22.4	5.7
	Trichoderma	24.4	5.8	18.6
	Basidiomycota/basidiomycetes	23.0	6.3	16.7
	Phanerochaete	17.8	4.9	12.9
	Hypocrea	14.1	3.0	11.1
	Enzymolysis	13.3	5.2	8.1
	Bacteria	11.9	15.1	–3.2
	Beta-glucosidase	9.6	3.9	5.7
	Enzyme kinetics	8.1	2.2	5.9
	Enzymology	6.7	6.3	0.4
	Microbiology		6.4	–6.4
	Enzymatic saccharification		5.5	–5.5
3	**Other Pretreatment**			
	Pretreatment	17.0	8.1	8.9
	Pretreatment	8.9	7.4	1.5
	Delignification	6.7	4.7	2.0
	pH	3.7	7.3	–3.6
4	**Other Processes**			
	Hydrolysis	57.8	30.1	27.7
	Saccharification	17.8	13.7	4.1
	Biodegradation	16.3	15.5	0.8
	Fermentation	16.3	13.6	2.7

(Continued)

TABLE 6.8 (*Continued*)
The Most-Prolific Keywords in Biological Pretreatments of the Biomass

No.	Keywords	Sample Papers (%)	Population Papers (%)	Surplus (%)
	Degradation	12.6	9.8	2.8
	Adsorption	8.9	3.5	5.4
	Oxidation	8.9	3.0	5.9
	Bioconversion	8.9	2.9	6.0
5	**Products**			
	Glucose	20.7	13.0	7.7
	Ethanol	20.7	9.9	10.8
	Polysaccharides	19.3	7.4	11.9
	Biofuel	16.3	9.9	6.4
	Sugar	15.6	10.2	5.4
	Xylose	8.9	3.8	5.1
	Bioethanol	5.9	5.4	0.5

Those related to the other pretreatments are pretreatments and pretreatments with 17% and 8% of the sample papers, respectively. Those related to other processes are hydrolysis, saccharification, biodegradation, and fermentation with 16%–58% of the sample papers each.

The prolific keywords related to the products of the biological pretreatments are glucose, ethanol, and polysaccharides with 19%–22% of the sample papers each. It is notable that bioethanol is used only by 6% of the sample papers.

6.3.10 The Most Prolific Research Fronts in Biological Pretreatments of the Biomass

Information about the research fronts for the sample papers in biological pretreatments of the biomass is given in Table 6.9. As Table 6.9 shows, there are three primary research fronts for this field: enzymatic pretreatments, fungal pretreatments, and enzymatic hydrolysis by the use of primarily fungal enzymes with 28%, 26%, 55% of the 135 HCPs. The other front is the microbial pretreatments of the biomass with 4% of the sample papers.

Information about the research fronts for the sample papers in biological pretreatments of the biomass with regard to the biomass used in these pretreatments is given in Table 6.10. As Table 6.10 shows, there are four primary research fronts for this field: biomass constituents, agricultural residues, biomass in general, and wood with 60%, 19%, 22%, and 10% of the 134 HCPs, respectively. The other research fronts are grass and algae with 2% and 1% of the sample papers, respectively.

Further, on the individual terms, the most prolific biomass are cellulose, lignin, wood, biomass, in general, lignocellulose, and to a lesser extent, lignocellulosic biomass, corn stover, straw, and sugarcane bagasse with 3%–27% of the sample papers each.

6.4 DISCUSSION

6.4.1 Introduction

The crude oil-based gasoline fuels have been widely used in the transportation sector since the 1920s. However, there have been great public concerns over the adverse environmental and human impact of these fuels. Hence, biomass-based bioethanol fuels have increasingly been used in blending gasoline fuels in the fuel cells, and in the biochemical production in a biorefinery context.

TABLE 6.9
The Most-Prolific Research Fronts for the Biological Pretreatments of the Biomass

No.	Research Fronts	N Paper % Sample
1	Enzymatic pretreatments	28
2	Fungal pretreatments	26
3	Enzymatic hydrolysis	55
4	Microbial pretreatments	4
	Sample	134

N paper (%) sample: the number of papers in the population sample of 134 papers.

TABLE 6.10
The Most-Prolific Research Fronts for the Biomass for the Biological Pretreatments of the Biomass

No.	Research Fronts	N Paper % Sample
1	**Biomass Constituents**	**60**
	Lignin	21
	Cellulose	27
	Lignocellulose	10
	Others	3
2	Wood	14
3	Grass	2
4	**Agricultural Residues**	**19**
	Corn stover	7
	Straw	7
	Other residues	1
	Agricultural residues in general	1
	Sugarcane bagasse	3
5	**Biomass in General**	**22**
	Biomass	12
	Lignocellulosic biomass	8
	Cellulosic biomass	1
6	Algae	1
	Sample	**134**

N paper (%) sample: the number of papers in the population sample of 134 papers.

The research in the fields of biological pretreatments of the biomass has also intensified in recent years in the research fronts of the enzymatic pretreatments, fungal pretreatments, enzymatic hydrolysis by enzymes, and microbial pretreatments.

However, it is essential to develop efficient incentive structures for the primary stakeholders to enhance the research in this field. The scientometric analysis has been used in this context to inform the primary stakeholders about the current state of the research in a selected research field.

As there have been no scientometric studies on the biological pretreatments of the biomass, this book chapter presents a scientometric study of the research in the biological pretreatments of the biomass. It examines the scientometric characteristics of both the sample and population data presenting scientometric characteristics of these both datasets in the order of documents, authors, publication years, institutions, funding bodies, source titles, countries, Scopus subject categories, keywords, and research fronts.

As a first step for the search of the relevant literature, the keywords were selected using the first most-cited 200 papers. The selected keyword list was optimized to obtain a representative sample of papers for the searched research field. This keyword list was provided in the Appendix for future replication studies.

As a second step, two sets of data were used for this study. First, a population sample of over 13,500 papers was used to examine the scientometric characteristics of the population data. Secondly, a sample of 135 most-cited papers, corresponding to 1% of the population dataset, was used to examine the scientometric characteristics of these citation classics.

The scientometric characteristics of these both sample and population datasets were presented in the order of documents, authors, publication years, institutions, funding bodies, source titles, countries, Scopus subject categories, keywords, and research fronts.

Lastly, the key scientometric findings for both datasets were discussed to highlight the research landscape for the biological pretreatments of the biomass. Additionally, a number of brief conclusions were drawn, and a number of relevant recommendations were made to enhance the future research landscape.

6.4.2 The Most Prolific Documents in the Biological Pretreatments of the Biomass

Both articles (together with conference papers) dominate the sample and population papers, while review papers (together with short surveys) dominate the sample papers (Table 6.1). Further, review papers and articles have a surplus and deficit, respectively. It is notable that the overpresentation of the reviews and short surveys by 24% is quite extraordinary, showing the public demand for such reviews by the major stakeholders.

Scopus differs from the Web of Science database in differentiating and showing articles and conference papers published in the journals separately. Similarly, Scopus differs from Web of Science database in introducing short surveys. Hence, the total number of articles and review papers in the sample dataset is 73% and 27%, respectively.

It is observed during the search process that there has been inconsistency in the classification of the documents in Scopus as well as in other databases such as Web of Science. This is especially relevant for the classification of papers as reviews or articles as the papers not involving a literature review may be erroneously classified as a review paper. There is also a case of review papers being classified as articles.

In this context, it would be helpful to provide a classification note for the published papers in the books and journals at the first instance. It would also be helpful to use the document types listed in Table 6.1 for this purpose. Book chapters may also be classified as articles or reviews as an additional classification to differentiate review chapters from the experimental chapters as it is done by the Web of Science. It would be further helpful to additionally classify the conference papers as articles or review papers as well as whether it is done in the Web of Science database.

6.4.3 The Most Prolific Authors in the Biological Pretreatments of the Biomass

There have been most-prolific 13 authors with at least 3% of the sample papers each as given in Table 6.2. These authors have shaped the development of the research in this field.

The most prolific authors Jack N. Saddler, Charles E. Wyman, Bernard Henrissat, Dan Cullen, Igor V. Grigoriev, and Pedro M. Coutinho are working primarily on the fungal pretreatments and enzymatic hydrolysis of the biomass by primarily fungal enzymes.

It is important to note the inconsistencies in indexing of the author names in Scopus and other databases. It is especially an issue for the names with more than two components such as 'Blake Sam de Hyun Wyman'. The probable outcomes are 'Wyman, B.S.D.H.', 'de Hyun Wyman, B.S.', or 'Hyun Wyman, B.S.D.'. The first choice is the gold standard of the publishing sector as the last word in the name is taken as the last name. In most of the academic databases such as PUBMED and EBSCO databases, this version is used predominantly. The second choice is a strong alternative, while the last choice is an undesired outcome as two last words are taken as the last name. It is a good practice to combine the words of the last name by a hyphen: 'Hyun-Wyman, B.S.D.'. It is notable that inconsistent indexing of the author names may cause substantial inefficiencies in the search process for the papers as well as allocating credit to the authors as there are different author entries for each outcome in the databases.

There are also inconsistencies in the shortening Chinese names. For example, 'Yuoyong Kim' is often shortened as 'Kim, Y.', 'Kim, Y.-Y.', and 'Kim Y.Y' as it is done in the Web of Science database as well. However, the gold standard in this case is 'Kim, Y' where the last word is taken as the last name and the first word is taken as a single forename. Nevertheless, the third option makes sense to differentiate the Chinese names efficiently: 'Kim, Y.Y.'. In most of the academic databases such as PUBMED and EBSCO, this first version is used predominantly. Therefore, there have been difficulties to locate papers for the Chinese authors. In such cases, the use of the unique author codes provided for each author by the Scopus database has been helpful.

There is also a difficulty in allowing credit for the authors especially for the authors with common names such as 'Zhang, Y', or 'Huang, Y,' or 'Zhu, Y' in conducting scientometric studies. These difficulties strongly influence the efficiency of the scientometric studies as well as allocating credit to the authors as there are the same author entries for different authors with the same name, e.g., 'Zhang Y' in the databases.

In this context, the coding of authors in Scopus database is a welcome innovation compared to the other databases such as Web of Science. In this process, Scopus allocates a unique number to each author in the database (Aman, 2018). However, there might still be substantial inefficiencies in this coding system especially for common names. For example, some of the papers for a certain author maybe allocated to another researcher with a different author code. It is possible that Scopus uses a number of software programs to differentiate the author names, and the program may not be false-proof (D'Angelo and van Eck, 2020).

In this context, it does not help that author names are not given in full in some journals and books. This makes it difficult to differentiate authors with common names and makes the scientometric studies further difficult in the author domain. Therefore, the author names should be given in all books and journals at the first instance. There is also a cultural issue where some authors do not use their full names in their papers. Instead, they use initials for their forenames: 'Wyman, H. J.' or just 'Wyman' instead of 'Wyman, Hyun Jae'.

There are also inconsistencies in naming of the authors with more than two components by the authors themselves in journal papers and book chapters. For example, 'Wyman, A.P.C.', 'Young, C.E.', and 'Simmons, S.J.' might be given as 'Wyman, A', 'Young, C', or 'Simmons, S' in the journals and books, respectively. This also makes the scientometric studies difficult in the author domain. Hence, contributing authors should use their name consistently in their publications.

The other critical issue regarding the author names is the inconsistencies in the spelling of the author names in the national spellings (e.g., Gonçalves, Übeyoğlü) rather than in the English spellings (e.g., Goncalves, Ubeyoglu) in Scopus database. Scopus differs from the Web of Science database and many other databases in this respect where the author names are given only in the English spellings. It is observed that national spellings of the author names do not help in conducting scientometric studies as well in allocating credits to the authors as sometimes there are the different author entries for the English and National spellings in the Scopus database.

In total, 13 institutions house these prolific authors. The most prolific country for the sample dataset is the USA. In total, seven countries house these authors.

The most-prolific research front is the fungal pretreatment of the biomass, followed by the enzymatic hydrolysis of the biomass by enzymes. It is notable that there are no authors related to both enzymatic and microbial pretreatments of the biomass.

It is also notable that there is significant gender deficit for the sample dataset as surprisingly only two of these top researchers are female. This finding is the most thought-provoking with strong public-policy implications. Hence, institutions, funding bodies, and policymakers should take efficient measures to reduce the gender deficit in this field as well as other scientific fields with strong gender deficit. In this context, it is worth to note the level of representation of the researchers from the minority groups in science on the basis of race, sexuality, age, and disability, besides the gender (Blankenship, 1993; Dirth and Branscombe, 2017; Konur, 2000, 2002a–c, 2006a, b, 2007a, b).

6.4.4 The Most Prolific Research Output by Years in the Biological Pretreatments of the Biomass

The research output observed between 1970 and 2022 is illustrated in Figure 6.1. This figure clearly shows that the bulk of the research papers in the population dataset were published primarily in the 2010s with 51% of the population dataset. Similarly, the bulk of the research papers in the sample dataset were published in the 2000s and to a lesser extent in 2010s.

These findings suggest that the most-prolific sample and population papers were primarily published in the last two decades. These are the thought-provoking findings as there has been no significant research in this field in the pre-2000s, but there has been significant research boom in the last two decades. In this context, the increasing public concerns about climate change (Change, 2007), greenhouse gas emissions (Carlson et al., 2017), and global warming (Kerr, 2007) have been certainly behind the boom in the research in this field in the last two decades.

Based on these findings, the size of the population papers is likely to more than double in the current decade, provided that the public concerns about climate change, greenhouse gas emissions, and global warming are translated efficiently to the research funding in this field.

6.4.5 The Most Prolific Institutions in the Biological Pretreatments of the Biomass

The most prolific eight institutions publishing papers on the biological pretreatments of the biomass with at least 3.7% of the sample papers each given in Table 6.3 have shaped the development of the research in this field.

The most prolific institutions are the USDA Forest Service, University of British Columbia, NREL, and Novozymes Inc., all from the USA.

The top country for these most prolific institutions is the USA and to a lesser extent is France. In total, only five countries house these top institutions.

On the other hand, the institutions with the most impact are Novozymes Inc., USDA Forest Service, and NREL while those with the least impact are CSIC and CNRS. These findings confirm the dominance of the US institutions in this research field.

6.4.6 The Most Prolific Funding Bodies in the Biological Pretreatments of the Biomass

The most prolific seven funding bodies funding at least 2.2% of the sample papers each are given in Table 6.4. It is notable that only 33% and 47% of the sample and population papers were funded.

The most prolific funding bodies are the US DOE and the European Commission. And the most prolific country for these top funding bodies is the USA. The funding bodies with the most and least citation impact are US DOE and the Office of Science, respectively.

These findings on the funding of the research in this field suggest that the level of the funding for the population papers, mostly in the last two decades, is highly intensive and it has been largely instrumental in enhancing the research in this field (Ebadi and Schiffauerova, 2016) in the light of North's institutional framework (North, 1991). However, considering the relatively low levels of funding, especially for the sample papers, there is ample room to enhance funding in this field.

6.4.7 The Most Prolific Source Titles in Biological Pretreatments of the Biomass

The most prolific 11 source titles publishing at least 2.2% of the sample papers each in biological pretreatments of the biomass have shaped the development of the research in this field (Table 6.5).

The most prolific source titles are Bioresource Technology and Biotechnology and Bioengineering. On the other hand, the source titles with the most and least impact are Biotechnology and Bioengineering and Biotechnology for Biofuels, respectively.

It is notable that these top source titles are related to the biomass, bioresources, biotechnology, and bioenergy. This finding suggests that the journals in this field have significantly shaped the development of the research in this field as they focus on the biological pretreatment aspects of bioethanol fuels.

6.4.8 The Most Prolific Countries in Biological Pretreatments of the Biomass

The most prolific 17 countries publishing at least 2.2% of the sample papers each have significantly shaped the development of the research in this field (Table 6.6).

The most prolific country is the USA with 52% of the sample papers. The other prolific countries are Canada, Spain, Sweden, Finland, the UK, and France. Further, eight European countries listed in Table 6.6 produce 47% and 22% of the sample and population papers, with a resulting 25% surplus.

On the other hand, the countries with the most and least citation impact are the USA and China, respectively. Additionally, it is notable that India, Japan, and Brazil were under-represented in the sample papers, while the Canada, Finland, Spain, and Sweden were over-represented.

The close examination of these findings suggests that the USA, Europe, and to a lesser extent, Canada, China, India, and Japan are the major producers of the research in this field. It is a fact that the USA has been a major player in science (Leydesdorff and Wagner, 2009; Leydesdorff et al., 2014). The USA has further developed a strong research infrastructure to support its corn- and grass-based bioethanol industry (Dwivedi et al., 2009).

However, China has been a rising star in scientific research in competition with the USA and Europe (Leydesdorff and Zhou, 2005). China is also a major player in this field as a major producer of bioethanol (Li and Chan-Halbrendt, 2009).

Next, Europe has been a persistent player in the scientific research in competition with both the USA and China (Leydesdorff, 2000). Europe has also been a persistent producer of bioethanol along with the USA and Brazil (Gnansounou, 2010).

6.4.9 The Most Prolific Scopus Subject Categories in Biological Pretreatments of the Biomass

The most prolific eight Scopus subject categories indexing at least 5% of the sample papers each, respectively, given in Table 6.7 have shaped the development of the research in this field.

The most prolific Scopus subject categories in the biological pretreatments of the biomass are Biochemistry, Genetics and Molecular Biology, and to a lesser extent, Chemical Engineering and Immunology and Microbiology. On the other hand, the Scopus subject categories with the most and least citation impact are the Biochemistry, Genetics and Molecular Biology, and Agricultural and Biological Sciences, respectively.

These findings are thought-provoking suggesting that the primary subject categories are Biochemistry, Genetics and Molecular Biology, Chemical Engineering, and Immunology and Microbiology. The other key finding is that social sciences are not well represented in both the sample and population papers, as in the most fields in bioethanol fuels. These findings are not surprising as the key research fronts in this field relate to the development and application of biological pretreatments of the biomass.

6.4.10 The Most Prolific Keywords in Biological Pretreatments of the Biomass

A limited number of keywords have shaped the development of the research in this field as shown in Table 6.8 and in the Appendix.

These keywords are grouped under the five headings: biomass, biological pretreatments, other pretreatments, other processes, and products of the biological pretreatments.

There are 12 keywords selected related to the biomass and biomass constituents: The prolific keywords are cellulose, lignin, biomass, and lignocellulose, and those related to the biological pretreatments of the biomass are cellulases, enzymes, fungi, and enzyme activity. Those related to the other pretreatments are pretreatments and pre-treatments and those related to other processes are hydrolysis, saccharification, biodegradation, and fermentation. Finally, those related to the products of the biological pretreatments are glucose, ethanol, and polysaccharides, while it is notable that bioethanol is used only by 6% of the sample papers.

These findings suggest that it is necessary to determine the keyword set carefully to locate the relevant research in each of these research fronts. Additionally, the size of the samples for each keyword highlights the intensity of the research in the relevant research areas.

6.4.11 The Most Prolific Research Fronts in Biological Pretreatments of the Biomass

As Table 6.9 shows, there are three primary research fronts for this research field: enzymatic pretreatments, fungal pretreatments, and enzymatic hydrolysis by the use of primarily fungal enzymes. The other research front is the microbial pretreatment of the biomass.

Information about the research fronts for the sample papers in biological pretreatments of the biomass with regard to the biomass or biomass constituents used in these pretreatments is given in Table 6.10. As Table 6.10 shows, there are four primary research fronts for this field: biomass constituents, agricultural residues, biomass, in general, and wood. The other research fronts are grass and algae. Further, on individual terms, the most prolific biomass are cellulose, lignin, wood, biomass, in general, lignocellulose, and to a lesser extent, lignocellulosic biomass, corn stover, straw, and sugarcane bagasse.

These findings are thought-provoking in seeking ways to increase the bioethanol yield through the biological pretreatments of the biomass at the global scale. It is clear that all of these research fronts with regard to both pretreatment types and biomass types have public importance and merit substantial funding and other incentives.

In the end, these most-cited papers in this field hint that the efficiency of bioethanol fuels and their derivatives could be optimized using the structure, processing, and property relationships of these biological pretreatments (Formela et al., 2016; Konur, 2018a, 2020b, 2021a–d; Konur and Matthews, 1989).

6.5 CONCLUSION AND FUTURE RESEARCH

The research on the biological pretreatments of the biomass has been mapped through a scientometric study of both sample and population datasets.

The critical issue in this study has been to obtain a representative sample of the research as in any other scientometric studies. Therefore, the keyword set has been carefully devised and optimized after a number of runs in the Scopus database. It is a representative sample of the wider population studies. The sheer size of this research field shows the public importance of the biological pretreatments of the biomass with over 13,500 papers.

The other issue has been the selection of a multidisciplinary database to carry out the scientometric study of the research in this field. For this purpose, Scopus database has been selected. The journal coverage of this database has been notably wider than that of Web of Science.

The key scientometric properties of the research in this field have been determined and discussed in this book chapter. It is evident that a limited number of documents, authors, institutions, publication periods, institutions, funding bodies, source titles, countries, Scopus subject categories, keywords, and research fronts have shaped the development of the research in this field.

There is ample scope to increase the efficiency of the scientometric studies in this field in the author and document domains by developing consistent policies and practices in both domains across all the academic databases. In this respect, it seems that authors, journals, and academic databases have a lot to do. Furthermore, the significant gender deficit as in most scientific fields emerges as a public policy issue. The potential deficits on the basis of age, race, disability, and sexuality need also to be explored in this field as in other scientific fields.

The research in this field has boomed in the last two decades possibly promoted by the public concerns on global warming, greenhouse gas emissions, and climate change. The institutions from the USA have mostly shaped the research in this field.

The relatively well-funding rate of 47% for the population papers suggests that funding in this field significantly enhanced the research in this field primarily in the last two decades, possibly more than doubling in the current decade. However, it is evident that there is ample room for more funding and other incentives to enhance the research in this field further as only 33% of the sample papers declared any funding.

The most-prolific source titles have been mostly indexed by the subject categories of Chemical Engineering, Biochemistry, Genetics and Molecular Biology, and Chemical Engineering as the focus of the sample papers has been on the development and utilization of the biological pretreatments of the biomass to increase the sugar and bioethanol yield at large.

The USA, Europe, and to a lesser extent, Canada, China, India, and Japan have been the major producers of the research in this field as the major producers and users of bioethanol fuels from different types of biomass such as corn, sugarcane, and grass as well as other types of biomass. It is evident that these countries have well-developed research infrastructure in bioethanol fuels and their derivatives.

The primary Scopus subject categories have been Biochemistry, Genetics and Molecular Biology, Chemical Engineering, and Environmental Science as the focus of the research has been on the development and utilization of the biological pretreatments of the biomass to increase the sugar and bioethanol yield at large. Further, social sciences are not well represented in both the sample and population papers, as in the most fields in bioethanol fuels. These findings are not surprising. This suggests that there is ample room for further research in the social sciences- and humanities-based interdisciplinary studies in this field.

It emerges that ethanol is more popular than bioethanol as a keyword with strong implications for the search strategy. In other words, the search strategy using only bioethanol keyword would not be much helpful.

These keywords are grouped under the five headings: biomass and biomass constituents, biological pretreatments, other pretreatments, other processes, and products of the biological pretreatments. These groups of keywords highlight the potential primary research fronts for these fields.

The three primary research fronts with regard to the biological pretreatments of the biomass are enzymatic pretreatments of the biomass, fungal pretreatments of the biomass, and enzymatic hydrolysis of the biomass with primarily fungal enzymes. The other research front is the microbial pretreatment of the biomass.

Similarly, there are four primary research fronts for the biomass or biomass constituents used for the biological pretreatments of the biomass: biomass constituents, agricultural residues, biomass in general, and wood. The other research fronts are grass and algae.

These findings are thought-provoking. The focus of these most-cited 135 papers as well as over 13,500 population papers is the development and utilization of biological pretreatments to increase the sugar and bioethanol yield. These studies highlight strong structure–processing–property relationships for biological pretreatments for bioethanol fuels and their derivatives.

Thus, the scientometric analysis has a great potential to gain valuable insights into the evolution of the research in this field as in other scientific fields.

It is recommended that further scientometric studies are carried out for the primary types of the biological pretreatments and the biomass used as well as about the other types of pretreatments. It is further recommended that reviews of the most-cited papers are carried out for each research front to complement these scientometric studies. Next, the scientometric studies of the hot papers in these primary fields are carried out.

ACKNOWLEDGMENT

The contribution of the highly cited researchers in the field of the biological pretreatments of the biomass has been gratefully acknowledged.

APPENDIX: THE KEYWORD SET FOR BIOLOGICAL PRETREATMENTS OF THE BIOMASS

(TITLE ("biological pretreat*" OR "biological treat*" OR "microbial pretreat*" OR "fungal pretreat*" OR enzym* OR fung* OR laccase* OR cellulosome* OR *enzyme OR bacter* OR hydrolase* OR cellobiohydrolase* OR glucosidase* OR xylanase* OR cellulase* OR trichoderma OR peroxidase* OR trametes OR pleurotus OR myceliophthora OR escherichia OR mucor OR rhizopus OR saccharomyces OR aspergillus OR esterase* OR microb* OR postia OR phanerochaete* OR oxidase* OR oudemansiella OR ceriporiopsis OR bifidobacterium OR lactobacillus OR candida OR echinodontium OR yeast* OR agaricomycetes OR hemicellulase* OR glucanase* OR endoglucanase* OR biosurfactant* OR pseudomonas OR mannanase* OR mannosidase* OR galactosidase* OR swollenin OR basidiomycete* OR pectinase* OR thielavia OR pycnoporus OR alcalase* OR ligninase* OR monooxygenase* OR neurospora OR hypocrea OR coniophora OR coriolus OR isozyme*) AND TITLE (pretreat* OR decay* OR *hydrolysis OR saccharification OR hydrolytic OR digestibili* OR disruption OR fractionation OR decomposition OR dissolution OR delignification OR depolymerization OR deconstruction OR disintegration OR *conversion OR recalcitrance OR *degradation OR *degradability OR *degrading OR extract* OR bioconversion OR ligninolysis) AND TITLE (*mannan OR waste* OR *grass OR grass* OR biomass* OR lignocellul* OR bagasse* OR cellulos* OR *wood OR woody OR *woods OR *algae OR *algal OR microalga OR olive OR chlamydomonas OR straw* OR stover* OR eucalyptus OR miscanthus OR pine OR chlorella OR lignin OR cellulolytic OR hemicellulose* OR xylan OR porphyridium OR nannochloropsis OR poplar* OR cedar OR pulp* OR "corn cob*" OR spruce OR beech OR oak OR pinus OR residue* OR willow* OR cypress OR sawdust OR prosopis OR "rice hull*" OR "*bean hull*" OR husk* OR birch OR cynara OR "*cane tops" OR "cotton stalk*" OR "corn stalk*" OR alfalfa OR "maize stem*" OR cornstalk* OR hyacinth OR salix OR substrate* OR corncob* OR "fruit bunch*" OR populus OR sage OR "orange peel*" OR "*weed stem*" OR bamboo* OR birch OR "*flower stalk*" OR "barley hull*" OR "maize stem*")) AND NOT (TITLE (sediment* OR {bacterial cellulose*} OR {microbial cellulose*} OR bioelectricity OR digestion OR *composites OR composite OR electrochemical* OR nano* OR anaerobic OR *methane OR methan* OR xylitol OR "microbial oil*" OR *methanation OR *hydrogen OR hydrogen* OR fiber* OR fibre* {glucose oxidase*} OR sludge OR biogas OR lipid* OR *diesel

OR fibril* OR butanol OR wastewater* OR "waste water*" OR human OR oil* OR {phenol degrad*} OR grassland* OR benzene OR "fuel cell*" OR dye* OR xenobiotic* OR trichloroethylene OR leaf* OR soil* OR hydrocarbon* OR advanced OR pectin OR *vanillic OR synthesis OR ddt OR "by-product*" OR snake OR juice* OR arsenic OR *capacitors OR extractive OR gels OR *leaching OR rna OR cadmium OR forest OR tmb3400 OR motor OR venom OR alcohol OR excret* OR bisphenol OR inhibit*) OR SRCTITLE (dairy OR phar* OR medi* OR cereal* OR nutrit* OR feed* OR brew* OR animal* OR zoo* OR lipid* OR soil* OR composite* OR atmosph* OR poult* OR postharv* OR botan* OR cellullar) OR SUBJAREA (medi OR phar OR vete OR heal OR neur OR eart)) AND (LIMIT-TO (SRCTYPE, "j") OR LIMIT-TO (SRCTYPE, "k") OR LIMIT-TO (SRCTYPE, "b")) AND (LIMIT-TO (DOCTYPE, "ar") OR LIMIT-TO (DOCTYPE, "cp") OR LIMIT-TO (DOCTYPE, "re") OR LIMIT-TO (DOCTYPE, "ch") OR LIMIT-TO (DOCTYPE, "no") OR LIMIT-TO (DOCTYPE, "le") OR LIMIT-TO (DOCTYPE, "sh") OR LIMIT-TO (DOCTYPE, "bk") OR LIMIT-TO (DOCTYPE, "ed")) AND (LIMIT-TO (LANGUAGE, "English"))

REFERENCES

Alvira, P., E. Tomas-Pejo, M. Ballesteros and M. J. Negro. 2010. Pretreatment technologies for an efficient bioethanol production process based on enzymatic hydrolysis: A review. *Bioresource Technology* 101:4851–4861.

Aman, V. 2018. Does the Scopus author ID suffice to track scientific international mobility? A case study based on Leibniz laureates. *Scientometrics* 117:705–720.

Angelici, C., B. M. Weckhuysen and P. C. A. Bruijnincx. 2013. Chemocatalytic conversion of ethanol into butadiene and other bulk chemicals. *ChemSusChem* 6:1595–1614.

Antolini, E. 2007. Catalysts for direct ethanol fuel cells. *Journal of Power Sources* 170:1–12.

Antolini, E. 2009. Palladium in fuel cell catalysis. *Energy and Environmental Science* 2:915–931.

Beaudry, C. and V. Lariviere. 2016. Which gender gap? Factors affecting researchers' scientific impact in science and medicine. *Research Policy* 45:1790–1817.

Blankenship, K. M. 1993. Bringing gender and race in: US employment discrimination policy. *Gender & Society* 7:204–226.

Brown, M. E. and M. C. Y. Chang. 2014. Exploring bacterial lignin degradation. *Current Opinion in Chemical Biology* 19:1–7.

Bugg, T. D. H., M. Ahmad, E. M. Hardiman and R. Singh. 2011. The emerging role for bacteria in lignin degradation and bio-product formation. *Current Opinion in Biotechnology* 22: 394–400.

Burnham, J. F. 2006. Scopus database: A review. *Biomedical Digital Libraries* 3:1–8.

Carlson, K. M., J. S. Gerber and D. Mueller, et al. 2017. Greenhouse gas emissions intensity of global croplands. *Nature Climate Change* 7:63–68.

Change, C. 2007. Climate change impacts, adaptation and vulnerability. *Science of the Total Environment* 326:95–112.

D'Angelo, C. A. and N. J. van Eck. 2020. Collecting large-scale publication data at the level of individual researchers: A practical proposal for author name disambiguation. *Scientometrics* 123:883–907.

Dashtban, M., H. Schraft and W. Qin. 2009. Fungal bioconversion of lignocellulosic residues; Opportunities & perspectives. *International Journal of Biological Sciences* 5:578–595.

Dirth, T. P. and N. R. Branscombe. 2017. Disability models affect disability policy support through awareness of structural discrimination. *Journal of Social Issues* 73:413–442.

Dwivedi, P., J. R. Alavalapati and P. Lal. 2009. Cellulosic ethanol production in the United States: Conversion technologies, current production status, economics, and emerging developments. *Energy for Sustainable Development* 13:174–182.

Ebadi, A. and A. Schiffauerova. 2016. How to boost scientific production? A statistical analysis of research funding and other influencing factors. *Scientometrics* 106:1093–1116.

Fernando, S., S. Adhikari, C. Chandrapal and M. Murali. 2006. Biorefineries: Current status, challenges, and future direction. *Energy & Fuels* 20:1727–1737.

Formela, K., A. Hejna, L. Piszczyk, M. R. Saeb and X. Colom. 2016. Processing and structure-property relationships of natural rubber/wheat bran biocomposites. *Cellulose* 23:3157–3175.

Garfield, E. 1955. Citation indexes for science. *Science* 122:108–111.

Gnansounou, E. 2010. Production and use of lignocellulosic bioethanol in Europe: Current situation and perspectives. *Bioresource Technology* 101:4842–4850.

Hahn-Hagerdal, B., M. Galbe, M. F. Gorwa-Grauslund, G. Liden and G. Zacchi. 2006. Bio-ethanol – The fuel of tomorrow from the residues of today. *Trends in Biotechnology* 24:549–556.

Hill, J., E. Nelson, D. Tilman, S. Polasky and D. Tiffany. 2006. Environmental, economic, and energetic costs and benefits of biodiesel and ethanol biofuels. *Proceedings of the National Academy of Sciences of the United States of America* 103:11206–11210.

Hill, J., S. Polasky and E. Nelson, et al. 2009. Climate change and health costs of air emissions from biofuels and gasoline. *Proceedings of the National Academy of Sciences of the United States of America* 106:2077–2082.

Hsieh, W. D., R. H. Chen, T. L. Wu and T. H. Lin. 2002. Engine performance and pollutant emission of an SI engine using ethanol-gasoline blended fuels. *Atmospheric Environment* 36:403–410.

Huang, H. J., S. Ramaswamy, U. W. Tschirner and B. V. Ramarao. 2008. A review of separation technologies in current and future biorefineries. *Separation and Purification Technology* 62:1–21.

Kerr, R. A. 2007. Global warming is changing the world. *Science* 316:188–190.

Konur, O. 2000. Creating enforceable civil rights for disabled students in higher education: An institutional theory perspective. *Disability & Society* 15:1041–1063.

Konur, O. 2002a. Access to nursing education by disabled students: Rights and duties of nursing programs. *Nurse Education Today* 22:364–374.

Konur, O. 2002b. Assessment of disabled students in higher education: Current public policy issues. *Assessment and Evaluation in Higher Education* 27:131–152.

Konur, O. 2002c. Access to employment by disabled people in the UK: Is the Disability Discrimination Act working? *International Journal of Discrimination and the Law* 5:247–279.

Konur, O. 2006a. Participation of children with dyslexia in compulsory education: Current public policy issues. *Dyslexia* 12:51–67.

Konur, O. 2006b. Teaching disabled students in Higher Education. *Teaching in Higher Education* 11:351–363.

Konur, O. 2007a. A judicial outcome analysis of the Disability Discrimination Act: A windfall for the employers? *Disability & Society* 22:187–204.

Konur, O. 2007b. Computer-assisted teaching and assessment of disabled students in higher education: The interface between academic standards and disability rights. *Journal of Computer Assisted Learning* 23:207–219.

Konur, O. 2011. The scientometric evaluation of the research on the algae and bio-energy. *Applied Energy* 88:3532–3540.

Konur, O. 2012a. Prof. Dr. Ayhan Demirbas' scientometric biography. *Energy Education Science and Technology Part A: Energy Science and Research* 28:727–738.

Konur, O. 2012b. The evaluation of the biogas research: A scientometric approach. *Energy Education Science and Technology Part A: Energy Science and Research* 29:1277–1292.

Konur, O. 2012c. The evaluation of the global energy and fuels research: A scientometric approach. *Energy Education Science and Technology Part A: Energy Science and Research* 30:613–628.

Konur, O. 2012d. The evaluation of the research on the biodiesel: A scientometric approach. *Energy Education Science and Technology Part A: Energy Science and Research* 28:1003–1014.

Konur, O. 2012e. The evaluation of the research on the bioethanol: A scientometric approach. *Energy Education Science and Technology Part A: Energy Science and Research* 28:1051–1064.

Konur, O. 2012f. The evaluation of the research on the biofuels: A scientometric approach. *Energy Education Science and Technology Part A: Energy Science and Research* 28:903–916.

Konur, O. 2012g. The evaluation of the research on the biohydrogen: A scientometric approach. *Energy Education Science and Technology Part A: Energy Science and Research* 29:323–338.

Konur, O. 2012h. The evaluation of the research on the microbial fuel cells: A scientometric approach. *Energy Education Science and Technology Part A: Energy Science and Research* 29:309–322.

Konur, O. 2012i. The scientometric evaluation of the research on the production of bioenergy from biomass. *Biomass and Bioenergy* 47:504–515.

Konur, O. 2015. Current state of research on algal bioethanol. In *Marine Bioenergy: Trends and Developments*, Eds. S. K. Kim and C. G. Lee, pp. 217–244. Boca Raton, FL: CRC Press.

Konur, O. 2018b. Bioenergy and biofuels science and technology: Scientometric overview and citation classics. In *Bioenergy and Biofuels*, Ed. O. Konur, pp. 3–63. Boca Raton: CRC Press.

Konur, O. 2019. Cyanobacterial bioenergy and biofuels science and technology: A scientometric overview. In *Cyanobacteria: From Basic Science to Applications*, Eds. A. K. Mishra, D. N. Tiwari and A. N. Rai, pp. 419–442. Amsterdam: Elsevier.

Konur, O. 2020a. The scientometric analysis of the research on the bioethanol production from green macroalgae. In *Handbook of Algal Science, Technology and Medicine*, Ed. O. Konur, pp. 385–401. London: Academic Press.

Konur, O., Ed. 2018a. *Bioenergy and Biofuels*. Boca Raton, FL: CRC Press.

Konur, O., Ed. 2020b. *Handbook of Algal Science, Technology and Medicine*. London: Academic Press.

Konur, O., Ed. 2021a. *Handbook of Biodiesel and Petrodiesel Fuels: Science, Technology, Health, and Environment*. Boca Raton, FL: CRC Press.

Konur, O., Ed. 2021b. *Handbook of Biodiesel and Petrodiesel Fuels: Science, Technology, Health, and Environment. Volume 1. Biodiesel Fuels: Science, Technology, Health, and Environment*. Boca Raton, FL: CRC Press.

Konur, O., Ed. 2021c. *Handbook of Biodiesel and Petrodiesel Fuels: Science, Technology, Health, and Environment. Volume 2. Biodiesel Fuels based on the Edible and Nonedible Feedstocks, Wastes, and Algae: Science, Technology, Health, and Environment*. Boca Raton, FL: CRC Press.

Konur, O., Ed. 2021d. *Handbook of Biodiesel and Petrodiesel Fuels: Science, Technology, Health, and Environment. Volume 3. Petrodiesel Fuels: Science, Technology, Health, and Environment*. Boca Raton, FL: CRC Press.

Konur, O. and F. L. Matthews. 1989. Effect of the properties of the constituents on the fatigue performance of composites: A review. *Composites* 20:317–328.

Leydesdorff, L. 2000. Is the European Union becoming a single publication system? *Scientometrics* 47:265–280.

Leydesdorff, L. and C. Wagner. 2009. Is the United States losing ground in science? A global perspective on the world science system. *Scientometrics* 78:23–36.

Leydesdorff, L., C. S. Wagner and L. Bornmann. 2014. The European Union, China, and the United States in the top-1% and top-10% layers of most-frequently cited publications: Competition and collaborations. *Journal of Informetrics* 8:606–617.

Leydesdorff, L. and P. Zhou. 2005. Are the contributions of China and Korea upsetting the world system of science? *Scientometrics* 63:617–630.

Li, S. Z. and C. Chan-Halbrendt. 2009. Ethanol production in (the) People's Republic of China: Potential and technologies. *Applied Energy* 86:S162–S169.

Lin, Y. and S. Tanaka. 2006. Ethanol fermentation from biomass resources: Current state and prospects. *Applied Microbiology and Biotechnology* 69:627–642.

Ma, X., L. Sun and C. Song. 2002. A new approach to deep desulfurization of gasoline, diesel fuel and jet fuel by selective adsorption for ultra-clean fuels and for fuel cell applications. *Catalysis Today* 77:107–116.

Morschbacker, A. 2009. Bio-ethanol based ethylene. *Polymer Reviews* 49:79–84.

Najafi, G., B. Ghobadian and T. Tavakoli, et al. 2009. Performance and exhaust emissions of a gasoline engine with ethanol blended gasoline fuels using artificial neural network. *Applied Energy* 86:630–639.

Newman, P. W. G. and J. R. Kenworthy. 1989. Gasoline consumption and cities: A comparison of U.S. cities with a global survey. *Journal of the American Planning Association* 55:24–37.

North, D. C. 1991. Institutions. *Journal of Economic Perspectives* 5:97–112.

Olsson, L. and B. Hahn-Hagerdal. 1996. Fermentation of lignocellulosic hydrolysates for ethanol production. *Enzyme and Microbial Technology* 18:312–331.

Sanchez, C. 2009. Lignocellulosic residues: Biodegradation and bioconversion by fungi. *Biotechnology Advances* 27:185–194.

Sanchez, O. J. and C. A. Cardona. 2008. Trends in biotechnological production of fuel ethanol from different feedstocks. *Bioresource Technology* 99:5270–5295.

Sindhu, R., P. Binod and A. Pandey. 2016. Biological pretreatment of lignocellulosic biomass – An overview. *Bioresource Technology* 199:76–82.

Sun, Y. and J. Cheng. 2002. Hydrolysis of lignocellulosic materials for ethanol production: A review. *Bioresource Technology* 83:1–11.

Taherzadeh, M. J. and K. Karimi. 2007. Enzyme-based hydrolysis processes for ethanol from lignocellulosic materials: A review. *Bioresources* 2:707–738.

Taherzadeh, M. J. and K. Karimi. 2008. Pretreatment of lignocellulosic wastes to improve ethanol and biogas production: A review. *International Journal of Molecular Sciences* 9:1621–1651.

Van Dyk, J. S. and B. I. Pletschke. 2012. A review of lignocellulose bioconversion using enzymatic hydrolysis and synergistic cooperation between enzymes – Factors affecting enzymes, conversion and synergy. *Biotechnology Advances* 30:1458–1480.

7 Biological Pretreatment of the Biomass

Review

Ozcan Konur
(Formerly) Ankara Yildirim Beyazit University

7.1 INTRODUCTION

The crude oil-based gasoline fuels (Ma et al., 2002; Newman and Kenworthy, 1989) have been widely used in the transportation sector since the 1920s. However, there have been great public concerns over the adverse environmental and human impact of these fuels (Hill et al., 2006, 2009). Hence, biomass-based bioethanol fuels (Alvira et al., 2010; Hill et al., 2006; Konur, 2012, 2015, 2020; Sun and Cheng, 2002) have increasingly been used in blending gasoline fuels (Hsieh et al., 2002; Najafi et al., 2009), in the fuel cells (Antolini, 2007, 2009), and in the biochemical production (Morschbacker, 2009) in a biorefinery context (Fernando et al., 2006; Huang et al., 2008). However, it is necessary to pretreat the biomass (Alvira et al., 2010; Taherzadeh and Karimi, 2008) to enhance the yield of the bioethanol (Hahn-Hagerdal et al., 2006; Sanchez and Cardona, 2008) prior to the bioethanol production through the hydrolysis (Alvira et al., 2010; Sun and Cheng, 2002; Taherzadeh and Karimi, 2007) and fermentation (Lin and Tanaka, 2006; Olsson and Hahn-Hagerdal, 1996) of the biomass and hydrolysates, respectively.

The research in the fields of biological pretreatments of the biomass has also intensified in recent years in the research fronts of the enzymatic pretreatments (Sindhu et al., 2016; van Dyk and Pletsche, 2012), fungal pretreatments (Dashtban et al., 2009; Sanchez, 2009), enzymatic hydrolysis by enzymes (Taherzadeh and Karimi, 2007; van Dyk and Pletsche, 2012), and microbial pretreatments (Brown and Chang, 2014; Bugg et al., 2011).

However, it is essential to develop efficient incentive structures (North, 1991) for the primary stakeholders to enhance the research in this field (Konur, 2000, 2002a–c, 2006a, b, 2007a, b). Although there has been a number of review papers on the biological pretreatment of the biomass (Brown and Chang, 2014; Bugg et al., 2011; Dashtban et al., 2009; Sanchez, 2009; Sindhu et al., 2016; Taherzadeh and Karimi, 2007; van Dyk and Pletsche, 2012), there has been no updated review of the most-cited 25 articles in this field.

Thus, this book chapter presents a review of the most-cited 25 articles in the field of the biological pretreatments of the biomass. Then, it discusses the key findings of these highly influential papers and comments on the future research priorities in this field.

7.2 MATERIALS AND METHODS

The search for this study was carried out using Scopus database (Burnham, 2006) in April 2022.

As a first step for the search of the relevant literature, the keywords were selected using all of the population papers. The selected keyword list was then optimized to obtain a representative sample of papers for the searched research field. This final keyword set was provided in the Appendix of Konur (2023) for future replication studies.

 DOI: 10.1201/9781003226536-10

As a second step, a sample dataset was used for this study. The first 25 articles with at least 424 citations each were selected for the review study. Key findings from each paper were taken from the abstracts of these papers and were discussed. Additionally, a number of brief conclusions were drawn, and a number of relevant recommendations were made to enhance the future research landscape.

7.3 RESULTS

The brief information about 25 most-cited papers with at least 424 citations each on the biological pretreatments of the biomass is given below. The primary research fronts are enzymatic pretreatments, fungal pretreatments, and enzymatic hydrolysis of the biomass with 9, 4, and 12 papers, respectively. It is notable that there has been no paper on the microbial pretreatment of the biomass in this dataset.

7.3.1 The Enzymatic Pretreatment of the Biomass

Bourbonnais and Paice (1990) studied the lignin biodegradation by laccases (*Trametes versicolor-Coriolus versicolor*) in a paper with 1,169 citations. They observed that veratryl alcohol (I) and 1-(3,4-dimethoxyphenyl)-2-(2-methoxyphenoxy)-propane-1,3-diol (III) were oxidized by laccase and mediator, 2,2′-azinobis(3-ethylbenzthiazoline-6-sulphonate) (ABTS) to give the α-carbonyl derivatives. The β-1 lignin model dimer, 1-(3,4-dimethoxyphenyl)-2-phenoxy-ethane-1,2-diol(II) was cleaved by laccase in the presence of the mediator to give veratraldehyde and benzaldehyde. They asserted that laccase was capable of oxidizing both phenolic and nonphenolic moieties of lignin but that the latter was dependent on the copresence of primary laccase substrates.

Floudas et al. (2012) studied the paleozoic origin of enzymatic lignin decomposition based on the analysis of 31 fungal genomes in a paper with 1,094 citations. They carried out a comparative analysis of 31 fungal genomes and found that lignin-degrading peroxidases expanded in the lineage leading to the ancestor of the Agaricomycetes, which was reconstructed as a white rot species, and then contracted in parallel lineages leading to brown rot and mycorrhizal species. Further, the origin of lignin degradation might have coincided with the sharp decrease in the rate of organic carbon burial around the end of the Carboniferous period.

Tien and Kirk (1983) studied the lignin-degrading enzyme from the *Phanerochaete chrysasporium* in a paper with 963 citations. They observed that this *P. chrysasporium* contained an enzyme that degraded lignin substructure model compounds as well as spruce and birch lignins. It had a molecular size of 42,000 Da and required hydrogen peroxide for activity.

Quinlan et al. (2011) studied the oxidative degradation of cellulose by a glycoside hydrolase (CAZy) GH61 enzyme focusing on cellulase enhancing factors in a paper with 634 citations. They showed that GH61 enzymes were a unique family of copper-dependent oxidases. Further, copper was needed for GH61 maximal activity and that the formation of cellodextrin and oxidized cellodextrin products by GH61 was enhanced in the presence of small molecule redox-active cofactors such as ascorbate and gallate. The active site of GH61 contained a type II copper and, uniquely, a methylated histidine in the copper's coordination sphere.

Glenn and Gold (1985) purified and characterized an extracellular Mn(II)-dependent peroxidase from the lignin-degrading *Phanerochaete chrysosporium* in a paper with 576 citations. They observed the presence of a heme prosthetic group and the enzyme contained one molecule of heme as iron protoporphyrin IX. The absorption maximum of the native enzyme (406 nm) shifted to 433 nm in the reduced enzyme and to 423 nm in the reduced-CO complex. Both CN- and N_3- readily bounded to the native enzyme, indicating an available coordination site and that the heme iron is high spin. The absorption spectrum of the H_2O_2 enzyme complex, maximum at 420 nm, was similar to that of horseradish peroxidase compound II. *P. chrysosporium* peroxidase activity was dependent on Mn(II), with maximal activity attained above 100 μm. The enzyme was also stimulated to varying degrees by α-hydroxy acids and protein. The peroxidase was capable of oxidizing NADH. Several of the substrates were oxidized by this Mn(II)-dependent peroxidase at considerably faster rates than those catalyzed by horseradish peroxidase. The enzyme rapidly oxidized Mn(II) to Mn(III). Inhibition of the oxidation of the substrate diammonium 2,2-azino-bis(3-ethyl-6-benzothiazolinesulfonate) (ABTS) by Na-pyrophosphate suggests that Mn(III) played a role in the enzyme mechanism.

Harris et al. (2010) studied the structure and function of glycoside hydrolase 61 (GH61) enzymes in a paper with 555 citations. They showed that certain GH61 proteins lacked measurable hydrolytic activity by themselves but in the presence of various divalent metal ions could significantly reduce the total protein loading required to hydrolyze lignocellulosic biomass. They also solved the structure of one highly active GH61 protein and found that it was devoid of conserved, closely juxtaposed acidic side chains that could serve as general proton donor and nucleophile/base in a canonical hydrolytic reaction. They concluded that the GH61 proteins were unlikely to be glycoside hydrolases. By incorporating the gene for one GH61 protein into a commercial *Trichoderma reesei* strain producing high levels of cellulolytic enzymes, they reduced by two-fold the total protein loading (and hence the cost) required to hydrolyze lignocellulosic biomass.

Bourbonnais et al. (1995) studied lignin the degradation by laccase isozymes from *Trametes versicolor* and the role of the mediator 2,2′-azinobis(3-ethylbenzthiazoline-6-sulfonate) (ABTS) in a paper with 504 citations. They observed that laccase II enzyme showed a very high sequence similarity to a laccase which was previously reported to depolymerize lignin. The reactivities of the laccase I and II on most of the substrates tested were similar, but there were some differences in the oxidation rate of polymeric substrates. They found that these two laccases produced similar qualitative effects on kraft lignin and residual lignin in kraft pulp, with no evidence of a marked preference for depolymerization by either enzyme. However, the presence of the mediator prevented and reversed the polymerization of kraft lignin by either laccase. Either laccase reduced the kappa number of pulp but only in the presence of the mediator.

Glenn et al. (1983) prepared an extracellular H_2O_2-requiring enzyme involved in lignin biodegradation by *Phanerochaete chrysosporium* in a paper with 482 citations. They observed that the enzyme system generated ethylene from 2-keto-4-thiomethyl butyric acid (KTBA) and oxidized a variety of lignin model compounds including the diarylpropane 1-(4′-ethoxy-3′-methoxyphenyl) 1,3-dihydroxy-2-(4″-methoxyphenyl)propane (I), a β-ether dimer 1-(4′-ethoxy-3′-methoxyphenyl) glycerol-β-guaiacyl ether (IV) and an olefin 1-(4′-ethoxy-3′-methoxy-phenyl)1,2-propene (VI). The products were equivalent to the metabolic products previously isolated from intact ligninolytic cultures. In addition, the enzyme system partially degraded 14C-ring labeled lignin. The enzyme was neither found in high nitrogen (N) cultures nor in cultures of a ligninolytic mutant strain, which is incapable of metabolizing lignin.

Phillips et al. (2011) studied the cellulose degradation using cellobiose dehydrogenase and a copper-dependent polysaccharide monooxygenase by *Neurospora crassa* in a paper with 448 citations. They observed that filamentous fungi used oxidative enzymes to cleave glycosidic bonds in cellulose. Deletion of *cdh-1*, the gene encoding the major cellobiose dehydrogenase of *N crassa*, reduced cellulase activity substantially, and addition of purified cellobiose dehydrogenases from *Myceliophthora thermophila* to the *Δcdh-1* strain resulted in a 1.6- to 2.0-fold stimulation in cellulase activity. They showed that cellobiose dehydrogenase enhanced cellulose degradation by coupling the oxidation of cellobiose to the reductive activation of copper-dependent polysaccharide monooxygenases (PMOs) that catalyzed the insertion of oxygen into C-H bonds adjacent to the glycosidic linkage. Their data supported a direct, enzyme-catalyzed oxidation of cellulose. This model for oxidative cellulose degradation might be widespread throughout the fungal kingdom. When added to mixtures of cellulases, these proteins enhanced cellulose saccharification, suggesting that they could be used to reduce the cost of biofuel production (Table 7.1).

TABLE 7.1
The Enzymatic Pretreatment of the Biomass

No.	Papers	Biomass	Enzymes	Parameters	Keywords	Lead Author	Affil.	Cits
1	Bourbonnais and Paice (1990)	Lignin	*Trametes versicolor (Coriolus versicolor)* laccase	Lignin biodegradation mechanisms	Laccase, substrate, lignin, biodegradation	Bourbonnais, Robert 7003682574	McGill Univ. Canada	1,169
2	Floudas et al. (2012)	Lignin	Fungi, peroxidase	Lignin biodegradation, genomic analysis	Lignin, decomposition, fungal, enzymatic	Grigoriev, Igor V. 25027225800	Joint Genome Inst. USA	1,049
3	Tien and Kirk (1983)	Lignin, spruce, birch	Fungi, *Phanerochaete chrysasporium,* enzyme	Lignin biodegradation	Lignin, degrading, enzyme, Phanerochaete	Tien, Ming 7006146774	USDA Forest Serv. USA	963
4	Quinlan et al. (2011)	Cellulose	Glycoside hydrolase	Glycoside hydrolase structure, cellulose degradation	Cellulose, degradation, metalloenzyme, biomass	Johansen, Katja S.* 36473579400	Copenhagen Univ. Denmark	634
5	Glenn and Gold (1985)	Lignin	*Phanerochaete chrysosporium* peroxidase	Peroxide purification and characterization, lignin degradation	Lignin, peroxidase, degrading, basidiomycete, Phanerochaete	Gold, Michael H. 7402444296	Oregon Hlth. Sci. Univ.	576
6	Harris et al. (2010)	Lignocellulosic biomass	Glycoside hydrolase	Glycoside hydrolase structure, biomass degradation	Lignocellulosic, biomass, hydrolysis, hydrolase	Harris, Paul V. 56305899400	Novozymes USA	555
7	Bourbonnais et al. (1995)	Lignin	*Trametes versicolor* laccase	Lignin degradation, mediator, enzyme activity	Lignin, laccase, Lignin, Trametes, depolymerization	Bourbonnais, Robert 7003682574	McGill Univ. Canada	504
8	Glenn et al. (1983)	Lignin	*Phanerochaete chrysosporium enzyme*	Enzyme preparation, lignin degradation	Lignin, enzyme, biodegradation, basidiomycete, Phanerochaete	Gold, Michael H. 7402444296	Oregon Hlth. Sci. Univ.	482
9	Phillips et al. (2011)	Cellulose	*Neurospora crassa* polysaccharide monooxygenase, cellobiose dehydrogenase	Cellulose degradation mechanisms	Cellulose, Neurospora, degradation	Marletta, Michael A. 7101716736	Univ. Calif. Berkeley USA	448

Prt, biological pretreatments; Other prts., other pretreatments; Na, nonavailable; Cits., number of citations received for each paper; *, female.

7.3.2 The Fungal Pretreatment of the Biomass

Martinez et al. (2008) provided a roadmap for constructing enhanced *Trichoderma reesei* strains as a source for cellulases and hemicellulases for the biomass degradation in a paper with 878 citations. They assembled 89 scaffolds to generate 34 Mbp of nearly contiguous *T. reesei* genome sequence comprising 9,129 predicted gene models. They observed that its genome encoded fewer cellulases and hemicellulases compared to any other sequenced fungus. Many *T. reesei* genes encoding carbohydrate-active enzymes were distributed nonrandomly in clusters that stood between regions of synteny with other Sordariomycetes. There was little mechanistic insight into its extraordinary capacity for protein secretion.

Pandey and Pitman (2003) studied wood (*Pinus sylvestris* L., *Fagus sylvatica* L.) degradation by *Coniophora puteana*, *Coriolus versicolor* (*Trichoderma cersicolor*), and *Phanerochaete chrysosporium* by a Fourier transform infrared (FTIR) method in a paper with 693 citations. In wood decayed by *C. puteana*, they observed that there was a progressive increase in lignin content relative to carbohydrate evident from increases in the relative height of lignin-associated bands (at 1,596, 1,505, 1,330, 1,230, and 1,122 cm^{-1} in beech and 1,596, 1,511, 1,268, and 1,220 cm^{-1} in pine) and a corresponding decrease in the intensities of carbohydrate bands (at 1,738, 1,375, 1,158, and 898 cm^{-1}). At higher weight losses, spectra for wood decayed by *C. puteana* had many similarities with that of Klason lignin isolated from wood. In contrast, wood decayed by *P. chrysosporium* showed selective-type decay with a reduction in peak heights associated with lignin relative to carbohydrates. Although weight losses in samples exposed to *C. versicolor* were high (45.5% and 39.8% for beech and Scots pine, respectively, after 12 weeks), simultaneous decay resulted in little change in the relative intensities of the lignin and carbohydrate bands, with only a slight preference for lignin.

Martinez et al. (2004) carried out a genomic analysis of wood-degrading *Phanerochaete chrysosporium* as a source for cellulases and hemicellulases for the biomass degradation in a paper with 668 citations. They sequenced the 30-million base-pair genome of *P. chrysosporium* strain RP78 using a whole genome shotgun approach. The *P. chrysosporium* genome revealed an impressive array of genes encoding secreted oxidases, peroxidases, and hydrolytic enzymes that cooperated in wood decay. This genome provided a high-quality draft sequence of a basidiomycete.

Martinez et al. (2009) carried out the genomic analysis of wood-degrading *Postia placenta,* as a source for cellulases and hemicellulases for the biomass degradation in a paper with 435 citations. They observed unique extracellular enzyme systems, including an unusual repertoire of extracellular glycoside hydrolases. Genes encoding exocellobiohydrolases and cellulose-binding domains, typical of cellulolytic microbes, were absent in this efficient cellulose-degrading fungus. When *P. placenta* was grown in medium containing cellulose as sole carbon source, transcripts corresponding to many hemicellulases and to a single putative β-1–4 endoglucanase were expressed at high levels relative to glucose-grown cultures. Also upregulated during growth on cellulose medium were putative iron reductases, quinone reductase, and structurally divergent oxidases potentially involved in extracellular generation of Fe(II) and H_2O_2. These observations were consistent with a biodegradative role for Fenton chemistry in which Fe(II) and H_2O_2 reacted to form hydroxyl radicals, highly reactive oxidants capable of depolymerizing cellulose (Table 7.2).

7.3.3 The Enzymatic Hydrolysis of the Biomass

Li et al. (2010) compared dilute acid and ionic liquid (IL) pretreatment of switchgrass in terms of delignification, enzymatic saccharification efficiency, and saccharide yields in a paper with 832 citations. For the IL pretreatment, they observed that switchgrass exhibited reduced cellulose crystallinity, increased surface area, and decreased lignin content compared to dilute acid pretreatment. Further, IL pretreatment enabled a significant enhancement in the rate of enzymatic hydrolysis of the cellulose component of switchgrass, with a rate increase of 16.7-fold, and a glucan yield of 96.0% obtained in 24 h.

TABLE 7.2
The Fungal Pretreatment of the Biomass

No.	Papers	Biomass	Prt.	Parameters	Keywords	Lead Author	Affil.	Cits
1	Martinez et al. (2008)	Biomass	*Trichoderma reesei* (*Hypocrea jecorina*)	Genomic analysis, biomass degradation	Biomass, degrading, fungus, Trichoderma, Hypocrea	Martinez, Diego 7202958664	Massachusetts Inst. Technol. USA	878
2	Pandey and Pitman (2003)	Wood, pine (*Pinus sylvestris* L.), beech (*Fagus sylvatica* L.)	*Coniophora puteana, Phanerochaete chrysosporium, Coriolus versicolor*	Wood degradation, spectroscopic study	Wood, decay, fungi	Pandey, Krishna K. 56252805800	Inst. Wood Sci. Technol. India	789
3	Martinez et al. (2004)	Wood	*Phanerochaete chrysosporium*	Genomic analysis, wood degradation	Lignocellulose, fungus, degrading	Cullen, Dan. 7202109135	USDA Forest Serv. USA	668
4	Martinez et al. (2009)	Wood	Postia placenta	Genomic analysis, cellulose degradation mechanisms	Lignocellulose, wood, decay, Postia, conversion	Cullen, Dan. 7202109135	USDA Forest Serv. USA	435

Prt, biological pretreatments; Other prts., other pretreatments; Na, nonavailable.

Lee et al. (2009) studied the enzymatic hydrolysis of lignin pretreated by the ILs in a paper with 795 citations. They used the IL, 1-ethyl-3 methylimidazolium acetate ([Emim]CH_3COO), as a pretreatment solvent to extract lignin from wood flour. They observed that the cellulose in the pretreated wood flour became far less crystalline without undergoing solubilization, while 40% of the lignin was removed, the cellulose crystallinity index dropped below 45, resulting in more than 90% of the cellulose in wood flour to be hydrolyzed by *Trichoderma viride* cellulase.

Eriksson et al. (2002) studied the enzymatic hydrolysis of the steam- and surfactant-pretreated spruce with cellulase of *Trichoderma reesei* in a paper with 739 citations. They found that nonionic surfactants were the most effective. The anionic and nonionic surfactants reduced enzyme adsorption to the spruce substrate. The approximate reduction of enzyme adsorption was from 90% adsorbed enzyme to 80% with surfactant addition. Surfactants had only a weak effect on cellulase temperature stability. They explained the improved conversion of lignocellulose with surfactant by the reduction of the unproductive enzyme adsorption to the lignin part of the substrate. This was due to hydrophobic interaction of surfactant with lignin on the lignocellulose surface, which released unspecifically bound enzyme.

Saha et al. (2005) studied the enzymatic hydrolysis of the acid-pretreated wheat straw for bioethanol production in a paper with 645 citations. They observed that the maximum yield of monomeric sugars from wheat straw (7.83%, w/v, dry solid-DS) by dilute H_2SO_4 (0.75%, v/v) pretreatment and enzymatic hydrolysis (45°C, pH 5.0, 72 h) using cellulase, β-glucosidase, xylanase, and esterase was 565 mg/g. Under this condition, there were no measurable quantities of furfural and hydroxymethylfurfural. The yield of ethanol (per liter) from acid pretreated enzyme saccharified wheat straw (78.3 g) hydrolysate by recombinant *Escherichia coli* strain FBR5 was 19 g with a yield of 0.24 g/g DS. Detoxification of the acid and enzyme-treated wheat straw hydrolysate by overliming reduced the fermentation time from 118 to 39 h in the case of separate hydrolysis and fermentation (SHF) (35°C, pH 6.5), and increased the ethanol yield from 13 to 17 g/L and decreased the fermentation time from 136 to 112 h in the case of simultaneous saccharification and fermentation (SSF) (35°C, pH 6.0).

Lloyd and Wyman (2005) studied the enzymatic hydrolysis of acid-pretreated corn stover in a paper with 616 citations. They observed that up to 15% of the total potential sugar in the substrate could be released as glucose during pretreatment and between 15% and 90+% of the xylose remaining in the solid residue could be recovered in subsequent enzymatic hydrolysis, depending on the enzyme loading. Glucose yields increased from as high as 56% of total maximum potential glucose plus xylose for just enzymatic digestion to 60% when glucose released in pretreatment was included. Xylose yields similarly increased from as high as 34% of total potential sugars for pretreatment alone to between 35% and 37% when credit was taken for xylose released in digestion. Yields were much lower if no acid was used. Overall, up to about 92.5% of the total sugars originally available in the corn stover used could be recovered for coupled dilute acid pretreatment and enzymatic hydrolysis. Enhanced hemicellulase activity could further improve xylose yields, particularly for low cellulase loadings.

Yang and Wyman (2004) studied the enzymatic hydrolysis of corn stover cellulose by batch and flow-through as well as water and acid pretreatment in a paper with 556 citations. They observed that increasing flow with just water enhanced the xylan dissolution rate, more than doubled total lignin removal, and increased cellulose digestibility. Furthermore, adding dilute sulfuric acid increased the rate of xylan removal for both batch and flow-through systems. Further, adding acid also increased the lignin removal rate with flow, but less lignin was left in solution when acid was added in batch. Although the enzymatic hydrolysis of pretreated cellulose was related to xylan removal, the digestibility was much better for flow-through compared with batch systems, for the same degree of xylan removal. Cellulose digestibility for flow-through reactors was related to lignin removal as well.

Hsu et al. (2010) studied enzymatic hydrolysis of acid-pretreated rice straw in a paper with 477 citations. They obtained a maximal sugar yield of 83% when the rice straw was pretreated with

1% (w/w) sulfuric acid with a reaction time of 1–5 min at 160°C or 180°C, followed by enzymatic hydrolysis. The complete release of sugar (xylose and glucose) increased the pore volume of the pretreated solid residues resulted in an efficiency of 70% for the enzymatic hydrolysis. The extra pore volume was generated by the release of acid-soluble lignin, and this resulted in the enzymatic hydrolysis being enhanced by nearly 10%. The increase in the crystallinity index of the pretreated rice straw was limited.

Ohgren et al. (2007) studied the enzymatic hydrolysis of the acid and steam pretreated corn stover in a paper with 466 citations. They observed that a near-theoretical glucose yield (96%–104%) could be obtained if xylanases were used to supplement cellulases during hydrolysis. Xylanases hydrolyzed residual hemicellulose, thereby improving the access of enzymes to cellulose. Under these conditions, xylose yields reached 70%–74%. When pretreatment severity was reduced by using autocatalysis instead of acid-catalyzed steam pretreatment, xylose yields were increased to 80%–86%. The overall glucose yield increased slightly due to delignification, but the overall xylose yield decreased due to hemicellulose loss in the delignification step.

Kristensen et al. (2009) studied the effect of the substrate concentration on the enzymatic hydrolysis of the lignocellulose in a paper with 462 citations. They observed the decreasing conversion at increasing substrate concentrations as a generic or intrinsic effect, describing a linear correlation from 5% to 30% initial total substrate content (w/w). Product inhibition by glucose and in particular cellobiose at the increased substrate concentrations at high substrate loading played a role but could not completely account for the decreasing conversion. Adsorption of cellulases decreased at increasing substrate concentrations. There was a strong correlation between the decreasing adsorption and conversion, indicating that the inhibition of cellulase adsorption to cellulose was causing the decrease in yield. They concluded that inhibition of enzyme adsorption by hydrolysis products was the main cause of the decreasing yields at increasing substrate concentrations in the enzymatic decomposition of cellulosic biomass.

Zhu et al. (2009) studied the enzymatic hydrolysis of the sulfite (SPORL)-and acid-pretreated and disk-refined spruce and red pine in a paper with 435 citations. They observed that after the sulfite pretreatment of spruce chips with 8%–10% bisulfite and 1.8%–3.7% sulfuric acid on oven dry (od) wood at 180°C for 30 min, more than 90% cellulose conversion of substrate was achieved with enzyme loading of about 14.6 FPU cellulase plus 22.5 CBU β-glucosidase per gram of od substrate after 48 h hydrolysis. Glucose yield from enzymatic hydrolysis of the substrate per 100 g of untreated o.d. spruce wood (glucan content 43%) was about 37 g. Hemicellulose removal was critical as lignin sulfonation for cellulose conversion in the SPORL process. Pretreatment altered the wood chips, which reduced the electric energy consumption for size reduction to about 19 Wh/kg o.d. untreated wood, or about 19 g glucose/Wh electricity. Furthermore, the SPORL produced low amounts of fermentation inhibitors, hydroxymethylfurfural (HMF) and furfural, of about 5 and 1 mg/g of untreated od wood, respectively. In addition, similar results were achieved when the SPORL was applied to red pine.

Grethlein (1985) studied the effect of pore size distribution on the rate of enzymatic hydrolysis of acid-pretreated hard and softwoods in a paper with 431 citations. They observed that regardless of the substrate, the initial rate of hydrolysis using cellulase from *Trichoderma reesei* was linearly correlated with the pore volume of the substrate accessible to a nominal diameter of 51 A, representative of the size of the cellulase. In contrast, crystallinity index had no relationship to the rate of hydrolysis.

Yang and Wyman (2006) studied the enzymatic hydrolysis of bovine serum albumin (BSA)-pretreated cellulose in lignin-containing substrates in a paper with 424 citations. They added cellulase and BSA to Avicel cellulose and solids containing 56% cellulose and 28% lignin from dilute sulfuric acid pretreatment of corn stover. They observed that little BSA was adsorbed on Avicel cellulose, while acid-pretreated corn stover solids adsorbed considerable amounts of this protein. On the other hand, cellulase was highly adsorbed on both substrates. Adding a 1% concentration of BSA to dilute acid pretreated corn stover prior to enzyme addition at 15 FPU/g cellulose enhanced filter paper activity in solution by about a factor of 2 and beta-glucosidase activity in solution by about a factor of 14. Overall, BSA treatment reduced adsorption of cellulase and particularly β-glucosidase on lignin. BSA treatment of pretreated corn stover solids prior to enzymatic hydrolysis increased 72h glucose yields from about 82% to about 92% at a cellulase loading of 15 FPU/g cellulose or achieved about the same yield at a loading of 7.5 FPU/g cellulose. They also observed similar improvements for enzymatic hydrolysis of ammonia fiber explosion (AFEX) pretreated corn stover and Douglas fir treated by SO_2 steam explosion and for simultaneous saccharification and fermentation (SSF) of BSA pretreated corn stover. In addition, BSA treatment prior to hydrolysis reduced the need for β-glucosidase supplementation of SSF (Table 7.3).

7.4 DISCUSSION

7.4.1 Introduction

The crude oil-based gasoline fuels have been widely used in the transportation sector since the 1920s. However, there have been great public concerns over the adverse environmental and human impact of these fuels. Hence, biomass-based bioethanol fuels have increasingly been used in blending gasoline and petrodiesel fuels, in the fuel cells, and in the biochemical production in a biorefinery context.

The research in the field of biological pretreatments of the biomass such as enzymatic and fungal pretreatments as well as enzymatic hydrolysis to increase the sugar and bioethanol yield has also intensified in recent years.

However, it is essential to develop efficient incentive structures for the primary stakeholders to enhance the research in this field. Although there have been a number of review papers for this field, there has been no review of the most-cited 25 articles in this field.

Thus, this book chapter presents a review of the most-cited 25 articles in this field. Then, it discusses the key findings of these highly influential papers and comments on the future research priorities in this field.

As a first step for the search of the relevant literature, the keywords were selected using the whole set of papers. The selected keyword list was optimized to obtain a representative sample of papers for the searched research field. This keyword list was provided in the Appendix for future replication studies.

As a second step, a sample dataset was used for this study. The first 25 articles with at least 424 citations each were selected for the review study. Key findings from each paper were taken from the abstracts of these papers and were discussed. Additionally, a number of brief conclusions were drawn, and a number of relevant recommendations were made to enhance the future research landscape.

There are three primary research fronts for these most-cited 25 papers: enzymatic pretreatments, enzymatic hydrolysis, and fungal pretreatments with 36%, 48%, and 16% of the 25 HCPs, respectively (Table 7.4). However, enzymatic pretreatments are over-represented in the reviewed papers compared to the 134 HCPs with 10% surplus. On the contrary, fungal pretreatments, enzymatic hydrolysis, and microbial pretreatments are under-represented by 10%, 7%, and 4%, respectively. Furthermore, there is no representation of microbial pretreatments in the reviewed papers.

Table 7.5 provides information on these biomass used in these 25 HCPs. There are four primary types of the biomass: biomass constituents, wood, agricultural residues, and biomass at large with 44%, 24%, 24%, and 12% of the 25 HCPs, respectively. Further, lignin and cellulose are the primary constituents with 28% and 16% of the 25 HCPs, respectively. Further, corn stover is the key agricultural residues with 12% of the 25 HCPs.

It is notable that the papers related to wood, corn stover, lignin, and agricultural residues are over-represented by 10%, 9%, 7%, and 5%, respectively. On the contrary, the papers related to the biomass constituents, cellulose, lignocellulose, and biomass are under-represented by 8%–15% of the 25 HCPs each.

TABLE 7.3
The Enzymatic Hydrolysis of the Biomass

No.	Papers	Biomass	Prt.	Other Prts.	Parameters	Keywords	Lead Author	Affil.	Cits
1	Li et al. (2010)	Switchgrass	Enzyme	IL, acid	Enzymatic hydrolysis	Switchgrass, biomass, recalcitrance, delignification, enzymatic, saccharification	Singh, Seema 35264950300	Joint Bioenergy Inst. USA	832
2	Lee et al. (2009)	Lignin	*Trichoderma viride* cellulase	IL	Enzymatic hydrolysis	Lignin, extraction, wood, cellulose, enzymatic, hydrolysis	Doherty, Thomas V. 57213283673	Rensselear Polytech. Inst. USA	795
3	Eriksson et al. (2002)	Spruce	*Trichoderma reesei* cellulase	Surfactants	Enzymatic hydrolysis, hydrolysis mechanisms	Enzymatic, lignocellulose, hydrolysis	Tjerneld, Folke 7006446969	Lund Univ. Sweden	739
4	Saha et al. (2005)	Wheat straw	β-glucosidase, xylanase and esterase	Acids	Enzymatic hydrolysis, sugar and ethanol yield	Acid, pretreatment, straw, enzymatic, saccharification	Saha, Badal C. 7202946302	USDA Agr. Res. Serv.	645
5	Lloyd and Wyman (2005)	Corn stover	Enzyme	Acids	Enzymatic hydrolysis, xylose and glucose yield	Corn stover, pretreatment, enzymatic, hydrolysis	Wyman, Charles E. 7004396809	Dartmouth Coll. USA	616
6	Yang and Wyman (2004)	Corn stover cellulose	Enzyme	Water, acids	Enzymatic hydrolysis, xylan and lignin degradation	Xylan, lignin, enzymatic, digestibility, corn stover, cellulose	Wyman, Charles E. 7004396809	Dartmouth Coll. USA	556
7	Hsu et al. (2010)	Rice straw	Enzyme	Acid	Enzymatic hydrolysis, sugar yield	Acid, pretreatment, rice straw, enzymatic, hydrolysis	Guo, Gia Luen 7402768046	Inst. Nucl. Ener. Res. Taiwan	477
8	Ohgren et al. (2007)	Corn stover	Xylanase, cellulase	Acid, steam	Hemicellulose and lignin degradation, enzymatic hydrolysis	Corn stover, hemicellulose, lignin, enzymatic, hydrolysis, pretreated	Zacchi, Guido 7006727748	Lund Univ. Sweden	466
9	Kristensen et al. (2009)	Cellulosic biomass	Cellulase	Na	Enzymatic hydrolysis, substrate concentration, sugar yield	Lignocellulose, enzymatic, hydrolysis	Jorgensen, Henning 7202554496	Copenhagen Univ. Denmark	462
10	Zhu et al. (2009)	Spruce, pine	Cellulase, β-glucosidase	Sulfite (SPORL), acid	Enzymatic hydrolysis, sugar yield	Sulfite, pretreatment, enzymatic, saccharification, spruce, pine	Zhu, Junyong 7405692678	USDA Forest Serv.	435
11	Grethlein (1985)	Hardwood, softwood	*Trichoderma reesei* cellulase	Acid	Enzymatic wood, cellulose degradation	Enzymatic, hydrolysis, substrate, cellulosic	Grethein, Hans E 7004179983	MBI Int. USA	431
12	Yang and Wyman (2006)	Corn stover cellulose	Enzyme, cellulase	Bovine serum albumin (BSA), acid	Enzymatic hydrolysis mechanisms, sugar yield	Enzymatic, hydrolysis, cellulose, lignin	Wyman, Charles E 7004396809	Dartmouth Coll. USA	424

Prt, biological pretreatments; Other parts., other pretreatments; Na, nonavailable.

TABLE 7.4
The Biological Pretreatment of the Biomass: Research Fronts

No.	Research Fronts	N Paper % Review	N Paper % Sample	Surplus
1	Enzymatic pretreatments	36	28	8
2	Fungal pretreatments	16	26	−10
3	Enzymatic hydrolysis	48	55	−7
4	Microbial pretreatments	0	4	−4
	Sample	25	134	

N paper (%) sample: the number of papers in the population sample of 134 papers. N paper (%) review: the number of papers in the sample of 25 most cited papers.

7.4.2 The Enzymatic Pretreatment of the Biomass

Bourbonnais and Paice (1990) studied the lignin biodegradation by laccases (*Trametes versicolor-Coriolus versicolor*) and asserted that laccase was capable of oxidizing both phenolic and non-phenolic moieties of lignin but that the latter was dependent on the copresence of primary laccase substrates. On the other hand, Bourbonnais et al. (1995) studied the lignin degradation by laccase isozymes from *Trametes versicolor* and the role of the mediator 2,2′-azinobis(3-ethylbenzthiazoline-6-sulfonate) (ABTS) and found that the presence of the mediator prevented and reversed the polymerization of kraft lignin by either laccase.

Floudas et al. (2012) studied the paleozoic origin of enzymatic lignin decomposition based on the analysis of 31 fungal genomes and found that lignin-degrading peroxidases expanded in the lineage leading to the ancestor of the Agaricomycetes. On the other hand, Tien and Kirk (1983) studied the lignin-degrading enzyme from the *Phanerochaete chrysasporium* and observed that this *P. chrysasporium* contained an enzyme that degraded lignin substructure model compounds as well as spruce and birch lignins.

Quinlan et al. (2011) studied the oxidative degradation of cellulose by a glycoside hydrolase (CAZy) GH61 enzyme and showed that GH61 enzymes were a unique family of copper-dependent oxidases. On the other hand, Glenn and Gold (1985) purified and characterized an extracellular Mn(II)-dependent peroxidase from the lignin-degrading *Phanerochaete chrysosporium* and found that Mn(III) played a role in the enzyme mechanism. Further, Harris et al. (2010) studied the structure and function of glycoside hydrolase 61 (GH61) enzymes and showed that certain GH61 proteins in the presence of various divalent metal ions could significantly reduce the total protein loading required to hydrolyze lignocellulosic biomass.

Glenn et al. (1983) prepared an extracellular H_2O_2-requiring enzyme involved in lignin biodegradation by *Phanerochaete chrysosporium* and found that the enzyme system partially degraded 14C-ring labeled lignin. On the other hand, Phillips et al. (2011) studied the cellulose degradation using cellobiose dehydrogenase and a copper-dependent polysaccharide monooxygenase by *Neurospora crassa* and found that when added to mixtures of cellulases, these proteins enhanced cellulose saccharification.

Overall, these HCPs clearly show that these enzymatic pretreatments degrade the biomass to improve the hydrolysis of the biomass and fermentation of the resulting hydrolysates with an improved sugar and bioethanol yield. It is notable that these enzymes are produced primarily from the fungal species.

7.4.3 The Fungal Pretreatment of the Biomass

Martinez et al. (2008) provided a roadmap for constructing enhanced *Trichoderma reesei* strains as a source for cellulases and hemicellulases for the biomass degradation and found that its genome encoded fewer cellulases and hemicellulases compared to any other sequenced fungus. On the other

TABLE 7.5
The Most Prolific Research Fronts for the Biomass for the Biological Pretreatments of the Biomass

No.	Research Fronts	N Paper % Review	N Paper % Sample	Surplus
1	**Biomass Constituents**	**44**	**60**	**−16**
	Lignin	28	21	7
	Cellulose	16	27	−11
	Lignocellulose	0	10	−10
	Others	0	3	−2
2	Wood	24	14	10
3	Grass	4	2	2
4	**Agricultural Residues**	**24**	**19**	**5**
	Corn stover	16	7	9
	Straw	8	7	1
	Other residues	0	1	−1
	Agricultural residues in general	0	1	−1
	Sugarcane bagasse	0	3	−3
5	**Biomass in General**	**12**	**22**	**−10**
	Biomass	4	12	−8
	Lignocellulosic biomass	4	8	−4
	Cellulosic biomass	4	1	3
6	Algae	0	1	−1
	Sample	**25**	**134**	

N paper (%) sample: the number of papers in the population sample of 134 papers. N paper (%) review: the number of papers in the sample of 25 most cited papers.

hand, Martinez et al. (2004) carried out a genomic analysis of wood-degrading *Phanerochaete chrysosporium* as a source for cellulases and hemicellulases for the biomass degradation and observed that the *P. chrysosporium* genome revealed an impressive array of genes encoding secreted oxidases, peroxidases, and hydrolytic enzymes that cooperated in wood decay. Further, Martinez et al. (2009) carried out the genomic analysis of wood-degrading *Postia placenta,* as a source for cellulases and hemicellulases for the biomass degradation and observed unique extracellular enzyme systems, including an unusual repertoire of extracellular glycoside hydrolases.

Pandey and Pitman (2003) studied wood (*Pinus sylvestris* L., *Fagus sylvatica* L.) degradation by *Coniophora puteana*, *Coriolus versicolor* (*Trichoderma cersicolor*), and *Phanerochaete chrysosporium* and found that there was a progressive increase in lignin content relative to carbohydrate evident.

Overall, these HCPs clearly show that the fungal pretreatments degrade the biomass to improve the hydrolysis of the biomass and fermentation of the resulting hydrolysates with an improved sugar and bioethanol yield.

7.4.4 The Enzymatic Hydrolysis of the Biomass

Li et al. (2010) compared dilute acid and IL pretreatment of switchgrass in terms of delignification, enzymatic saccharification efficiency, and saccharide yields and observed that IL pretreatment enabled a significant enhancement in the rate of enzyme hydrolysis of the cellulose component of switchgrass. On the other hand, Lee et al. (2009) studied the enzymatic hydrolysis of lignin pretreated by the ILs and observed more than 90% of the cellulose in wood flour to be hydrolyzed by *Trichoderma viride* cellulase.

Eriksson et al. (2002) studied the enzymatic hydrolysis of the steam- and surfactant-pretreated spruce with cellulase of *Trichoderma reesei* and observed that the approximate reduction of enzyme adsorption was from 90% adsorbed enzyme to 80% with surfactant addition. On the other hand, Saha et al. (2005) studied the enzymatic hydrolysis of the acid-pretreated wheat straw for bioethanol production in a paper with 645 citations. They observed that the maximum yield of monomeric sugars from wheat straw by dilute H_2SO_4 pretreatment and enzymatic hydrolysis using cellulase, β-glucosidase, xylanase, and esterase was 565 mg/g.

Lloyd and Wyman (2005) studied the enzymatic hydrolysis of acid-pretreated corn stover and observed that up to about 92.5% of the total sugars originally available in the corn stover used could be recovered for coupled dilute acid pretreatment and enzymatic hydrolysis. On the other hand, Yang and Wyman (2004) studied the enzymatic hydrolysis of corn stover cellulose by batch and flow-through as well as water and acid pretreatment and observed that, the digestibility was much better for flow-through compared with batch systems, for the same degree of xylan removal. Further, Yang and Wyman (2006) studied the enzymatic hydrolysis of bovine serum albumin (BSA)-pretreated cellulose and observed that cellulase was highly adsorbed on both substrates.

Hsu et al. (2010) studied enzymatic hydrolysis of acid-pretreated rice straw and obtained a maximal sugar yield of 83% when the rice straw was pretreated with 1% (w/w) sulfuric acid, followed by enzymatic hydrolysis. On the other hand, Ohgren et al. (2007) studied the enzymatic hydrolysis of the acid and steam pretreated corn stover and observed that a near-theoretical glucose yield could be obtained if xylanases were used to supplement cellulases during hydrolysis.

Kristensen et al. (2009) studied the effect of the substrate concentration on the enzymatic hydrolysis of the lignocellulose and observed that there was a strong correlation between the decreasing adsorption and conversion. On the other hand, Zhu et al. (2009) studied the enzymatic hydrolysis of the sulfite (SPORL)- and acid-pretreated and disk-refined spruce and red pine and observed that more than 90% cellulose conversion of substrate was achieved with enzyme loading of about cellulase plus β-glucosidase. Further, Grethlein (1985) studied the effect of pore size distribution on the rate of enzymatic hydrolysis of acid-pretreated hard and softwoods and observed that regardless of the substrate, the initial rate of hydrolysis using cellulase from *Trichoderma reesei* was linearly correlated with the pore volume of the substrate.

Overall, these HCPs clearly show that these enzymes degrade the biomass to improve the hydrolysis of the biomass and fermentation of the resulting hydrolysates with an improved sugar and bioethanol yield. For this purpose, hydrolysis and fermentation conditions were optimized. A number of the biomass and enzymes are used for this purpose.

ACKNOWLEDGMENT

The contribution of the highly cited researchers in the field of the biological pretreatment of the biomass has been gratefully acknowledged.

REFERENCES

Alvira, P., E. Tomas-Pejo, M. Ballesteros and M. J. Negro. 2010. Pretreatment technologies for an efficient bioethanol production process based on enzymatic hydrolysis: A review. *Bioresource Technology* 101:4851–4861.

Antolini, E. 2007. Catalysts for direct ethanol fuel cells. *Journal of Power Sources* 170:1–12.

Antolini, E. 2009. Palladium in fuel cell catalysis. *Energy and Environmental Science* 2:915–931.

Bourbonnais, R. and M. G. Paice. 1990. Oxidation of non-phenolic substrates. An expanded role for laccase in lignin biodegradation. *FEBS Letters* 267:99–102.

Bourbonnais, R., M. G. Paice, I.D. Reid, P. Lanthier and M. Yaguchi. 1995. Lignin oxidation by laccase isozymes from *Trametes versicolor* and role of the mediator 2,2'-azinobis(3-ethylbenzthiazoline-6-sulfonate) in kraft lignin depolymerization. *Applied and Environmental Microbiology* 61:1876–1880.

Brown, M. E. and M.C. Y. Chang. 2014. Exploring bacterial lignin degradation. *Current Opinion in Chemical Biology* 19:1–7.

Bugg, T. D. H, M. Ahmad, E. M. Hardiman and R. Singh. 2011. The emerging role for bacteria in lignin degradation and bio-product formation. *Current Opinion in Biotechnology* 22: 394–400.

Burnham, J. F. 2006. Scopus database: A review. *Biomedical Digital Libraries* 3:1–8.

Dashtban, M., H. Schraft and W. Qin. 2009. Fungal bioconversion of lignocellulosic residues; Opportunities & perspectives. *International Journal of Biological Sciences* 5:578–595.

Eriksson, T., J. Borjesson and F. Tjerneld. 2002. Mechanism of surfactant effect in enzymatic hydrolysis of lignocellulose. *Enzyme and Microbial Technology* 31:353–364.

Fernando, S., S. Adhikari, C. Chandrapal and M. Murali. 2006. Biorefineries: Current status, challenges, and future direction. *Energy & Fuels* 20:1727–1737.

Floudas, D., M. Binder and R. Riley, et al. 2012. The paleozoic origin of enzymatic lignin decomposition reconstructed from 31 fungal genomes. *Science* 336:1715–1719.

Glenn, J. K. and M. H. Gold. 1985. Purification and characterization of an extracellular Mn(II)-dependent peroxidase from the lignin-degrading basidiomycete, Phanerochaete chrysosporium. *Archives of Biochemistry and Biophysics* 24:329–341.

Glenn, J. K., M. A. Morgan, M. B. Mayfield, M. Kuwahara and M. H. Gold. 1983. An extracellular H_2O_2-requiring enzyme preparation involved in lignin biodegradation by the white rot basidiomycete *Phanerochaete chrysosporium. Biochemical and Biophysical Research Communications* 114:1077–1083.

Grethlein, H. E. 1985. The effect of pore size distribution on the rate of enzymatic hydrolysis of cellulosic substrates. *Bio/Technology* 3:155–160.

Hahn-Hagerdal, B., M. Galbe, M. F. Gorwa-Grauslund, G. Liden and G. Zacchi. 2006. Bio-ethanol – The fuel of tomorrow from the residues of today. *Trends in Biotechnology* 24:549–556.

Harris, P. V., D. Welner and K. C. McFarland, et al. 2010 Stimulation of lignocellulosic biomass hydrolysis by proteins of glycoside hydrolase family 61: Structure and function of a large, enigmatic family. *Biochemistry* 49:3305–3316.

Hill, J., E. Nelson, D. Tilman, S. Polasky and D. Tiffany. 2006. Environmental, economic, and energetic costs and benefits of biodiesel and ethanol biofuels. *Proceedings of the National Academy of Sciences of the United States of America* 103:11206–11210.

Hill, J., S. Polasky and E. Nelson, et al. 2009. Climate change and health costs of air emissions from biofuels and gasoline. *Proceedings of the National Academy of Sciences of the United States of America* 106:2077–2082.

Hsieh, W. D., R. H. Chen, T. L. Wu and T. H. Lin. 2002. Engine performance and pollutant emission of an SI engine using ethanol-gasoline blended fuels. *Atmospheric Environment* 36:403–410.

Hsu, T. C., G. L. Guo, W. H. Chen and W. S. Hwang. 2010. Effect of dilute acid pretreatment of rice straw on structural properties and enzymatic hydrolysis. *Bioresource Technology* 101:4907–4913.

Huang, H. J., S. Ramaswamy, U. W. Tschirner and B. V. Ramarao. 2008. A review of separation technologies in current and future biorefineries. *Separation and Purification Technology* 62:1–21.

Konur, O. 2000. Creating enforceable civil rights for disabled students in higher education: An institutional theory perspective. *Disability & Society* 15:1041–1063.

Konur, O. 2002a. Access to nursing education by disabled students: Rights and duties of nursing programs. *Nurse Education Today* 22:364–374.

Konur, O. 2002b. Assessment of disabled students in higher education: Current public policy issues. *Assessment and Evaluation in Higher Education* 27:131–52.

Konur, O. 2002c. Access to employment by disabled people in the UK: Is the Disability Discrimination Act working? *International Journal of Discrimination and the Law* 5:247–279.

Konur, O. 2006a. Participation of children with dyslexia in compulsory education: Current public policy issues. *Dyslexia* 12:51–67.

Konur, O. 2006b. Teaching disabled students in Higher Education. *Teaching in Higher Education* 11:351–363.

Konur, O. 2007a. A judicial outcome analysis of the Disability Discrimination Act: A windfall for the employers? *Disability & Society* 22:187–204.

Konur, O. 2007b. Computer-assisted teaching and assessment of disabled students in higher education: The interface between academic standards and disability rights. *Journal of Computer Assisted Learning* 23:207–219.

Konur, O. 2012. The evaluation of the research on the bioethanol: A scientometric approach. *Energy Education Science and Technology Part A: Energy Science and Research* 28:1051–1064.

Konur, O. 2015. Current state of research on algal bioethanol. In *Marine Bioenergy: Trends and Developments*, Eds. S. K. Kim and C. G. Lee, pp. 217–244. Boca Raton, FL: CRC Press.

Konur, O. 2020. The scientometric analysis of the research on the bioethanol production from green macroalgae. In *Handbook of Algal Science, Technology and Medicine*, Ed. O. Konur, pp. 385–401. London: Academic Press.

Konur, O. 2023. Biological pretreatment of the biomass: Scientometric study. In *Bioethanol Fuel Production Processes. I: Biomass Pretreatments. Handbook of Bioethanol Fuels Volume 1*, Ed. O. Konur. Boca Raton, FL: CRC Press.

Kristensen, J. B., C. Felby and H. Jorgensen. 2009. Yield-determining factors in high-solids enzymatic hydrolysis of lignocellulose. *Biotechnology for Biofuels* 2:11.

Lee, S. H., T. V. Doherty, R. J. Linhardt and J. S. Dordick. 2009. Ionic liquid-mediated selective extraction of lignin from wood leading to enhanced enzymatic cellulose hydrolysis. *Biotechnology and Bioengineering* 102:1368–1376.

Li, C., B. Knierim and C. Manisseri, et al. 2010. Comparison of dilute acid and ionic liquid pretreatment of switchgrass: Biomass recalcitrance, delignification and enzymatic saccharification. *Bioresource Technology* 101:4900–4906.

Lin, Y. and S. Tanaka. 2006. Ethanol fermentation from biomass resources: Current state and prospects. *Applied Microbiology and Biotechnology* 69:627–642.

Lloyd, T. A. and C. E. Wyman. 2005. Combined sugar yields for dilute sulfuric acid pretreatment of corn stover followed by enzymatic hydrolysis of the remaining solids. *Bioresource Technology* 96:1967–1977.

Ma, X., L. Sun and C. Song. 2002. A new approach to deep desulfurization of gasoline, diesel fuel and jet fuel by selective adsorption for ultra-clean fuels and for fuel cell applications. *Catalysis Today* 77:107–116.

Martinez, D., J. Challacombe and I. Morgenstern, et al. 2009. Genome, transcriptome, and secretome analysis of wood decay fungus *Postia placenta* supports unique mechanisms of lignocellulose conversion. *Proceedings of the National Academy of Sciences of the United States of America* 106:1954–1959.

Martinez, D., L. F. Larrondo and N. Putnam, et al. 2004. Genome sequence of the lignocellulose degrading fungus *Phanerochaete chrysosporium* strain RP78. *Nature Biotechnology* 22:695–700.

Martinez, D., R. M. Berka and B. Henrissat, et al. 2008. Genome sequencing and analysis of the biomass-degrading fungus *Trichoderma reesei* (syn. *Hypocrea jecorina*). *Nature Biotechnology* 26:553–560.

Morschbacker, A. 2009. Bio-ethanol based ethylene. *Polymer Reviews* 49:79–84.

Najafi, G., B. Ghobadian and T. Tavakoli, et al. 2009. Performance and exhaust emissions of a gasoline engine with ethanol blended gasoline fuels using artificial neural network. *Applied Energy* 86:630–639.

Newman, P. W. G. and J. R. Kenworthy. 1989. Gasoline consumption and cities: A comparison of U.S. cities with a global survey. *Journal of the American Planning Association* 55:24–37.

North, D. C. 1991. Institutions. *Journal of Economic Perspectives* 5:97–112.

Ohgren, K., R. Bura, J. Saddler and G. Zacchi. 2007. Effect of hemicellulose and lignin removal on enzymatic hydrolysis of steam pretreated corn stover. *Bioresource Technology* 98:2503–2510.

Olsson, L. and B. Hahn-Hagerdal. 1996. Fermentation of lignocellulosic hydrolysates for ethanol production. *Enzyme and Microbial Technology* 18:312–331.

Pandey, K. K. and A. J. Pitman. 2003. FTIR studies of the changes in wood chemistry following decay by brown-rot and white-rot fungi. *International Biodeterioration and Biodegradation* 52:151–160.

Phillips, C. M., W. T. Beeson, J. H. Cate and M. A. Marletta. 2011. Cellobiose dehydrogenase and a copper-dependent polysaccharide monooxygenase potentiate cellulose degradation by *Neurospora crassa*. *ACS Chemical Biology* 6:1399–1406.

Quinlan, R. J., M. D. Sweeney and L. L. Leggio, et al. 2011. Insights into the oxidative degradation of cellulose by a copper metalloenzyme that exploits biomass component. *Proceedings of the National Academy of Sciences of the United States of America* 108:15079–15084.

Saha, B. C., L. B. Iten, M. A. Cotta and Y. W. Wu. 2005. Dilute acid pretreatment, enzymatic saccharification and fermentation of wheat straw to ethanol. *Process Biochemistry* 40:3693–3700.

Sanchez, C. 2009. Lignocellulosic residues: Biodegradation and bioconversion by fungi. *Biotechnology Advances* 27:185–194.

Sanchez, O. J. and C. A. Cardona. 2008. Trends in biotechnological production of fuel ethanol from different feedstocks. *Bioresource Technology* 99:5270–5295.

Sindhu, R., P. Binod and A. Pandey. 2016. Biological pretreatment of lignocellulosic biomass – An overview. *Bioresource Technology* 199:76–82.

Sun, Y. and J. Cheng. 2002. Hydrolysis of lignocellulosic materials for ethanol production: A review. *Bioresource Technology* 83:1–11.

Taherzadeh, M. J. and K. Karimi. 2007. Enzyme-based hydrolysis processes for ethanol from lignocellulosic materials: A review. *Bioresources* 2:707–738.

Taherzadeh, M. J. and K. Karimi. 2008. Pretreatment of lignocellulosic wastes to improve ethanol and biogas production: A review. *International Journal of Molecular Sciences* 9:1621–1651.

Tien, M. and Kirk, T. K. 1983. Lignin-degrading enzyme from the hymenomycete *Phanerochaete chrysasporium* burds. *Science* 221:661–663.

Van Dyk, J. S. and B. I. Pletschke. 2012. A review of lignocellulose bioconversion using enzymatic hydrolysis and synergistic cooperation between enzymes-Factors affecting enzymes, conversion and synergy. *Biotechnology Advances* 30:1458–1480.

Yang, B. and C. E. Wyman. 2004. Effect of xylan and lignin removal by batch and flowthrough pretreatment on the enzymatic digestibility of corn stover cellulose. *Biotechnology and Bioengineering* 86:88–98.

Yang, B. and C. E. Wyman. 2006. BSA treatment to enhance enzymatic hydrolysis of cellulose in lignin containing substrates. *Biotechnology and Bioengineering* 94:611–617.

Zhu, J. Y., X. J. Pan, G. S. Wang and R. Gleisner. 2009. Sulfite pretreatment (SPORL) for robust enzymatic saccharification of spruce and red pine. *Bioresource Technology* 100:2411–2418.

8 Lignocellulolytic Enzymes for Biofuel Production

Sonali Thakur
Lovely Professional University

Richa Arora
Punjab Agricultural University

Ajay Kumar
Sam Higginbottom University of Agriculture, Technology and Sciences

Amit Pratush
Arboreal Bio-Innovations Pvt. Ltd.

ABBREVIATIONS

AA	Auxiliary activities
AAO	Aryl-alcohol oxidase
BGL	β-glucosidase
CAZymes	Carbohydrate-active enzymes
CBM	Carbohydrate-binding modules
CD	Catalytic domains
CDH	Cellobiose dehydrogenase
CE	Carbohydrate esterases
GH	Glycoside hydrolases
GLO	Glyoxal oxidase
LCB	Lignocellulosic biomass
LiP	Lignin peroxidase
MnP	Manganese peroxidase
ODC	Oxalate decarboxylase
PL	Polysaccharide lyases
SBH	Soybean hulls
VP	Versatile peroxidase

8.1 INTRODUCTION

Attention to cellulosic biological commodities as a substitute for fossil-based materials has increased over the past decade. This growing research can be accredited mainly due to the huge financial costs as well as the environmental problems raised due to fossil-based materials. Lignocellulosic biomass (LCB) is one of the most abundant, nominal effort, limitless, and cost-effective feedstocks for the derivation of biochemicals (including biofuels) due to its rich cellulosic properties (Obeng et al., 2017).

Generally, product development from LCB has three key stages (i) pretreatment step which uncovers the cellulose and hemicellulose by eliminating lignin and other waxy materials, (ii) saccharification which converts celluloses and hemicelluloses into fermentable sugars by the use of

DOI: 10.1201/9781003226536-11

hydrolytic enzymes, and (iii) fermentation process which further forms various renewable products from hydrolyzed sugars (Gunny et al., 2014). However, it has been seen that geometrical confirmation of cellulose (solid-occupied crystal-like arrangement) having cross-linkages with hemicelluloses and lignin further slows down its degradation and makes it overpriced (Arora et al., 2015).

Despite extensive investigations of cellulosic material bioconversions, a few other technical blockages also exist. Efficient and economical enzymatic hydrolysis of recalcitrant LCB is among the major constraints. For optimum hydrolysis, the contact between cellulase enzyme and their corresponding substrate must be in a proper manner and higher ratio (Orencio-Trejo et al., 2016). Improvement in a bioprocess for using inexpensive media to produce cellulases is still a challenge (Li et al., 2016a). Furthermore, the cellulase-producing microorganisms also produce some unwanted metabolites which reduce the product yield; thus, to overcome this problem, these organisms need special cultivation conditions. In this regard, alkaliphilic/acidophilic thermozymes produced by thermophilic and alkaliphilic/acidophilic microbes having high catalytic activity are the main focus of the current article. Literature studies concluded that under stressful environmental conditions, these thermozymes are more stable, which further enhance productivity and play a key role in economic development (Gomes et al., 2016).

Currently, fungus-based cellulases are major commercial products. Thermophilic bacteria have numerous benefits in the hydrolysis of lignocellulosic materials such as producing highly active thermostable enzymes and higher solubility of reactants and products which further increase the reaction velocities and make the hydrolysis reaction shorter. In such reactions, productivity is much higher with a minimum risk of contamination (Bhalla et al., 2013). Hence, to find out attractive strategies for enzymatic saccharification in a biorefinery, there is a necessity to search for cellulases of bacterial origin (Sriariyanun et al., 2016). The current chapter is focused on lignocellulolytic enzyme-producing microbes, their production mechanism, and hyper-enzyme production for efficient saccharification of LCB.

8.2 LIGNOCELLULOLYTIC ENZYMES: AN OVERVIEW

The combined activity of all lignocellulases—cellulases, hemicellulases, and ligninases (lignolytic compounds)—is needed for the compelling demolition of LCB. Therefore, in the literature, different studies focusing on lignocellulose-degrading enzymes, their arrangement, and their mechanism have been described in detail (Bornscheuer et al., 2014; Wu et al., 2014). Herein, the three most important lignocellulolytic enzymes such as cellulases, hemicellulases, and ligninases are discussed in detail.

8.2.1 Cellulases

Cellulases comprise the subunit's endoglucanases and exoglucanases which in addition comprehend cellobiohydrolases, cellodextrinases, and β-glucosidases. Cellulases are divided into two major categories: non-complex and complex cellulases (Patel et al., 2019). Several anaerobic microorganisms such as *Clostridium thermocellum* produce complex cellulases, although non-complex cellulases are produced by aerobic fungi and bacteria named *Trichoderma reesei* and *Bacillus megaterium* (Himmel et al., 2018; Singhania et al., 2017), respectively. Gupta and Verma (2015) demonstrate three different types of extracellular cellulolytic enzymes, which are produced by the fungus *T. reesei,* and these enzymes are made up of two cellobiohydrolases, two β-glucosidases (BGL), and five endoglucanases.

Furthermore, it has been seen that various fungal and bacterial species can interrupt plant cell wall structure which comprises polysaccharides by producing enough amount of complex enzymes named carbohydrate-active enzymes (CAZymes) (Munir and Levin, 2016). The detailed CAZyme structure shows the presence of cellulases, hemicellulases, and associated glycoside hydrolases, as well as carbohydrate esterases, polysaccharide lyases, catalysts with carbohydrate-binding modules and auxiliary activities. Different types of cellulases secreted by bacteria, fungi, and actinomycetes are tabulated in Table 8.1.

TABLE 8.1
Enzyme Activity of Cellulase-Producing Microbes on Different Substrates

Microorganisms	Enzyme Produced	Reaction Mixture	Substrate(s)	Enzyme Activity	References
		Bacteria			
Bacillus cereus	Cellulase	pH-7, temperature-30°C, time-48 h	*Acacia Arabica* *Bauhinia forficata* *Cassia surattensis* *Peltophorum pterocarpum*	0.440 IU/mL/min 0.213 IU/mL/min 0.187 IU/mL/min 0.190 IU/mL/min	Patagundi et al. (2014)
Bacillus subtilis	Cellulase	pH-7, temperature-30°C, time-48 h	*Acacia Arabica* *Bauhinia forficata* *Cassia surattensis* *Peltophorum pterocarpum*	0.357 IU/mL/min 0.201 IU/mL/min 0.189 IU/mL/min 0.178 IU/mL/min	Paudel and Qin (2015)
Bacillus thuringiensis	Cellulase	pH-7, temperature-30°C, time-48 h	*Acacia Arabica* *Bauhinia forficata* *Cassia surattensis* *Peltophorum pterocarpum*	0.334 IU/mL/min 0.219 IU/mL/min 0.203 IU/mL/min 0.193 IU/mL/min	Patagundi et al. (2014)
Bacillus sp. K1	Carboxy methyl cellulase	pH-6, temperature-50°C, time- 72 h	Carboxymethyl cellulose	5.21±0.21 U/mL	Paudel and Qin (2015)
Streptomyces sp. MDS	Endoglucanase Exoglucanase Cellobiase	pH-5–8, temperature-50°C–80°C, moisture content- 1:1.5	Rice waste biomass	132.6±1.75 U/gds 14.6±0.62 U/gds 125.6±1.75 U/gds	Saratale et al. (2017)
		Fungi			
Aspergillus niger	Cellulase	pH- 4–7, temperature-20°C–50°C, time 2–8 days	Saw dust, paper cellulose	3.9 IU	Devi and Kumar (2012)
Trichoderma reesei	Cellulase	Temperature-30°C, time-72 h, agitation- 200 rpm	Rice straw	70.3%	Kogo et al. (2017)
Humicola insolens	Cellulase	Temperature-30°C, time- 72 h, agitation- 200 rpm	Rice straw	23.5%	Kogo et al. (2017)
Trichoderma viride	Cellulase	pH- 5.0, temperature-30°C, time- 120 h	Forest waste	1.76 U/mL	Kumar et al. (2014)
Myceliophthora thermophila JCP 1-4	Endoglucanase *B*-glucosidase	pH- 5, temperature-45°C, time- 96 h, moisture content- 70%	Sugarcane bagasse	357.51 U/g 45.42 U/g	de Cassia Pereira et al. (2015)
		Actinomycetes			
Streptomyces thermoalkaliphilus	Carboxymethyl cellulase	pH- 9.0, temperature 50°C , time- 3 days	Carboxymethyl cellulose	19.5±1.5 IU/mL	Wu et al. (2018)
Streptomyces sp. NEAE-D	Carboxymethyl cellulase	pH- 6.5, temperature-35°C, time- 5 days	Carboxymethyl cellulose	70.99%	El-Naggar and Abdelwahed (2012)
Streptomyces sp.	Carboxymethyl cellulase	pH- 7.0, temperature-37°C, time- day 5	Carboxymethyl cellulose	1.388 µmol/mL/min	Deepthi et al. (2012)
Streptomyces griseorubens	Carboxymethyl cellulase	pH- 7, temperature-45°C and time- 6 days	Carboxymethyl cellulose	4.5 mg/mL	Prasad et al. (2013)

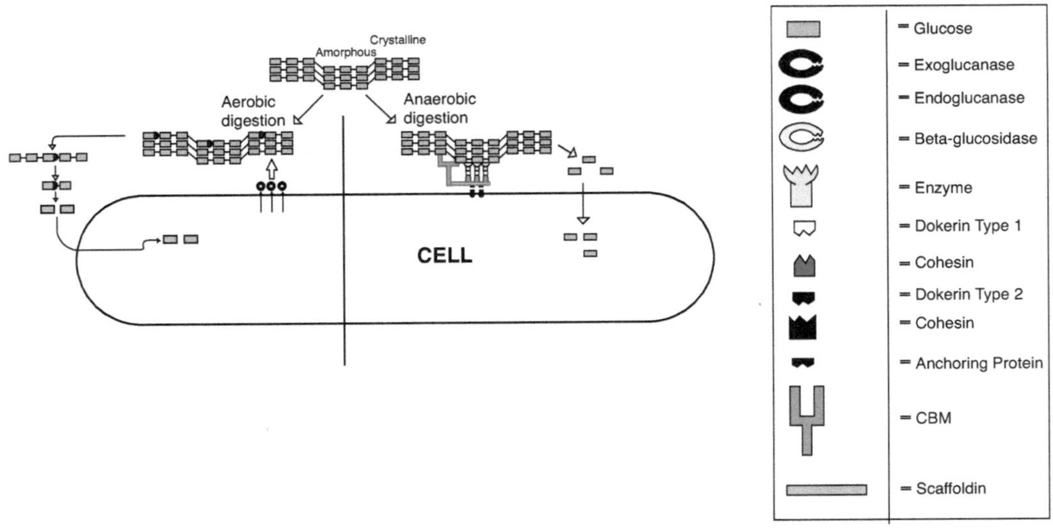

FIGURE 8.1 Aerobic and anaerobic hydrolysis by cellulases.

Besides this, certain aerobic microbial species such as *Pseudomonas, Thermomonospora, Cellulomonas,* and *Microbispora* which weaken the cellulases are extensively investigated (Tiwari and Verma, 2019). Investigation reports elaborate on the cellulase reaction mechanism both in aerobic and anaerobic conditions. In aerobic conditions, cellulose is converted into H_2O and CO_2, whereas in anaerobic conditions, it is converted into ethanol and hydrogen. In addition, *C. thermocellum*, thermophilic cellulose-degrading bacteria, is widely employed in industries for a variety of applications. In ruminants, the most significant plant cell wall degrading bacteria are *Ruminococcus albus*, *Ruminococcus flavefaciens,* and *Fibrobacter succinogenes.* These microbes are well recognized to harbor a whole well-developed system of cellulase enzymes (Munir et al., 2015). A diagrammatical representation of the cellulose hydrolysis mechanism of both aerobic and anaerobic bacteria is shown in Figure 8.1.

The microscopic studies of anaerobic bacteria show the presence of tiny outgrowths in the cell envelope. These outgrowths spread and form filamentous extensions which further bind the substrate and cells together. This whole structure developed by anaerobic bacteria is known as cellulosome. These cellulosomes can act as means of communication and help disperse the dissoluble products to the cell from the cellulose substrate that is existing external to the cell. Alternatively, the detachment of cellulosomes is possible in the future growth stages which facilitates the hydrolysis of each substrate (Smith et al., 2017).

8.2.2 Hemicellulases

Hemicellulases are made up of a large portion of secluded proteins that have catalytic domains, Carbohydrate-binding modules (CMBs), and active modules to promote the slicing of glycosidic bonds or side groups like esterified acid. The hemicellulose mainly consists of β-(1, 4) glycosidic bonds, and most of the time, hemicellulases share relative actions with the cellulases due to these consistent bonds. The complex structure of the hemicellulose substrate is made up of various simple hydrolyzable polymers, for example, pentoses (like xylose and arabinose), hexoses (like glucose and galactose), and sugar acids. The principal constituents of hemicellulose—xylans, horticultural plants, and the xylooligomers—deceptively have a cellulase inhibitory property (Harris et al., 2014). For instance, the ester bonds of ferulic acid or acetate in plant cell walls have hydrolyzed feruloyl esterases or acetyl whereas the glycosidic bonds are broken by enzymes like mannanases, α-glucuronidases, α-d-galactosidases, and α-arabinofuranosidases (Obeng et al., 2017). Depolymerizarion of the xylan spine is catalyzed by endo-β-1,4-xylanase and produces xylo-oligosaccharides, which are further hydrolyzed

TABLE 8.2
Enzyme Activity of Hemicellulase-Producing Microbes on Different Substrates

Microorganism	Enzyme Produced	Reaction Mixture	Substrate(s)	Enzyme Activity	References
Bacteria					
Cellulosimicrobium sp. strain HY-13	B-mannanase	pH-7, agitation-180 rpm, time-48 h, temperature-37°C	Bean gum Guar gum	7,109 IU/mg 5,158 IU/mg	Ham et al. (2011)
Acinetobacter junii F6-02	Xylanase	pH 7.0, temperature-35°C, agitation-150 rpm, time-3 days	Xylan	28.5 U/mL	Lo et al. (2010)
Bosea sp. FBZP-16	Endoxylanases	Temperature-28°C, time-4 weeks, agitation-150 rpm	Xylan	0.2–32 U/min mL	Houfani et al. (2017)
Fungi					
Trichoderma asperellum KIF125	b-Xylosidase	Time-96 h, temperature-28°C, pH-4.8, agitation-300 rpm	PNP-b-D-xylopyranoside	2.04 U/mg	Inoue et al. (2016)
Aspergillus niger	Endo b-1–4 xylanase	Temperature-50°C, pH-7	Birch wood xylan	664.4 U/mg	Do et al. (2013)
Malbranchea flava	Xylanase	Temperature 35°C–50°C, pH -4.0–10.0, moisture content 55%–80%	Sorghum straw Rice straw Paper waste	7,091 U/gds 3,700 U/gds 1,421 U/gds	Sharma et al. (2016)
Lichtheimia ramosa	Xylanase β-xylosidase	Agitation-300 rpm, temperature -50°C, time-48 h, pH-5.5	Sugarcane bagasse	34 U/g 199.2 U/g	Garcia et al. (2018)
Actinomycetes					
Streptomyces sp.	Xylanase	Temperature -45°C, time-144 h, agitation-120 rpm	Sugarcane bagasse	92.40 U/mL	de Queiroz Brito-Cunha et al. (2013)
Amycolatopsis sp. GDS	Xylanase	Temperature -30°C, time-6 days	CMC	69.21 U/mL	Kshirsagar et al. (2016)
Streptomyces sp. K37	Xylanase	pH-7.0, temperature-40°C–50°C, time-72 h	xylan	4.337 U/mg	Nour El-Dein et al. (2014)

by β-xylosidase. Whereas several glycosidases play an active role in the release of adjacent groups, the most common glycosidases are α-1-arabinofuranosidase, β-glucuronidases, ferulic, p-coumaric acid esterase, and acetyl xylan esterase (Chakdar et al., 2016). Microbes which are commonly reported to produce hemicellulase along with their reaction conditions and substrates are tabulated in Table 8.2.

8.2.3 Ligninase

Lignin is an organic polymer that is found as a structural material in most plants. It helps in the formation of cell walls, wood, and bark of the plant. Core enzyme systems that degrade lignin are versatile peroxidase, manganese peroxidase, lignin peroxidase), and laccase enzyme (usually found in the white-rot fungi) (Janusz et al., 2013). Besides this, a couple of enzymes are also isolated from fungus, e.g., oxalate decarboxylase (Makela et al., 2010), aryl-alcohol oxidase (Carro et al., 2016), cellobiose dehydrogenase (Zhang et al., 2011), P450 monooxygenase (Syed and Yadav, 2012), glyoxal oxidase (Daou and Faulds, 2017), pyranose 2-oxidase (Salaheddin et al., 2010), multicopper oxidase (Hilden et al., 2013), and manganese superoxide dismutase (Manavalan et al., 2015). In Table 8.3, a list of some ligninase-producing microorganisms along with their substrate and growth conditions is given.

TABLE 8.3
Enzyme Activity of Ligninase-Producing Microbes on Different Substrates

Microorganism	Enzyme Produced	Reaction Mixture	Substrate(s)	Enzyme Activity	References
		Bacteria			
Serratia liquefaciens	Lignin peroxidase (LiP)	Temperature - 30°C, time- 2 days	Pulp and paper mill effluent	18.1 IU/mL	Haq et al. (2016)
Bacillus sp. CS-1	Laccase	Temperature – 37°C, pH- 8, agitation-120 rpm, time- 3 days.	Rice straw	80%	Chang et al. (2014)
Burkholderia sp.H1	Lignin peroxidase (LiP)	pH 7.0, temperature - 37°, agitation -150 rpm, time- 48 h	Rotten wood	5.677 U/L	Yang et al. (2017)
Rheinheimera sp.	Laccase	Temperature -37°C, pH- 7, time- 48 h, agitation- 150 rpm	Soybean meal	60.50 nkat/mL	Sharma et al. (2017)
Lysinibacillus sp.	Laccase	Temperature -37°C, pH- 7, time- 72 h, agitation- 160 rpm	Soybean meal	455.23±9.9 nkats/g	Sharma et al. (2017)
		Fungi			
Pleurotus sajor-caju	Laccase (Lac), lignin peroxidase (LiP) and manganese peroxidase (MnP)	Temperature - 30 °C, time- 35–40 days	Oil palm fiber waste Banana residues Cotton stalks	0.83 U/mL 0.65 U/mL 0.64 U/mL	Thiribhuvanamala et al. (2017)
Phanerochaete chrysosporium-IBL-03	Lignin peroxidase (LiP) and manganese peroxidase	pH 4.5, temperature 35°C, time- 6 days	Wheat straw	993.9 U/mL 1,215.15 U/mL	Tiwari and Verma (2019)
Podoscypha elegans	Laccase Lignin peroxidase Manganese peroxidase	Temperature – 27°C, time- 7 days	Ligninolytic enzyme screening medium	1,013 U/L 1,509 U/L 14,230 U/L	Agrawal et al. (2017)
		Actinomycetes			
Streptomyces psammoticus	Manganese peroxidase	pH- 5–8.5, temperature- 20°C–50°C	2,4-dichlorophenol	3.06 U/mL	Niladevi and Prema (2005)
Streptomyces griseorubens ssr38	LiP (lignin peroxidase)	Temperature- 37°C, time- 10 and 21 days	Paddy straw	3.04 IU/mL	Saritha et al. (2013)
Streptomyces griseorubens ssr38	Laccase	Temperature- 37°C, time- 10 and 21 days		1.55 U/mL	Saritha et al. (2013)

8.3 ENZYME COCKTAIL

Despite various new advances and recent developments in the area of biorefining, lignocellulosic saccharification remains the foremost widespread and cost-effective procedure. An enzyme cocktail of *Podospora anserina* was produced after growing on soybean hulls (SBH) that seemingly showed a tremendously unusual ability to liberate reducing sugars out of all studied plant feedstocks which were used as biomass. SBHs bounded in numerous and very diverse enzyme groups targeted on plant cell wall polymers were exposed in the thorough analysis of proteomics of the culture and were precisely rich in enzymes, for example, xylan, mannan, and some other enzymes that are known to act on pectin (Makela et al., 2017). Hardwoods are pretreated explicitly with a raised-up amount of lignin, with a proficiently fabricated laccase saccharification by HoloMix. After saccharification for 24 h by HoloMix, 70%–100% of sugar yield was achieved (Malgas et al., 2017). A detailed overview of recently published reports of enzyme cocktails for enhanced saccharification is given in Table 8.4. There are several parameters such as temperature, pH, and type of lignocellulolytic material which can greatly affect the saccharification of lignocellulosic materials. Under extreme environmental conditions like high temperatures, extremely high or low salt concentrations, and pH or in the presence of some different organic solvents, enzymes isolated from prokaryotic organisms give the best results with higher conversion rates. Such properties of these enzymes make them robust for rough or harsh industrial procedures (Yu and Li, 2015).

8.4 ENZYME RECYCLING

The enzyme recycling process is usually carried out by three procedures. These processes are (i) hydrolysate, (ii) after fermentation-generated broth, and (iii) stillage (which is a crucial outcome from the distillation column) (Jorgensen and Pinelo, 2017). The main constituent on which the whole result depends is the type of biorefinery, which is being utilized for process and product development. The normal ethanol purification process also interrupts the enzyme working or reutilization (Zabed et al., 2016). There are two points that should be taken into consideration when selecting the methods to apply for enzyme reutilizing. First, different compounds in the stream block the enzymes or growth of microorganisms that are used in performing the fermentation process, will these compounds come together or will they be capable of exclusion? Second is the physical and chemical pressure that an enzyme is able to bear. A substantial suppression of cellulase enzymes could be caused by sugars and sugar oligomers, and these sugars will exist if the hydrolyzed slurry is recycled (Jorgensen and Pinelo, 2017). Likewise, plenty of compounds, for instance, organic acids, phenolic acids, 5-hydroxymethylfurfural, and furfural formed pretreatment can confine the enzymes as well as the microorganisms that perform fermentation. These inhibitors can be removed using membrane filters or immobilized enzyme sideways; with the presence of the utilization step, these unwanted compounds get excluded from enzymes (Nilsson et al., 2016). Lindedam et al. (2013) explained that thermostability is currently an issue in commercial cellulase systems, as the maximum cellulolytic measure was missing after purifying; only 34%–40% of the cumulative cellulolytic activity was recovered. Aiming at efficient reuse of cellulases, all other enzymes in both liquid and solid phases must be used. Reuse of fluid and solid (basically whole slurry) has been done (Baeyens et al., 2015; Haven et al., 2015). Orsteda Fredericia, Denmark-based company, has verified this process for sale, the preliminary solid procedure for the loading of 20% in correlation to reprocessing of fermentation broth (20%) via the hydrolysis phase (Adjei-Gyapong et al., 2012). In this fashion, the utilization of enzymes is decreased to 5%. Currently, this technique is confirmed to be successful at an industrial level in the utmost appropriate environment for high solid loading and total enzyme utilization. Experiments on a pilot scale have led to the suggestion that there is an option to lessen the concentration of the dry material by hydrolyzation and fermentation sideways with sustaining the overall ethanol concentration and enzyme recovery. The main ambiguity of this method is to save the enzyme for a limited time in the steam process laterally with microorganism growth accumulation or inhibitory enzymes (Adjei-Gyapong et al., 2016; Jorgensen and Pinelo, 2017).

TABLE 8.4
Amount of Sugars Released Using the Cocktail of Enzymes by Different Microbes Using Different Substrates

Enzyme Cocktail	Microorganism	Substrate	Amount Released	References
Cocktail of xylan-, mannan- and pectin-acting enzyme	*Podospora anserina*	Soybean hulls	1.2 g/L	Makela et al. (2017)
Cocktail of xylan-, mannan- and pectin-acting enzyme	*Podospora anserine*	Wheat straw	0.3 g/L	Makela et al. (2017)
Cocktail of xylan-, mannan- and pectin-acting enzyme	*Podospora anserine*	Cotton seed hulls	0.7 g/L	Makela et al. (2017)
HoloMix cocktail (CelMix+XynMix)	*Selenomonas ruminantium, Trichoderma longibrachiatum, Geobacillus stearothermophilus, Agaricus bisporus, Aspergillus niger, Trichoderma viride*	Hardwood biomass	70%–100%	Malgas et al. (2017)
TrCel7A (8%–40%), TrCel7B (40%–67%), TrCel6A (0%–20%)	*Trichoderma reesei*	Sugarcane	>80%	Chylenski et al. (2017)
TrCel7A (18%–35%), TrCel7B (49%–67%), TrCel6A (0%–11%)	*Trichoderma reesei*	Norway spruce	>90%	Chylenski et al. (2017)
ISN-CLEAs (cellulase, xylanase, β-1,3-glucanase)	*Trichoderma citrinoviride AUKAR04*	Sugarcane bagasse pulp	86% (glucose) and 73.3% (xylose)	Periyasamy et al. (2018)
ISN-CLEAs (cellulase, xylanase, β-1,3-glucanase)	*Trichoderma citrinoviride AUKAR04*	Eucalyptus pulp	88% (glucose) and 76.7% (xylose)	Periyasamy et al. (2018)
Multifect CL	*Chrysoporthe cubensis*	Sugarcane bagasse	5.30 g/L	Maitan-Alfenas et al. (2015)
Multifect XL	*Chrysoporthe cubensis*	Sugarcane bagasse	8.33 g/L	Maitan-Alfenas et al. (2015)
SmChiA (38%), SmCHB (28%), SmChiC (17%), SmLPMO10A (5.0%), SmCHB (12%)	*Serratia marcescens*	Chitin	71%	Mekasha et al. (2017)
SmchiA (38%), SmChiB (26%), SmChiC (25%), SmLPMO10A (2%), SmCHB (76%)	*Serratia marcescens*	PTS	76%	Mekasha et al. (2017)

8.5 PROCESS ECONOMICS

The preliminary stage in bioconversion of lignocellulose is the pretreatment step which is economically unfriendly and makes the whole process expensive. Further, physical pretreatments on lignocellulose require a high temperature which is generated by hydrothermal or steam explosion processes using electrical energy. This whole temperature-rising procedure increases the process cost. Even during the chemical pretreatment method, chemicals such as H_2SO_4 (for acidic pretreatment), and NaOH and NH_3 (for alkali pretreatment) are costly chemicals (Ghosh et al., 2017). It has been estimated that biorefineries associated with second generation ethanol production recover about 28% of the total expense of the enzymatic hydrolysis process. This spending is enormously large and jeopardizes the competition in the market; in consequence, to improve the yield and decline expenditures, the properties of natural enzymes entail to be moderated. At all times with reduced costs in mind, the improvement of mechanisms that allow for the reuse of enzymes and low load must also be given attention (Silva et al., 2018). The four chief groups alienated for the expense of the entire procedure showed next: feedstock expenditures (varies by shipping of feedstocks, change in season, geological location, and the situation of demand–supply); investment cost (this takes in the cost of land, infrastructures, all expenditures of machines, needed equipment, and their installation); expenditure on the operation (includes maintenance, power supplies, cultures for process, essential chemicals, enzymes, cost of manual labor and machinery repairing, insurance, taxes, etc.); and economic revenues from by-products (Arora et al., 2015).

8.6 FUTURE PROSPECTS

There are still huge opportunities and potentials left for lignocellulosic studies which are mainly dependent on the metagenomic methods to deliver the potential enzymes that are eventually required in the rapid industrial advancement of sustainable bioethanol production from LCB (Batista-Garcia et al., 2016). The plentiful availability and regenerative nature of lignocellulose make it a very competent, suitable, and economically viable biomass for the production of several industrially beneficial products. Lignocellulolytic biocatalysts are, therefore, substantial in the developments connected with bioconversion modification of LCB (Arevalo-Gallegos et al., 2017). At present, most lignocellulases that are commercially active are of fungal or bacterial origin. Actinomycetes are relatively not as much discovered in regard to the hydrolysis potential of the biomass. The research can be straightforwardly set for an explanation for finding the novel actinomycetes which produce lignocellulases (Shivlata and Tulasi, 2015). Moreover, many feedstocks express characteristic dissimilarity in their chemical conformation, and enzyme making is imagined to be enhanced for various biomasses. It should be kept in mind that the production of lignocellulases that use almost all microbial sources is very expensive (Li et al., 2016b). Nevertheless, making some attempts in diverse ways can also be made for dropping the price of the manufacture of these lignocellulase enzymes via utilizing highly potent actinomycete enzyme systems with a wider array of tolerance to demonstrate activity in different ecological situations. Moreover, with the help of genetic engineering techniques, the enzyme system can be simply created. Similarly, the research can be extended gradually and steadily to grade up the standards of manufacturing industries for their succeeding implementation in profitable processes (Patel et al., 2019).

8.7 CONCLUSIONS

Lignocellulose is widely present in plant cells that are considered essential plentiful biomasses on this planet, and it is known to represent about 50% of the yearly worldwide production of organic materials. Likewise, it is assessed that it can be used proficiently for the production of biofuels as discussed in this chapter. There are varieties of microbes that secrete significant enzymes that have the competence to disrupt lignocellulosic substrates very proficiently and effectively. Aerobic bacteria like *Cellulomonas*, *Pseudomonas,* etc., are known to break down the cellulose present in biomass. Other than bacteria, there are fungi, for instance, *Penicillium oxalicum*, *Antrodia cinnamomea*, and *Myceliophthora fergusii* that are known to secrete an enzyme that is known as hemicellulase whereas actinomycetes, for example, *Streptomyces*, *Micromonospora*, and *Actinomadura* are well recognized for releasing the cellulase enzyme. However, another fact that must be taken into account is that with respect to the bioconversion of lignocelluloses, actinomycetes are less studied. Hence, a lignocellulosic breakdown is mainly reliant on enzymes that are of fungal and bacterial origin and not from actinomycetes. Hydrolysis mechanisms of lignin and hemicellulose are trailed by enzymes that are included in the bioconversion process. Henceforth, it can be straightforwardly determined that fungi and bacteria are extremely important in the production of industrial enzymes. Lignocellulolytic enzymes (named hemicellulases, cellulases, and ligninases) are crucial portions of industries related to food, for instance in vegetable oil processing, vegetable juice and fruit juice factories, fermentation, baking industries, and winemaking industries. Also, cocktails of different enzymes enhanced the lignocellulosic saccharification very effectively and efficiently. Irrespective of these significant and unusual characteristics and technical and research advancement, bioconversion of LCB is still an enormously luxurious procedure, and to tackle this, new ways should be invented.

REFERENCES

Adjei-Gyapong, T., R. C. Abaidoo, E. Yeboah, H. Hauggaard-Nielsen and C. Quansah. 2012. Assessment of second generation biofuel residuals as soil amendment for soil fertility improvement. In *Advanced Biofuels in a Biorefinery Approach*, Ed. H. Jorgensen. Copenhagen: University of Copenhagen.

Agrawal, N., P. Verma, R. S. Singh and S. K. Shahi. 2017. Ligninolytic enzyme production by white rot fungi *Podoscypha elegans* strain FTG4. *International Journal of Current Microbiology and Applied Sciences* 6:2757–2764.

Arevalo-Gallegos, A., Z. Ahmad, M. Asgher, R. Parra-Saldivar and H. M. Iqbal. 2017. Lignocellulose: A sustainable material to produce value-added products with a zero waste approach-a review. *International Journal of Biological Macromolecules* 99:308–318.

Arora, R., S. Behera and S. Kumar. 2015. Bioprospecting thermophilic/thermotolerant microbes for production of lignocellulosic ethanol: A future perspective. *Renewable and Sustainable Energy Reviews* 51:699–717.

Baeyens, J., Q. Kang and L. Appels, et al. 2015. Challenges and opportunities in improving the production of bio-ethanol. *Progress in Energy and Combustion Science* 47:60–88.

Batista-Garcia, R. A., M. del Rayo Sanchez-Carbente and P. Talia, et al. 2016. From lignocellulosic metagenomes to lignocellulolytic genes: Trends, challenges and future prospects. *Biofuels, Bioproducts and Biorefining* 10:864–882.

Bhalla, A., N. Bansal, S. Kumar, K. M. Bischoff and R. K. Sani. 2013. Improved lignocellulose conversion to biofuels with thermophilic bacteria and thermostable enzymes. *Bioresource Technology* 128:751–759.

Bornscheuer, U., K. Buchholz and J. Seibel. 2014. Enzymatic degradation of (ligno) cellulose. *Angewandte Chemie International Edition* 53:10876–10893.

Carro, J., A. Serrano, P. Ferreira and A. T. Martinez. 2016. Fungal aryl-alcohol oxidase in lignocellulose degradation and bioconversion. In *Microbial Enzymes in Bioconversions of Biomass, Biofuel and Biorefinery Technologies, Volume 3*, Ed. V. Gupta, pp. 301–322. Cham: Springer.

Chakdar, H., M. Kumar and K. Pandiyan, K. et al. 2016. Bacterial xylanases: Biology to biotechnology. *3 Biotech* 6:150.

Chang, Y. C., D. Choi, K. Takamizawa and S. Kikuchi. 2014. Isolation of *Bacillus* sp. strains capable of decomposing alkali lignin and their application in combination with lactic acid bacteria for enhancing cellulase performance. *Bioresource Technology* 152:429–436.

Chylenski, P., Z. Forsberg and J. Stahlberg, et al. 2017. Development of minimal enzyme cocktails for hydrolysis of sulfite-pulped lignocellulosic biomass. *Journal of Biotechnology* 246:16–23.

Daou, M. and C. B. Faulds. 2017. Glyoxal oxidases: Their nature and properties. *World Journal of Microbiology & Biotechnology* 33:87.

De Cassia Pereira, J., N. P. Marques and A. Rodrigues, A., et al. 2015. Thermophilic fungi as new sources for production of cellulases and xylanases with potential use in sugarcane bagasse saccharification. *Journal of Applied Microbiology* 118:928–939.

de Queiroz Brito-Cunha, C. C., I. T. N. de Campos, F. P. de Faria, F. P. and L. A. M. Bataus, 2013. Screening and xylanase production by *Streptomyces* sp. grown on lignocellulosic wastes. *Applied Biochemistry and Biotechnology* 170:598–608.

Deepthi, M. K., M. S. Sudhakar and M. N. Devamma. 2012. Isolation and screening of *Streptomyces* sp. from Coringa mangrove soils for enzyme production and antimicrobial activity. *International Journal of Pharmaceutical, Chemical and Biological Sciences* 2:110–116.

Devi, M. C. and M. S. Kumar. 2012. Production, optimization and partial purification of cellulase by *Aspergillus niger* fermented with paper and timber sawmill industrial wastes. *Journal of Microbiology and Biotechnology Research* 2:120–128.

Do, T. T., D. T. Quyen and T. N. Nguyen. 2013. Molecular characterization of a glycosyl hydrolase family 10 xylanase from *Aspergillus niger*. *Protein Expression and Purification* 92:196–202.

El-Naggar, N. E. A. and N. A. Abdelwahed. 2012. Optimization of process parameters for the production of alkali-tolerant carboxymethylcellulase by newly isolated *Streptomyces* sp. strain NEAE-D. *African Journal of Biotechnology* 11:1185–1196.

Garcia, N. F. L., F. R. da Silva Santos and D. A. Bocchini, et al. 2018. Catalytic properties of cellulases and hemicellulases produced by *Lichtheimia ramosa*: Potential for sugarcane bagasse saccharification. *Industrial Crops and Products* 122:49–56.

Ghosh, S., R. Chowdhury and P. Bhattacharya. 2017. Sustainability of cereal straws for the fermentative production of second-generation biofuels: A review of the efficiency and economics of biochemical pretreatment processes. *Applied Energy* 198:284–298.

Gomes, E., A. R. de Souza and G. L. Orjuela, et al. 2016. Applications and benefits of thermophilic microorganisms and their enzymes for industrial biotechnology. In *Gene Expression Systems in Fungi: Advancements and Applications*, Eds. M. Schmoll and C. Dattenbock, pp. 459–492. Cham: Springer.

Gunny, A. A. N., D. Arbain, R. E. Gumba, B. C. Jong and P. Jamal. 2014. Potential halophilic cellulases for *in situ* enzymatic saccharification of ionic liquids pretreated lignocelluloses. *Bioresource Technology* 155:177–181.

Gupta, A. and J. P. Verma. 2015. Sustainable bio-ethanol production from agro-residues: A review. *Renewable and Sustainable Energy Reviews* 41:550–567.

Ham, S. J., H. J. Lee and Y. J. Kim, et al. 2011. A highly active endo-β-1, 4-mannanase produced by *Cellulosimicrobium* sp. strain HY-13, a hemicellulolytic bacterium in the gut of *Eisenia fetida*. *Enzyme and Microbial Technology* 48:365–370.

Haq, I., S. Kumar, V. Kumari, S. K. Singh and A. Raj. 2016. Evaluation of bioremediation potentiality of ligninolytic *Serratia liquefaciens* for detoxification of pulp and paper mill effluent. *Journal of Hazardous Materials* 305:190–199.

Harris, P. V., F. Xu, F., N. E. Kreel, C. Kang and S. Fukuyama. 2014. New enzyme insights drive advances in commercial ethanol production. *Current Opinion in Chemical Biology* 19:162–170.

Haven, M. O., J. Lindedam and M. D. Jeppesen. 2015. Continuous recycling of enzymes during production of lignocellulosic bioethanol in demonstration scale. *Applied Energy* 159:188–195.

Hilden, K., M. R. Makela, P. Lankinen and T. Lundell. 2013. *Agaricus bisporus* and related *Agaricus* species on lignocellulose: Production of manganese peroxidase and multicopper oxidases. *Fungal Genetics and Biology* 55:32–41.

Himmel, M. E., W. S. Adney, J. O. Baker, R. A. Nieves and S. R. Thomas. 2018. Cellulases: Structure, function, and applications. In *Handbook on Bioethanol*, Ed. C. Wyman, pp. 143–161. Abingdon: Routledge.

Houfani, A. A., T. Vetrovsky, P. Baldrian and S. Benallaoua. 2017. Efficient screening of potential cellulases and hemicellulases produced by *Bosea* sp. FBZP-16 using the combination of enzyme assays and genome analysis. *World Journal of Microbiology & Biotechnology* 33:29.

Inoue, H., C. Kitao, S. Yano and S. Sawayama. 2016. Production of β-xylosidase from *Trichoderma asperellum* KIF125 and its application in efficient hydrolysis of pretreated rice straw with fungal cellulase. *World Journal of Microbiology & Biotechnology* 32:186.

Janusz, G., K. H. Kucharzyk, A. Pawlik, M. Staszczak and A. J. Paszczynski. 2013. Fungal laccase, manganese peroxidase and lignin peroxidase: Gene expression and regulation. *Enzyme and Microbial Technology* 52:1–12.

Jorgensen, H. and M. Pinelo. 2017. Enzyme recycling in lignocellulosic biorefineries. *Biofuels, Bioproducts and Biorefining* 11:150–167.

Kogo, T., Y. Yoshida and K. Koganei. 2017. Production of rice straw hydrolysis enzymes by the fungi *Trichoderma reesei* and *Humicola insolens* using rice straw as a carbon source. *Bioresource Technology* 233:67–73.

Kshirsagar, S. D., G. D. Saratale, R. G. Saratale, S.P. Govindwar and M. K. Oh. 2016. An isolated *Amycolatopsis* sp. GDS for cellulase and xylanase production using agricultural waste biomass. *Journal of Applied Microbiology* 120:112–125.

Kumar, U., A. Tapwal, P. Kalkal, S. Varghese and S. Chandra. 2014. Isolation and screening of cellulase producing fungi from forest waste. *International Journal of Pharmaceutical and Biological Science Archive* 5:56–59.

Li, K., J. C. Qin, C. G. Liu and F. W. Bai. 2016a. Optimization of pretreatment, enzymatic hydrolysis and fermentation for more efficient ethanol production by Jerusalem artichoke stalk. *Bioresource Technology* 221:188–194.

Li, Y., C. Liu, F. Bai and X. Zhao. 2016b. Overproduction of cellulase by *Trichoderma reesei* RUT C30 through batch-feeding of synthesized low-cost sugar mixture. *Bioresource Technology* 216:503–510.

Lindedam, J., M. O. Haven, P. Chylenski, H. Jorgensen and C. Felby. 2013. Recycling cellulases for cellulosic ethanol production at industrial relevant conditions: Potential and temperature dependency at high solid processes. *Bioresource Technology* 148:180–188.

Lo, Y. C., W. C. Lu, C. Y. Chen, W. M. Chen and J. S. Chang. 2010. Characterization and high-level production of xylanase from an indigenous cellulolytic bacterium *Acinetobacter junii* F6-02 from southern Taiwan soil. *Biochemical Engineering Journal* 53:77–84.

Maitan-Alfenas, G. P., E. M. Visser and R. F. Alfenas. 2015. The influence of pretreatment methods on saccharification of sugarcane bagasse by an enzyme extract from *Chrysoporthe cubensis* and commercial cocktails: A comparative study. *Bioresource Technology* 192:670–676.

Makela, M. R., K. Hilden and T. K. Lundell. 2010. Oxalate decarboxylase: Biotechnological update and prevalence of the enzyme in filamentous fungi. *Applied Microbiology and Biotechnology* 87:801–814.

Makela, M. R., O. Bouzid and D. Robl. 2017. Cultivation of *Podospora anserina* on soybean hulls results in an efficient enzyme cocktail for plant biomass hydrolysis. *New Biotechnology* 37:162–171.

Malgas, S., R. Chandra, J. S. Van Dyk, J. N. Saddler and B. I. Pletschke. 2017. Formulation of an optimized synergistic enzyme cocktail, HoloMix, for effective degradation of various pre-treated hardwoods. *Bioresource Technology* 245:52–65.

Manavalan, T., A. Manavalan and K. Heese. 2015. Characterization of lignocellulolytic enzymes from white-rot fungi. *Current Opinion in Microbiology* 70:485–498.

Mekasha, S., I. R. Byman and C. Lynch. 2017. Development of enzyme cocktails for complete saccharification of chitin using mono-component enzymes from *Serratia marcescens*. *Process Biochemistry* 56:132–138.

Munir, N., M. Asgher and I. M. Tahir, et al. 2015. Utilization of agro-wastes for production of ligninolytic enzymes in liquid state fermentation by *Phanerochaete chrysosporium*-IBL-03. *International Journal of Chemical and Biological Sciences* 7:9–14.

Munir, R. and D. B. Levin. 2016. Enzyme systems of anaerobes for biomass conversion. *Advances in Biochemical Engineering/Biotechnology* 156:113–138.

Niladevi, K. N. and P. Prema. 2005. Mangrove actinomycetes as the source of ligninolytic enzymes. *Actinomycetologica* 19:40–47.

Nilsson, R. L. K., M. Holmgren, B. Madavi, R. T. Nilsson and A. Sellstedt. 2016. Adaptability of *Trametes versicolor* to the lignocellulosic inhibitors furfural, HMF, phenol and levulinic acid during ethanol fermentation. *Biomass and Bioenergy* 90:95–100.

Nour El-Dein, M. M., A. E. A. Shereif, F. A. Mansour, M. I. Abou-Dobara and A. S. Ball. 2014. Optimization of xylanase and peroxidase production from *Streptomyces* sp. K37. *Journal of BioScience and Biotechnology* 3:29–42.

Obeng, E. M., S. N. N. Adam and C. Budiman, et al. 2017. Lignocellulases: A review of emerging and developing enzymes, systems, and practices. *Bioresources and Bioprocessing* 4:16.

Orencio-Trejo, M., S. de la Torre-Zavala, A. Rodriguez-Garcia, H. Aviles-Arnaut and A. Gastelum-Arellanez. 2016. Assessing the performance of bacterial cellulases: The use of *Bacillus* and *Paenibacillus* strains as enzyme sources for lignocellulose saccharification. *BioEnergy Research* 9:1023–1033.

Patagundi, B. I., Shivasharan, C. T. and Kaliwal, B. B. 2014. Isolation and characterization of cellulase producing bacteria from soil. *International Journal of Current Microbiology and Applied Sciences* 3:59–69.

Patel, A. K., R. R. Singhania, S. J. Sim and A. Pandey. 2019. Thermostable cellulases: Current status and perspectives. *Bioresource Technology* 279:385–392.

Paudel, Y. P. and W. Qin. 2015. Characterization of novel cellulase-producing bacteria isolated from rotting wood samples. *Applied Biochemistry and Biotechnology* 177:1186–1198.

Periyasamy, K., L. Santhalembi and G. Mortha, et al. 2018. Bioconversion of lignocellulosic biomass to fermentable sugars by immobilized magnetic cellulolytic enzyme cocktails. *Langmuir* 34:6546–6555.

Prasad, P., T. Singh and S. Bedi. 2013. *Characterization of the cellulolytic enzyme produced by* Streptomyces griseorubens (Accession No. AB184139) isolated from Indian soil. *Journal of King Saud University – Science* 25:245–250.

Salaheddin, C., Y. Takakura and M. Tsunashima. 2010. Characterization of recombinant pyranose oxidase from the cultivated mycorrhizal basidiomycete *Lyophyllum shimeji* (hon-shimeji). *Microbial Cell Factories* 9:57.

Saratale, G. D., R. G. Saratale and G. S. Ghodake. 2017. Solid state fermentative lignocellulolytic enzymes production, characterization and its application in the saccharification of rice waste biomass for ethanol production: An integrated biotechnological approach. *Journal of the Taiwan Institute of Chemical Engineers* 76:51–58.

Saritha, M., A. Arora, S. Singh and L. Nain. 2013. *Streptomyces griseorubens* mediated delignification of paddy straw for improved enzymatic saccharification yields. *Bioresource Technology* 135:12–17.

Sharma, A., V. Gupta and M. Khan, et al. 2017. Flavonoid-rich agro-industrial residues for enhanced bacterial laccase production by submerged and solid-state fermentation. *3 Biotech* 7:200.

Sharma, M., C. Mahajan, M. S. Bhatti and B. S. Chadha. 2016. Profiling and production of hemicellulases by thermophilic fungus Malbrancheaflava and the role of xylanases in improved bioconversion of pretreated lignocellulosics to ethanol. *3 Biotech* 6:30.

Shivlata, L. and S. Tulasi. 2015. Thermophilic and alkaliphilic Actinobacteria: Biology and potential applications. *Frontiers in Microbiology* 6:1014.

Silva, C. O., R. P. Vaz and E. X. Filho. 2018. Bringing plant cell wall-degrading enzymes into the lignocellulosic biorefinery concept. *Biofuels, Bioproducts and Biorefining* 12:277–289.

Singhania, R. R., M. Adsul, A. Pandey and A. K. Patel. 2017. Cellulases. In *Current Developments in Biotechnology and Bioengineering*, Eds. C. Larroche, M. Sanroman, G. Du and A. Pandey, pp. 73–101. Amsterdam: Elsevier.

Smith, S. P., E. A. Bayer and M. Czjzek. 2017. Continually emerging mechanistic complexity of the multi-enzyme cellulosome complex. *Current Opinion in Structural Biology* 44:151–160.

Sriariyanun, M., P. Tantayotai, P. Yasurin, P. Pornwongthong and K. Cheenkachorn. 2016. Production, purification and characterization of an ionic liquid tolerant cellulase from *Bacillus* sp. isolated from rice paddy field soil. *Electronic Journal of Biotechnology* 19:23–28.

Syed, K. and J. S. Yadav. 2012. P450 monooxygenases (P450ome) of the model white rot fungus *Phanerochaete chrysosporium. Critical Reviews in Microbiology* 38:339–363.

Thiribhuvanamala, G., G. Kalaiselvi, S. Parthasarathy and B. Anusha. 2017. Induction of lignolytic enzyme activities in different agro residues by the white rot fungi, *Pleurotus sajar-caju. International Journal of Chemical Studies* 5:89–94.

Tiwari, S. and T. Verma. 2019. Cellulose as a potential feedstock for cellulose enzyme production. In *Approaches to Enhance Industrial Production of Fungal Cellulases*, Eds. M. Srivastava, N. Srivastava, P. W. Ramteke and P. K. Mishra, pp. 89–116. Cham: Springer.

Wu, H., B. Liu and X. Ou, et al. 2018. *Streptomyces thermoalkaliphilus* sp. nov., an alkaline cellulase producing thermophilic actinomycete isolated from tropical rainforest soil. *Antonie van Leeuwenhoek* 111:413–422.

Wu, Z., H. Hao, Y. Tu, et al. 2014. Diverse cell wall composition and varied biomass digestibility in wheat straw for bioenergy feedstock. *Biomass and Bioenergy* 70:347–355.

Yang, C. X., T. Wang and L. N. Gao, et al. 2017. Isolation, identification and characterization of lignin-degrading bacteria from Qinling, China. *Journal of Applied Microbiology* 123:1447–1460.

Yu, H. Y. and X. Li. 2015. Alkali-stable cellulase from a halophilic isolate, *Gracilibacillus* sp. SK1 and its application in lignocellulosic saccharification for ethanol production. *Biomass and Bioenergy* 81:19–25.

Zabed, H., J. N. Sahu, A. N. Boyce and G. Faruq. 2016. Fuel ethanol production from lignocellulosic biomass: An overview on feedstocks and technological approaches. *Renewable and Sustainable Energy Reviews* 66:751–774.

Zhang, R., Z. Fan and T. Kasuga. 2011. Expression of cellobiose dehydrogenase from *Neurospora crassa* in *Pichia pastoris* and its purification and characterization. *Protein Expression and Purification* 75:63–69.

Chemical Biomass Pretreatments for Bioethanol Production

9 Chemical Pretreatment of Biomass

Scientometric Study

Ozcan Konur
(Formerly) Ankara Yildirim Beyazit University

9.1 INTRODUCTION

Crude oil-based gasoline fuels (Ma et al., 2002; Newman and Kenworthy, 1989) have been widely used in the transportation sector since the 1920s. However, there have been great public concerns over the adverse environmental and human impact of these fuels (Hill et al., 2006, 2009). Hence, biomass-based bioethanol fuels (Alvira et al., 2010; Hill et al., 2006; Konur, 2012e, 2015, 2020a; Sun and Cheng, 2002) have increasingly been used in blending gasoline fuels (Hsieh et al., 2002; Najafi et al., 2009), in fuel cells (Antolini, 2007, 2009), and in biochemical production (Angelici et al., 2013; Morschbacker, 2009) in a biorefinery context (Fernando et al., 2006; Huang et al., 2008).

Research in the field of chemical pretreatments of biomass has also intensified in recent years. Acid (Huang and Fu, 2013; Mussatto and Roberto, 2004), ionic liquid (IL) (Brandt et al., 2013; Zhu et al., 2006), alkali (Kim et al., 2016; Knill and Kennedy, 2002), ammonia (Balan et al., 2009; Kim et al., 2003), organic solvent (Shuai and Luterbacher, 2016; Zhao et al., 2009), hydrogen peroxide (Glenn et al., 1983; Sun et al., 2000), deep eutectic solvent (DES) (Satlewal et al., 2018; Zhang et al., 2016), sub- and supercritical water (Sasaki et al., 1998, 2000), surfactant (Eriksson et al., 2002; Helle et al., 1992), ozone (Garcia-Cubero et al., 2009; Vidal and Molinier, 1988), sulfite (Shuai et al., 2010; Zhu et al., 2009), and supercritical carbon dioxide (Kim and Hong, 2001; Zheng et al., 1998) pretreatments of the biomass have been widely researched to increase the sugar and bioethanol yield in this context in recent years.

However, it is essential to develop efficient incentive structures (North, 1991) for the primary stakeholders to enhance research in this field (Konur, 2000, 2002a–c, 2006a, b, 2007a, b). A scientometric analysis has been used in this context to inform the primary stakeholders about the current state of research in a selected research field (Garfield, 1955; Konur, 2011, 2012a–i, 2015, 2018b, 2019, 2020a).

As there have been no scientometric studies on chemical pretreatments of biomass as of April 2022, this chapter presents a scientometric study of research in chemical pretreatments of biomass. It examines the scientometric characteristics of both the sample and population data presenting scientometric characteristics of these both datasets in the order of documents, authors, publication years, institutions, funding bodies, source titles, countries, Scopus subject categories, keywords, and research fronts, as well as the highly cited 25 papers.

9.2 MATERIALS AND METHODS

Search for this study was carried out using the Scopus database (Burnham, 2006) in April 2022.

As a first step for the search of the relevant literature, keywords were selected using the first most-cited 200 papers. The selected keyword list was optimized to obtain a representative sample of papers for the searched research field. This keyword list was provided in the Appendix for future replication studies.

DOI: 10.1201/9781003226536-13

As a second step, two sets of data were used for this study. First, a population sample of over 9,000 papers was used to examine the scientometric characteristics of the population data. Secondly, a sample of 182 most-cited papers, corresponding to 2% of the population papers, was used to examine the scientometric characteristics of these citation classics.

Scientometric characteristics of these both sample and population datasets were presented in the order of documents, authors, publication years, institutions, funding bodies, source titles, countries, Scopus subject categories, keywords, and research fronts.

Lastly, the key scientometric findings for both datasets were discussed to highlight the research landscape for chemical pretreatments of biomass. Additionally, a number of brief conclusions were drawn and a number of relevant recommendations were made to enhance the future research landscape.

9.3 RESULTS

9.3.1 The Most Prolific Documents in the Chemical Pretreatments of Biomass

Information on the types of documents for both datasets is given in Table 9.1. The articles and conference papers dominate both the sample and population papers as they are under-represented in the sample papers. Further, review papers and short surveys have a surplus as they are over-represented in the sample papers.

It is further noted that 97% of the population papers was published in journals while 2% and 1% of them were published in book series and books, respectively. On the contrary, all the sample papers were published in journals.

9.3.2 The Most Prolific Authors in the Chemical Pretreatments of the Biomass

Information about the most prolific 11 authors with at least 2.2% of sample papers each is given in Table 9.2.

The most prolific author is Blake A. Simmons of Lawrence Berkeley National Laboratory with 4.4% of sample papers working primarily on IL pretreatment of the biomass. Tae Hyun Kim, Seema Singh, and Yoon Y. Lee follow this top author with 2.7%–3.3% of sample papers each working on ammonia, IL, and ammonia pretreatments of the biomass, respectively.

The most prolific institution for the sample dataset is the Imperial College of the United Kingdom with three authors. Lawrence Berkeley National Laboratory follows this top institution with two authors. In total, eight institutions house these prolific authors.

The most prolific country for the sample dataset is the United States with six authors while the United Kingdom has three authors. In total, four countries house these authors.

The most-prolific research front is the IL pretreatment of the biomass with five authors. The other fronts are acid and ammonia pretreatments of the biomass with three authors each.

On the other hand, there is a significant gender deficit (Beaudry and Lariviere, 2016) for the sample dataset as surprisingly only two of these top researchers are female.

Additionally, there are other authors with a relatively low citation impact and with 0.4%–1.0% of the population papers each: Runcang Sun, Arthur J. Ragauskas, Feng Xu, Venkatesh Balan, Qiang Yong, Charles E. Wyman, Ying Xu, Herbert Sixta, Yunqiao Pu, Keikhosro Karimi, and Zhongshun Yuan.

9.3.3 The Most Prolific Research Output by Years in Chemical Pretreatments of the Biomass

Information about papers published between 1970 and 2022 is given in Figure 9.1. This figure clearly shows that the bulk of research papers in the population dataset was published primarily in

TABLE 9.1
Documents of Chemical Pretreatments of the Biomass

Documents	Sample Dataset (%)	Population Dataset (%)	Surplus (%)
Article	85.7	92.7	–7.0
Review	10.4	2.1	8.3
Conference paper	3.3	3.7	–0.4
Short survey	0.5	0.1	0.4
Book chapter	0.0	0.9	–0.9
Book	0.0	0.1	–0.1
Editorial	0.0	0.1	–0.1
Letter	0.0	0.1	–0.1
Note	0.0	0.1	–0.1
Sample size	182	9,082	

Sample dataset: the number of papers (%) in a set of 182 highly cited papers. Population dataset: the number of papers (%) in a set of 9,082 population papers.

TABLE 9.2
The Most-Prolific Authors in Chemical Pretreatments of the Biomass

No.	Author Name	Author Code	Sample Papers %	Population Papers %	Institution	Country	HI	N	Res. Front
1	Simmons, Blake A.	7102183263	4.4	0.9	Lawrence Berkeley Natl. Lab.	USA	74	440	ILs
2	Kim, Tae Hyun	57210847338	3.3	0.4	Hanyang Univ.	S. Korea	24	88	Ammonia
3	Singh, Seema*	35264950300	2.7	0.7	Lawrence Berkeley Natl. Lab.	USA	55	175	ILs
4	Lee, Yoon Y.	8948274900	2.7	0.3	Auburn Univ.	USA	44	102	Ammonia
5	Zhu, Junyong	7405692678	2.2	2.3	USDA Forest Serv.	USA	61	303	Acid, alkali, sulfite
6	Dale, Bruce E.	7201511969	2.2	0.6	Michigan State Univ.	USA	89	427	Ammonia
7	Welton, Tom	7003503272	2.2	0.6	Imperial coll.	UK	73	186	ILs
8	Taherzadeh, Muhammad J.	6701407496	2.2	0.3	Univ. Boras	Sweden	64	403	Acid
9	Hallett, Jason P.	7102331746	2.2	0.2	Imperial Coll.	UK	46	174	ILs
10	Himmel, Michael E.	7007125552	2.2	0.2	Natl. Renew. Ener. Lab.	USA	73	421	Acid
11	Brandt, Agnieszka*	35785816800	2.2	0.1	Imperial Coll.	UK	22	39	ILs

Author code: the unique code given by Scopus to authors. Sample papers: the number of papers authored in the sample dataset. Population papers: the number of papers authored in the population dataset.

HI, H-index; N, number of papers published by each author; *, female.

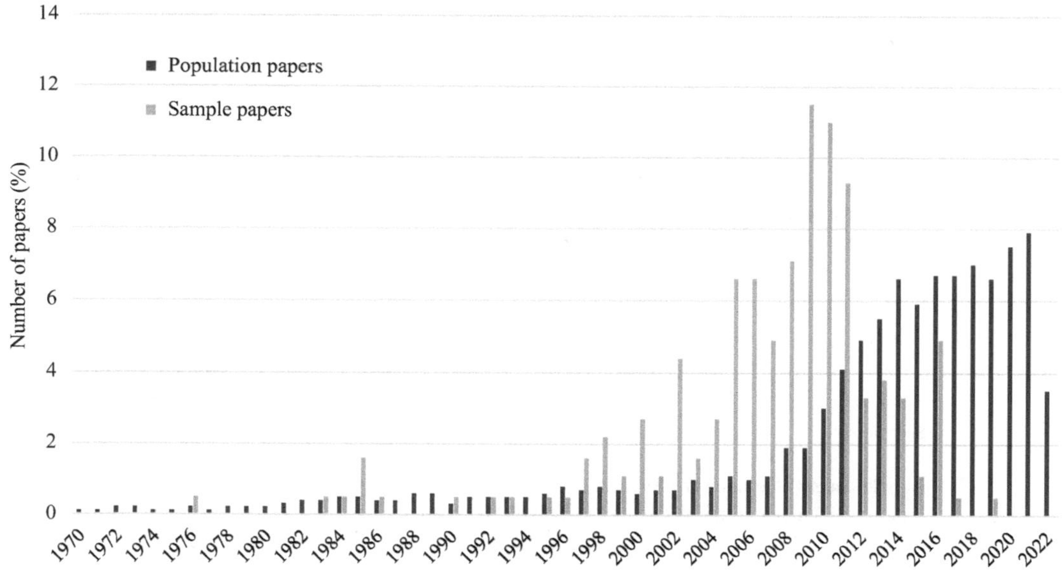

FIGURE 9.1 Research output by years regarding chemical pretreatments of the biomass.

the 2010s with 57% of the population dataset. The publication rates for the 2020s, 2000s, 1990s, 1980s, and 1970s were 15%, 11%, 6%, 4%, and 2% respectively. Additionally, 2% of the population papers were published between 1892 and 1969.

Similarly, the bulk of research papers in the sample dataset were published in the 2010s and 2000s with 38% and 49% of the sample dataset, respectively. The publication rates for the 1990s, 1980s, and 1970s were 7%, 4%, and 2% of the sample papers, respectively.

The most prolific publication year for the population dataset was 2021 with 8% while 73% of the population papers were published between 2011 and 2022. Similarly, 57% of sample papers were published between 2005 and 2011 while the most prolific publication years were 2009 and 2011 with 12% and 9% of sample papers, respectively.

9.3.4 The Most Prolific Institutions in the Chemical Pretreatments of the Biomass

Information about the most prolific 11 institutions publishing papers on the chemical pretreatments of the biomass with at least 2.7% of sample papers each is given in Table 9.3.

The most prolific institutions are the Lawrence Berkeley National Laboratory, National Renewable Energy Laboratory, and Agricultural Research Service (ARS) of the US Department of Agriculture (USDA) with 5.5% of sample papers each. The Forest Service of the USDA and Sandia Laboratory follow these top institutions with 3.8% of sample papers each.

The top country for these most prolific institutions is the United States with nine institutions. In total, three countries house these top institutions.

On the other hand, the institution with the most impact is the USDA ARS with 4.4% of the surplus. Similarly, the institution with the least impact is the Chinese Academy of Sciences with a 1% deficit.

Additionally, there are other institutions with a relatively low citation impact and with 0.9%–3.2% of the population papers each: South China University of Technology, Nanjing Forestry University, Beijing Forestry University, University of Sao Paulo, Qilu University of Technology, Michigan State University, and the State University of Campinas.

TABLE 9.3
The Most Prolific Institutions in Chemical Pretreatments of the Biomass

No.	Institutions	Country	Sample Papers (%)	Population Papers (%)	Surplus (%)
1	Natl. Renew. Ener. Lab.	USA	5.5	1.3	4.2
2	Lawrence Berkeley Natl. Lab.	USA	5.5	1.3	4.2
3	USDA Agr. Res. Serv.	USA	5.5	0.8	4.7
4	USDA Forest Serv.	USA	3.8	1.1	2.7
5	Sandia Natl. Lab.	USA	3.8	0.9	2.9
6	Chin. Acad. Sci.	China	2.7	3.5	–0.8
7	Oak Ridge Natl. Lab.	USA	2.7	1.2	1.5
8	Univ. Wisconsin-Madison	USA	2.7	1.0	1.7
9	Auburn Univ.	USA	2.7	0.6	2.1
10	Georgia Inst. Technol.	USA	2.7	0.6	2.1
11	Imperial Coll.	UK	2.7	0.3	2.4

TABLE 9.4
The Most Prolific Funding Bodies in Chemical Pretreatments of the Biomass

No.	Funding Bodies	Country	Sample Paper No. (%)	Population Paper No. (%)	Surplus (%)
1	US Dept. Ener.	USA	2.7	3.0	–0.3
2	Eur. Commis.	EU	2.7	1.4	1.3
3	Natl. Natr. Sci. Found. China	China	2.2	13.7	–11.5
4	Natl. Sci. Found.	USA	1.6	1.7	–0.1
5	Off Sci.	USA	1.6	1.6	0.0
6	US Dept. Agr.	USA	1.6	1.0	0.6
7	Seventh Framework Prog.	EU	1.6	0.4	1.2
8	Eng. Phys. Sci. Res. Counc.	UK	1.6	0.4	1.2

9.3.5 The Most Prolific Funding Bodies in the Chemical Pretreatments of the Biomass

Information about the most prolific eight funding bodies funding at least 1.6% of sample papers each is given in Table 9.4. Only 27% and 51% of sample and population papers were funded.

The most prolific funding body is the US Department of Energy and European Commission with 2.7% of sample papers each. The National Natural Science Foundation of China follows them with 2.2% of sample papers.

The most prolific country for these top funding bodies is the United States with four funding bodies while Europe has two funding bodies. In total, three countries and the EU house these top funding bodies.

The funding body with the most citation impact is the European Commission with a 1.3% surplus. Similarly, the funding body with the least citation impact is the National Natural Science Foundation of China with a 12% deficit.

The other funding bodies with a relatively low citation impact and with 1.1%–2.8% of the population papers each are the Ministry of Science and Technology of China, National Council for Scientific and Technological Development of Brazil, National Key Research and Development Program of China, Ministry of Education of China, Fundamental Research Funds for the Central Universities of China, Higher Education Personnel Improvement Coordination of Brazil, Ministry of Education, Culture, Sports, Science and Technology of Japan, Research Support Foundation of the State of Sao Paulo of Brazil, National Research Foundation of South Korea, and Ministry of Finance of Japan.

9.3.6 The Most Prolific Source Titles in Chemical Pretreatments of the Biomass

Information about the most prolific 12 source titles publishing at least 2.2% of the sample papers each in chemical pretreatments of the biomass is given in Table 9.5.

The most prolific source title is Bioresource Technology with 20% of sample papers, followed by Green Chemistry and Biotechnology and Bioengineering with 12% of sample papers each.

On the other hand, the source title with the most impact is Biotechnology and Bioengineering with an 11% surplus, closely followed by Green Chemistry and Bioresource Technology with 10% and 9% surpluses, respectively. Similarly, the source title with the least impact is Industrial Crops and Products with a 1% deficit.

The other source titles with a relatively low citation impact with 1.2%–2.8% of the population paper each are Bioresources, Cellulose, ACS Sustainable Chemistry and Engineering, Biomass Conversion and Biorefinery, Industrial and Engineering Chemistry Research, Carbohydrate Polymers, RSC Advances, and Waste and Biomass Valorization.

9.3.7 The Most Prolific Countries in Chemical Pretreatments of the Biomass

Information about the most prolific 15 countries publishing at least 2.2% of sample papers each in chemical pretreatments of the biomass is given in Table 9.6.

The most prolific country is the United States with 49% of sample papers, followed by China, Japan, and the United Kingdom with 14%, 7%, and 6% of sample papers, respectively. The other prolific countries are Canada, Sweden, and Spain with 4%–5% of sample papers each. Further, six European countries listed in Table 9.6 produce 23% and 14% of sample and population papers, respectively, with a 9% surplus.

On the other hand, the country with the most citation impact is the United States with a 30% surplus. Similarly, the country with the least citation impact is China with a 14% deficit. India, Brazil, and South Korea are the other countries with a 2%–4% deficit each.

Additionally, there are other countries with a relatively low citation impact and with 1.1%–2.9% of sample papers each: Malaysia, Finland, France, Thailand, Iran, Turkey, Italy, Portugal, Russian Federation, Taiwan, and Indonesia.

TABLE 9.5
The Most Prolific Source Titles in Chemical Pretreatments of the Biomass

No.	Source Titles	Sample Papers (%)	Population Papers (%)	Surplus (%)
1	Bioresource Technology	19.8	11.1	8.7
2	Green Chemistry	12.1	1.9	10.2
3	Biotechnology and Bioengineering	12.1	1.1	11.0
4	Chemsuschem	2.7	0.5	2.2
5	Biotechnology Progress	2.7	0.5	2.2
6	Industrial Crops and Products	2.2	3.3	−1.1
7	Biomass and Bioenergy	2.2	2.3	−0.1
8	Applied Biochemistry and Biotechnology	2.2	1.9	0.3
9	Biotechnology for Biofuels	2.2	1.4	0.8
10	Process Biochemistry	2.2	0.8	1.4
11	Applied Biochemistry and Biotechnology Part A	2.2	0.7	1.5
12	Enzyme and Microbial Technology	2.2	0.5	1.7

TABLE 9.6
The Most Prolific Countries in Chemical Pretreatments of the Biomass

No.	Countries	Sample Papers (%)	Population Papers (%)	Surplus (%)
1	USA	48.9	19.1	29.8
2	China	13.7	27.5	−13.8
3	Japan	7.1	6.0	1.1
4	UK	6.0	3.0	3.0
5	Canada	4.9	3.6	1.3
6	Sweden	4.9	2.9	2.0
7	Spain	4.4	3.9	0.5
8	Australia	3.3	1.8	1.5
9	Netherlands	3.3	1.2	2.1
10	Germany	2.7	2.3	0.4
11	Mexico	2.7	1.1	1.6
12	India	2.2	6.6	−4.4
13	Brazil	2.2	5.0	−2.8
14	South Korea	2.2	4.3	−2.1
15	Denmark	2.2	1.0	1.2

TABLE 9.7
The Most Prolific Scopus Subject Categories in Chemical Pretreatments of the Biomass

No.	Scopus Subject Categories	Sample Papers (%)	Population Papers (%)	Surplus (%)
1	Chemical Engineering	61.5	44.1	17.4
2	Environmental Science	46.7	34.5	12.2
3	Biochemistry, Genetics, and Molecular Biology	35.7	20.0	15.7
4	Energy	32.4	31.6	0.8
5	Immunology and Microbiology	25.8	10.7	15.1
6	Chemistry	12.6	27.1	−14.5
7	Materials Science	11.0	16.3	−5.3
8	Agricultural and Biological Sciences	7.7	15.8	−8.1
9	Engineering	6.0	12.2	−6.2
10	Physics and Astronomy	4.4	3.4	1.0

9.3.8 The Most Prolific Scopus Subject Categories in Chemical Pretreatments of the Biomass

Information about the most prolific ten Scopus subject categories indexing at least 4% of sample papers each is given in Table 9.7.

The most prolific Scopus subject category in chemical pretreatments of biomass is Chemical Engineering with 62% of sample papers, followed by Environmental Science with 47% of sample papers. The other prolific subject categories are Biochemistry, Genetics and Molecular Biology, Energy, and Immunology and Microbiology with 26%–36% of sample papers each.

On the other hand, the Scopus subject category with the most citation impact is Chemical Engineering with a 17% surplus. Biochemistry, Genetics and Molecular Biology, Immunology and Microbiology, and Environmental Science follow it with a 12%–16% surplus each.

Similarly, the Scopus subject category with the least citation impact is Chemistry with a 15% deficit. Agricultural and Biological Sciences, Engineering, and Materials Science follow it with a 5%–8% deficit each.

9.3.9 The Most Prolific Keywords in Chemical Pretreatments of the Biomass

Information about the keywords used in at least 8% or 5% of sample or population papers, respectively, is given in Table 9.8. For this purpose, keywords related to the keyword set given in the Appendix are selected from a list of the most prolific keyword set provided by the Scopus database.

TABLE 9.8
The Most Prolific Keywords in Chemical Pretreatments of the Biomass

No.	Keywords	Sample Papers (%)	Population Papers (%)	Surplus (%)
1	**Biomass**			
	Cellulose	62.6	35.3	27.3
	Lignin	47.3	29.2	18.1
	Biomass	39.0	23.8	15.2
	Lignocellulose	26.4	9.5	16.9
	Hemicellulose	21.4	9.1	12.3
	Zea mays	14.8	5.1	9.7
	Lignocellulosic biomass	11.0	9.2	1.8
	Wood	8.2	6.5	1.7
	Xylan	7.7	3.4	4.3
	Corn	7.7		7.7
	Straw	7.1	5.4	1.7
	Bagasse	6.0	6.8	−0.8
	Corn stover	10.4	3.1	7.3
2	**Chemical Pretreatment**			
	ILs	33.0	17.2	15.8
	Sulfuric acids	24.2	9.7	14.5
	Solvents	14.8	13.4	1.4
	Alcohol	13.2	7.0	6.2
	Acids	9.3	7.0	2.3
	Water	9.3	5.3	4
	Ammonia	8.8	3.1	5.7
	Alkalinity	6.0	5.5	0.5
	pH	4.9	7.0	−2.1
	Sodium hydroxide	4.4	6.6	−2.2
	Chlorine compounds		5.6	−5.6
3	**Other Processes**			
	Hydrolysis	59.9	30.1	29.8
	Enzymatic hydrolysis	27.5	18.2	9.2
	Fermentation	23.6	16.0	7.6
	Pretreatment	23.1	13.3	9.8
	Enzyme activity	22.5	11.3	11.2

(Continued)

TABLE 9.8 (*Continued*)
The Most Prolific Keywords in Chemical Pretreatments of the Biomass

No.	Keywords	Sample Papers (%)	Population Papers (%)	Surplus (%)
	Enzymes	21.4	7.8	13.6
	Cellulase	21.4	7.2	14.2
	Saccharification	18.1	12.2	5.9
	Pretreatment	17.0	15.3	1.7
	Dissolution	12.6	6.9	5.7
	Enzymolysis	11.5	4.7	6.8
	Delignification	10.4	8.5	1.9
	Degradation	9.3	6.0	3.3
	Extraction		8.9	–8.9
4	**Products**			
	Sugar	26.9	14.0	12.9
	Glucose	25.8	15.7	10.1
	Ethanol	24.2	17.0	7.2
	Biofuel	19.2	9.6	9.6
	Xylose	13.2	6.5	6.7
	Furfural	7.7	2.7	5.0
	Bioethanol	4.9	6.9	–2

These keywords are grouped into four headings: biomass, chemical pretreatments, other processes, and products of chemical pretreatments.

There are 13 keywords selected related to the biomass and biomass constituents: The prolific keywords are cellulose, lignin, biomass, lignocellulose, and hemicellulose with 21%–63% of sample papers each.

The prolific keywords related to chemical pretreatments of the biomass are ILs, sulfuric acids, solvents, and alcohol with 13%–33% of sample papers each while those related to other processes are hydrolysis, enzymatic hydrolysis, fermentation, pretreatment, enzyme activity, enzymes, and cellulases with 21%–63% of sample papers each.

The prolific keywords related to the products of chemical pretreatments are sugar, glucose, ethanol, biofuel, and xylose with 13%–27% of sample papers each.

9.3.10 The Most Prolific Research Fronts in Chemical Pretreatments of the Biomass

Information about the research fronts for the sample papers in chemical pretreatments of biomass is given in Table 9.9. As Table 9.9 shows, there are three primary research fronts for this field: acid pretreatments, IL pretreatments, and alkaline pretreatments with 37%, 25%, and 18% of the 182 highly cited papers (HCPs), respectively.

The other prolific research fronts are ammonia, organic solvent, solvent, hydrogen peroxide, and DES pretreatments with 3%–7% of sample papers each.

Information about the research fronts for the sample papers in chemical pretreatments of the biomass with regard to the biomass used in these pretreatments is given in Table 9.10. As Table 9.10 shows, there are three primary research fronts for this field: biomass constituents, agricultural residues, and biomass in general with 46%, 33%, and 19% of the 182 HCPs, respectively. The other research fronts are wood, grass, and other biomasses such as algae and water hyacinth with 10%, 8%, and 1% of sample papers, respectively.

TABLE 9.9
The Most Prolific Research Fronts for Chemical Pretreatments of the Biomass

No.	Research Fronts	N Paper % Sample
1	Acid pretreatments	36.8
2	IL pretreatments	24.7
3	Alkaline pretreatments	18.1
4	Ammonia pretreatments	7.1
5	Organic solvent pretreatments	4.4
6	Solvent pretreatments	3.8
8	**Minor Chemical Pretreatments**	**15.8**
	Hydrogen peroxide pretreatments	3.3
	Deep eutectic solvent pretreatments	3.3
	Supercritical water pretreatments	1.6
	Chemical pretreatments in general	1.6
	Wet oxidation pretreatments	1.6
	Surfactant pretreatments	1.1
	Ozone pretreatments	1.1
	Sulfite pretreatments	1.1
	Supercritical carbon dioxide pretreatments	1.1

N paper (%) sample: the number of papers in the population sample of 182 papers.

TABLE 9.10
The Most Prolific Research Fronts for Chemical Pretreatments of the Biomass

No.	Research Fronts	N Paper % Sample
1	**Biomass Constituents**	**46.2**
	Cellulose	29.7
	Lignin	7.7
	Hemicellulose	4.4
	Lignocellulose	3.3
	Cellobiose	1.1
2	Wood	9.9
3	Grass	7.7
4	**Agricultural Residues**	**33.0**
	Wheat straw	5.5
	Corn stover	10.4
	Rice straw	6.6
	Sugarcane bagasse	4.4
	Other residues	6.0
5	**Biomass in General**	**19.2**
	Biomass	8.8
	Lignocellulosic biomass	10.4
6	Other biomass	1.1

N paper (%) sample: the number of papers in the population sample of 182 papers.

9.4 DISCUSSION

9.4.1 Introduction

Crude oil-based gasoline fuels have been widely used in the transportation sector since the 1920s. However, there have been great public concerns over the adverse environmental and human impact of these fuels. Hence, biomass-based bioethanol fuels have increasingly been used in blending gasoline fuels in fuel cells and in biochemical production in a biorefinery context.

Research in the field of chemical pretreatments of biomass such as acid, IL, alkali, and ammonia pretreatments to increase the sugar and bioethanol yield has also intensified in recent years.

However, it is essential to develop efficient incentive structures for the primary stakeholders to enhance research in this field. A scientometric analysis has been used in this context to inform the primary stakeholders about the current state of the research in a selected research field.

As there have been no scientometric studies on the chemical pretreatments of biomass, this chapter presents a scientometric study of research in the chemical pretreatments of biomass. It examines the scientometric characteristics of both the sample and population data presenting the scientometric characteristics of these both datasets in the order of documents, authors, publication years, institutions, funding bodies, source titles, countries, Scopus subject categories, keywords, and research fronts.

As a first step for search of the relevant literature, keywords were selected using the first most-cited 200 papers. The selected keyword list was optimized to obtain a representative sample of papers for the searched research field. This keyword list was provided in the Appendix for future replication studies.

As a second step, two sets of data were used for this study. First, a population sample of over 9,082 papers was used to examine the scientometric characteristics of the population data. Secondly, a sample of 182 most-cited papers corresponding to 2% of the population dataset was used to examine the scientometric characteristics of these citation classics.

The scientometric characteristics of these both sample and population datasets were presented in the order of documents, authors, publication years, institutions, funding bodies, source titles, countries, Scopus subject categories, keywords, and research fronts.

Lastly, the key scientometric findings for both datasets were discussed to highlight the research landscape for chemical pretreatments of biomass. Additionally, a number of brief conclusions were drawn and a number of relevant recommendations were made to enhance the future research landscape.

9.4.2 The Most Prolific Documents in Chemical Pretreatments of the Biomass

Both articles (together with conference papers) and review papers (together with short surveys) dominate the sample and population papers (Table 9.1). Further, review papers and articles have a surplus and deficit, respectively.

Scopus differs from the Web of Science database in differentiating and showing articles and conference papers published in the journals separately. Similarly, Scopus differs from the Web of Science database in introducing short surveys. Hence, the total number of articles and review papers in the sample dataset is 89% and 11%, respectively.

It is observed during the search process that there has been inconsistency in the classification of documents in Scopus as well as in other databases such as Web of Science. This is especially relevant for the classification of papers as reviews or articles as the papers not involving a literature review may be erroneously classified as a review paper. There is also a case of review papers being classified as articles. For example, the percentage of the review papers and short surveys was found manually as 12.1%, compared to 10.7%.

In this context, it would be helpful to provide a classification note for the published papers in books and journals in the first instance. It would also be helpful to use the document types listed in Table 9.1 for this purpose. Book chapters may also be classified as articles or reviews as an additional classification to differentiate review chapters from the experimental chapters as done by the Web of Science. It would be further helpful to additionally classify the conference papers as articles or review papers as it is done in the Web of Science database.

9.4.3 The Most Prolific Authors in Chemical Pretreatments of the Biomass

There have been the most-prolific 11 authors with at least 2.2% of sample papers each as given in Table 9.2. These authors have shaped the development of research in this field.

The most prolific authors are Blake A. Simmons, Tae Hyun Kim, Seema Singh, and Yoon Y. Lee working primarily on ammonia and IL pretreatments of the biomass.

It is important to note the inconsistencies in indexing the author names in Scopus and other databases. It is especially an issue for names with more than two components such as 'Blake Sam de Hyun Kim'. The probable outcomes are 'Kim, B.S.D.H.', 'de Hyun Kim, B.S.', or 'Hyun Kim, B.S.D.'. The first choice is the gold standard of the publishing sector as the last word in the name is taken as the last name. In most academic databases such as PUBMED and EBSCO databases, this version is used predominantly. The second choice is a strong alternative with an increasing trend while the last choice is an undesired outcome as two last words are taken as the last name. It is a good practice combining the words of the last name with a hyphen: 'Hyun-Kim, B.S.D.'. It is noted that inconsistent indexing of author names may cause substantial inefficiencies in the search process for papers as well as allocating credit to authors as there are different author entries for each outcome in the databases.

There are also inconsistencies in shortening Chinese names. For example, 'Yuoyong Zhang' is often shortened as 'Zhang, Y.', 'Zhang, Y.-Y.', and 'Zhang Y.Y' as it is done in the Web of Science database as well. However, the gold standard, in this case, is 'Zhang Y' where the last word is taken as the last name and the first word is taken as a single forename. In most academic databases such as PUBMED and EBSCO, this first version is used predominantly. However, there is an increasing trend to use the third option 'Zhang, Y.Y.' to differentiate Chinese names efficiently. Therefore, there have been difficulties to locate papers for Chinese authors. In such cases, the use of unique author codes provided for each author by the Scopus database has been helpful.

There is also a difficulty in allowing credit for the authors especially for authors with common names such as 'Zhang, W', or 'Huang, W' or 'Zhu, W' in conducting scientometric studies. These difficulties strongly influence the efficiency of scientometric studies as well as allocating credit to authors as there are the same author entries for different authors with the same name, e.g., 'Zhang W' in the databases.

In this context, the coding of authors in the Scopus database is a welcome innovation compared to other databases such as Web of Science. In this process, Scopus allocates a unique number to each author in the database (Aman, 2018). However, there might still be substantial inefficiencies in this coding system, especially for common names. For example, some papers for a certain author may be allocated to another researcher with a different author code. It is possible that Scopus uses a number of software programs to differentiate author names and the program may not be false-proof (D'Angelo and van Eck, 2020).

In this context, it does not help that author names are not given in full in some journals and books. This makes it difficult to differentiate authors with common names and makes the scientometric studies further difficult in the author domain. Therefore, author names should be given in all books and journals in the first instance. There is also a cultural issue where some authors do not use their full names in their papers. Instead, they use initials for their forenames: 'Kim, H.J.' or just 'Kim' instead of 'Kim, Hyun Jae'.

There are also inconsistencies in naming the authors with more than two components by the authors themselves in journal papers and book chapters. For example, 'Kim, A.P.C.', 'Simmons,

C.E.', and 'Blake, S.J.' might be given as 'Kim, A', 'Simmons, C', or 'Blake, S' in the journals and books, respectively. This also makes the scientometric studies difficult in the author domain. Hence, contributing authors should use their names consistently in their publications.

The other critical issue regarding the author names is the inconsistencies in the spelling of author names in national spellings (e.g., Gonçalves, Üböğloş) rather than in English spellings (e.g., Goncalves, Uboglos) in Scopus database. Scopus differs from the Web of Science database and many other databases in this respect where the author names are given only in English spellings. It is observed that national spellings of author names do not help in conducting scientometric studies as well as in allocating credits to authors as sometimes there are different author entries for English and National spellings in the Scopus database.

The most prolific institutions for the sample dataset are the Imperial College and Lawrence Berkeley National Laboratory. In total, eight institutions house these prolific authors. The most prolific countries for the sample dataset are the United States and the United Kingdom. In total, four countries house these authors.

The most-prolific research front is the IL pretreatment of the biomass with five authors. The other fronts are acid and ammonia treatments of the biomass with three authors each.

It is also noted that there is a significant gender deficit for the sample dataset as surprisingly only two of these top researchers are female. This finding is the most thought-provoking with strong public policy implications. Hence, institutions, funding bodies, and policymakers should take efficient measures to reduce the gender deficit in this field as well as other scientific fields with a strong gender deficit. In this context, it is worth noting the level of representation of researchers from minority groups in science on the basis of race, sexuality, age, and disability, besides gender (Blankenship, 1993; Dirth and Branscombe, 2017; Konur, 2000, 2002a–c, 2006a, b, 2007a, b).

9.4.4 The Most Prolific Research Output by Years in Chemical Pretreatments of the Biomass

The research output observed between 1970 and 2022 is illustrated in Figure 9.1. This figure clearly shows that the bulk of research papers in the population dataset was published primarily in the 2010s. Similarly, the bulk of research papers in the sample dataset was published in the last two decades.

These findings suggest that the most-prolific sample and population papers were primarily published in the last two decades. These are thought-provoking findings as there has been no significant research in this field in the pre-2000s, but there has been a significant research boom in the last two decades. In this context, the increasing public concerns about climate change (Change, 2007), greenhouse gas emissions (Carlson et al., 2017), and global warming (Kerr, 2007) have been certainly behind the boom in the research in this field in the last two decades.

Based on these findings, the size of the population papers is likely to more than double in the current decade, provided that the public concerns about climate change, greenhouse gas emissions, and global warming are translated efficiently into research funding in this field.

9.4.5 The Most Prolific Institutions in Chemical Pretreatments of the Biomass

The most prolific 11 institutions publishing papers on chemical pretreatments of the biomass with at least 2.7% of sample papers each given in Table 9.3 have shaped the development of research in this field.

The most prolific institutions are the Lawrence Berkeley National Laboratory, National Renewable Energy Laboratory, USDA ARS, USDA Forest Service, and Sandia Laboratory, all from the United States.

The top country for these most prolific institutions is the United States with nine institutions. In total, only three countries house these top institutions.

On the other hand, the institutions with the most and least impact are USDA ARS and the Chinese Academy of Sciences, respectively. These findings confirm the dominance of the US institutions in research field.

9.4.6 The Most Prolific Funding Bodies in Chemical Pretreatments of the Biomass

The most prolific eight funding bodies that are funding at least 1.6% of sample papers each are given in Table 9.4. It is noted that only 27% and 51% of sample and population papers were funded.

The most prolific funding bodies are the US Department of Energy, the European Commission, and the National Natural Science Foundation of China.

The most prolific countries for these top funding bodies are the United States and European Union (EU). In total, three countries and the EU house these top funding bodies. The funding bodies with the most and least citation impact are the European Commission and the National Natural Science Foundation of China, respectively.

These findings on the funding of research in this field suggest that the level of funding, mostly in the last two decades, is highly intensive for the population papers, and it has been largely instrumental in enhancing research in this field (Ebadi and Schiffauerova, 2016) in the light of North's institutional framework (North, 1991). However, considering the relatively low levels of funding especially for sample papers, there is ample room to enhance funding in this field.

9.4.7 The Most Prolific Source Titles in Chemical Pretreatments of the Biomass

The most prolific 12 source titles publishing at least 2.2% of sample papers each in chemical pretreatments of the biomass have shaped the development of research in this field (Table 9.5).

The most prolific source titles are Bioresource Technology, Green Chemistry, and Biotechnology and Bioengineering. On the other hand, the source titles with the most and least impact are Biotechnology and Bioengineering and Industrial Crops and Products, respectively.

It is noted that these top source titles are related to biomass, bioresources, biotechnology, and bioenergy. This finding suggests that the journals in this field have significantly shaped the development of research in this field as they focus on the chemical pretreatment aspects of bioethanol fuels.

9.4.8 The Most Prolific Countries in Chemical Pretreatments of the Biomass

The most prolific 15 countries publishing at least 2.2% of sample papers each have significantly shaped the development of research in this field (Table 9.6).

The most prolific country is the United States with 49% of sample papers. The other prolific countries are China, Japan, the United Kingdom, Canada, Sweden, and Spain. On the other hand, the countries with the most and least citation impact are the United States and China, respectively. Further, six European countries listed in Table 9.6 produce 23% and 14% of sample and population papers, respectively, with a 9% surplus. Additionally, it is noted that India, Brazil, and South Korea were under-represented in sample papers while the United Kingdom, Netherlands, and Sweden were over-represented.

Close examination of these findings suggests that the United States, Europe as a whole, China, and to a lesser extent Japan and Australia are the major producers of research in this field. It is a fact that the United States has been a major player in science (Leydesdorff and Wagner, 2009; Leydesdorff et al., 2014). The United States has further developed a strong research infrastructure to support its corn- and grass-based bioethanol industry (Vadas et al., 2008).

However, China has been a rising star in scientific research in competition with the United States and Europe (Leydesdorff and Zhou, 2005). China is also a major player in this field as a major producer of bioethanol fuels (Li and Chan-Halbrendt, 2009).

Next, Europe as a whole has been a persistent player in scientific research in competition with both the United States and China (Leydesdorff, 2000). Europe has also been a persistent producer of bioethanol along with the United States and Brazil (Gnansounou, 2010).

Additionally, Brazil has also been a persistent player in scientific research at a moderate level (Glanzel et al., 2006). Brazil has also developed a strong research infrastructure to support its biomass-based bioethanol industry (Soccol et al., 2010).

9.4.9 The Most Prolific Scopus Subject Categories in Chemical Pretreatments of the Biomass

The most prolific ten Scopus subject categories indexing at least 4% of sample papers each, respectively, given in Table 9.7 have shaped the development of research in this field.

The most prolific Scopus subject categories in chemical pretreatments of biomass are Chemical Engineering and Environmental Science. On the other hand, the Scopus subject categories with the most and least citation impact are Chemical Engineering and Chemistry, respectively.

These findings are thought provoking suggesting that the primary subject categories are Chemical Engineering and Environmental Science. The other key finding is that social sciences are not well represented in both the sample and population papers, as in most fields in bioethanol fuels. These findings are not surprising as the key research fronts in this field relate to the development and application of chemical pretreatments of biomass.

9.4.10 The Most Prolific Keywords in Chemical Pretreatments of the Biomass

A limited number of keywords have shaped the development of research in this field as shown in Table 9.8 and the Appendix.

These keywords are grouped into four headings: biomass and biomass constituents, chemical pretreatments, other processes, and products of chemical pretreatments.

There are 13 keywords selected related to biomass: The prolific keywords are cellulose, lignin, biomass, lignocellulose, and hemicellulose.

The prolific keywords related to chemical pretreatments of the biomass are ILs, sulfuric acids, solvents, and alcohol while those related to other processes are hydrolysis, enzymatic hydrolysis, fermentation, pretreatment, enzyme activity, enzymes, and cellulases. Further, the prolific keywords related to the products of chemical pretreatments are sugar, glucose, ethanol, biofuel, and xylose.

These findings suggest that it is necessary to determine the keyword set carefully to locate the relevant research in each of these research fronts. Additionally, the size of the samples for each keyword highlights the intensity of research in the relevant research areas.

9.4.11 The Most Prolific Research Fronts in Chemical Pretreatments of the Biomass

As Table 9.9 shows, there are three primary research fronts for this research field: acid pretreatments, IL pretreatments, and alkaline pretreatments. The other prolific research fronts are ammonia pretreatments, organic solvent pretreatments, solvent pretreatments, hydrogen peroxide pretreatments, and DES pretreatments.

Information about the research fronts for sample papers in chemical pretreatments of the biomass with regard to the biomass or biomass constituents used in these pretreatments is given in Table 9.10. As Table 9.10 shows, there are three primary research fronts for this field: biomass constituents, agricultural residues, and biomass in general. The other research fronts are wood, grass, and other biomasses such as algae and water hyacinth.

These findings are thought provoking in seeking ways to increase bioethanol yield through chemical pretreatments of the biomass at the global scale. It is clear that all these research fronts with regard to both pretreatment types and biomass types have public importance and merit substantial funding and other incentives.

In the end, these most-cited papers in this field hint that the efficiency of bioethanol fuels and their derivatives could be optimized using the structure, processing, and property relationships of these chemical pretreatments (Formela et al., 2016; Konur, 2018a, 2020b, 2021a–d; Konur and Matthews, 1989).

9.5 CONCLUSION AND FUTURE RESEARCH

Research on chemical pretreatments of biomass has been mapped through a scientometric study of both sample and population datasets.

The critical issue in this study has been to obtain a representative sample of research as in any other scientometric studies. Therefore, the keyword set has been carefully devised and optimized after a number of runs in the Scopus database. It is a representative sample of wider population studies.

The other issue has been the selection of a multidisciplinary database to carry out the scientometric study of research in this field. For this purpose, the Scopus database has been selected. The journal coverage of this database has been notably wider than that of the Web of Science.

The key scientometric properties of research in this field have been determined and discussed in this chapter. It is evident that a limited number of documents, authors, institutions, publication periods, institutions, funding bodies, source titles, countries, Scopus subject categories, keywords, and research fronts as well as HCPs have shaped the development of research in this field.

There is ample scope to increase the efficiency of scientometric studies in this field in the author and document domains by developing consistent policies and practices in both domains across all academic databases. In this respect, it seems that authors, journals, and academic databases have a lot to do. Furthermore, the significant gender deficit as in most scientific fields emerges as a public policy issue. The potential deficits on the basis of age, race, disability, and sexuality need also to be explored in this field as in other scientific fields.

Research in this field has boomed in the last two decades possibly promoted by public concerns about global warming, greenhouse gas emissions, and climate change. Institutions from the United States have mostly shaped the research in this field.

A relatively well funding rate of 51% for population papers suggests that funding in this field significantly enhanced research in this field primarily in the last two decades, possibly more than doubling in the current decade. However, it is evident that there is ample room for more funding and other incentives to enhance research in this field further as only 27% of sample papers declared any funding.

The most-prolific source titles have been mostly indexed by the subject categories of Chemical Engineering and Environmental Science as the focus of sample papers has been on the development and utilization of chemical pretreatments of the biomass to increase the sugar and bioethanol yield at large.

The United States, Europe, and China and to a lesser extent Japan and Australia have been the major producers of research in this field as the major producers and users of bioethanol fuels from different types of biomass such as corn, sugarcane, grass, as well as other types of biomass. It is evident that these countries have well-developed research infrastructure in bioethanol fuels and their derivatives.

The primary Scopus subject categories have been Chemical Engineering and Environmental Science as the focus of research has been on the development and utilization of chemical pretreatments of the biomass to increase the sugar and bioethanol yield at large. Further, social sciences are not well represented in both the sample and population papers, as in most fields in bioethanol fuels. These findings are not surprising.

It emerges that ethanol is more popular than bioethanol as a keyword with strong implications for the search strategy. In other words, the search strategy using only the bioethanol keyword would not be much helpful.

These keywords are grouped into four headings: biomass and biomass constituents, chemical pretreatment, other processes, and products of chemical pretreatments. These groups of keywords highlight the potential primary research fronts for these fields.

The three primary research fronts with regard to chemical pretreatments of the biomass are acid pretreatments, IL pretreatments, and alkaline pretreatments. The other prolific research fronts are ammonia pretreatments, organic solvent pretreatments, other solvent pretreatments, hydrogen peroxide pretreatments, and DES pretreatments.

Similarly, there are three primary research fronts for the biomass or biomass constituents used for chemical pretreatments of the biomass: biomass constituents, agricultural residues, and biomass in general. The other research fronts are wood, grass, and other biomass such as algae and water hyacinth.

These findings are thought-provoking. The focus of these most-cited 182 papers as well as 9,082 population papers is the development and utilization of chemical pretreatments to increase the sugar and bioethanol yield. Thus, these studies highlight strong structure-processing-property relationships for chemical pretreatments for bioethanol fuels and their derivatives.

Thus, the scientometric analysis has a great potential to gain valuable insights into the evolution of research in this field as in other scientific fields.

It is recommended that further scientometric studies are carried out for the primary types of chemical pretreatments and the biomass used as well as about the other types of pretreatments. It is further recommended that reviews of the most-cited papers are carried out for each research front to complement these scientometric studies. Next, scientometric studies of hot papers in these primary fields are carried out.

ACKNOWLEDGMENTS

The contribution of highly cited researchers in the field of chemical pretreatments of biomass has been gratefully acknowledged.

APPENDIX: THE KEYWORD SET FOR CHEMICAL PRETREATMENTS OF THE BIOMASS

(TITLE ("*chemical treat*" OR "*chemical pretreat*" OR acid* OR h2so4 OR "h 2so 4" OR "hydrogen peroxide*" OR h2o2 OR "h 2o 2" OR "alkaline peroxide*" OR "chemo-mechanical pretreat*" OR "thermo-chemical pretreat*" OR "thermochemical pretreat*" OR hcl OR salt* OR sulfite* OR sporl OR ph OR lime* OR "sulphur dioxide" OR "sulfur dioxide" OR so2 OR "so 2" OR alkali* OR *alkali OR naoh* OR "sodium hydroxide*" OR koh OR caoh* OR "ca(oh)*" OR "calcium hydroxide" OR "sodium methoxide" OR base OR "ionic liquid*" OR solvent* OR *solvent OR *methylimidazolium OR "methyl imidazolium" OR dissolv* OR *organosolv OR *organosolve OR ammonia OR nh3 OR ozonation OR ozonolysis OR ozone OR "wet oxidation" OR "air oxidation" OR flowthrough OR "supercritical co2" OR "carbon dioxide explosion" OR "co2 explosion" OR "supercritical carbon dioxide" OR "supercritical co2" OR methylmorpholine OR nmmo* OR bmim* OR surfactant* OR afex* OR ammonolysis OR "recycled percolation" OR "physicochemical pretreat*" OR "physicochemical treat*" OR "physico-chemical pretreat*" OR "supercritical water" OR "subcritical water" OR fecl3 OR fenton) AND TITLE (ethanol OR bioethanol OR pretreat* OR "pre-treat*" OR *hydrolysis OR saccharification OR ferment* OR hydrolytic OR digestibili* OR disruption OR fractionation OR decomposition OR dissolution OR delignification OR depolymerization OR sugar OR sugars OR *oligosaccharides OR glucose OR deconstruction OR xylose OR disintegration OR *conversion OR recalcitrance OR hydrolyzate* OR hydrolysate* OR *degradation OR extract* OR detoxification) AND (TITLE (waste* OR *grass OR grass* OR biomass* OR lignocellul* OR bagasse* OR cellulos* OR *wood OR woody OR *woods OR *algae OR *algal OR microalga OR chlamydomonas OR straw* OR stover* OR eucalyptus OR miscanthus OR pine OR chlorella OR lignin OR hemicellulose* OR porphyridium OR nannochloropsis OR poplar* OR cedar OR "corn cob*" OR spruce OR beech OR oak OR pinus OR residue* OR willow* OR cypress OR sawdust OR prosopis OR "rice hull*" OR "*bean hull*" OR husk* OR birch OR cynara OR "*cane tops" OR "cotton stalk*" OR "corn stalk*" OR alfalfa OR "maize stem*" OR cornstalk* OR hyacinth OR salix OR corncob* OR "fruit bunch*" OR populus OR sage OR "orange peel*" OR "*weed stem*" OR bamboo* OR birch OR "*flower stalk*" OR "barley hull*" OR "maize stem*") OR SRCTITLE (cellulose OR biomass OR sugar))) AND NOT (TITLE (digestion OR *composites OR composite OR lactic OR polysaccharide* OR wastewater* OR pyroly* OR nano* OR anaerobic OR *methane OR methan* OR *methanation OR *hydrogen OR hydrogen* OR fiber* OR fibre* OR levulinic OR sludge OR furan* OR dye* OR aromatic* OR *furfural OR polyol* OR amino OR bovine OR docosahexaen* OR fatty OR nucleic OR protease OR *porous OR adhesive* OR seed* OR radioactive* OR phar* OR keton* OR "value added" OR nuclear OR phenol* OR carotene OR corrosion OR "activated carbon" or hexitol* OR soil* OR biogas OR carbohydrate* OR cytokinin* OR magic OR label* OR earth OR oxidase* OR lipid* OR walnut OR strawberr* OR antioxidant* OR *diesel OR tandem) OR SRCTITLE (dairy OR phar* OR medi* OR cereal* OR nutrit* OR feed* OR brew* OR animal* OR zoo* OR lipid* OR soil* OR composite* OR atmosph*) OR SUBJAREA (medi OR phar OR vete OR heal OR neur OR eart)) AND (LIMIT-TO (SRCTYPE, "j") OR LIMIT-TO (SRCTYPE, "k") OR LIMIT-TO (SRCTYPE, "b")) AND (LIMIT-TO (DOCTYPE, "ar") OR LIMIT-TO (DOCTYPE, "cp") OR LIMIT-TO (DOCTYPE, "re") OR LIMIT-TO (DOCTYPE, "ch") OR LIMIT-TO (DOCTYPE, "le") OR LIMIT-TO (DOCTYPE, "bk") OR LIMIT-TO (DOCTYPE, "no") OR LIMIT-TO (DOCTYPE, "ed") OR LIMIT-TO (DOCTYPE, "sh") OR LIMIT-TO (DOCTYPE, "cr")) AND (LIMIT-TO (LANGUAGE, "English"))

REFERENCES

Alvira, P., E. Tomas-Pejo, M. Ballesteros and M. J. Negro. 2010. Pretreatment technologies for an efficient bioethanol production process based on enzymatic hydrolysis: A review. *Bioresource Technology* 101:4851–4861.

Aman, V. 2018. Does the Scopus author ID suffice to track scientific international mobility? A case study based on Leibniz laureates. *Scientometrics* 117:705–720.

Angelici, C., B. M. Weckhuysen and P. C. A. Bruijnincx. 2013. Chemocatalytic conversion of ethanol into butadiene and other bulk chemicals. *ChemSusChem* 6:1595–1614.

Antolini, E. 2007. Catalysts for direct ethanol fuel cells. *Journal of Power Sources* 170:1–12.

Antolini, E. 2009. Palladium in fuel cell catalysis. *Energy and Environmental Science* 2:915–931.

Balan, V., B. Bals, S. P. Chundawat, D. Marshall and B. E. Dale. 2009. Lignocellulosic biomass pretreatment using AFEX. *Methods in Molecular Biology* 581:61–77.

Beaudry, C. and V. Lariviere. 2016. Which gender gap? Factors affecting researchers' scientific impact in science and medicine. *Research Policy* 45:1790–1817.

Blankenship, K. M. 1993. Bringing gender and race in: US employment discrimination policy. *Gender & Society* 7:204–226.

Brandt, A., J. Grasvik, J. P. Hallett and T. Welton. 2013. Deconstruction of lignocellulosic biomass with ionic liquids. *Green Chemistry* 15:550–583.

Burnham, J. F. 2006. Scopus database: A review. *Biomedical Digital Libraries* 3:1–8.

Carlson, K. M., J. S. Gerber and D. Mueller, et al. 2017. Greenhouse gas emissions intensity of global croplands. *Nature Climate Change* 7:63–68.

Change, C. 2007. Climate change impacts, adaptation and vulnerability. *Science of the Total Environment* 326:95–112.

D'Angelo, C. A. and N. J. van Eck. 2020. Collecting large-scale publication data at the level of individual researchers: A practical proposal for author name disambiguation. *Scientometrics* 123:883–907.

Dirth, T. P. and N. R. Branscombe. 2017. Disability models affect disability policy support through awareness of structural discrimination. *Journal of Social Issues* 73:413–442.

Ebadi, A. and A. Schiffauerova. 2016. How to boost scientific production? A statistical analysis of research funding and other influencing factors. *Scientometrics* 106:1093–1116.

Eriksson, T., J. Borjesson and F. Tjerneld. 2002. Mechanism of surfactant effect in enzymatic hydrolysis of lignocellulose. *Enzyme and Microbial Technology* 31:353–364.

Fernando, S., S. Adhikari, C. Chandrapal and M. Murali. 2006. Biorefineries: Current status, challenges, and future direction. *Energy & Fuels* 20:1727–1737.

Formela, K., A. Hejna, L. Piszczyk, M. R. Saeb and X. Colom. 2016. Processing and structure-property relationships of natural rubber/wheat bran biocomposites. *Cellulose* 23:3157–3175.

Garcia-Cubero, M. T., G. González-Benito, I. Indacoechea, M. Coca and S. Bolado. 2009. Effect of ozonolysis pretreatment on enzymatic digestibility of wheat and rye straw. *Bioresource Technology* 100:1608–1613.

Garfield, E. 1955. Citation indexes for science. *Science* 122:108–111.

Glanzel, W., J. Leta and B. Thijs. 2006. Science in Brazil. Part 1: A macro-level comparative study. *Scientometrics* 67:67–86.

Glenn, J. K., M. A. Morgan, M. B. Mayfield, M. Kuwahara and M. H. Gold. 1983. An extracellular H_2O_2-requiring enzyme preparation involved in lignin biodegradation by the white rot basidiomycete *Phanerochaete chrysosporium*. *Biochemical and Biophysical Research Communications* 114:1077–1083.

Gnansounou, E. 2010. Production and use of lignocellulosic bioethanol in Europe: Current situation and perspectives. *Bioresource Technology* 101:4842–4850.

Helle, S. S., S. J. B. Duff and D. G. Cooper. 1993. Effect of surfactants on cellulose hydrolysis. *Biotechnology and Bioengineering* 42:611–617.

Hill, J., E. Nelson, D. Tilman, S. Polasky and D. Tiffany. 2006. Environmental, economic, and energetic costs and benefits of biodiesel and ethanol biofuels. *Proceedings of the National Academy of Sciences of the United States of America* 103:11206–11210.

Hill, J., S. Polasky and E. Nelson, et al. 2009. Climate change and health costs of air emissions from biofuels and gasoline. *Proceedings of the National Academy of Sciences of the United States of America* 106:2077–2082.

Hsieh, W. D., R. H. Chen, T. L. Wu and T. H. Lin. 2002. Engine performance and pollutant emission of an SI engine using ethanol-gasoline blended fuels. *Atmospheric Environment* 36:403–410.

Huang, H. J., S. Ramaswamy, U. W. Tschirner and B. V. Ramarao. 2008. A review of separation technologies in current and future biorefineries. *Separation and Purification Technology* 62:1–21.

Huang, Y. B. and Y. Fu. 2013. Hydrolysis of cellulose to glucose by solid acid catalysts. *Green Chemistry* 15:1095–1111.

Kerr, R. A. 2007. Global warming is changing the world. *Science* 316:188–190.

Kim, J. S., Y. Y. Lee and T. H. Kim. 2016. A review on alkaline pretreatment technology for bioconversion of lignocellulosic biomass. *Bioresource Technology* 199:42–48.

Kim, K. H. and J. Hong. 2001. Supercritical CO_2 pretreatment of lignocellulose enhances enzymatic cellulose hydrolysis. *Bioresource Technology* 77:139–144.

Kim, T. H., J. S. Kim, C. Sunwoo and Y. Y., Lee. 2003. Pretreatment of corn stover by aqueous ammonia. *Bioresource Technology* 90:39–47.

Knill, C. J. and J. F. Kennedy. 2002. Degradation of cellulose under alkaline conditions. *Carbohydrate Polymers* 51:281–300.

Konur, O. 2000. Creating enforceable civil rights for disabled students in higher education: An institutional theory perspective. *Disability & Society* 15:1041–1063.

Konur, O. 2002a. Access to nursing education by disabled students: Rights and duties of nursing programs. *Nurse Education Today* 22:364–374.

Konur, O. 2002b. Assessment of disabled students in higher education: Current public policy issues. *Assessment and Evaluation in Higher Education* 27:131–152.

Konur, O. 2002c. Access to employment by disabled people in the UK: Is the Disability Discrimination Act working? *International Journal of Discrimination and the Law* 5:247–279.

Konur, O. 2006a. Participation of children with dyslexia in compulsory education: Current public policy issues. *Dyslexia* 12:51–67.

Konur, O. 2006b. Teaching disabled students in Higher Education. *Teaching in Higher Education* 11:351–363.

Konur, O. 2007a. A judicial outcome analysis of the Disability Discrimination Act: A windfall for the employers? *Disability & Society* 22:187–204.

Konur, O. 2007b. Computer-assisted teaching and assessment of disabled students in higher education: The interface between academic standards and disability rights. *Journal of Computer Assisted Learning* 23:207–219.

Konur, O. 2011. The scientometric evaluation of the research on the algae and bio-energy. *Applied Energy* 88:3532–3540.

Konur, O. 2012a. Prof. Dr. Ayhan Demirbas' scientometric biography. *Energy Education Science and Technology Part A: Energy Science and Research* 28:727–738.

Konur, O. 2012b. The evaluation of the biogas research: A scientometric approach. *Energy Education Science and Technology Part A: Energy Science and Research* 29:1277–1292.

Konur, O. 2012c. The evaluation of the global energy and fuels research: A scientometric approach. *Energy Education Science and Technology Part A: Energy Science and Research* 30:613–628.

Konur, O. 2012d. The evaluation of the research on the biodiesel: A scientometric approach. *Energy Education Science and Technology Part A: Energy Science and Research* 28:1003–1014.

Konur, O. 2012e. The evaluation of the research on the bioethanol: A scientometric approach. *Energy Education Science and Technology Part A: Energy Science and Research* 28:1051–1064.

Konur, O. 2012f. The evaluation of the research on the biofuels: A scientometric approach. *Energy Education Science and Technology Part A: Energy Science and Research* 28:903–916.

Konur, O. 2012g. The evaluation of the research on the biohydrogen: A scientometric approach. *Energy Education Science and Technology Part A: Energy Science and Research* 29:323–338.

Konur, O. 2012h. The evaluation of the research on the microbial fuel cells: A scientometric approach. *Energy Education Science and Technology Part A: Energy Science and Research* 29:309–322.

Konur, O. 2012i. The scientometric evaluation of the research on the production of bioenergy from biomass. *Biomass and Bioenergy* 47:504–515.

Konur, O. 2015. Current state of research on algal bioethanol. In *Marine Bioenergy: Trends and Developments*, Eds. S. K. Kim and C. G. Lee, pp. 217–244. Boca Raton, FL: CRC Press.

Konur, O. 2018b. Bioenergy and biofuels science and technology: Scientometric overview and citation classics. In *Bioenergy and Biofuels*, Ed. O. Konur, pp. 3–63. Boca Raton: CRC Press.

Konur, O. 2019. Cyanobacterial bioenergy and biofuels science and technology: A scientometric overview. In *Cyanobacteria: From Basic Science to Applications*, Eds. A. K. Mishra, D. N. Tiwari and A. N. Rai, pp. 419–442. Amsterdam: Elsevier.

Konur, O. 2020a. The scientometric analysis of the research on the bioethanol production from green macroalgae. In *Handbook of Algal Science, Technology and Medicine*, Ed. O. Konur, pp. 385–401. London: Academic Press.

Konur, O., Ed. 2018a. *Bioenergy and Biofuels*. Boca Raton, FL: CRC Press.

Konur, O., Ed. 2020b. *Handbook of Algal Science, Technology and Medicine*. London: Academic Press.

Konur, O., Ed. 2021a. *Handbook of Biodiesel and Petrodiesel Fuels: Science, Technology, Health, and Environment*. Boca Raton, FL: CRC Press.

Konur, O., Ed. 2021b. *Handbook of Biodiesel and Petrodiesel Fuels: Science, Technology, Health, and Environment. Volume 1. Biodiesel Fuels: Science, Technology, Health, and Environment*. Boca Raton, FL: CRC Press.

Konur, O., Ed. 2021c. *Handbook of Biodiesel and Petrodiesel Fuels: Science, Technology, Health, and Environment. Volume 2. Biodiesel Fuels based on the Edible and Nonedible Feedstocks, Wastes, and Algae: Science, Technology, Health, and Environment*. Boca Raton, FL: CRC Press.

Konur, O., Ed. 2021d. *Handbook of Biodiesel and Petrodiesel Fuels: Science, Technology, Health, and Environment. Volume 3. Petrodiesel Fuels: Science, Technology, Health, and Environment*. Boca Raton, FL: CRC Press.

Konur, O. and F. L. Matthews. 1989. Effect of the properties of the constituents on the fatigue performance of composites: A review. *Composites* 20:317–328.

Leydesdorff, L. 2000. Is the European Union becoming a single publication system? *Scientometrics* 47:265–280.

Leydesdorff, L. and C. Wagner. 2009. Is the United States losing ground in science? A global perspective on the world science system. *Scientometrics* 78:23–36.

Leydesdorff, L., C. S. Wagner and L. Bornmann. 2014. The European Union, China, and the United States in the top-1% and top-10% layers of most-frequently cited publications: Competition and collaborations. *Journal of Informetrics* 8:606–617.

Leydesdorff, L. and P. Zhou. 2005. Are the contributions of China and Korea upsetting the world system of science? *Scientometrics* 63:617–630.

Li, S. Z. and C. Chan-Halbrendt. 2009. Ethanol production in (the) People's Republic of China: Potential and technologies. *Applied Energy* 86:S162–S169.

Ma, X., L. Sun and C. Song. 2002. A new approach to deep desulfurization of gasoline, diesel fuel and jet fuel by selective adsorption for ultra-clean fuels and for fuel cell applications. *Catalysis Today* 77:107–116.

Morschbacker, A. 2009. Bio-ethanol based ethylene. *Polymer Reviews* 49:79–84.

Mussatto, S. I. and I. C. Roberto. 2004. Alternatives for detoxification of diluted-acid lignocellulosic hydrolyzates for use in fermentative processes: A review. *Bioresource Technology* 93:1–10.

Najafi, G., B. Ghobadian and T. Tavakoli, et al. 2009. Performance and exhaust emissions of a gasoline engine with ethanol blended gasoline fuels using artificial neural network. *Applied Energy* 86:630–639.

Newman, P. W. G. and J. R. Kenworthy. 1989. Gasoline consumption and cities: A comparison of U.S. cities with a global survey. *Journal of the American Planning Association* 55:24–37.

North, D. C. 1991. Institutions. *Journal of Economic Perspectives* 5:97–112.

Sasaki, M., B. Kabyemela and R. Malaluan, et al. 1998. Cellulose hydrolysis in subcritical and supercritical water. *Journal of Supercritical Fluids* 13:261–268.

Sasaki, M., Z. Fang, Y. Fukushima, T. Adschiri and K. Arai. 2000. Dissolution and hydrolysis of cellulose in subcritical and supercritical water. *Industrial and Engineering Chemistry Research* 39:2883–2890.

Satlewal, A., R. Agrawal, S. Bhagia, J. Sangoro and A. J. Ragauskas. 2018. Natural deep eutectic solvents for lignocellulosic biomass pretreatment: Recent developments, challenges and novel opportunities. *Biotechnology Advances* 36:2032–2050.

Shuai, L. and J. Luterbacher. 2016. Organic solvent effects in biomass conversion reactions. *ChemSusChem* 9:133–155.

Shuai, L., Q. Yang and J. Y. Zhu, et al. 2010. Comparative study of SPORL and dilute-acid pretreatments of spruce for cellulosic ethanol production. *Bioresource Technology* 101:3106–3114.

Soccol, C. R., L. P. de Souza Vandenberghe and A. B. P. Medeiros, et al. 2010. Bioethanol from lignocelluloses: Status and perspectives in Brazil. *Bioresource Technology* 101:4820–4825.

Sun, R. C., J. Tomkinson, P. L. Ma and S. F. Liang. 2000. Comparative study of hemicelluloses from rice straw by alkali and hydrogen peroxide treatments. *Carbohydrate Polymers* 42:111–122.

Sun, Y. and J. Cheng. 2002. Hydrolysis of lignocellulosic materials for ethanol production: A review. *Bioresource Technology* 83:1–11.

Vadas, P. A., K. H. Barnett and D. J. Undersander 2008. Economics and energy of ethanol production from alfalfa, corn, and switchgrass in the Upper Midwest, USA. *Bioenergy Research* 1:44–55.

Vidal, P. F. and J. Molinier. 1988. Ozonolysis of lignin – Improvement of *in vitro* digestibility of poplar sawdust. *Biomass* 16:1–17.

Zhang, K., Z. Pei and D. Wang. 2016. Organic solvent pretreatment of lignocellulosic biomass for biofuels and biochemicals: A review. *Bioresource Technology* 199:21–33.

Zhao, X., K. Cheng and D. Liu. 2009. Organosolv pretreatment of lignocellulosic biomass for enzymatic hydrolysis. *Applied Microbiology and Biotechnology* 82:815–827.

Zheng, Y., H. M. Lin and G. T. Tsao. 1998. Pretreatment for cellulose hydrolysis by carbon dioxide explosion. *Biotechnology Progress* 14:890–896.

Zhu, J. Y., X. J. Pan, G. S. Wang and R. Gleisner. 2009. Sulfite pretreatment (SPORL) for robust enzymatic saccharification of spruce and red pine. *Bioresource Technology* 100:2411–2418.

Zhu, S., Y. Wu and Q. Chen, et al. 2006. Dissolution of cellulose with ionic liquids and its application: A mini-review. *Green Chemistry* 8:325–327.

10 Chemical Pretreatment of the Biomass

Review

Ozcan Konur
(Formerly) Ankara Yildirim Beyazit University

10.1 INTRODUCTION

The crude oil-based gasoline fuels (Ma et al., 2002; Newman and Kenworthy, 1989) have been widely used in the transportation sector since the 1920s. However, there have been great public concerns over the adverse environmental and human impact of these fuels (Hill et al., 2006, 2009). Hence, biomass-based bioethanol fuels (Alvira et al., 2010; Hill et al., 2006; Konur, 2012, 2015, 2020a; Sun and Cheng, 2002) have increasingly been used in blending gasoline fuels (Hsieh et al., 2002; Najafi et al., 2009), in the fuel cells (Antolini, 2007, 2009), and in the biochemical production (Angelici et al., 2013; Morschbacker, 2009) in a biorefinery context (Fernando et al., 2006; Huang et al., 2008).

However, it is necessary to pretreat the biomass (Alvira et al., 2010; Taherzadeh and Karimi, 2008) to enhance the yield of the bioethanol (Hahn-Hagerdal et al., 2006; Sanchez and Cardona, 2008) prior to the bioethanol production through the hydrolysis (Alvira et al., 2010; Sun and Cheng, 2002) and fermentation (Lin and Tanaka, 2006; Olsson and Hahn-Hagerdal, 1996) of the biomass and hydrolysates, respectively.

The research in the fields of chemical pretreatments of the biomass has also intensified in recent years. Acid (Huang and Fu, 2013; Rinaldi and Schuth, 2009), ionic liquid (IL) (Brandt et al., 2013; Zhu et al., 2006), alkali (Kim et al., 2016; Knill and Kennedy, 2002), ammonia (Balan et al., 2009; Kim et al., 2003), organic solvent (Shuai and Luterbacher, 2016; Zhao et al., 2009), hydrogen peroxide (Glenn et al., 1983; Sun et al., 2000), deep eutectic solvent (DES) (Satlewal et al., 2018; Zhang et al., 2016), sub- and supercritical water (Sasaki et al., 1998, 2000), surfactant (Eriksson et al., 2002; Helle et al., 1993), ozone (O_3) (Garcia-Cubero et al., 2009; Vidal and Molinier, 1988), sulfite (Shuai et al., 2010; Zhu et al., 2009), and supercritical carbon dioxide (CO_2) (Kim and Hong, 2001; Zheng et al., 1998) pretreatments of the biomass have been widely researched to increase the sugar and bioethanol yield in this context in recent years.

However, it is essential to develop efficient incentive structures (North, 1991) for the primary stakeholders to enhance the research in this field (Konur, 2000, 2002a–c, 2006a, b, 2007a, b). Although there have been a number of review papers on the chemical pretreatment of the biomass (Balan et al., 2009; Brandt et al., 2013; Huang and Fu, 2013; Kim et al., 2016; Knill and Kennedy, 2002; Rinaldi and Schuth, 2009; Satlewal et al., 2018; Shuai and Luterbacher, 2016; Zhao et al., 2009; Zhu et al., 2006), there has been no updated review of the most-cited 25 articles in this field.

Thus, this book chapter presents a review of the most-cited 25 articles in the field of the chemical pretreatments of the biomass. Then, it discusses the key findings of these highly influential papers and comments on the future research priorities in this field.

DOI: 10.1201/9781003226536-14

10.2 MATERIALS AND METHODS

The search for this study was carried out using Scopus database (Burnham, 2006) in April 2022.

As a first step for the search of the relevant literature, the keywords were selected using all of the population papers. The selected keyword list was optimized to obtain a representative sample of papers for the searched research field. This final keyword set was provided in the Appendix of Konur (2023) for future replication studies.

As a second step, a sample dataset was used for this study. The first 25 articles with at least 477 citations each were selected for the review study. Key findings from each paper were taken from the abstracts of these papers and were discussed. Additionally, a number of brief conclusions were drawn, and a number of relevant recommendations were made to enhance the future research landscape.

10.3 RESULTS

The brief information about 25 most-cited papers with at least 477 citations each on the chemical pretreatments of the biomass is given below. The primary research fronts are acid, IL, and other chemical pretreatments with 10, 9, and 9 papers, respectively.

10.3.1 The Acid Pretreatment of the Biomass

Roman and Winter (2004) studied the impact of sulfate groups from sulfuric acid (H_2SO_4) hydrolysis on the thermal degradation behavior of bacterial cellulose in a paper with 966 citations. They observed that the sulfate group content increased with acid concentration, acid-to-cellulose ratio, and hydrolysis time. Even at low levels, the sulfate groups caused a significant decrease in degradation temperatures and an increase in char fraction confirming that the sulfate groups acted as flame retardants. There was thermal separation of the degradation reactions by the sulfate groups into low- and high-temperature processes. The activation energies were lower at larger amounts of sulfate groups showing a catalytic effect on the degradation reactions. They recommended that for high thermostability in the cellulose crystals, low acid concentrations, small acid-to-cellulose ratios, and short hydrolysis times should be used.

Larsson et al. (1999) studied the impact of the severity of dilute H_2SO_4 hydrolysis of spruce on sugar yield and on the fermentability of the hydrolysate by *Saccharomyces cerevisiae* in a paper with 863 citations. When the severity of the hydrolysis (CS) conditions increased, they observed that the yield of fermentable sugars increased to a maximum between CS 2.0–2.7 for mannose, and 3.0–3.4 for glucose above which it decreased. The decrease in the yield of monosaccharides coincided with the maximum concentrations of furfural and 5-hydroxymethylfurfural (5-HMF). With the further increase in CS, the concentrations of furfural and 5-HMF decreased, while the formation of formic acid and levulinic acid increased. The yield of ethanol decreased at approximately CS 3, while the volumetric productivity decreased at lower CS. Ethanol yield and volumetric productivity decreased with increasing concentrations of acetic acid, formic acid, and levulinic acid. Furfural and 5-HMF decreased the volumetric productivity but did not influence the final yield of ethanol. They asserted that the inhibition observed in hydrolysates produced in higher CS could not be fully explained by the effect of the by-products furfural, 5-HMF, acetic acid, formic acid, and levulinic acid.

Li et al. (2010) compared the efficiency of the dilute acid and IL pretreatment of switchgrass in terms of the delignification, saccharification efficiency, and saccharide yields in a paper with 832 citations. When subject to IL pretreatment, they observed that switchgrass exhibited reduced cellulose crystallinity, increased surface area, and decreased lignin content compared to dilute acid pretreatment. IL pretreatment enabled a significant enhancement in the rate of enzymatic hydrolysis of the cellulose component of switchgrass, with a rate increase of 16.7-fold and a glucan yield of 96.0% obtained in 24 h. They asserted that IL pretreatment might offer unique advantages compared to the dilute acid pretreatment process for switchgrass.

Silverstein et al. (2007) compared H_2SO_4, sodium hydroxide (NaOH), hydrogen peroxide (H_2O_2), and ozone (O_3) pretreatments for improving saccharification of cotton stalks in a paper with 652 citations. They pretreated ground cotton stalks at a solid loading of 10% (w/v) with H_2SO_4, NaOH, and H_2O_2 at concentrations of 0.5%, 1%, and 2% (w/v) at the treatment temperatures of 90°C and 121°C at 15 psi for residence times of 30, 60, and 90 min. They performed O_3 pretreatment at 4°C with constant sparging of stalks in water. Solids from H_2SO_4, NaOH, and H_2O_2 pretreatments (at 2%, 60 min, 121°C/15 psi) showed significant lignin degradation and/or high sugar availability and hence were hydrolyzed by Celluclast 1.5 L and Novozym 188 at 50°C. H_2SO_4 pretreatment resulted in the highest xylan reduction (95.23% for 2% acid, 90 min, 121°C/15 psi) but the lowest cellulose to glucose conversion during hydrolysis (23.85%). NaOH pretreatment resulted in the highest level of delignification (65.63% for 2% NaOH, 90 min, 121°C/15 psi) and cellulose conversion (60.8%). H_2O_2 pretreatment resulted in significantly lower delignification (maximum of 29.51% for 2%, 30 min, 121°C/15 psi) and cellulose conversion (49.8%) than NaOH pretreatment but had a higher cellulose conversion than H_2SO_4 pretreatment. O_3 did not cause any significant changes in lignin, xylan, or glucan contents over time.

Saha et al. (2005) studied the dilute acid pretreatment, enzymatic saccharification, and fermentation of wheat straw in a paper with 635 citations. They observed that the maximum yield of monomeric sugars from wheat straw (7.83%, w/v, dry solid-DS) by dilute H_2SO_4 (0.75%, v/v) pretreatment and enzymatic saccharification (45°C, pH 5.0, 72 h) using cellulase, β-glucosidase, xylanase, and esterase was 565 mg/g. Under this condition, no measurable quantities of furfural and hydroxymethylfurfural were produced. The yield of ethanol (per liter) from acid-pretreated enzyme-saccharified wheat straw (78.3 g) hydrolysate by recombinant *Escherichia coli* strain FBR5 was 19 g with a yield of 0.24 g/g DS. Detoxification of the acid and enzyme-treated wheat straw hydrolysate by overliming reduced the fermentation time from 118 to 39 h in the case of separate hydrolysis and fermentation (SHF) (35°C, pH 6.5), and increased the ethanol yield from 13 to 17 g/L and decreased the fermentation time from 136 to 112 h in the case of simultaneous saccharification and fermentation (SSF) (35°C, pH 6.0).

Lloyd and Wyman (2005) evaluated the combined sugar yields for dilute H_2SO_4 pretreatment of corn stover followed by enzymatic hydrolysis of the remaining solids in a paper with 616 citations. They observed that up to 15% of the total potential sugar in the substrate could be released as glucose during pretreatment and between 15% and 90+% of the xylose remaining in the solid residue could be recovered in subsequent enzymatic hydrolysis, depending on the enzyme loading. Glucose yields increased from as high as 56% of total maximum potential glucose plus xylose for just enzymatic digestion to 60% when glucose released in pretreatment was included. Xylose yields similarly increased from as high as 34% of total potential sugars for pretreatment alone to between 35% and 37% when credit was taken for xylose released in digestion. Yields were much lower if no acid was used. Overall, up to about 92.5% of the total sugars originally available in the corn stover used could be recovered for coupled dilute acid pretreatment and enzymatic hydrolysis. These results also suggest that enhanced hemicellulase activity could further improve xylose yields, particularly for low cellulase loadings.

Yang and Wyman (2004) studied the effect of xylan and lignin removal by batch and flow-through pretreatment using water and H_2SO_4 on the enzymatic digestibility of corn stover cellulose in a paper with 556 citations. They observed that increasing flow with just water enhanced the xylan dissolution rate, more than doubled total lignin removal, and increased cellulose digestibility. Furthermore, adding dilute H_2SO_4 increased the rate of xylan removal for both batch and flow-through systems. Further, adding acid also increased the lignin removal rate with flow, but less lignin was left in solution when acid was added in batch. Although the enzymatic hydrolysis of pretreated cellulose was related to xylan removal, the digestibility was much better for flow-through compared with batch systems, for the same degree of xylan removal. Cellulose digestibility for flow-through reactors was related to lignin removal as well. Altering lignin also affected the enzymatic digestibility of corn stover.

Donohoe et al. (2008) visualized lignin coalescence and migration through maize cell walls following acid pretreatment in a paper with 545 citations. They observed droplet morphologies that appeared on and within cell walls of pretreated biomass, as well as the specific ultrastructural regions that accumulate the droplets. These droplets contained lignin. They observed that acid pretreatments reaching temperatures above the range for lignin phase transition caused lignins to coalesce into larger molten bodies that migrated within and out of the cell wall, and could redeposit on the surface of plant cell walls. They asserted that this decompartmentalization and relocalization of lignins was likely to be at least as important as lignin removal in the quest to improve the digestibility of biomass for sugars and ethanol production.

Onda et al. (2008) studied the selective hydrolysis of cellulose into glucose over solid acid catalysts in a paper with 509 citations. They showed solid acid catalysis for the hydrolysis of cellulose with β-1,4-glycosidic bonds into glucose selectively higher than 90 C-%. Among the solid acid catalysts tested, such as the H-form zeolite catalysts and the sulfated and sulfonated catalysts, they observed that a sulfonated activated-carbon catalyst showed a remarkably high yield of glucose, which was due to the high hydrothermal stability and the excellent catalytic property attributed to the strong acid sites of SO_3H functional groups and the hydrophobic planes.

Rinaldi et al. (2008) studied the depolymerization of cellulose using solid acid catalysts in ILs in a paper with 491 citations. They observed that solid acids acted as powerful catalysts for the hydrolysis of cellulose dissolved in an IL. Cellulose underwent selective depolymerization, yielding cellulose oligomers (cellooligomers) without any substantial formation of side products. They then hydrolyzed wood using this methodology.

Hsu et al. (2010) studied the impact of dilute acid pretreatment of rice straw on structural properties and enzymatic hydrolysis in a paper with 477 citations. They obtained a maximal sugar yield of 83% when the rice straw was pretreated with 1% (w/w) H_2SO_4 with a reaction time of 1–5 min at 160°C or 180°C, followed by enzymatic hydrolysis. The complete release of sugar (xylose and glucose) increased the pore volume of the pretreated solid residues resulted in an efficiency of 70% for the enzymatic hydrolysis. The extra pore volume was generated by the release of acid-soluble lignin, and this resulted in the enzymatic hydrolysis being enhanced by nearly 10%. The increase in the crystallinity index of the pretreated rice straw was limited (Table 10.1).

10.3.2 The Ionic Liquid Pretreatment of the Biomass

Kilpelainen et al. (2007) studied the dissolution of wood in imidazolium-based ILs under gentle conditions in a paper with 817 citations. A variety of ILs can only partially dissolve wood chips, whereas ILs such as 1-butyl-3-methylimidazolium chloride and 1-allyl-3-methylimidazolium chloride have good solvating power for Norway spruce sawdust and Norway spruce and Southern pine thermomechanical pulp fibers. Despite the fact that the obtained solutions were not fully clear, these ILs provided solutions which permitted the complete acetylation of the wood. Alternatively, transparent amber solutions of wood could be obtained when the dissolution of the same lignocellulosic samples was attempted in 1-benzyl-3-methylimidazolium chloride. This realization was based on a designed augmented interaction of the aromatic character of the cation of the IL with the lignin in the wood. After dissolution, wood can be regenerated as an amorphous mixture of its original components. The cellulose of the regenerated wood can be efficiently digested to glucose by a cellulase enzymatic hydrolysis treatment.

Sun et al. (2009) studied the complete dissolution and partial delignification of wood (pine and oak) in the IL 1-ethyl-3-methylimidazolium acetate in a paper with 807 times. They found that both softwood (southern yellow pine) and hardwood (red oak) could be completely dissolved in the IL 1-ethyl-3-methylimidazolium acetate ([C_2mim]OAc) after mild grinding. They showed that [C_2mim]OAc] was a better solvent for wood than 1-butyl-3-methylimidazolium chloride ([C_4mim] Cl) and that variables such as type of wood, initial wood load, particle size, etc., affected dissolution and dissolution rates; for example, red oak dissolved better and faster than southern yellow

TABLE 10.1
The Acid Pretreatment of the Biomass

No.	Papers	Biomass	Prt.	Other Prts.	Parameters	Keywords	Lead Author	Affil.	Cits
1	Roman and Winter (2004)	Cellulose	Acids	Na	Acid concentration, acid-to-cellulose ratio, sulfate group content	Hydrolysis, cellulose, degradation	Winter, William T. 7102526472	McGill Univ. Canada	966
2	Larsson et al. (1999)	Softwood	Acids	Enzymes	Sugar yield, fermentability, ethanol yield, hydrolysis severity	Acid, hydrolysis, fermentation, softwood	Hahn-Hagerdal, Barbel* 7005389381	Lund Univ. Sweden	863
3	Li et al. (2010)	Switchgrass, cellulose	Acids, ILs	Enzymes	Delignification, saccharification efficiency, and saccharide yields	Acids, ILs, pretreatment, switchgrass, biomass, delignification, saccharification, recalcitrance	Singh, Seema* 35264950300	Lawrence Livermore Natl. Lab. USA	832
4	Silverstein et al. (2007)	Cotton stalks	Acids, alkali, ozone, hydrogen peroxide	Enzymes	Pretreatment type, concentration treatment temperature, time	Chemical pretreatment, saccharification, cotton stalks	Sharma-Shivappa, Ratna R.* 16231216900	NC State Univ. USA	652
5	Saha et al. (2005)	Wheat straw	Acids	Enzymes	Dissolution severity, sugar yield, ethanol yield	Acids, pretreatment, saccharification, ethanol, straw, fermentation	Saha, Badal C. 7202946302	USDA Agr. Res. Serv. USA	645
6	Lloyd and Wyman (2005)	Corn stover	Acids	Na	Sugar yield, acid concentration	Corn stover, acids, sugar, pretreatment, hydrolysis	Wyman, Charles E. 7004396809	Univ. Calif. Riverside USA	616
7	Yang and Wyman (2004)	Corn stover, cellulose	Acids, water	Enzymes	Dissolution severity, pretreatment type, system type	Pretreatment, digestibility, lignin, corn stover, cellulose, flow-through	Wyman, Charles E. 7004396809	Univ. Calif. Riverside USA	556
8	Donohoe et al. (2008)	Maize stems	Acids	Na	Dissolution mechanisms and morphology	Lignin, maize, thermochemical pretreatment	Donohoe, Bryon S. 36688509400	Natl. Renew. Ener. Lab. USA	545
9	Onda et al. (2008)	Cellulose	Acids	Na	Acid-type, dissolution severity, sugar yield	Cellulose, hydrolysis, glucose	Onda, Ayumu 56689677300	Kochi Univ. Japan	509
10	Rinaldi et al. (2008)	Cellulose	ILs, Acids	Na	Dissolution severity	Depolymerization, cellulose, ILs	Rinaldi, Roberto 7101680610	Imperial coll. UK	491
11	Hsu et al. (2010)	Rice straw	acids	Enzyme	Dissolution severity and sugar yield	Straw, acids, pretreatment, hydrolysis	Guo, Gia Luen 7402768046	Inst. Nucl. Ener. Res. Taiwan	477

Prt, chemical pretreatments; Other prts., other pretreatments; Na, nonavailable; Cits., number of citations received for each paper; *, female.

pine. Carbohydrate-free lignin and cellulose-rich materials could be obtained by using the proper reconstitution solvents (e.g., acetone/water 1:1 v/v), and approximately 26.1% and 34.9% reductions of lignin content in the reconstituted cellulose-rich materials (from pine and oak, respectively) were achieved in one dissolution/reconstitution cycle. For pine, 59% of the holocellulose (i.e., the sum of cellulose and hemicellulose) in the original wood could be recovered in the cellulose-rich reconstituted material, whereas 31% and 38% of the original lignin were recovered, respectively, as carbohydrate-free lignin and as carbohydrate-bonded lignin in the cellulose-rich material.

Lee et al. (2009) studied the IL-mediated selective extraction of lignin from wood leading to enhanced enzymatic cellulose hydrolysis in a paper with 795 citations. They used the IL, 1-ethyl-3-methylimidazolium acetate ([C_2mim]OAc) as a pretreatment solvent to extract lignin from wood flour. They observed that the cellulose in the pretreated wood flour became far less crystalline without undergoing solubilization. When 40% of the lignin was removed, the cellulose crystallinity index dropped below 45, resulting in larger than 90% of the cellulose in wood flour to be hydrolyzed by *Trichoderma viride* cellulase. [C_2mim]OAc was easily reused, thereby resulting in a highly concentrated solution of chemically unmodified lignin.

Fukaya et al. (2008) studied the cellulose dissolution with polar ILs under mild conditions in a paper with 658 citations. They obtained a series of alkylimidazolium salts containing dimethyl phosphate, methyl methylphosphonate, or methyl phosphonate prepared by a facile, one-pot procedure as room-temperature ILs, which has the potential to solubilize cellulose under mild conditions. They observed that especially N-ethyl-N′-methylimidazolium methylphosphonate enabled the preparation of 10 wt% cellulose solution by keeping it at 45°C for 30 min with stirring and rendered soluble 2–4 wt% cellulose without pretreatments and heating.

Remsing et al. (2006) studied the mechanism of cellulose dissolution in the IL 1-n-butyl-3-methylimidazolium chloride ([C_4mim]Cl) in a paper with 645 citations. ^{13}C and $^{35/37}Cl$ NMR relaxation measurements on several model systems demonstrated that the solvation of cellulose by the [C_4mim]Cl involved hydrogen bonding between the carbohydrate hydroxyl protons and the IL chloride ions in a 1:1 stoichiometry.

Kosan et al. (2008) studied the dissolution and formation of cellulose with a number of ILs and used the 1-N-butyl-3-methylimidazolium chloride, the 1-ethyl-3-methylimidazolium chloride, the 1-N-butyl-2,3-dimethylimidazolium chloride, the 1-N-butyl-3-methylimidazolium acetate, and the 1-ethyl-3-methylimidazolium acetate.

Dadi et al. (2006) studied the enhancement of cellulose saccharification kinetics using an IL pretreatment in a paper with 488 citations. They disrupted the cellulose structure using the IL, 1-n-butyl-3-methylimidazolium chloride, in a cellulose regeneration strategy which accelerated the subsequent hydrolysis reaction. The initial enzymatic hydrolysis rates were approximately 50-fold higher for regenerated cellulose as compared to untreated cellulose (Avicel PH-101) as measured by a soluble reducing sugar assay (Table 10.2).

10.3.3 The Other Chemical Pretreatments of the Biomass

Eriksson et al. (2002) studied the mechanism of surfactant effect in enzymatic hydrolysis of steam-pretreated spruce in a paper with 739 citations. They found that nonionic surfactants were most effective. Studies of adsorption of the dominating cellulase of *Trichoderma reesei*, Cel7A (CBHI), during hydrolysis showed that the anionic and nonionic surfactants reduced enzyme adsorption to the lignocellulose substrate. The approximate reduction of enzyme adsorption was from 90% adsorbed enzyme to 80% with surfactant addition. Surfactants had only a weak effect on cellulase temperature stability. They explained the improved conversion of lignocellulose with surfactant by the reduction of the unproductive enzyme adsorption to the lignin part of the substrate. This was due to hydrophobic interaction of surfactant with lignin on the lignocellulose surface, which released unspecifically bound enzyme.

TABLE 10.2
The IL Pretreatment of the Biomass

No.	Papers	Biomass	Prt.	Other Prts.	Parameters	Keywords	Lead Author	Affil.	Cits
1	Li et al. (2010)	Switchgrass, cellulose	Acids, ILs	Enzymes	Delignification, saccharification efficiency and saccharide yields	Acids, ILs, pretreatment, switchgrass, biomass, delignification, saccharification, recalcitrance	Singh, Seema* 35264950300	Lawrence Livermore Natl. Lab. USA	832
2	Kilpelainen et al. (2007)	Wood	ILs	Enzymes	IL types, dissolution severity, solubility	Wood, ILs, dissolution	Kilpelainen, Ilkka 7006830888	Univ. Helsinki Finland	807
3	Sun et al. (2009)	Wood, lignin	ILs	Microwave, ultrasound	IL types, dissolution severity, solubility, wood type	Dissolution, delignification, wood, ILs	Rogers, Robin D. 35474829200	Univ. Alabama USA	807
4	Lee et al. (2009)	Wood, lignin, cellulose	ILs	Enzymes	Dissolution severity	ILs, wood, lignin, cellulose, hydrolysis	Doherty, Thomas V. 57213283673	Rensselaer Polytech. Inst. USA	795
5	Fukaya et al. (2008)	Cellulose	ILs	Na	IL type, dissolution severity, solubility	Cellulose, dissolution, ILs	Ohno, Hiroyuki 7403244652	Tokyo Univ. Japan	658
6	Remsing et al. (2006)	Cellulose	ILs	Na	Dissolution mechanisms	Cellulose, dissolution, ILs,	Rogers, Robin D. 35474829200	Univ. Alabama USA	645
7	Kosan et al. (2008)	Cellulose	ILs	Na	IL type and dissolution severity	Dissolution, ILs, cellulose	Kosan, Birgit* 6602356689	Thuringian Inst. Text. Plast. Res. Germany	511
8	Rinaldi et al. (2008)	Cellulose	ILs, acids	Na	Dissolution severity	Depolymerization, cellulose, ILs	Rinaldi, Roberto 7101680610	Imperial coll. UK	491
9	Dadi et al. (2006)	Cellulose	ILs	Enzyme	Dissolution severity, cellulose regeneration	Cellulose, saccharification, ILs, pretreatment	Schall, Constance A.* 6603671396	Univ. Toledo	488

Prt, chemical pretreatments; Other prts., other pretreatments; Na, nonavailable; *, female.

Cai and Zhang (2005) studied the rapid dissolution of cellulose in LiOH/urea and NaOH/urea aqueous solutions in a paper with 715 citations. They evaluated the dissolution behavior and solubility of cellulose. They found that cellulose having viscosity-average molecular weight ($\bar{M}_n$) of 11.4×10^4 and 37.2×10^4 could be dissolved, respectively, in 7% NaOH/12% urea and 4.2% LiOH/12% urea aqueous solutions precooled to −10°C within 2 min, whereas all of them could not be dissolved in KOH/urea aqueous solution. The dissolution power of the solvent systems was in the order of LiOH/urea >NaOH/urea≫KOH/urea aqueous solution. LiOH/urea and NaOH/urea aqueous solutions as nonderivatizing solvents broke the intra- and intermolecular hydrogen bonding of cellulose and prevented the approach toward each other of the cellulose molecules, leading to the good dispersion of cellulose to form an actual solution.

Sasaki et al. (2000) studied the dissolution and hydrolysis of cellulose in subcritical and supercritical water in a paper with 599 citations. They performed decomposition experiments of microcrystalline cellulose in subcritical and supercritical water (25 MPa, 320°C–400°C, and 0.05–10.0 s). At 400°C, they obtained hydrolysis products mainly, while in 320°C–350°C water, aqueous decomposition products of glucose were the main products. Below 350°C, the cellulose decomposition rate was slower than the glucose and cellobiose decomposition rates, while above 350°C, the cellulose hydrolysis rate drastically increased and became higher than the glucose and cellobiose decomposition rates. Below 280°C, cellulose particles became gradually smaller with increasing reaction time, but at high temperatures (300°C–320°C), cellulose particles disappeared with increasing transparency and much more rapidly than expected from the lower temperature results. They asserted that cellulose hydrolysis at high temperature took place with dissolution in water. This was probably because of the cleavage of intra- and intermolecular hydrogen linkages in the cellulose crystal. Thus, a homogeneous atmosphere was formed in supercritical water, and this resulted in the drastic increase of the cellulose decomposition rate above 350°C.

Sasaki et al. (1998) studied cellulose hydrolysis in subcritical and supercritical water in a paper with 577 citations. They observed that hydrolysis product yields (around 75%) in supercritical water were much higher than those in subcritical water. At a low temperature region, the glucose or oligomer conversion rate was much faster than the hydrolysis rate of cellulose. Thus, even if the hydrolysis products, such as glucose or oligomers, were formed, their further decomposition rapidly took place and thus high yields of hydrolysis products could not be obtained. However, around the critical point, the hydrolysis rate jumped to more than an order of magnitude higher level and became faster than the glucose or oligomer decomposition rate. This was the reason why they obtained a high yield of hydrolysis products in supercritical water.

Kim et al. (2003) studied the pretreatment of corn stover by aqueous ammonia in a paper with 496 citations. They pretreated corn stover with aqueous ammonia in a flow-through column reactor, a process termed ammonia-recycled percolation (ARP). This method was highly effective in delignifying of the biomass, reducing the lignin content by 70%–85%. Most lignin removal occurred within the first 20 min of the process. The ARP process solubilized 40%–60% of the hemicellulose but left the cellulose intact. The solubilized carbohydrate existed in oligomeric form. Carbohydrate decomposition during the pretreatment was insignificant. Corn stover pretreated for 90 min exhibited enzymatic digestibility of 99% with 60 FPU/g of glucan enzyme loading and 92.5% with 10 FPU/g of glucan. The digestibility of ARP pretreated corn stover was substantially higher than that of α-cellulose. The enzymatic digestibility was related to the removal of lignin and hemicellulose, perhaps due to increased surface area and porosity. The biomass structure was deformed, and its fibers were exposed by the pretreatment. The crystallinity index increased with pretreatment reflecting removal of the amorphous portion of biomass. The crystalline structure of the cellulose in the biomass, however, was not changed by the ARP treatment.

Pan et al. (2005) studied biorefining of softwoods using ethanol organosolv pulping in a paper with 484 citations. They prepared pulps with residual lignin ranging from 6.4% to 27.4% (w/w) from mixed softwoods using a proprietary biorefining technology (the Lignol process) based on aqueous ethanol organosolv extraction. They observed that all pulps were readily hydrolyzed without further delignification. More than 90% of the cellulose in low lignin pulps (≤18.4% residual lignin) was hydrolyzed to glucose in 48 h using an enzyme loading of 20 filter paper units/g cellulose. Cellulose in a high lignin pulp (27.4% residual lignin) was hydrolyzed to >90% conversion within 48 h using 40 filter paper units/g. The pulps performed well in both sequential and simultaneous saccharification and fermentation (SSF) trials indicating an absence of metabolic inhibitors. Lignin extracted during organosolv pulping of softwood was a suitable feedstock for the production of lignin-based adhesives and other products due to its high purity, low molecular weight, and abundance of reactive groups. Further, additional co-products may be derived from the hemicellulose sugars and furfural recovered from the water-soluble stream.

Glenn et al. (1983) studied an extracellular H_2O_2-requiring enzyme preparation involved in lignin biodegradation by the white rot basidiomycete *Phanerochaete chrysosporium* in a paper with 482 citations. They found a H_2O_2-requiring enzyme system in the extracellular medium of ligninolytic cultures of *Phanerochaete chrysosporium*. The enzyme system generated ethylene from 2-keto-4-thiomethyl butyric acid (KTBA) and oxidized a variety of lignin model compounds including the diarylpropane 1-(4′-ethoxy-3′-methoxyphenyl) 1,3-dihydroxy-2-(4″-methoxyphenyl) propane (I), a β-ether dimer 1-(4′-ethoxy-3′-methoxyphenyl)glycerol-β-guaiacyl ether (IV), and an olefin 1-(4′-ethoxy-3′-methoxy-phenyl)1,2-propene (VI). The products were equivalent to the metabolic products previously isolated from intact ligninolytic cultures. In addition, the enzyme system partially degraded 14C-ring labeled lignin (Table 10.3).

10.4 DISCUSSION

10.4.1 Introduction

The crude oil-based gasoline fuels have been widely used in the transportation sector since the 1920s. However, there have been great public concerns over the adverse environmental and human impact of these fuels. Hence, biomass-based bioethanol fuels have increasingly been used in blending gasoline and petrodiesel fuels, in the fuel cells, and in the biochemical production in a biorefinery context.

The research in the field of chemical pretreatments of the biomass such as acid, IL, alkali, and ammonia pretreatments to increase the sugar and bioethanol yield has also intensified in recent years.

However, it is essential to develop efficient incentive structures for the primary stakeholders to enhance the research in this field. Although there have been a number of review papers for this field, there has been no review of the most-cited 25 articles in this field.

Thus, this book chapter presents a review of the most-cited 25 articles in this field. Then, it discusses the key findings of these highly influential papers and comments on the future research priorities in this field.

As a first step for the search of the relevant literature, the keywords were selected using the whole set of papers. The selected keyword list was optimized to obtain a representative sample of papers for the searched research field. This keyword list was provided in the Appendix of Konur (2023) for future replication studies.

As a second step, a sample dataset was used for this study. The first 25 articles with at least 477 citations each were selected for the review study. Key findings from each paper were taken from the abstracts of these papers and were discussed. Additionally, a number of brief conclusions were drawn, and a number of relevant recommendations were made to enhance the future research landscape.

As Table 10.4 shows, there are two primary research fronts for these most-cited 25 papers: acid and IL pretreatments with 37% and 25% of the 25 HCPs, respectively. The other research fronts are alkali, ammonia, organic solvent, hydrogen peroxide, supercritical water, surfactant, and ozone pretreatment with 4%–12% of these HCPs each.

TABLE 10.3
The Other Chemical Pretreatments of the Biomass

No.	Papers	Biomass	Prt.	Other Prts.	Parameters	Keywords	Lead Author	Affil.	Cits
1	Eriksson et al. (2002)	Lignocellulose	Surfactants	Enzymes	Dissolution severity, surfactant-type, enzyme stability	Surfactants, lignocellulose, hydrolysis	Tjerneld, Folke 7006446969	Lund Univ. Sweden	739
2	Cai and Zhang (2005)	Cellulose	Alkali	Na	Alkali-type, dissolution severity, solubility	NaOH, dissolution, cellulose	Zhang, Lina 55917992100	Wuhan Univ. China	715
3	Silverstein et al. (2007)	Cotton stalks	Acids, alkali, ozone, hydrogen peroxide	Enzymes	Pretreatment type, concentration treatment temperature, time	Chemical pretreatment, saccharification, cotton stalks	Sharma-Shivappa, Ratna R. 16231216900	NC State Univ. USA	652
4	Sasaki et al. (2000)	Cellulose	Supercritical water	Na	Dissolution severity, temperature	Dissolution, hydrolysis, cellulose, supercritical water	Arai, Kunio 7403965625	Tohoku Univ. Japan	599
5	Sasaki et al. (1998)	Cellulose	Supercritical water	Na	Dissolution severity, temperature	Hydrolysis, cellulose, supercritical water	Arai, Kunio 7403965625	Tohoku Univ. Japan	577
6	Yang and Wyman (2004)	Corn stover, cellulose	Acids, water	Enzymes	Dissolution severity, pretreatment type, system type	Pretreatment, digestibility, lignin, corn stover, cellulose, flow-through	Wyman, Charles E. 7004396809	Univ. Calif. Riverside USA	556
7	Kim et al. (2003)	Corn stover	Ammonia	Enzymes	Dissolution severity and sugar yield	Corn stover, pretreatment, ammonia	Lee, Yoon Y. 8948274900	Auburn Univ. USA	496
8	Pan et al. (2005)	Softwood	Organic solvents	Enzymes	Dissolution severity and sugar yield	Softwood, organosolv, ethanol	Pan Xuejun 57203296000	Univ. British Columbia Canada	484
9	Glenn et al. (1983)	Lignin	Hydrogen peroxide	Enzyme	Enzyme-type and dissolution severity	H_2O_2 and biodegradation	Gold, MH 7402444296	Oregon Health Sci. Univ. USA	484

Prt, chemical pretreatments; Other prts., other pretreatments; Na, nonavailable.

TABLE 10.4
The Chemical Pretreatment of the Biomass: Thematic Research Fronts

No.	Research Fronts	N Paper % Review	N Paper % Sample	Surplus
1	Acid pretreatments	44.0	36.8	7.2
2	IL pretreatments	36.0	24.7	11.3
3	Supercritical water pretreatments	12.0	1.6	10.4
4	Alkali pretreatments	8.0	18.1	−10.1
5	Hydrogen peroxide pretreatments	8.0	3.3	4.7
6	**Minor Research Fronts**			
	Ammonia pretreatments	4.0	7.1	−3.1
	Organic solvent pretreatments	4.0	4.4	−0.4
	Ozone pretreatments	4.0	1.1	2.9
	Surfactant pretreatments	4.0	1.1	2.9
	Missing research fronts			
	Solvent pretreatments	0.0	3.8	−3.8
7	**Deep Eutectic Solvent Pretreatments**	**0.0**	**3.3**	**−3.3**
	Chemical pretreatments in general	0.0	1.6	−1.6
	Wet oxidation pretreatments	0.0	1.6	−1.6
	Sulfite pretreatments	0.0	1.1	−1.1
15	Supercritical carbon dioxide pretreatments	0.0	1.1	−1.1
	Sample	25	182	

N paper (%) sample: the number of papers in the population sample of 182 papers. N paper (%) review: the number of papers in the sample of 25 most cited papers.

However, IL, supercritical water, and acid pretreatments are over-represented in the reviewed papers compared to the 182 HCPs with 7%–11% surplus each. On the contrary, alkaline, other solvent, ammonia, and deep eutectic solvent pretreatments are under-represented by 3%–10% each. Furthermore, a number of minor research fronts are not represented at all in the set of the reviewed papers: other solvents, deep eutectic solvent, chemical pretreatments in general, wet oxidation, sulfite, and supercritical carbon dioxide pretreatments.

Table 10.5 provides information on the biomass used in these 25 HCPs. There are three primary types of the biomass: biomass constituents, wood, and agricultural residues with 72%, 20%, and 28% of the 25 HCPs, respectively. Further, cellulose and lignin are the primary constituents with 56% and 12% of the 25 HCPs, respectively. Further, corn stover is the key agricultural residues with 12% of the 25 HCPs.

It is notable that the papers related to biomass constituents, cellulose, and wood are over-represented by 26%, 26%, and 10%, respectively. On the contrary, the papers related to the biomass in general, lignocellulosic biomass, and biomass are under-represented by 19%, 10%, and 9% of the 25 HCPs, respectively, as there are no papers for these research fronts among the 25 HCPs. The papers related to the agricultural residues are also under-represented by 5% among the 25 HCPs.

10.4.2 The Acid Pretreatment of the Biomass

Roman and Winter (2004) studied the impact of sulfate groups from H_2SO_4 hydrolysis on the thermal degradation behavior of bacterial cellulose and recommended that for high thermostability in the crystals, low acid concentrations, small acid-to-cellulose ratios, and short hydrolysis times should be used. On the other hand, Larsson et al. (1999) studied the impact of the severity of dilute

TABLE 10.5
The Most Prolific Research Fronts for the Biomass for the Chemical Pretreatments of the Biomass

No.	Research Fronts	N Paper % Review	N Paper % Sample	Surplus
1	**Biomass Constituents**	**72.0**	**46.2**	**25.8**
	Cellulose	56.0	29.7	26.3
	Lignin	12.0	7.7	4.3
	Lignocellulose	4.0	3.3	0.7
	Hemicellulose	0.0	4.4	–4.4
	Cellobiose	0.0	1.1	–1.1
2	Wood	20.0	9.9	10.1
3	Grass	4.0	7.7	–3.7
4	**Agricultural Residues**	**28.0**	**33.0**	**–5.0**
	Corn stover	12.0	10.4	1.6
	Other residues	8.0	6.0	2
	Rice straw	4.0	6.6	–2.6
	Wheat straw	4.0	5.5	–1.5
	Sugarcane bagasse	0.0	4.4	–4.4
5	**Biomass in General**	**0.0**	**19.2**	**–19.2**
	Biomass	0.0	8.8	–8.8
	Lignocellulosic biomass	0.0	10.4	–10.4
6	Other biomass	0.0	1.1	–1.1
	Sample	25	182	

N paper (%) sample: the number of papers in the population sample of 182 papers. N paper (%) review: the number of papers in the sample of 25 most cited papers.

H_2SO_4 hydrolysis of spruce on sugar yield and on the ethanol yield and volumetric productivity of the hydrolysate by *Saccharomyces cerevisiae* and asserted that the inhibition observed in hydrolysates produced in higher hydrolysis severity could not be fully explained by the effect of the by-products such as furfural, 5-HMF, acetic acid, formic acid, and levulinic acid.

Li et al. (2010) compared the efficiency of the dilute acid and IL pretreatment of switchgrass in terms of the delignification, saccharification efficiency, and saccharide yields, and asserted that IL pretreatment may offer unique advantages compared to the dilute acid pretreatment process for switchgrass. On the other hand, Silverstein et al. (2007) compared H_2SO_4, NaOH, H_2O_2, and O_3 pretreatments for improving saccharification of cotton stalks and found that these pretreatments had different impact on the delignification of the biomass and its saccharification.

Saha et al. (2005) studied the dilute acid pretreatment, enzymatic saccharification, and fermentation of wheat straw to ethanol and observed that this pretreatment improved the sugar and bioethanol yield of the biomass. On the other hand, Lloyd and Wyman (2005) evaluated the combined sugar yields for dilute H_2SO_4 pretreatment of corn stover followed by enzymatic hydrolysis of the remaining solids and observed that this pretreatment improved the sugar yield (92.5% of the total sugars) of the biomass.

Yang and Wyman (2004) studied the effect of xylan and lignin removal by batch and flow-through pretreatment using water and H_2SO_4 and on the enzymatic digestibility of corn stover cellulose and found that these pretreatments improved the sugar yield of the biomass. On the other hand, Donohoe et al. (2008) visualized lignin coalescence and migration through maize cell walls following acid pretreatment and asserted that the decompartmentalization and relocalization of lignins was likely to be at least as important as lignin removal in the quest to improve the digestibility of biomass for sugars and ethanol production.

Onda et al. (2008) studied the selective hydrolysis of cellulose into glucose over solid acid catalysts and observed that a sulfonated activated-carbon catalyst showed a remarkably high yield of glucose. On the other hand, Rinaldi et al. (2008) studied the depolymerization of cellulose using solid catalysts in ILs and observed that solid acids acted as powerful catalysts for the hydrolysis of cellulose dissolved in an IL. Further, Hsu et al. (2010) studied the impact of dilute acid pretreatment of rice straw on structural properties and obtained a maximal sugar yield of 83% when the rice straw was pretreated with 1% (w/w) H_2SO_4 with a reaction time of 1–5 min at 160°C or 180°C, followed by enzymatic hydrolysis.

Overall, these HCPs clearly show that the acids (H_2SO_4 and solid acids) improve the hydrolysis of the biomass and fermentation of the resulting hydrolysates with an improved sugar and bioethanol yield. For this purpose, hydrolysis and fermentation conditions were optimized.

10.4.3 The IL Pretreatment of the Biomass

Kilpelainen et al. (2007) studied the dissolution of wood in imidazolium-based ILs under gentle conditions and found that ILs such as 1-butyl-3-methylimidazolium chloride and 1-allyl-3-methylimidazolium chloride have good solvating power for Norway spruce sawdust and Norway spruce and Southern pine thermomechanical pulp fibers. On the other hand, Sun et al. (2009) studied the complete dissolution and partial delignification of wood in the IL 1-ethyl-3-methylimidazolium acetate ([C_2mim]OAc) and found that [C_2mim]OAc was a better solvent for wood than 1-butyl-3-methylimidazolium chloride ([C_4mim]Cl).

Lee et al. (2009) studied the IL ([C_2mim]OAc)-mediated selective extraction of lignin from wood leading to enhanced enzymatic cellulose hydrolysis and found that this IL was much efficient for the lignin extraction and cellulose hydrolysis. On the other hand, Fukaya et al. (2008) studied the cellulose dissolution with polar ILs under mild conditions and found that N-ethyl-N′-methylimidazolium methylphosphonate enabled the preparation of 10 wt% cellulose solution by keeping it at 45°C for 30 min with stirring and rendered soluble 2–4 wt% cellulose without pretreatments and heating.

Remsing et al. (2006) studied the mechanism of cellulose dissolution in the IL 1-n-butyl-3-methylimidazolium chloride ([C_4mim]Cl) and observed that the solvation of cellulose by the [C_4mim]Cl involved hydrogen bonding between the carbohydrate hydroxyl protons and the IL chloride ions in a 1:1 stoichiometry. On the other hand, Kosan et al. (2008) studied the dissolution and forming of cellulose with ILs in a paper with 511 citations. They arranged the preparation of the dopes starting from slurry of cellulose in the aqueous IL by removing the water at elevated temperature, vacuum, and high shearing rates. As ILs, they sued the 1-N-butyl-3-methylimidazolium chloride, the 1-ethyl-3-methylimidazolium chloride, the 1-N-butyl-2,3-dimethylimidazolium chloride, the 1-N-butyl-3-methylimidazolium acetate, and the 1-ethyl-3-methylimidazolium acetate. In addition, they compared these results with cellulose solutions in N-methyl-morpholine-N-oxide monohydrate.

Rinaldi et al. (2008) studied the depolymerization of cellulose using solid acid catalysts in ILs and observed that solid acids acted as powerful catalysts for the hydrolysis of cellulose dissolved in an IL. Further, Dadi et al. (2006) studied the enhancement of cellulose saccharification kinetics using an IL pretreatment step and found that the initial enzymatic hydrolysis rates were approximately 50-fold higher for regenerated cellulose as compared to untreated cellulose (Avicel PH-101). On the other hand, Li et al. (2010) compared the efficiency of the dilute acid and IL pretreatment of switchgrass in terms of the delignification, saccharification efficiency, and saccharide yields and asserted that IL pretreatment may offer unique advantages compared to the dilute acid pretreatment process for switchgrass.

Overall, these HCPs clearly show that ILs improve the hydrolysis of the biomass and fermentation of the resulting hydrolysates with an improved sugar and bioethanol yield. For this purpose, hydrolysis and fermentation conditions were optimized.

10.4.4 The Other Chemical Pretreatments of the Biomass

Cai and Zhang (2005) studied the rapid dissolution of cellulose in LiOH/urea and NaOH/urea aqueous solutions and found that the dissolution power of the solvent systems was in the order of LiOH/urea >NaOH/urea ≫ KOH/urea aqueous solution. On the other hand, Silverstein et al. (2007) compared H_2SO_4, NaOH, H_2O_2, and O_3 pretreatments for improving saccharification of cotton stalks and found that these pretreatments had different impact on the delignification of the biomass and its saccharification.

Sasaki et al. (2000) studied the dissolution and hydrolysis of cellulose in subcritical and supercritical water and asserted that cellulose hydrolysis at high temperature took place with dissolution in water. On the other hand, Sasaki et al. (1998) studied cellulose hydrolysis in subcritical and supercritical water and found that hydrolysis product yields in supercritical water were much higher than those in subcritical water. Further, Yang and Wyman (2004) studied the effect of xylan and lignin removal by batch and flow-through pretreatment using water and H_2SO_4 and on the enzymatic digestibility of corn stover cellulose and found that these pretreatments improved the sugar yield of the biomass.

Kim et al. (2003) studied the pretreatment of corn stover by aqueous ammonia and found that this pretreatment was highly effective in delignifying of the biomass. On the other hand, Pan et al. (2005) studied biorefining of softwoods using ethanol organosolv pulping and found that all pulps were readily hydrolyzed without further delignification.

Glenn et al. (1983) studied an extracellular H_2O_2-requiring enzyme preparation and found a hydrogen peroxide-requiring enzyme system in the extracellular medium of ligninolytic cultures of *Phanerochaete chrysosporium*. On the other hand, Eriksson et al. (2002) studied the mechanism of surfactant effect in enzymatic hydrolysis of spruce in a paper and found that nonionic surfactants were most effective.

Overall, these HCPs clearly show that these other chemical pretreatments improve the hydrolysis of the biomass and fermentation of the resulting hydrolysates with an improved sugar and bioethanol yield. For this purpose, hydrolysis and fermentation conditions were optimized.

10.5 CONCLUSION AND FUTURE RESEARCH

The brief information about the key research fronts covered by the 25 most-cited papers with at least 477 citations each is given under four headings: acid, IL, and other chemical pretreatments.

The usual characteristics of these highly cited papers are that these chemical pretreatments are effective in disrupting the biomass constituents such as cellulose, lignin, and hemicellulose of these feedstocks.

The key findings on these research fronts should be read in the light of the increasing public concerns about climate change, GHG emissions, and global warming as these concerns have been certainly behind the boom in the research on the bioethanol production as an alternative to crude oil-based gasoline and diesel fuels in the last decades.

As Table 10.4 shows, there are two primary research fronts for these most-cited 25 papers: acid and IL pretreatments. The other research fronts are alkaline, ammonia, organic solvent, hydrogen peroxide, supercritical water, surfactant, and ozone pretreatments. Furthermore, a number of minor research fronts are not represented at all in the set of the reviewed papers.

Table 10.5 provides information on the biomass used in these 25 HCPs. There are three primary types of the biomass: biomass constituents, wood, and agricultural residues. Further, cellulose and lignin are the primary constituents while corn stover is the key agricultural residue.

These studies emphasize the importance of proper incentive structures for the efficient development and application of the chemical pretreatments to enhance sugar and bioethanol yield of the biomass after the hydrolysis of the biomass and the following fermentation of the resulting hydrolysates in the light of North's institutional framework (North, 1991). In this context, the major producers and users of bioethanol fuels such as the USA, China, Japan, Europe, and Brazil had developed strong incentive structures for the effective development and application of chemical pretreatments for bioethanol and sugar production.

It is recommended that such review studies are performed for the primary research fronts of acid, IL, and other chemical pretreatments as well as the biomass constituents, wood, grass, and agricultural residues. Further, such studies should also be carried out for the other pretreatments of the biomass.

ACKNOWLEDGMENTS

The contribution of the highly cited researchers in the field of the chemical pretreatment of the biomass has been gratefully acknowledged.

REFERENCES

Alvira, P., E. Tomas-Pejo, M. Ballesteros and M. J. Negro. 2010. Pretreatment technologies for an efficient bioethanol production process based on enzymatic hydrolysis: A review. *Bioresource Technology* 101:4851–4861.

Angelici, C., B. M. Weckhuysen and P. C. A. Bruijnincx. 2013. Chemocatalytic conversion of ethanol into butadiene and other bulk chemicals. *ChemSusChem* 6:1595–1614.

Antolini, E. 2007. Catalysts for direct ethanol fuel cells. *Journal of Power Sources* 170:1–12.

Antolini, E. 2009. Palladium in fuel cell catalysis. *Energy and Environmental Science* 2:915–931.

Balan, V., B. Bals, S. P. Chundawat, D. Marshall and B. E. Dale. 2009. Lignocellulosic biomass pretreatment using AFEX. *Methods in Molecular Biology* 581:61–77.

Brandt, A., J. Grasvik, J. P. Hallett and T. Welton. 2013. Deconstruction of lignocellulosic biomass with ionic liquids. *Green Chemistry* 15:550–583.

Burnham, J. F. 2006. Scopus database: A review. *Biomedical Digital Libraries* 3:1–8.

Cai, J. and L. Zhang. 2005. Rapid dissolution of cellulose in LiOH/urea and NaOH/urea aqueous solutions. *Macromolecular Bioscience* 5:539–548.

Dadi, A. P., S. Varanasi and C. A. Schall. 2006. Enhancement of cellulose saccharification kinetics using an ionic liquid pretreatment step. *Biotechnology and Bioengineering* 95:904–910.

Donohoe, B. S., S. R. Decker, M. P. Tucker, M. E. Himmel and T. B. Vinzant. 2008. Visualizing lignin coalescence and migration through maize cell walls following thermochemical pretreatment. *Biotechnology and Bioengineering* 101:913–925.

Eriksson, T., J. Borjesson and F. Tjerneld. 2002. Mechanism of surfactant effect in enzymatic hydrolysis of lignocellulose. *Enzyme and Microbial Technology* 31:353–364.

Fernando, S., S. Adhikari, C. Chandrapal and M. Murali. 2006. Biorefineries: Current status, challenges, and future direction. *Energy & Fuels* 20:1727–1737.

Fukaya, Y., K. Hayashi, M. Wada and H. Ohno. 2008. Cellulose dissolution with polar ionic liquids under mild conditions: Required factors for anions. *Green Chemistry* 10:44–46.

Garcia-Cubero, M. T., G. Gonzalez-Benito, I. Indacoechea, M. Coca and S. Bolado. 2009. Effect of ozonolysis pretreatment on enzymatic digestibility of wheat and rye straw. *Bioresource Technology* 100:1608–1613.

Glenn, J. K., M. A. Morgan, M. B. Mayfield, M. Kuwahara and M. H. Gold. 1983. An extracellular H_2O_2-requiring enzyme preparation involved in lignin biodegradation by the white rot basidiomycete *Phanerochaete chrysosporium*. *Biochemical and Biophysical Research Communications* 114:1077–1083.

Hahn-Hagerdal, B., M. Galbe, M. F. Gorwa-Grauslund, G. Liden and G. Zacchi. 2006. Bio-ethanol – The fuel of tomorrow from the residues of today. *Trends in Biotechnology* 24:549–556.

Helle, S. S., S. J. B. Duff and D. G. Cooper. 1993. Effect of surfactants on cellulose hydrolysis. *Biotechnology and Bioengineering* 42:611–617.

Hill, J., E. Nelson, D. Tilman, S. Polasky and D. Tiffany. 2006. Environmental, economic, and energetic costs and benefits of biodiesel and ethanol biofuels. *Proceedings of the National Academy of Sciences of the United States of America* 103:11206–11210.

Hill, J., S. Polasky and E. Nelson, et al. 2009. Climate change and health costs of air emissions from biofuels and gasoline. *Proceedings of the National Academy of Sciences of the United States of America* 106:2077–2082.

Hsieh, W. D., R. H. Chen, T. L. Wu and T. H. Lin. 2002. Engine performance and pollutant emission of an SI engine using ethanol-gasoline blended fuels. *Atmospheric Environment* 36:403–410.

Hsu, T. C., G. L. Guo and W. H. Chen. 2010. Effect of dilute acid pretreatment of rice straw on structural properties and enzymatic hydrolysis. *Bioresource Technology* 101:4907–4913.

Huang, H. J., S. Ramaswamy, U. W. Tschirner and B. V. Ramarao. 2008. A review of separation technologies in current and future biorefineries. *Separation and Purification Technology* 62:1–21.

Huang, Y. B. and Y. Fu. 2013. Hydrolysis of cellulose to glucose by solid acid catalysts. *Green Chemistry* 15:1095–1111.

Kilpelainen, I., H. Xie and A. King, et al. 2007. Dissolution of wood in ionic liquids. *Journal of Agricultural and Food Chemistry* 55:9142–9148.

Kim, J. S., Y. Y. Lee and T. H. Kim. 2016. A review on alkaline pretreatment technology for bioconversion of lignocellulosic biomass. *Bioresource Technology* 199:42–48.

Kim, K. H. and J. Hong. 2001. Supercritical CO_2 pretreatment of lignocellulose enhances enzymatic cellulose hydrolysis. *Bioresource Technology* 77:139–144.

Kim, T. H., J. S. Kim, C. Sunwoo and Y. Y., Lee. 2003. Pretreatment of corn stover by aqueous ammonia. *Bioresource Technology* 90:39–47.

Knill, C. J. and J. F. Kennedy. 2002. Degradation of cellulose under alkaline conditions. *Carbohydrate Polymers* 51:281–300.

Konur, O. 2000. Creating enforceable civil rights for disabled students in higher education: An institutional theory perspective. *Disability & Society* 15:1041–1063.

Konur, O. 2002a. Access to nursing education by disabled students: Rights and duties of nursing programs. *Nurse Education Today* 22:364–374.

Konur, O. 2002b. Assessment of disabled students in higher education: Current public policy issues. *Assessment and Evaluation in Higher Education* 27:131–152.

Konur, O. 2002c. Access to employment by disabled people in the UK: Is the Disability Discrimination Act working? *International Journal of Discrimination and the Law* 5:247–279.

Konur, O. 2006a. Participation of children with dyslexia in compulsory education: Current public policy issues. *Dyslexia* 12:51–67.

Konur, O. 2006b. Teaching disabled students in Higher Education. *Teaching in Higher Education* 11:351–363.

Konur, O. 2007a. A judicial outcome analysis of the Disability Discrimination Act: A windfall for the employers? *Disability & Society* 22:187–204.

Konur, O. 2007b. Computer-assisted teaching and assessment of disabled students in higher education: The interface between academic standards and disability rights. *Journal of Computer Assisted Learning* 23:207–219.

Konur, O. 2012. The evaluation of the research on the bioethanol: A scientometric approach. *Energy Education Science and Technology Part A: Energy Science and Research* 28:1051–1064.

Konur, O. 2015. Current state of research on algal bioethanol. In *Marine Bioenergy: Trends and Developments*, Eds. S. K. Kim and C. G. Lee, pp. 217–244. Boca Raton, FL: CRC Press.

Konur, O. 2020a. The scientometric analysis of the research on the bioethanol production from green macroalgae. In *Handbook of Algal Science, Technology and Medicine*, Ed. O. Konur, pp. 385–401. London: Academic Press.

Konur, O. 2023. Chemical pretreatment of the biomass: Scientometric study. In *Bioethanol Fuel Production Processes. I: Biomass Pretreatments. Handbook of Bioethanol Fuels*, Ed. O. Konur. Boca Raton, FL: CRC Press.

Kosan, B., C. Michels and F. Meister. 2008. Dissolution and forming of cellulose with ionic liquids. *Cellulose* 15:59–66.

Larsson, S., E. Palmqvist and B. Hahn-Hagerdal, et al. 1999. The generation of fermentation inhibitors during dilute acid hydrolysis of softwood. *Enzyme and Microbial Technology* 24:151–159.

Lee, S. H., T. V. Doherty, R. J. Linhardt and J. S. Dordick. 2009. Ionic liquid-mediated selective extraction of lignin from wood leading to enhanced enzymatic cellulose hydrolysis. *Biotechnology and Bioengineering* 102:1368–1376.

Li, C., B. Knierim and C. Manisseri, et al. 2010. Comparison of dilute acid and ionic liquid pretreatment of switchgrass: Biomass recalcitrance, delignification and enzymatic saccharification. *Bioresource Technology* 101:4900–4906.

Lin, Y. and S. Tanaka. 2006. Ethanol fermentation from biomass resources: Current state and prospects. *Applied Microbiology and Biotechnology* 69:627–642.

Lloyd, T. A. and C. E. Wyman. 2005. Combined sugar yields for dilute sulfuric acid pretreatment of corn stover followed by enzymatic hydrolysis of the remaining solids. *Bioresource Technology* 96:1967–1977.

Ma, X., L. Sun and C. Song. 2002. A new approach to deep desulfurization of gasoline, diesel fuel and jet fuel by selective adsorption for ultra-clean fuels and for fuel cell applications. *Catalysis Today* 77:107–116.

Morschbacker, A. 2009. Bio-ethanol based ethylene. *Polymer Reviews* 49:79–84.

Najafi, G., B. Ghobadian and T. Tavakoli, et al. 2009. Performance and exhaust emissions of a gasoline engine with ethanol blended gasoline fuels using artificial neural network. *Applied Energy* 86:630–639.

Newman, P. W. G. and J. R. Kenworthy. 1989. Gasoline consumption and cities: A comparison of U.S. cities with a global survey. *Journal of the American Planning Association* 55:24–37.

North, D. C. 1991. Institutions. *Journal of Economic Perspectives* 5:97–112.

Olsson, L. and B. Hahn-Hagerdal. 1996. Fermentation of lignocellulosic hydrolysates for ethanol production. *Enzyme and Microbial Technology* 18:312–331.

Onda, A., T. Ochi and K. Yanagisawa. 2008. Selective hydrolysis of cellulose into glucose over solid acid catalysts. *Green Chemistry* 10:1033–1037.

Pan, X., C. Arato and N. Gilkes, et al. 2005. Biorefining of softwoods using ethanol organosolv pulping: Preliminary evaluation of process streams for manufacture of fuel-grade ethanol and co-products. *Biotechnology and Bioengineering* 90:473–481.

Remsing, R. C., R. P. Swatloski, R. D. Rogers and G. Moyna. 2006. Mechanism of cellulose dissolution in the ionic liquid 1-n-butyl-3- methylimidazolium chloride: A^{13}C and$^{35/37}$Cl NMR relaxation study on model systems. *Chemical Communications* 2006:1271–1273.

Rinaldi, R. and F. Schuth. 2009. Acid hydrolysis of cellulose as the entry point into biorefinery schemes. *ChemSusChem* 2:1096–1107.

Rinaldi, R., R. Palkovits and F. Schuth. 2008. Depolymerization of cellulose using solid catalysts in ionic liquids. *Angewandte Chemie – International Edition* 47:8047–8050.

Roman, M. and W. T. Winter. 2004. Effect of sulfate groups from sulfuric acid hydrolysis on the thermal degradation behavior of bacterial cellulose. *Biomacromolecules* 5:1671–1677.

Saha, B. C., L. B. Iten, M. A. Cotta and Y. V. Wu. 2005. Dilute acid pretreatment, enzymatic saccharification and fermentation of wheat straw to ethanol. *Process Biochemistry* 40:3693–3700.

Sanchez, O. J. and C. A. Cardona. 2008. Trends in biotechnological production of fuel ethanol from different feedstocks. *Bioresource Technology* 99:5270–5295.

Sasaki, M., B. Kabyemela and R. Malaluan, et al. 1998. Cellulose hydrolysis in subcritical and supercritical water. *Journal of Supercritical Fluids* 13:261–268.

Sasaki, M., Z. Fang, Y. Fukushima, T. Adschiri and K. Arai. 2000. Dissolution and hydrolysis of cellulose in subcritical and supercritical water. *Industrial and Engineering Chemistry Research* 39:2883–2890.

Satlewal, A., R. Agrawal, S. Bhagia, J. Sangoro and A. J. Ragauskas. 2018. Natural deep eutectic solvents for lignocellulosic biomass pretreatment: Recent developments, challenges and novel opportunities. *Biotechnology Advances* 36:2032–2050.

Shuai, L. and J. Luterbacher. 2016. Organic solvent effects in biomass conversion reactions. *ChemSusChem* 9:133–155.

Shuai, L., Q. Yang and J. Y. Zhu, et al. 2010. Comparative study of SPORL and dilute-acid pretreatments of spruce for cellulosic ethanol production. *Bioresource Technology* 101:3106–3114.

Silverstein, R. A., Y. Chen, R. R. Sharma-Shivappa, M. D. Boyette and J. Osborne. 2007. A comparison of chemical pretreatment methods for improving saccharification of cotton stalks. *Bioresource Technology* 98:3000–3011.

Sun, N., M. Rahman and Y. Qin, et al. 2009. Complete dissolution and partial delignification of wood in the ionic liquid 1-ethyl-3-methylimidazolium acetate. *Green Chemistry* 11:646–65.

Sun, R. C., J. Tomkinson, P. L. Ma and S. F. Liang. 2000. Comparative study of hemicelluloses from rice straw by alkali and hydrogen peroxide treatments. *Carbohydrate Polymers* 42:111–122.

Sun, Y. and J. Cheng. 2002. Hydrolysis of lignocellulosic materials for ethanol production: A review. *Bioresource Technology* 83:1–11.

Taherzadeh, M. J. and K. Karimi. 2008. Pretreatment of lignocellulosic wastes to improve ethanol and biogas production: A review. *International Journal of Molecular Sciences* 9:1621–1651.

Vidal, P. F. and J. Molinier. 1988. Ozonolysis of lignin – Improvement of *in vitro* digestibility of poplar sawdust. *Biomass* 16:1–17.

Yang, B. and C. E. Wyman. 2004. Effect of xylan and lignin removal by batch and flowthrough pretreatment on the enzymatic digestibility of corn stover cellulose. *Biotechnology and Bioengineering* 86:88–98.

Zhang, C. W., S. Q. Xia, P. S. Ma. 2016. Facile pretreatment of lignocellulosic biomass using deep eutectic solvents. *Bioresource Technology* 219:1–5.

Zhao, X., K. Cheng and D. Liu. 2009. Organosolv pretreatment of lignocellulosic biomass for enzymatic hydrolysis. *Applied Microbiology and Biotechnology* 82:815–827.

Zheng, Y., H. M. Lin and G. T. Tsao. 1998. Pretreatment for cellulose hydrolysis by carbon dioxide explosion. *Biotechnology Progress* 14:890–896.

Zhu, J. Y., X. J. Pan, G. S. Wang and R. Gleisner. 2009. Sulfite pretreatment (SPORL) for robust enzymatic saccharification of spruce and red pine. *Bioresource Technology* 100:2411–2418.

Zhu, S., Y. Wu and Q. Chen, et al. 2006. Dissolution of cellulose with ionic liquids and its application: A minireview. *Green Chemistry* 8:325–327.

11 Advances of Bioethanol Production Using Ionic Liquids

Amal A.M. Elgharbawy and Nor Azrini Nadiha Azmi
International Islamic University Malaysia

Muhammad Moniruzzaman
Universiti Teknologi PETRONAS

ABBREVIATIONS

[AM][Cl]	1-allyl-3-methylimidazolium chloride
[BM]	1-butyl-3-methylimidazolium
[BMPy]	1-butyl-3-methylpyridinium
[Ch][Lys]	cholinium lysinate
[DEP]	diethyl phosphate
[EM][Ac]	1-ethyl-3-methylimidazolium acetate
[HEAM]	bis-2-hydroxyethylammonium
HMF	5-hydroxymethylfurfural
ILs	ionic liquids
[OM]	1-octyl-3-methylimidazolium
OPEFB	oil palm empty fruit bunches
[TEA][HSO_4]	triethylammonium hydrogen sulfate
[TMA][$MeSO_4$]	tris (2-hydroxyethyl) methylammonium-methylsulfate

11.1 INTRODUCTION

The production of bioethanol and biodiesel fuels from lignocellulosic biomass to replace conventional liquid fuels like diesel and petroleum has sparked a lot of interest as they are capable to minimize the emissions of greenhouse gases by utilizing a renewable feedstock. Furthermore, interest in the synthesis of green materials from lignocellulosic biomass conversion is rapidly increasing (Halder et al., 2019). The growing worldwide demand for fossil fuels, as well as their widespread usage, which has resulted in environmental damage, has been a major source of worry. More precisely, between 2007 and 2035, global energy consumption is expected to rise by 49% owing to the population increment, societal pressure, and economic expansion. Alternative energy supplies need to be developed to replace fossil fuels in order to provide energy security and environmental sustainability. Bioethanol is the most intriguing biofuel since it may be used as a fuel replacement in compatible automobile engines or blended up to 30% with gasoline without requiring any alterations to the car engine (Cheah et al., 2020). Biomass consists of three main components: cellulose (40%–80%), hemicellulose (15%–30%), and lignin (10%–25%) of the total mass (Singh et al., 2017). Cellulose, which accounts for nearly all of the total mass of biomass, is the most prevalent organic polymer that plants use to form their cell walls. It is also the most common substrate for fermentation. The polymer is made up of D-glucose connected by -1,4-glycosidic linkages. Hemicellulose is a branching heteropolymer composed of pentoses (such as L-arabinose and D-xylose), hexoses (such as D-mannose, D-galactose, and D-glucose), and sugar acids (like D-glucuronic acid). Because of the diversity of sugars, a wide range of enzymes are necessary to hydrolyze these polymers

DOI: 10.1201/9781003226536-15

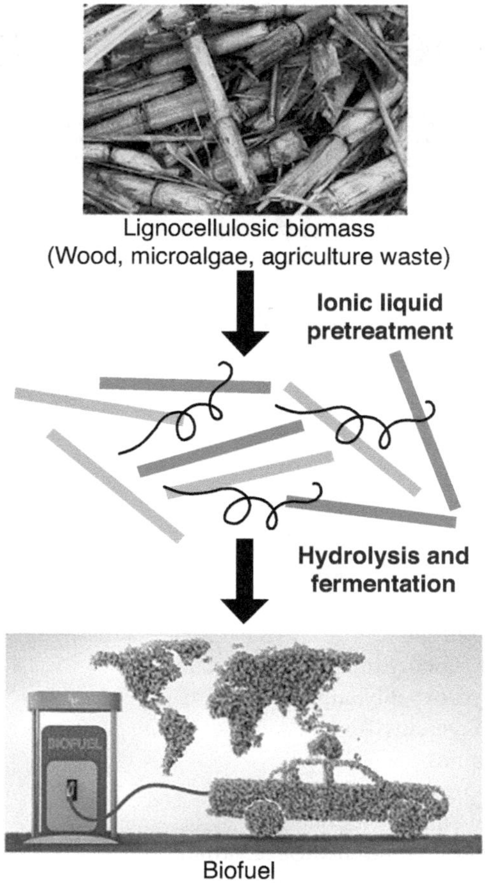

FIGURE 11.1 Typical scheme of ionic liquids (ILs) treatment of biomass.

into monomers (Limayem and Ricke, 2012) that are able to undergo fermentation. Regardless, the mentioned merits make the biomass one of the sustainable resources to replace feedstocks used in bioethanol production, such as corn.

There are many methods adapted in the production of bioethanol fuels from biomass, including chemical pretreatment (acid or alkaline treatment) and physical pretreatment (such as high pressure and temperature). However, those methods require intensive work and cost and cause more damage to the environment (Paul and Chakraborty, 2018). One of the methods that has developed throughout the years is the employment of ionic liquids (ILs) in the processing of biomass. Ionic liquid (IL) used in the pretreatment of lignocellulose has recently received a lot of interest. These compounds have been widely used because of their increased thermal stabilities, reduced vapor pressures, and environmental-friendly features, which limit the production of harmful gases. The common lignocellulosic biomass pretreatment by ILs for the generation of biofuel is depicted in Figure 11.1. ILs often used for lignocellulose pretreatment include imidazolium-based ILs and pyridinium-based ILs (Cheah et al., 2020).

11.2 BIOMASS IN BIOETHANOL PRODUCTION

Biomass is made up of agricultural, forestry, and industrial waste, as well as by-products, deposits, and waste. Biodegradable organic components of urban and industrial solid wastes are also included, as well as liquids and gases generated from organic material breakdown. Biomass is unique in that it may be retained and kept as gases, liquids, or solids, compared to other sources of renewable energy. As a result, it can be used to generate electricity as well as mechanical and heat power. Consequently, if alternate energy sources could be produced by biomass, the usage of fossil fuel and emissions of greenhouse gases might be minimized. In addition, biomass volatility may result in the emergence of a new business that supports reforestation, leading in a more stable climate and a boosted economy (Saga et al., 2008; Wulfhorst et al., 2015). Biomass resources may be separated into their constituent parts, such as carbohydrates, proteins, and lipids, using a variety of technologies. The plant that produces lignocellulose-containing raw components might be an excellent candidate for the concept of biorefinery, in which cellulose and hemicellulose make accessible hydrolysates (fermentable sugars) while lignin generates target chemicals (e.g., pesticides, resins, polymers, and other components) (Busic et al., 2018). Some examples of biomass used for biofuels production include corn stover, barley straw, poplar sawdust, eucalyptus, municipal, forestry, and greening waste, Napier grass, and Deenanath grass (Duque et al., 2021; Kumar and Reetu, 2015).

The raw materials utilized in the production of bioethanol may be categorized into cellulose components, sugars, and starches. Cellulose (waste of pulp and paper mills, wood and agricultural residues) must be hydrolyzed to sugars first, usually with the use of cellulolytic enzymes or acids before it undergoes the fermentation. Sugars (like molasses, sweet sorghum juice, or cane juice) can be fermented directly to ethanol. Starches (root crops, potatoes, cassava, and maize) must undergo a hydrolysis process by enzymes into fermentable sugars (Franks et al., 2006). In this chapter, the focus will be on the lignocellulosic biomass as a feedstock.

11.3 IMPORTANCE OF BIOETHANOL AS A BIOFUEL

The United States and Brazil have been adamant in pushing the indigenous production of bioethanol fuels. The first generation bioethanol is generated from sugar feedstocks like molasses (in India), cane juice (in Brazil), or starch-rich materials like maize (in the US). These raw materials, on the other hand, are interfering with food materials, causing inadequate supply to satisfy the increasing fuel demand, and could have a severe impact on biodiversity. This might even lead to deforestation in order to have more farmland (Kumar and Reetu, 2015).

Bioethanol is commonly regarded as a one-of-a-kind transportation fuel with significant environmental, economic, and strategic benefits. Bioethanol, which may be made from the agricultural plants' biomass, provides potential for small-holder farmers to increase their revenue. Farmers may be able to produce income-generating energy crops while also fulfilling their communal food needs. Ethanol derived from biomass is the only liquid transportation fuel that does not contribute to greenhouse gas emissions. Ethanol represents a closed carbon dioxide cycle since it is used by plants to manufacture cellulose during photosynthesis; therefore, when it is burned, the emitted carbon dioxide is recycled back into the plant material. Ethanol can be combined with gasoline up to a 10% concentration in the United States. This type of gasoline is known as E10, and it does not need any substantial technological changes to current infrastructure or car engines. Higher blends (e.g., 15% ethanol in gasoline) need minor vehicle modifications and a waiver of hydrocarbon emission limitations. By adding ethanol (10%) to gasoline, the fine particulate matter (PM) emissions decrease by 36%, reaching 65% in big displacement vehicles (Ramos et al., 2016).

11.4 IONIC LIQUIDS FOR BIOMASS PROCESSING

Low-melting organic salts known as ionic liquids (ILs) have the possibility to be employed as industrial solvents for pretreatment of lignocellulose. They have a number of advantages over organic solvents, including low vapor pressure and nonflammability (Chambon et al., 2018). It is recognized by now that ILs are molten salts in liquid form under 100°C. Because of their ionic composition, they have a low vapor pressure, making them preferable to traditional solvents (Van Osch et al., 2017). Also known as designer solvents, IL ions may be easily altered (there are 10^{18} theoretical possibilities of anions and cations) to achieve desired qualities. In addition, biocompatibility, solvation, and hydrophobicity are among the important features attributed to ILs for their possible applicability. ILs may be classified into three generations based on the intricacy of these stated attributes (Huang et al., 2019). The first generation ILs can be classified based on the characterization of their physical parameters (melting point, viscosity, and thermostability). The second generation IL groups are classed based on their chemical characteristics (reactivity, chirality, and solvation). Finally, the third generation IL groups possess biological activities (emollient and antimicrobial activities) (Egorova et al., 2017; Elgharbawy et al., 2020).

Furthermore, ILs are classed as either aprotic or protic, with the protic subset readily created by mixing a Bronsted acid with a Bronsted base in a proton transfer single step. Generally, the most thoroughly investigated IL is primarily aprotic ILs (APILs) that dissolve cellulose and require a lot of synthesis steps, hence the expensive production. The ILs known as 1-ethyl-3-methylimidazolium acetate ([EM][Ac]) release remarkable amounts of sugar from nearly any feedstock (Kimon et al., 2011; Olivier-Bourbigou et al., 2010). In recent times, the processing of lignocellulosic biomass employing ILs (aprotic or protic) has been investigated extensively, and this process is known as ion solvolysis. The biomass can be immediately solubilized by pure or neat ILs. In addition, a mixture containing an alternate solvent can be processed, allowing dissolution of chemicals (organic or inorganic). Total raw biomass or its constituents such as hemicellulose and cellulose can be dissolved by ILs. Therefore, the employment of ILs in biomass processing has promoted their usage in many other applications (Radhakrishnan et al., 2021).

11.4.1 Ionic Liquids as Solvents for Enzymatic Biomass Pretreatment and Hydrolysis

ILs often used for lignocellulose pretreatment include pyridinium- and imidazolium-; [EM][Ac], [EM][Cl], and [BM][Cl]. Overall, pyridinium+, imidazolium+, morpholinium+, phosphonium+, or ammonium+-based cations and anions are commonly found in ILs that are efficiently dissolving cellulose. Because of their capacity to create strong hydrogen bonds with hydroxyl groups, these IL groups have been proven to be useful. In the pretreatment of biomass using an IL, lignin and hemicellulose are swiftly eliminated, while efficiently decreasing the crystalline structure of the cellulose (Li et al., 2010). However, it has been noticed that adding water to those ILs results in poor performance most of the times. Efforts, therefore, have been made to further investigate this issue. For instance, the use of bionic liquid cholinium lysinate with 50% water demonstrated a good pretreatment performance with more than 80% glucose yield at 90°C, 6 h of pretreatment, and biomass loadings of 5%. But the group showed the instability and the high cost of the amino acid APIL utilized, limiting its economic uses (Hou et al., 2013).

These disadvantages have fueled research toward the creation of moisture-tolerant, conveniently produced, and low-cost protic ionic liquids (PILs) with comparable performance (Reis et al., 2017). Through successful simultaneous saccharification and fermentation (SSF) of sugarcane bagasse, an efficient technique for cofermenting xylose and glucose in IL was demonstrated. By two engineered xylose-assimilating *Saccharomyces cerevisiae* yeast strains, the fermentation successfully converted pure xylose to ethanol in ILs. ILs, namely, 1-butyl-3-methylpyridinium chloride, provided a greater saccharification efficiency of the substrate, resulting in 2.3 g/L glucose and 0.7 g/L xylose. In the absence of IL, only 26.7% ethanol by the reference strain were produced compared to 84% ethanol with the presence of IL (Amoah et al., 2019). The yield of ethanol from triticale straw increased with the addition of [EM][Ac] from 1.60 g/L without pretreatment to 10.64 g/L after pretreatment with the IL using enzymatic hydrolysis from *Aspergillus niger* and *Trichoderma reesei*. However, regeneration step was implemented in this study (Smuga-Kogut et al., 2019b).

The utilization of low-cost ILs is a potential approach for lignocellulosic biomass processing into bioethanol. For example, Ziaei-Rad et al. (2021) investigated the use of triethylammonium hydrogen sulfate ([TEA][HSO_4,]), a low-cost IL for wheat straw pretreatment. In extracting carbohydrates and lignin from the source, the pretreatment was examined under circumstances defined by 30% water, 130°C, and high solid-to-solvent load ratio of 1:5 g/g. After 48 h of fermentation, the production of ethanol and yield of pretreated biomass (3 h) were 43.1 g/L and 84.34% of the maximum theoretical yield, respectively, was realized. The obtained data proved unequivocally that low-cost IL pretreatment resulted in greater ethanol output due to effective delignification. The procedure established in this study demonstrated the technique's economic feasibility at the industrial scale (Ziaei-Rad et al., 2021). The same IL was used by Nurdin et al. (2021) in the transformation of oil palm empty fruit bunches (OPEFB). Based on the optimization process performed by the group, 120°C was the ideal temperature for the pretreatment. The cellulose content was 45.84 wt% interestingly, the performance of the recovered [TEA][HSO_4] for reuse has been tested, and the findings were promising (Figure 11.2).

Fungal delignification can also be attainable technique for pretreating biomass for bioethanol with some modifications to enhance its efficiency. Laccase was employed in the pretreatment of rice straw in [AM][Cl] and a surfactant (TritonX-100), which resulted in enhanced lignin removal to 18.49% and 31.79%, respectively. After pretreatment with laccase + [AM][Cl], both surface area and cellulose crystallinity were significantly changed (Chang et al., 2021). Another study tried to predict the operating conditions to achieve the highest scarification yield, done by Chuetor et al. (2022). According to the findings, pure methanol is the best antisolvent alternative for maximizing sugar production. Following the projected pretreatment condition, the recyclability of [EM][Ac] combined with methanol was tested up to five times. Even after five recycles, fermentation experiments revealed that recycled [EM][Ac] produced 89% more ethanol than untreated rice straw. Wheat straw was hydrolyzed into reducing sugars via one-pot process using [BMPy][Cl] coupled with nine different metal chlorides. Reducing sugar of 70% was achieved with 67% (w/w) lignin separated from the straw. Among the ILs coupled with ions, 86% of $[BMPy]^+CoCl_3^-$ was recycled (Naz et al., 2020). Those studies indicate the possibility of recycling ILs in ethanol production, consequently lowering industrial production costs (see Figure 11.3). Table 11.1 presents a summary of some recent studies that used ILs in bioethanol production.

Microwave-assisted heating (MAH) is one favorable approach used in IL pretreatment for bioethanol production as it reduces the pretreatment time remarkably. Rice straw was used in MAH-IL process with [Ch][Ac] catalyzed by SiO_2–OSO_3H at high pressure and 340 W for 6.5 min and generated around 30% of sugar after hydrolysis with *Accellerase 1500* (Sakdaronnarong et al., 2017). Another study used [EM][Ac] and [BM][Ac] with MAH on cotton for 12 s, 150°C, and 50 W to generate 15.6 g/L and 17.1 total reducing sugar (Ha et al., 2011). Lin et al. disclosed the greatest saccharification rate of 52.9% with the IL, 4-(3-methyl-1-imidazolio)1-butanesulfonic acid hydrogen

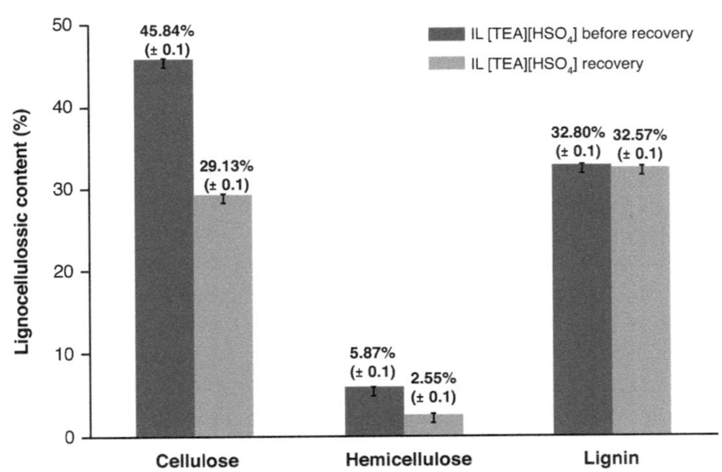

FIGURE 11.2 Comparison of pretreatment results between the neat and recovered [TEA][HSO_4] in the pretreatment of OPEFB. (Adapted from Nurdin et al. (2021). Reproduced with permission from the publisher under a Creative Commons Attribution 4.0 International License.)

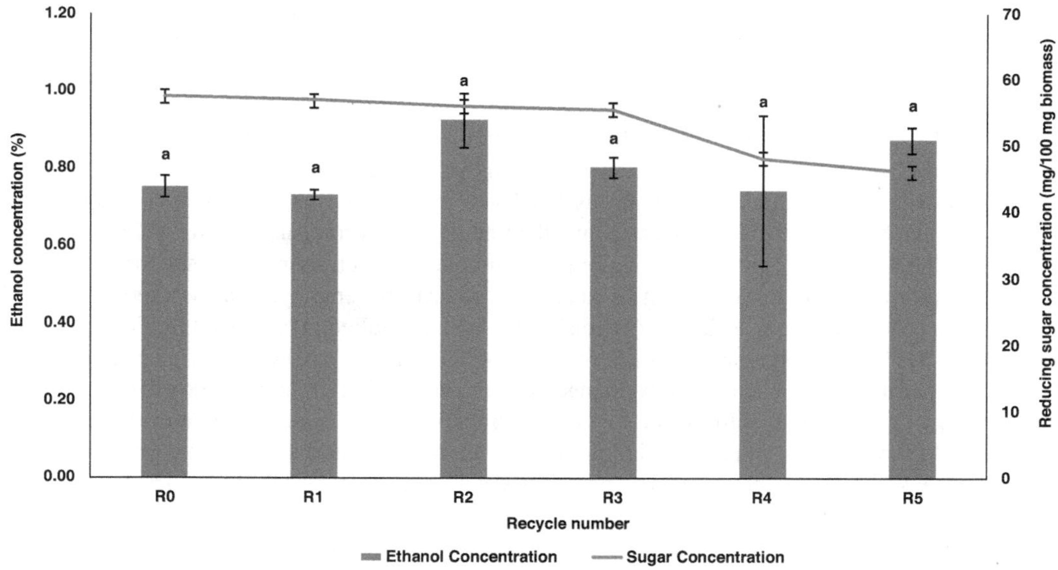

FIGURE 11.3 Ethanol concentration and reducing sugar yield obtained from enzymatic hydrolysis of pretreated rice straw by recycled [EM][Ac] and methanol as an antisolvent. Small alphabets indicate the results of ANOVA analysis, and different alphabet means significantly different ($p<0.05$). (Adapted from Chuetor et al. (2022) and reproduced with permission from the publisher under the terms and conditions of the Creative Commons Attribution (CC BY) license.)

sulfate prepared with microwave heating (Lin et al., 2021). According to the findings, sugar extraction in cellulosic feedstock might be replaced by IL dissolution with MAH pretreatment followed by enzymatic saccharification. Those studies also suggest the use of MAH aids in the reduction of by-products that inhibit bioethanol generation during the fermentation process, such as ferulic acid and 4-hydroxycinnamic acid.

Though many processes are still in the pilot stage, it is possible that the current progress may spark interest in the industrial context and lead to greater research into enhancing scalability for other continuous process biomass pretreatment (Ocreto et al., 2021). Also, a unique study tested eight ILs, and the greatest saccharification yield of 95.5% was found in ethyl ammonium nitrate, while among hydrophobic deep eutectic solvents, menthol:lactic acid (Me:LA) yielded the highest saccharification yield of 85.7% with no requirements of high temperature or other pretreatment of the microalgae *Scenedesmus sp.* for hydrolysis, suggesting a solvent system which could potentially decrease the cost of bioethanol from microalgae (Kulshrestha et al., 2021).

When sugarcane bagasse (SCB) was pretreated with IL [BM][Cl] and a surfactant (PEG-8000), the *sugar yield* increased by 16.5% during enzymatic saccharification. The combined effects of IL, surfactant, and ultrasound on the ultrastructure of biomass were severe and distinctive, resulting in increased saccharification and enzyme accessibility to biomass and higher sugar output (Sharma et al., 2019). Cellulose was ultrasonically processed in IL showing that combining IL and ultrasonication during mild ultrasonic pretreatment increased the susceptibility and accessibility of the amorphous region in cellulose. Meanwhile, the crystalline portion of cellulose remained unaffected, which resulted in higher reactivity of the amorphous region allowing acid-catalyzed hydrolysis to proceed smoothly (Pang et al., 2018). Enzymatic hydrolysis of wheat straw and bagasse pretreated with [BM][Ac]-ultrasound yielded the highest glucose yields at 28 kHz and the highest acid hydrolysis yields at 50 kHz. Moreover, in terms of reducing sugar yields, the combination of [BM][Ac] and ultrasound outperforms [BM][Cl] (Yu et al., 2018).

TABLE 11.1
Summary of Some Recent Studies that used ILs in Bioethanol Production

ILs Used	Biomass	Ethanol Yield	Main Findings	References
[EM][Ac]	Rice straw	5.9 g/L	Conditions: 5% solid loading, 128.4°C, 71.83 min; able to recycle the IL five times	Chuetor et al. (2022)
[TEA][HSO_4]	Wheat straw	43.1 g/L	80% delignification rate and 64.45% hemicellulose removal rate.	Ziaei-Rad et al. (2021)
[EM][Ac]	Rye straw	27.54 g/L	When treated for 2 h, 120°C, 10 mL IL/10 g dry mass rye straw, enzymatic hydrolysis for 72 h, 50°C, and the yield of reducing sugars was the highest.	Smuga-Kogut et al. (2017)
[BM][Cl]	Softwood, hardwood, sugarcane bagasse, rice straw.	21.86–29.56 g/L	The pretreatment with IL increased the cellulose digestibility of hardwood and softwood by 29 and 20 times, respectively, and most of the cellulose in rice straw and sugarcane bagasse was converted into glucose; bioethanol can be efficiently produced using a wide range of materials that have been pretreated with an acidified [BM][Cl] solution.	Trinh et al. (2019)
Choline acetate [Ch][Ac] IL	Paperboard mill sludge (PMS)	11.21 ± 0.54 g/L	Removal efficiency of hemicellulose and lignin of 36.38 ± 4.51, 17.42 ± 1.19, respectively, and cellulose recovery of 82.17 ± 4.28%; pretreatment with 10% (w/w) ratio of [Ch][Ac] IL/PMS for 1 h, at 120°C showed hemicellulose and lignin removal efficiency of 5.05 ± 0.52 and 14.71 ± 1.22%, respectively, with 89.19 ± 5.62% cellulose recovery.	Farghaly et al. (2017)
[EM][Ac]	Guinea grass (*Panicum maximum*)	21.93–23.11 g/L	The chemical defined mean (CDM) and hydrolyzed samples produced ethanol concentration of (23.11 ± 1.08) and (21.93 ± 1.18) g/L, respectively; the enzymatic and acid hydrolysis stages (157°C, 30 min, and biomass concentration [relative to total mass = 14%]) were maximized in these IL pretreatment conditions.	Odorico et al. (2018)
[EM][Ac]	Triticale straw	10.64 g/L	The yield of ethanol from triticale straw increased with employment of IL from 1.60 g/L after processing without pretreatment to 10.64 g/L after pretreatment; enzymatic hydrolysis of triticale straw purified with IL produced the highest concentration of reducing sugars.	Michalczyk et al. (2015)
[EM][Ac]	Spruce and oak sawdust	53.8%–54.5%	After the [EM][Ac]-pretreatment at 45°C, ethanolic yields were 2.6–3.9 higher; eco-friendly pretreatment strategy using IL at low temperature.	Alayoubi et al. (2020)
[BM][Ac]	Mixed softwood	8.7 g/L of ethanol	The optimal pretreatment conditions: 100°C, 15 h; Bioethanol production from mixed softwood that has been pretreated with IL is economically feasible.	Trinh et al. (2019)
[EM][Ac] and [EM][Cl]	Fireweed, common broom, hay, goldenrod biomass.	18.5 g/L	The sample from grass mixture (42.42 g/L) and fireweed (37.68 g/L) had the highest amount of reducing sugars; the lignocellulose from fireweed treated with [EM][Ac] has the highest concentration of bioethanol.	Smuga-Kogut et al., (2019a)
[BM][Cl]	Sunflower stalk	0.078 g of ethanol per g of SFS biomass	Combination of IL and sodium hydroxide resulted in easier enzymatic saccharification and higher sugar and ethanol yields.	Nargotra et al. (2018)

(Continued)

TABLE 11.1 (*Continued*)
Summary of Some Recent Studies that used ILs in Bioethanol Production

ILs Used	Biomass	Ethanol Yield	Main Findings	References
[BM][Ac] [EM][Ac]	Sawdust of *Eucalyptus sp.*, *Pinus sp.*	0.03–0.04 g ethanol/g biomass	0.16–0.19 g glucose/g biomass	Tura et al. (2018)
[EM][Ac]	Barley straw	22.9 g ethanol/100 g pretreated barley straw.	The pretreatment with IL greatly improves the enzymatic hydrolysis, shortens the time to only 24 h; the IL treatment resulted in a higher hydrolysis yield, with a glucose yield close to 100%.	Zappi et al. (2019)
[Ch][Ac]	Empty palm fruit bunch (EFB)	0.275 g ethanol/g EFB	Enzymatic hydrolysis by the cellulase and IL mixture increased the sugar yield from 0.058 g/g of EFB in the crude sample (untreated) to 0.283 g/g (24 h treatment) and 0.62 g/g (48 h treatment) in IL-E system.	Elgharbawy et al. (2018)
[Ch][Ac], [Ch][Lys]), and cholinium aspartate $[Ch]_2[Asp]$	Lignocellulosic biomass	41.1 g/L	Using diluted bioderived ILs, high-gravity bioethanol can be produced in a single pot; the cost of cellulosic ethanol generated has decreased by 40%, and local pressures on water resources and waste management infrastructure have decreased.	Xu et al. (2016)

Aside from their applications in biomass pretreatment as a crucial step for fermentation, ILs have been also introduced for ethanol distillation. A study demonstrated the possibility of extracting sugar from carob pods utilizing a four-stage counter-current system, with sugar losses of less than 2%. After 20 h of alcoholic fermentation with *S. cerevisiae*, aqueous extracts with a sugar content of 168 g/L yielded the highest ethanol concentration of 95 g/L, regardless of the microorganism's initial concentration. The ILs tetradecyl (trihexyl)phosphonium chloride [TDTP][Cl] and methyltr ioctylammonium{bis(trifluoromethyl)sulfonyl}imide [MTOA][NTf_2] were viable prospects for the establishment of *in situ* extraction–fermentation procedures, based on biocompatibility and ethanol extraction experiment (Sanchez-Segado et al., 2019). Figure 11.4 presents the separation process of bioethanol by extractive distillation to acquire anhydrous ethanol. Extractive distillation, azeotropic distillation, and membrane separation are among few of the technologies available to deal with the azeotropic ethanol.

11.4.2 Ionic Liquids and Nanotechnology Integration for Bioethanol Production

Metal nanoparticles (NPs) stabilized by ILs (NP@ILs) are appealing catalysts for biomass conversion because the catalytic system's characteristics can be tailor-made by selecting a metal NP and suitable IL combination (Luska et al., 2015). Second generation biomass conversion including the conversion of substance derived from lignocellulose has used NP@IL catalysts. Cr(0) NPs in [EM]

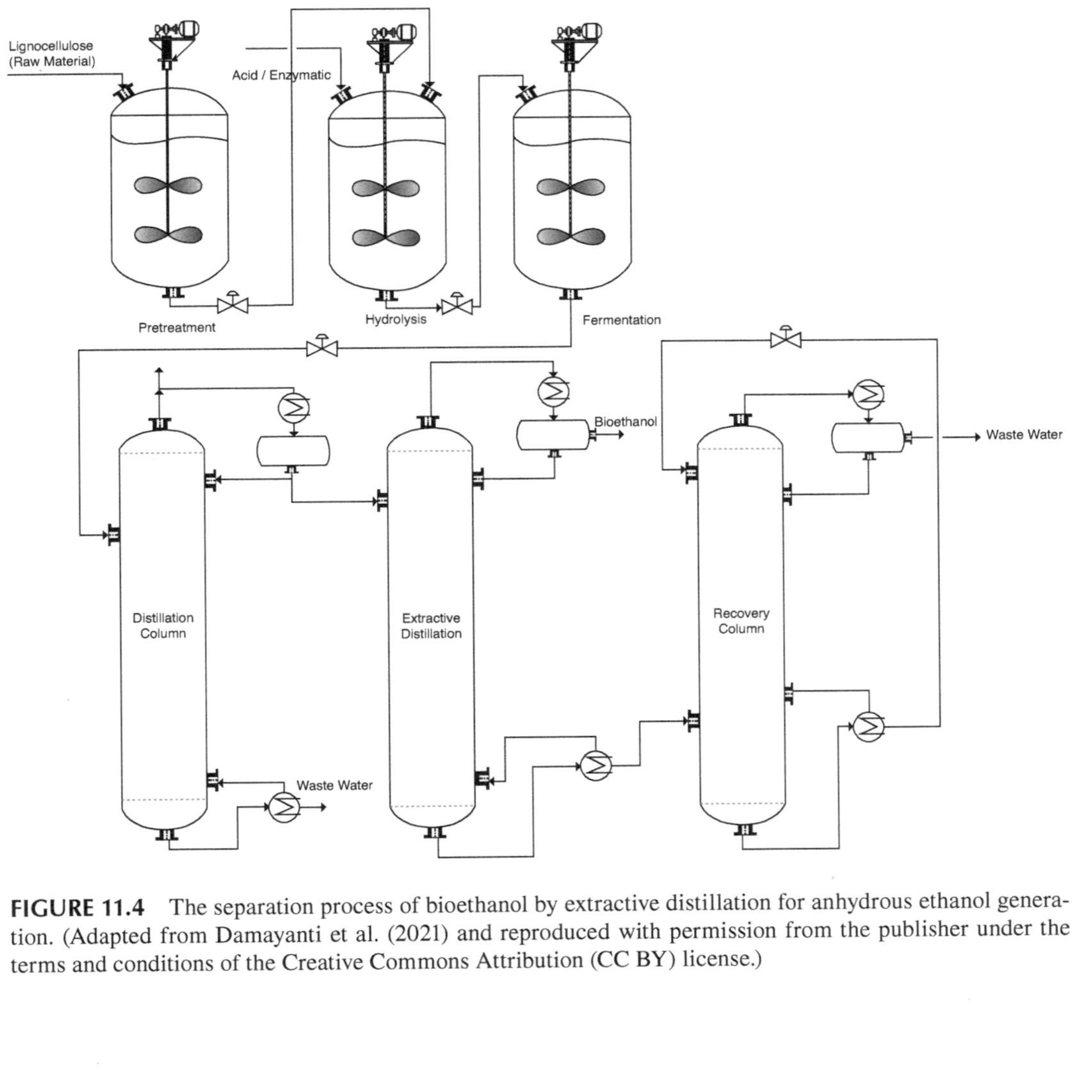

FIGURE 11.4 The separation process of bioethanol by extractive distillation for anhydrous ethanol generation. (Adapted from Damayanti et al. (2021) and reproduced with permission from the publisher under the terms and conditions of the Creative Commons Attribution (CC BY) license.)

[Cl] were used to catalyze the glucose dehydration to 5-hydroxymethylfurfural (HMF). The Cr NPs sized at 2.3 nm made by decomposing $Cr(CO)_6$ in [EM][Cl] yielded 49% HMF (He et al., 2013).

IL supported silica, [BMIM][Cl] and SiO_2, were used in the conversion of cellulose to 77% of reducing sugar (Xu et al., 2015). Polysulfone (PSF) grafted with IL, 1-(propyl-3-sulfonate) imidazolium hydrosulfate, was used to process inulin biomass (Jiang et al., 2021a). The hydrolysis of inulin generated 88.22% displaying promising potential of PSF-IL in biomass-conversion applications. The same researchers investigated cellulose hydrolysis using the same approach with high solid loading and reduction of two-third in energy consumption (Jiang et al., 2021b). Zirconia NPs and mesoporous titania were used to investigate the conversion of cellulose to glucose and HMF in [EM][Cl] system, resulting in 18.2% HMF and 12.9% glucose yields (Kuo et al., 2013), the very crucial step that leads to bioethanol production.

Furthermore, immobilization of in-house IL-stable enzyme by covalent bonding on magnetic NP functionalized with amino-propyl triethoxysilan has formed active nanobiocatalysts. The biomass of sugarcane tops was treated with [TMA][$MeSO_4$]. The hydrolysis with the IL-tolerant enzymes (cellulase and xylanase from fungus isolate, *Aspergillus flavus* PN3) yielded 181.18 mg reducing sugar/g biomass (Sharma et al., 2021). Nanocellulose was also formed via hydrolysis with ILs [EM][Ac] and [AM][Cl] by treating microcrystalline celluloses using enzymatic pretreatment. The enzyme-pretreated cellulose had a cellulose II structure and a particle size of about 200 nm compared to the IL-treated samples which demonstrated 30–40 nm diameter and low crystallinity (Babicka et al., 2021).

Rai et al. (2019) recently evaluated the role of several nanobiocatalysts in biomass hydrolysis and hypothesized that a range of nanomaterials may be employed to immobilize various enzymes responsible for the degradation of polymers found in lignocellulose. In addition, magnetic nanobiocatalysts were more promising. Because nanobiocatalysts can be employed repeatedly for more than one hydrolysis reaction, it is thought that using magnetic NPs for purposes such as the functionalization and the immobilization of enzymes would be a game-changing alternative to existing approaches (Ingle et al., 2019).

11.4.3 Effect of Ionic Liquids on Fermentation

Small amounts of residual IL can hinder fermentative microorganisms downstream, such as *Saccharomyces cerevisiae*, even though biomass is routinely washed after IL pretreatment (Sitepu et al., 2014). When assessing the effect of ILs on *Saccharomyces cerevisiae*, it was thought that imidazolium ILs target mitochondria because their chemical genomic profiles are similar to that of the mitochondrial membrane. Furthermore, deletions of genes encoding mitochondrial proteins result in an increase in sensitivity towards ILs. The effects of ILs on mitochondrial protein levels were validated using high-throughput chemical proteomics. Imidazolium IL, like valinomycin, caused aberrant mitochondrial morphology and changed polarization of mitochondrial membrane potential (Dickinson et al., 2016).

Despite that, some ILs have shown low or no toxicity towards microorganisms. For instance, Rigual et al. (2020) investigated the approach of one-pot bioproduction process. One-pot IL-based methods are a promising strategy for reducing water usage (by eliminating biomass washing), reducing inhibitory by-products and lowering the cost. The researchers used [HEAM][Ac] and [Ch][Lys] for the pretreatment of eucalyptus and pine at 11% of the ILs, where [HEAM][Ac] gave the highest digestibility of 75% in eucalyptus. Moreover, when the growth of the fungal strain *Rhodosporidium toruloides* was tested in hydrolysates, the results demonstrated the growth ability of the latter in the presence of 5% (w/w) of [HEAM][Ac] (Rigual et al., 2020). This outcome supports the hypothesis that the presence of small concentrations of IL reduces the inhibitory compounds that impact the fermentation of microorganism.

Another study previously done by Elgharbawy et al. (2018) has proven the ability of *S. cerevisiae* to grow in presence of 10% of [Ch][Ac] and yielded 87.94% of the bioethanol theoretical yield

(Elgharbawy et al., 2018). In another study, [BM][Ac] was the most effective IL in direct ethanol fermentation of biomass in a process incorporated with IL and cellulase-displaying yeast was. After 96h of fermentation, the ethanol yield of the [BM][Cl] pretreated bagasse reached 73.4% of the theoretical yield (Yamada et al., 2017). The effects of [BM][Cl] on the ethanol fermentation of *S. cerevisiae* AY93161 was examined using microcalorimetry. The examination was done at different [BM][Cl] concentrations from 0.001 to 5g/L. It was found that IL presence resulted in an increase of the maximum specific produced heat rate value P_m (from 2.08 to 7.06 mWl/g) and lowering the maximum specific growth rate μm (from 0.226 to 0.105 h^{-1}). This led to a reduction of ethanol from 40.3 to 25.1 g/L and final yeast concentration from 3.85 to 2.39g/L (Huang et al., 2016).

Furthermore, Tura et al. (2018) discovered that ethanol could be produced from *Pinus sp.* and *Eucalyptus sp.* pretreated with IL and fermented by *S. cerevisiae* and *S. pombe* yeasts. However, different ethanol yields were observed depending on the IL utilized in the pretreatment of the biomass. They have, therefore, suggested that the presence of residual IL interfered with the metabolism of *S. pombe* more than *S. cerevisiae*, with [EM][Ac] pretreated samples showing less interference with the production of ethanol by both yeasts tested (Tura et al., 2018).

Sitepu et al. evaluated the growth of 80 strains with [EM][Ac], which are responsible to hydrolyze the polysaccharides in plant cell wall and generate ethanol. They have found that the tolerance range was 1%–5% of the IL for the strains understuy, with the best strains named *Zygoascus, Yarrowia, Hyphopichia, Debaryomyces, Wickerhamomyces, Galactomyces, Naumovozyma, Clavispora, Kazachstania,* and *Wickerhamomyces* (Sitepu et al., 2014). The metabolite concetration could provide an evidence of the effect of ILs on yeast growth. For instance, in an attempt to identify IL-tolerant yeasts to 0%, 2%, 4% of [EM][Ac] and [EM][Cl] using 45 strains, at the species and strain level, the impacts of $[EM]^+$ exposure on lipid accumulation and cell mass production differed. For six strains, the acetate-IL reduced lipid synthesis and cell biomass more than the $[Cl]^-$. Lipid output in *Yarrowia lipolytica* UCDFST 51–30 was not significantly different, while it dropped by 5%–65% in other yeasts in the study. The effects of an equimolar concentration of the chloried-IL were substantially milder, decreasing to 25% and 66% rise in lipid production. *Vanrija humicola* UCDFST 10–1004 and UCDFST 12–717 produced the largest lipid yields (8.3 and 7.9g/L, respectively). These findings showed that in the presence of [EM][Cl] at quantities seen in lignocellulosic hydrolysates, significant lipid synthesis occurs (Sitepu et al., 2017).

Some studies reported the mechanism of action of ILs on cells. For example, after incubation in [OM][Cl], permeabilization of the cell membrane has been observed in *Saccharomyces cerevisiae.* In the same study, it was discovered that [OM][Cl] could induce permeability in large unilamellar vesicles (Cook et al., 2019) (see Figure 11.5). Another study reported that in the presence of 200mM ([EM][DEP]) and [EM][Cl], and 100mM [EM][Ac], a strain of *S. cerevisiae* can maintain its original viability (Nakashima et al., 2011).

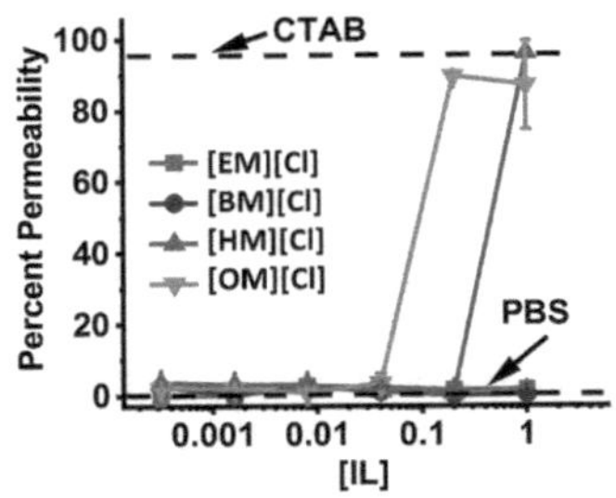

FIGURE 11.5 Effect of IL chain length on *S. cerevisiae* permeability. Cells with positive fluorescence at 575nm were considered permeable. Exposure to cetyltrimethylammonium bromide (CTAB) (102μM) yielded 94% permeable cells vs. PBS treatment 4% (dotted lines). (Adapted from Cook et al. (2019) and reproduced with permission from the publisher under the terms and conditions of the Creative Commons Attribution (CC BY) license.)

11.5 CONCLUSIONS AND OUTLOOK

Biological, physical, and chemical pretreatments can be used to handle lignocellulose before it is converted to bioethanol in the fermentation process. ILs have played an important role in the processing of biomass to undergo hydrolysis and fermentation. The employment of green solvents will minimize the usage of harmful substances in hydrolysis of biomass while also reducing the coproduction of inhibitory materials in the fermentation process in a sustainable manner. Despite the enormous progress in the IL field, more targeted research into cost-effective, eco-friendly techniques of enhancing IL recovery is required at the laboratory bench and pilot plant stages in order to discover and solve the hurdles impeding commercial usage of ILs for biomass pretreatment. ILs could potentially be used in biomass pretreatment to create valuable products with less health and environmental consequences, and this advantage should be taken into consideration while developing a new model or a novel design. Studies investigating the compatibility of ILs with enzymes and microbes simultaneously are still scarce, and some studies are contradicted in terms of toxicity level of ILs. More research into the biocompatibility of protic acidic ILs are anticipated. Moreover, there is lack of studies on tolerant microorganisms that can operate in high concentrations of ILs. The majority of prior attempts concentrated on the resistance of enzymes and microorganisms toward imidazolium-based ILs. However, with the advent of low-cost ILs, it is desirable to refocus research on these ILs. Additionally, nanomaterials treated or functionalized with ILs are game-changing due to their recyclability as well as the ability to prevent or reduce enzymatic inhibition in the hydrolysis of cellulose.

ACKNOWLEDGMENT

This research is funded by the Fundamental Research Grant Scheme, MoHE-Malaysia [FRGS/1/2021/STG02/UIAM/03/1].

REFERENCES

Alayoubi, R., N. Mehmood and E. Husson, et al. 2020. Low temperature ionic liquid pretreatment of lignocellulosic biomass to enhance bioethanol yield. *Renewable Energy* 145:1808–1816.

Amoah, J., K. Ogura, Q. Schmetz, A. Kondo and C. Ogino. 2019. Co-fermentation of xylose and glucose from ionic liquid pretreated sugar cane bagasse for bioethanol production using engineered xylose assimilating yeast. *Biomass and Bioenergy* 128:105283.

Babicka, M., M. Wozniak, K. Szentner, et al. 2021. Nanocellulose production using ionic liquids with enzymatic pretreatment. *Materials* 14:3264.

Busic, A., N. Mardetko and S. Kundas, et al. 2018. Bioethanol production from renewable raw materials and its separation and purification: A review. *Food Technology and Biotechnology* 56:289–311.

Chambon, C. L., T. Y. Mkhize and P. Reddy, et al. 2018. Pretreatment of South African sugarcane bagasse using a low-cost protic ionic liquid: A comparison of whole, depithed, fibrous and pith bagasse fractions. *Biotechnology for Biofuels and Bioproducts* 11:247.

Chang, K. L., C. H. Liu and P. Phitsuwan, et al. 2021. Enhancement of biological pretreatment on biomass by an ionic liquid or surfactant. *Catalysts* 11:1274.

Cheah, W. Y., R. Sankaran and P. L. Show, et al. 2020. Pretreatment methods for lignocellulosic biofuels production: Current advances, challenges and future prospects. *Biofuel Research Journal* 7:1115–1127.

Chuetor, S., E. J. Panakkal and T. Ruensodsai, et al. 2022. Improvement of enzymatic saccharification and ethanol production from rice straw using recycled ionic liquid : The effect of anti-solvent mixture. *Bioengineering* 9:115

Cook, K., K. Tarnawsky and A. J. Swinton, et al. 2019. Correlating lipid membrane permeabilities of imidazolium ionic liquids with their cytotoxicities on yeast, bacterial, and mammalian cells. *Biomolecules* 9:251.

Damayanti, D., D. Supriyadi and D. Amelia, et al. 2021. Conversion of lignocellulose for bioethanol production, applied in bio-polyethylene terephthalate. *Polymers* 13:2886.

Dickinson, Q., S. Bottoms and L. Hinchman, et al. 2016. Mechanism of imidazolium ionic liquids toxicity in *Saccharomyces cerevisiae* and rational engineering of a tolerant, xylose-fermenting strain. *Microbial Cell Factories* 15:17.

Duque, A., C. Alvarez, P. Domenech, P. Manzanares and A. D. Moreno. 2021. Advanced bioethanol production: From novel raw materials to integrated biorefineries. *Processes* 9:206.

Egorova, K. S., E. G. Gordeev and V. P. Ananikov. 2017. Biological Activity of ionic liquids and their application in pharmaceutics and medicine. *Chemical Reviews* 117:7132–7189.

Elgharbawy, A. A., M. Z. Alam, M. Moniruzzaman, N. A. Kabbashi and P. Jamal. 2018. Chemical and structural changes of pretreated empty fruit bunch (EFB) in ionic liquid-cellulase compatible system for fermentability to bioethanol. *3 Biotech* 8:236.

Elgharbawy, A. A. M., M. Moniruzzaman and M. Goto. 2020. Recent advances of enzymatic reactions in ionic liquids: Part II. *Biochemical Engineering Journal* 154:107426.

Farghaly, A., M. Elsamadony, S. Ookawara and A. Tawfik. 2017. Bioethanol production from paperboard mill sludge using acid-catalyzed bio-derived choline acetate ionic liquid pretreatment followed by fermentation process. *Energy Conversion and Management* 145:255–264.

Franks, C. D., G. B. Burow and J. J. Burke. 2006. A comparison of U.S. and Chinese sorghum germplasm for early season cold tolerance. *Crop Science* 46:1371–1376.

Ha, S. H., N. L. Mai, G. An and Y. M. Koo. 2011. Microwave-assisted pretreatment of cellulose in ionic liquid for accelerated enzymatic hydrolysis. *Bioresource Technology* 102:1214–1219.

Halder, P., S. Kundu and S. Patel, et al. 2019. Progress on the pre-treatment of lignocellulosic biomass employing ionic liquids. *Renewable and Sustainable Energy Reviews* 105:268–292.

He, J., Y. Zhang and E. Y. X. Chen. 2013. Chromium(0) nanoparticles as effective catalyst for the conversion of glucose into 5-hydroxymethylfurfural. *ChemSusChem* 6:61–64.

Hou, X. D., N. Li and M. H. Zong. 2013. Facile and simple pretreatment of sugar cane bagasse without size reduction using renewable ionic liquids – Water mixtures. *ACS Sustainable Chemistry & Engineering* 1:519–526.

Huang, W., J. Jin and L. Feng, et al. 2016. Microcalorimetric investigation of the effect of the ionic liquid 1-butyl-3-methylimidazolium chloride on the fermentation of *Saccharomyces cerevisiae* AY93161 for lignocellulosic ethanol production. *Open Biotechnology Journal* 10:391–397.

Huang, W., X. Wu and J. Qi, et al. 2019. Ionic liquids: Green and tailor-made solvents in drug delivery. *Drug Discovery Today* 25:901–908.

Ingle, A. P., A. K. Chandel, F. A. F. Antunes, M. Rai and S. S. da Silva. 2019. New trends in application of nanotechnology for the pretreatment of lignocellulosic biomass. *Biofuels, Bioproducts and Biorefining* 13:776–788.

Jiang, H., T. Wang, S. Li and Z. P. Zhao. 2021a. Fabrication of porous polymer membrane from polysulfone grafted with acid ionic liquid and the catalytic property for inulin hydrolysis. *Journal of Membrane Science* 618:118742.

Jiang, X., R. Zhai and M. Jin. 2021b. Increased mixing intensity is not necessary for more efficient cellulose hydrolysis at high solid loading. *Bioresource Technology* 329:124911.

Kimon, K. S., E. L. Alan and D. W. O. Sinclair. 2011. Enhanced saccharification kinetics of sugarcane bagasse pretreated in 1-butyl-3-methylimidazolium chloride at high temperature and without complete dissolution. *Bioresource Technology* 102:9325–9329.

Kulshrestha, A., I. Pancha, S. Mishra and A. Kumar. 2021. Deep eutectic solvents and ionic liquid assisted hydrolysis of microalgal biomass: A promising approach towards sustainable biofuel production. *Journal of Molecular Liquids* 335:116264.

Kumar, J. and S. Reetu. 2015. Lignocellulosic agriculture wastes as biomass feedstocks for second-generation bioethanol production : Concepts and recent developments. *3 Biotech* 5:337–353.

Kuo, I. J., N. Suzuki, Y. Yamauchi and K. C. W. Wu. 2013. Cellulose-to-HMF conversion using crystalline mesoporous titania and zirconia nanocatalysts in ionic liquid systems. *RSC Advances* 3:2028–2034.

Li, C., B. Knierim and C. Manisseri, et al. 2010. Comparison of dilute acid and ionic liquid pretreatment of switchgrass: Biomass recalcitrance, delignification and enzymatic saccharification. *Bioresource Technology* 101:4900–4906.

Limayem, A. and S. C. Ricke. 2012. Lignocellulosic biomass for bioethanol production: Current perspectives, potential issues and future prospects. *Progress in Energy and Combustion Science* 38:449–467.

Lin, Y. C., S. Shangdiar, S. C. Chen, C. A. Cho and A. Datov. 2021. A study on different parameters affecting the saccharification rate of *Typha orientalis* pretreated with ionic liquids (ILs) and microwave irradiation for bioethanol production by using response surface methodology. *Catalysis Communications* 150:106265.

Luska, K. L., P. Migowski and W. Leitner. 2015. Ionic liquid-stabilized nanoparticles as catalysts for the conversion of biomass. *Green Chemistry* 17:3195–3206.

Michalczyk, A., A. Cieniecka-Roslonkiewicz and M. Cholewinska. 2015. Application of ionic liquids in the ultrasound-assisted extraction of antimicrobial compounds from the bark of *Cinnamomum cassia*. *Journal of the Chilean Chemical Society* 60:2698–2703.

Nakashima, K., K. Yamaguchi and N. Taniguchi, et al. 2011. Direct bioethanol production from cellulose by the combination of cellulase-displaying yeast and ionic liquid pretreatment. *Green Chemistry* 13:2948–2953.

Nargotra, P., V. Sharma, M. Gupta, S. Kour and B. K. Bajaj. 2018. Application of ionic liquid and alkali pretreatment for enhancing saccharification of sunflower stalk biomass for potential biofuel-ethanol production. *Bioresource Technology* 267:560–568.

Naz, S., M. Uroos, A. M. Asim, N. Muhammad and F. U. Shah. 2020. One-pot deconstruction and conversion of lignocellulose into reducing sugars by pyridinium-based ionic liquid-metal salt system. *Frontiers in Chemistry* 8:236.

Nurdin, M., H. Abimanyu and H. Putriani, et al. 2021. Optimization of OPEFB lignocellulose transformation process through ionic liquid [TEA][HSO_4] based pretreatment. *Scientific Reports* 11:11338.

Ocreto, J. B., W. H. Chen, and A. T. Ubando, et al. 2021. A critical review on second-and third-generation bioethanol production using microwaved-assisted heating (MAH) pretreatment. *Renewable and Sustainable Energy Reviews* 152:111679.

Odorico, F. H., A. de Araujo Morandim-Giannetti, A. C. Lucarini and R. B. Torres. 2018. Pretreatment of Guinea grass (*Panicum maximum*) with the ionic liquid 1-ethyl-3-methyl imidazolium acetate for efficient hydrolysis and bioethanol production. *Cellulose* 25:2997–3009.

Olivier-Bourbigou, H., L. Magna and D. Morvan. 2010. Ionic liquids and catalysis: Recent progress from knowledge to applications. *Applied Catalysis A: General* 373:1–56.

Pang, Z., P. Wang and C. Dong. 2018. Ultrasonic pretreatment of cellulose in ionic liquid for efficient preparation of cellulose nanocrystals. *Cellulose* 25:7053–7064.

Paul, S. K. and S. Chakraborty. 2018. Microwave-assisted ionic liquid-mediated rapid catalytic conversion of non-edible lignocellulosic Sunn hemp fibres to biofuels. *Bioresource Technology* 253:85–93.

Radhakrishnan, R., P. Patra, M. Das and A. Ghosh. 2021. Recent advancements in the ionic liquid mediated lignin valorization for the production of renewable materials and value-added chemicals. *Renewable and Sustainable Energy Reviews* 149:111368.

Rai, M., A. P. Ingle and R. Pandit, et al. 2019. Emerging role of nanobiocatalysts in hydrolysis of lignocellulosic biomass leading to sustainable bioethanol production. *Catalysis Reviews* 61:1–26.

Ramos, J. L., M. Valdivia, F. Garcia-Lorente and A. Segura. 2016. Benefits and perspectives on the use of biofuels. *Microbial Biotechnology* 9:436–440.

Reis, C. L. B., L. M. A. e Silva, L.M.A. and T. H.S. Rodrigues, et al. 2017. Pretreatment of cashew apple bagasse using protic ionic liquids: Enhanced enzymatic hydrolysis. *Bioresource Technology* 224:694–701.

Rigual, V., G. Papa and A. Rodriguez, et al. 2020. Evaluating protic ionic liquid for woody biomass one-pot pretreatment+saccharification, followed by Rhodosporidium toruloides cultivation. *ACS Sustainable Chemistry & Engineering* 8:782–791.

Saga, K., S. Yokoyama, K. Imou and Y. Kaizu. 2008. A comparative study of the effect of CO_2 emission reduction by several bioenergy production systems. *International Energy Journal* 9:383.

Sakdaronnarong, C., K. Jiratanakittiwat, T. Tangkitthanasakul and N. Laosiripojana. 2017. Ionosolv pretreatment of sugarcane bagasse and rice straw assisted by catalytic hydrothermal and microwave heating for biorefining. *Food and Bioproducts Processing* 105:104–116.

Sanchez-Segado, S., M. J. Salar-Garcia and V. M. Ortiz-Martinez, et al. 2019. Evaluation of ionic liquids as in situ extraction agents during the alcoholic fermentation of carob pod extracts. *Fermentation* 5:90.

Sharma, S., P. Nargotra and V. Sharma, et al. 2021. Nanobiocatalysts for efficacious bioconversion of ionic liquid pretreated sugarcane tops biomass to biofuel. *Bioresource Technology* 333:125191.

Sharma, V., P. Nargotra and B. K. Bajaj. 2019. Ultrasound and surfactant assisted ionic liquid pretreatment of sugarcane bagasse for enhancing saccharification using enzymes from an ionic liquid tolerant *Aspergillus assiutensis* VS34. *Bioresource Technology* 285:121319.

Singh, Y. D., P. Mahanta and U. Bora. 2017. Comprehensive characterization of lignocellulosic biomass through proximate, ultimate and compositional analysis for bioenergy production. *Renewable Energy* 103:490–500.

Sitepu, I. R., L. A. Garay and L. Enriquez, et al. 2017. 1-Ethyl-3-methylimidazolium tolerance and intracellular lipid accumulation of 38 oleaginous yeast species. *Applied Microbiology and Biotechnology* 101:8621–8631.

Sitepu, I. R., S. Shi, and B.A. Simmons, et al. 2014. Yeast tolerance to the ionic liquid 1-ethyl-3-methylimidazolium acetate. *FEMS Yeast Research* 14:1286–1294.

Smuga-Kogut, M., B. Walendzik and D. Szymanowska-Powalowska, et al. 2019b. Comparison of bioethanol preparation from triticale straw using the ionic liquid and sulfate methods. *Energies* 12:1155.

Smuga-Kogut, M., K. Zgorska and T. Kogut, et al. 2017. The use of ionic liquid pretreatment of rye straw for bioethanol production. *Fuel* 191:266–274.

Smuga-Kogut, M., T. Piskier, B. Walendzik and D. Szymanowska-Powałowska. 2019a. Assessment of wasteland derived biomass for bioethanol production. *Electronic Journal of Biotechnology* 41:1–8.

Trinh, L. T. P., Y. J. Lee, C. S. Park and H. J. Bae. 2019. Aqueous acidified ionic liquid pretreatment for bioethanol production and concentration of produced ethanol by pervaporation. *Journal of Industrial and Engineering Chemistry* 69:57–65.

Tura, A., S. Montipo, S., R. C. Fontana, D. P. Dillon and M. Camassola. 2018. Ethanol production from sugar liberated from *Pinus* sp. and *Eucalyptus* sp. biomass pretreated by ionic liquids. *Brazilian Journal of Chemical Engineering* 35:467–475.

Van Osch, D. J. G. P., L. J. B. M. Kollau and A. van den Bruinhorst, et al. 2017. Ionic liquids and deep eutectic solvents for lignocellulosic biomass fractionation. *Physical Chemistry Chemical Physics* 19, 2636–2665.

Wulfhorst, H., N. Harwardt and H. Giese, et al. 2015. *Fuels from Biomass: An Interdisciplinary Approach. Notes on Numerical Fluid Mechanics and Multidisciplinary Design v.129*, Eds. M. Klaas, S. Pischinger and W. Schroder. Heidelberg, Berlin: Springer.

Xu, F., J. Sun, and N. V. S. N. M. Konda, et al. 2016. Transforming biomass conversion with ionic liquids: Process intensification and the development of a high-gravity, one-pot process for the production of cellulosic ethanol. *Energy and Environmental Science* 9:1042–1049.

Xu, H., H. Zhao and H. Song, et al. 2015. Functionalized ionic liquids supported on silica as mild and effective heterogeneous catalysts for dehydration of biomass to furan derivatives. *Journal of Molecular Catalysis A: Chemical* 410:235–241.

Yamada, R., K. Nakashima and N. Asai-Nakashima, et al. 2017. Direct ethanol production from ionic liquid-pretreated lignocellulosic biomass by cellulase-displaying yeasts. *Applied Biochemistry and Biotechnology* 182:229–237.

Yu, X., X. Bao and C. Zhou, et al. 2018. Ultrasound-ionic liquid enhanced enzymatic and acid hydrolysis of biomass cellulose. *Ultrasonics Sonochemistry* 41:410–418.

Zappi, D., S. Gabriele and L. Gontrani, et al. 2019. Biologically friendly room temperature ionic liquids and nanomaterials for the development of innovative enzymatic biosensors: Part II. *Talanta* 194:26–31.

Ziaei-Rad, Z., J. Fooladi, M. Pazouki and S. N. Gummadi. 2021. Lignocellulosic biomass pre-treatment using low-cost ionic liquid for bioethanol production: An economically viable method for wheat straw fractionation. *Biomass and Bioenergy* 151:106140.

12 Emerging Solvents for Biomass Pretreatment and their Sustainability

Yi Zhang
Zhejiang University
Westlake University

Lei Wang
Westlake University
Westlake Institute for Advanced Study

12.1 INTRODUCTION

The growing consumption of fossil fuels and increasing concerns about greenhouse gas emissions have attracted researchers' and industrial attentions to develop green and renewable energy and chemicals. Lignocellulosic biomass is a promising candidate to produce sustainable energy via biorefining processes. Typical lignocellulosic biomass is mainly composed of 30%–60% cellulose, 20%–40% hemicellulose, and 15%–25% lignin (Dahadha et al., 2017). However, due to complex interlinks among these components, many barriers need to be overcome to effectively utilize lignocellulosic biomass. It is necessary to conduct a proper pretreatment to separate cellulose, hemicellulose, and lignin before their separate uses for value-added products, such as liquid fuels, bio-based chemicals, and degradable materials (Rastogi and Shrivastava, 2017).

Various pretreatment methods have been studied extensively, including biological, physical, chemical, and their combinations (Ong and Wu, 2020). Among these conventional methods, organic solvents-based pretreatment could hydrolyze hemicellulose efficiently (Zhao et al., 2009), produce high-purity cellulose with only minor degradation, and allow solvent recycling via distillation (Kim and Pan, 2010). However, the large quantities of organic solvents may introduce secondary pollution due to their volatility, flammability, and toxicity. For instance, the volatility of organic solvent may cause volatile organic compounds (VOCs) emitted to air (Byrne et al., 2016); dipolar aprotic solvents could dissolve a broad range of substrates and accelerate reactions, but these characteristics likely come along the reproductive toxicity as consequences (Ashcroft et al., 2015); and the hydrophobicity of hydrocarbon solvent implies its tendency to bioaccumulate in the organism and lead to aquatic toxicity eventually (Parkerton et al., 2021). In general, the inevitable utilization of organic solvent may eventually result in unwanted exposure of harmful and toxic substances to the environment and humans. Therefore, regulations, such as Registration, Evaluation, Authorization and Restriction of Chemicals launched by the European Union, have been applied to govern organic solvent usage and control their emissions.

Conventional organic solvents face challenges to meet the requirements for performance, reduction in energy, water, solvent consumption, and carbon emissions simultaneously (Benvenutti et al., 2019; Chemat et al., 2017). Emerging solvents, a substitute for conventional organic solvents, have been applied in lignocellulosic biomass pretreatment with their outstanding abilities. Ionic liquids (ILs), deep eutectic solvents (DESs), and some polar aprotic solvents are claimed as green solvents available for biomass pretreatment. However, the definition of greenness is a complex, weighted, multi-criteria, and sometimes vague concept, which should be supported by a comprehensive assessment.

DOI: 10.1201/9781003226536-16

Existing researches on green solvents normally only focus on one aspect regarding greenness, for example, their low volatility or bio-based origin. Moreover, the toxicity of solvents might be underestimated due to the assessment concerning their precursors instead of their own emissions (Lima et al., 2021). Hence, it is necessary to dissect these presumed green solvents from a more holistic point of view, including life cycle thinking and comprehensive impact indicators.

This chapter aims to review recent emerging solvents used in lignocellulosic biorefinery research field with the focus on their sustainability assessments, to introduce current emerging solvents for lignocellulosic biomass pretreatment (Section 12.2), to discuss the criteria green solvents shall meet (Section 12.3), and to provide a critical review on sustainability assessments of green solvents (Section 12.4).

12.2 EMERGING SOLVENTS

In general, four main groups are covered in this chapter: bio-based solvents, ILs, DESs, and supercritical fluids (SCFs).

12.2.1 Bio-based Solvents

As mentioned in the beginning, biomass feedstock is a promising resource to produce chemicals, mitigating fossil resources depletion and global warming; significant investments have been set up to explore the renewable biomass utilization for producing chemicals (Ogmundarson et al., 2020). Recent analysis predicted that globally market share of bio-based chemicals will increase to 22% in 2025 (Biddy et al., 2016). Bio-based solvents are defined as solvents produced from biomass feedstock (Vovers et al., 2017). As a typical branch of bio-based chemicals, most of the bio-based solvents show the advantages of being biodegradable except for their inherent characteristics of low-carbon footprint and no fossil resources depletion (Gu and Jerome, 2013).

Bio-based solvents could be classified according to their origins, for example, biodiesel fuel, carbohydrate, and lignin, or classified based on their polarities, an important solvent property along with equilibria, solubility, cleaning, or extraction efficiency (Jin et al., 2017). Typical examples sorted by polarities of bio-based solvents are as follows: (i) strongly polar solvents such as γ-valerolactone (GVL), dihydrolevoglucosenone (Cyrene™), and dimethyl isosorbide (DMI); (ii) 'borderline' polar solvents include 2-methyltetrahydrofurfuran (2-MeTHF), glycerol and its derivatives, carbohydrates, and gluconic acid; and (iii) weakly polar solvents such as *p*-cymene, limonene, and terpinene (Gao et al., 2020). Figure 12.1 adopted from Clark et al. (2015) illustrates bio-based solvents could be produced within either of the petrochemical or biorefinery models. Quite many of them have been used in biomass pretreatment research such as 2-MeTHF, GVL, Cyrene™, and DMI, demonstrating the concept of 'closed-loop' biorefinery (Li et al., 2017; Luterbacher et al., 2015; Meng et al., 2020; Yang et al., 2022). However, besides their bio-based origin, their commercialization is also a crucial concern for the future sustainable requirement. Though it seems advantageous in the laboratory, it may not be technically, economically, or environmentally feasible at an industrial scale.

12.2.2 ILs

ILs are salts comprising weak interactions of organic cations associated with organic or inorganic anions, melting point of which is below 100°C (Lei et al., 2017). The first IL of ethylammonium nitrate ($[EtNH_3][NO_3]$) was discovered by Paul Walden in 1914, who was searching for liquid molten salts to suit his equipment (Walden, 1914). However, ILs as innovative solvents have sparked a surge of interest until they are successfully used as electrolytes, establishing 'green chemistry' as an emerging scientific field (Shamshina et al., 2019). As an alternative to traditional volatile organic solvents, ILs benefit from their unique advantages: negligible vapor pressure, low melting point, nonflammability, good thermal stability, high conductivity, wide electrochemical window, and catalytic properties (Jindal and Sablok, 2015). They can be formed by different cation–anion combinations, resulting

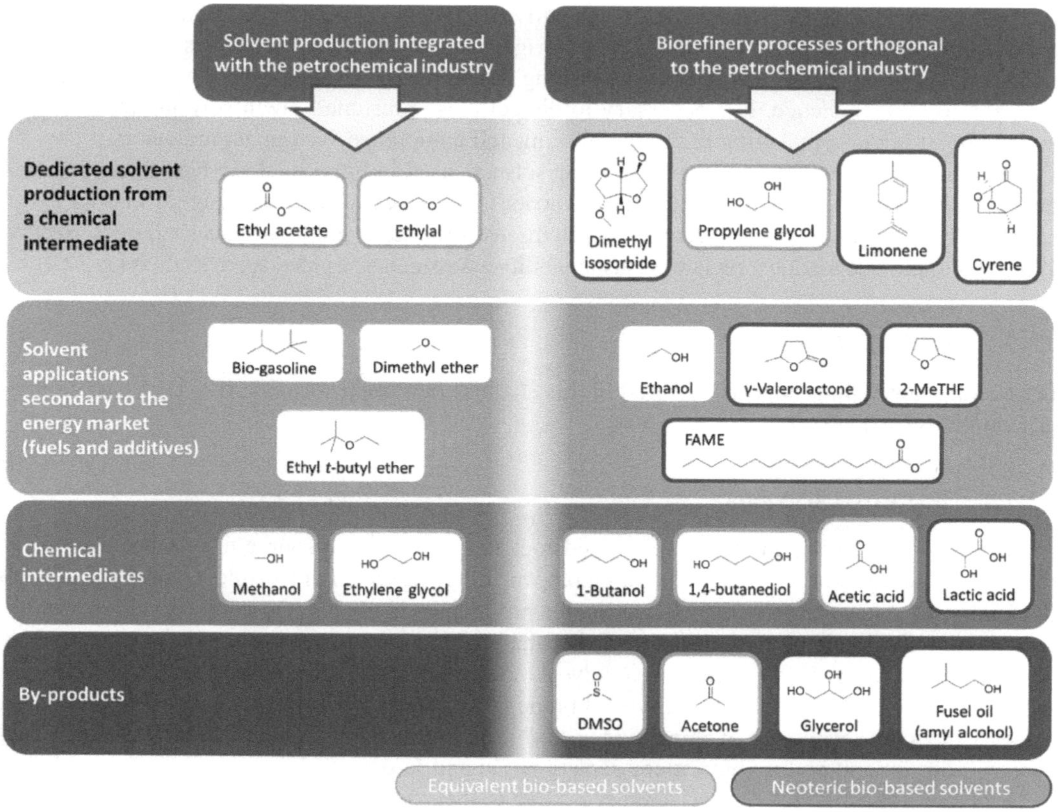

FIGURE 12.1 The relationship between different bio-based solvents according to their origin and their place in the chemical industry hierarchy.

Source: Clark et al. (2015).

in tunable physicochemical properties via appropriate designs (Hessel et al., 2022). According to the statistics, the possible combinations could reach up to 10^{18} theoretically (Chatel et al., 2017). According to the chemical varieties, ILs could be divided into several types (Lei et al., 2017), for example, room-temperature ILs (RTILs), poly-ionic liquids, task-specific ILs, and low-temperature ILs (Borra et al., 2007). Among them, RTILs are the most used ILs which consist of bulky and asymmetric organic cations such as 1-alkyl-3-methylimidazolium, 1-alkylpyridinium, N-methyl-N-alkylpyrrolidinium, and ammonium ions, while anions are employed with a wide range of halides, inorganic anions (tetrafluoroborate, hexafluorophosphate, and bis-trifluoromethanesulfonimide), and large organic anions (bistriflimide, triflate, and tosylate). There have been quite a few critical reviews on performance of ILs in lignocellulosic biomass pretreatment or biorefinery for audiences' interests (Amini et al., 2021; Roy and Chundawat, 2022; Usmani et al., 2020). Despite the advantages mentioned above, their high cost, viscous nature, and recycling hinder the large-scale adoption and require further research on commercialization. Moreover, these reviews focus more on technical and economic pros and cons rather than their sustainability as a remaining gap.

12.2.3 DESs

DES, introduced as an alternative to ILs, is a hypoeutectic mixture composed of hydrogen bond donor (HBD) and hydrogen bond acceptor (HBA). The first DES was coined by Abbott et al. to describe the formation of a liquid eutectic mixture (melting point 12°C) starting from two solid materials of choline chloride [ChCl] and urea with high melting points of 302°C and 133°C, respectively (Abbott et al., 2003, 2004). DESs are usually obtained by the complexation of a quaternary ammonium salt with an HBD that has the ability to form a complex with the halide anion of the ammonium salt (Smith et al., 2014). The charge delocalization occurs through the hydrogen bonding between HBA and HBD, which is responsible for the decrease in the melting point of the mixture compared to those of the individual constituent components (Abbott et al., 2001; Hansen et al., 2020). DESs can be represented by a general formula: $Cat^{+}X^{-}zY$ where Cat^{+} is in principle any ammonium, phosphonium, or sulfonium cation, and X is Lewis base, generally a halide anion. The complex anionic species are formed between X^{-} and either a Lewis or Brønsted acid Y. z refers to the number of Y molecules that interact with the anion (Smith et al., 2014).

According to the general formula, DESs could be divided into four types (Table 12.1). Type 1 eutectics are analogous to metal halide/imidazolium salt systems, which combine a quaternary ammonium salt and a metal chloride; however, nonhydrated metal halides are generally limited to suitable melting point (Yang et al., 2016). Type 2 eutectics use hydrated metal halides as HBD has a wider range. In addition, most of hydrated metal salts are inexpensive and air/moisture insensitive, enabling their wide usage on a large production scale appropriately (Smith et al., 2014). Type 3 eutectics, generated by ChCl and HBDs like amides, carboxylic acids, and alcohols, are generally inexpensive and biodegradable (Abbott et al., 2006). Types 4 eutectics are mainly composed of inorganic transition metal cations which can be incorporated into eutectics under ambient temperature (Abbott et al., 2007). As an emerging group in DES, natural deep eutectic solvents (NADESs) are composed of two or more compounds that are generally bio-based primary metabolites, that is, organic acids, sugars, alcohols, amines, and aminoacids (Dai et al., 2013). NADESs are firstly reported from a hypothesis of third-type fluids, which is separated from water and lipids and formed by the large amount of metabolites in cells (Choi et al., 2011). In other words, when these bio-based primary metabolites are mixed in a proper ratio, they change the state from solid to liquid and then NADESs generated.

DESs are acknowledged to share similar characteristics and properties to ILs, for example, good thermal stability, low volatility, negligible vapor pressures, and tunable characteristics by changing the composition and molar ratios of the HBA and HBD. Up to now, choline chloride is the most commonly used HBA for producing DESs, due to the low toxicity of cholinium (Mbous et al., 2017). Despite the above-mentioned advantages and lower cost compared to ILs, there are still some drawbacks such as their poor thermal stabilities and high viscosities, in general, therefore difficulties in recycling due to their decomposition at higher temperatures (Chen and Mu, 2019).

12.2.4 SCFs

SCFs are substances above their critical temperature and pressure. At these conditions, a fluid only exists in a homogeneous phase condition whose properties such as diffusivity, viscosity, and density are between those of gases and liquids (Ibanez et al., 2016). The properties of SCFs can be changed by adjusting operation conditions of temperature and/or pressure, or in the absence of modifiers such as catalysts and/or cosolvents (Escobar et al., 2020). The most common SCFs highlighted are carbon dioxide, ammonia, water, and hydrocarbons such as ethane, propane, ethylene, and propylene. SCFs can be applied for biomass valorization as a pretreatment technique or as a reactive extraction procedure to yield value-added coproducts (Escobar et al., 2020). For example, at 390°C and 250 bar, supercritical water could yield 61 and 71 wt% C_6 and C_5 sugars from beet pulp (Martínez et al., 2018). Zhao et al. (2019) realized the pretreatment of corn stover, corn cob, and sorghum stalk using supercritical CO_2 at the pressure of 175–220 bar and 50°C–80°C with the consequential enzymatic hydrolysis sugar yields at 50%–60%. In addition, SCFs could also be applied in the presence of catalysts to produce several carbohydrate-based value-added products such as furfural, furfuryl alcohol, levulinic acid, ethyl levulinate, and butyl levulinate (Escobar et al., 2020). Overall, SCF technologies demonstrate advantages in utilizing these solvents and direct fractionation of lignocellulose to fuels without costly consumption of chemicals or downstream products separation.

TABLE 12.1
General Formula for the Classification of DESs (Yang et al., 2016; Yuan et al., 2022)

Type	General Formula	Terms
1	$Cat^{+}X^{-}zMCl_x$; M=Zn, Sn, Fe, Al, Ga, In	Metal salt+organic salt
2	$Cat^{+}X^{-}zMCl_x \cdot yH_2O$; M=Cr, Co, Cu, Ni, Fe	Hydrate metal salt+organic salt
3	$Cat^{+}X^{-}zRZ$; Z=$CONH_2$, COOH, OH	HBD+organic salt
4	$MCl_x + RZ = MCl_{x-1}{}^{+}RZ + MCl_{x+1}{}^{-}$; M=Al/Zn, Z=$CONH_2$, OH	Zinc/aluminum chloride+HBD

However, their feasibility associated with high temperature and pressure operating conditions shall be evaluated thoroughly from both economic and environmental points of view.

12.3 CRITERIA OF GREEN SOLVENTS

Under the 'green chemistry' prompt, numerous researches have been published to proclaim the use of 'green solvents' or 'emerging process with green solvents' in different research fields. These green solvents are usually referred to as 'emerging solvents', 'benign solvents', 'eco-friendly solvents', 'benign solvents', 'cleaner solvents', etc. (Winterton, 2021). Despite various terminologies, these solvents are coincidently expected to achieve a 'green' purpose.

How shall green solvent be defined? Characteristics associated with green solvents should be renewable, sustainable, nontoxic, recyclable, and scalable; however, the solvent is usually labeled as green when they only meet one of these characteristics. Moreover, existing researches usually describe green solvents with their high chemical stability and low volatility, claiming the elimination of chemical loss as environmental benefits or their bio-based origins (Rogers and Seddon, 2003). Nevertheless, the definition of green solvent should entail more implications. Beyond the characteristics and origins of green solvents themselves, it should broaden the disciplinary boundaries from chemistry to environmental science and engineering, toxicology, industrial ecology, public policy, and other adjacent fields (Anastas and Zimmerman, 2019). Overall, we here to summarize the definition of green solvent as a substitute for traditional solvent, which is important to solvent property improvement, environment conservation, organism safe and health protection, final processing optimization, energy efficiency promotion, and economic saving (Gao et al., 2020).

12.4 COMPREHENSIVE ASSESSMENT OF GREEN SOLVENTS

12.4.1 State-of-Art Selection Methods for Green Solvents

Atom economy and E-factor are the oldest green metrics proposed in 1991 and 1992, respectively (Sheldon, 1992; Trost, 1991). Atom economy is equal to the ratio of the mass of desired product to the total mass of products. E-factor equals to the ratio between the mass of waste to the mass of product. However, these two methods take only the yield into account, allocating the same weighting to different coproduct and/or waste. Environmental, health, and safety (EHS) ranking therefore was developed to identify the potential hazardous solvent with health and safety criteria in 1985. Some well-known pharmaceutical companies and specialty institutes have published selection guides of solvents according to their EHS ranking, such as AstraZeneca, GSK, Pfizer, and the ACS Green Chemistry Institute Pharmaceutical Roundtable (Clarke et al., 2018; Diorazio et al., 2016; Henderson et al., 2011). The EHS ranking provides a powerful strategy to select appropriate solvents in the chemical production process, supporting sustainable development. However, the EHS method mainly focuses on the use stage of solvents while neglecting the impacts associated with their entire life cycles, including production, recycling, and disposal. Moreover, most of the above-mentioned emerging solvents usually haven't been listed in the EHS selection guides, despite the critical role they play in sustainable and green chemistry.

Life cycle assessment (LCA) is a comprehensive method designed to assess the various environmental impacts of all stages of a product's life cycle, from raw materials extraction through materials processing to distribution, use, and disposal or recycling. As a result, the systematic inclusion of the LCA method is essential during the selection of a solvent or potential alternative to conventional solvents. The first comprehensive framework proposed by Capello et al. (2007) includes the EHS method and the LCA method for assessing the 'greenness' of solvent; however, cumulative energy demand (CED) is the only impact category used as impact category for LCA methodology. Other studies are trying to develop multi-impacts integrated index or interactive tools to facilitate solvent selection (Diorazio et al., 2016; Fadel and Tarabieh, 2019). Nevertheless, although these frameworks include LCA, most of them focus on traditional solvents and ignore the data deficiencies of emerging solvents, let alone the uncertainty.

12.4.2 LCA of Emerging Solvents

LCA quantifies the environmental impacts by compiling all the relevant inputs and outputs for each of these life cycle stages (Figure 12.2) through four steps, including (i) goal and scope definition, (ii) life cycle inventory (LCI), (iii) life cycle impact assessment (LCIA), and (iv) results interpretation (ISO, 2006a, b).

Goal and scope definition: This process is to point out the questions that need to be addressed in study. This step should define the functional unit, set the system boundaries, and describe the important assumptions and constraints. A cradle-to-grave LCA should be conducted to avoid impacts shifting, including raw material extraction, materials processing, product manufacture distribution, use, and end-of-life (EoL) options.

Life cycle inventory (LCI): The main step of LCI is to collect data according to the determined system boundary, that is, to organize, count, and analyze the input–output data of resources, energy, resource, and waste emissions during each life cycle stage. Primary data from the industrial processes are recommended for LCA, which can be obtained from manufacturers, environmental product declarations, company sustainability publications, industry associations, and LCI databases. However, emerging solvents are a relatively emerging topic of compounds, and therefore, there are not many primary data from industrial-scale production available. Thus, there is a need to upscale the lab-scale data with assistance from process simulation modeling, such as Aspen Plus™, CHEMCAD®, and advanced simulation library (Parvatker et al., 2018).

Life cycle impact assessment (LCIA): LCIA is the translation of inventories to their environmental or human health effects via characterizations. The LCIA methods available are CML IA-baseline, ReCiPe, Ecoindicator, IMPACT, IPCC, etc. Among them, CML IA-baseline is the most widely used method because of its good applicability and comprehensive coverage of indicators. However, characterization factors (CFs) for impacts conversion are lacking for most emerging solvents, especially for (eco)toxicity categories, hindering the completeness of the comprehensive assessment.

Results interpretation: Finally, interpretation is an extraction or translation from the LCA results to the scientific standpoint. Interpretation can help identify the environmental hot spot and therefore indicate the scientific standpoint and give suggestions for decision-making. Sensitivity and uncertainty analyses are also suggested by ISO standards to improve the interpretation of LCIA results.

Even though a large number of publications introduce wide applications of the emerging solvents mentioned above, the number of studies assessing the 'greenness' over their whole life cycle is surprisingly small. Especially for DESs, there are still no LCA studies to our best knowledge. Some of researches upscaled the results to industrial scale with the help of process simulation software

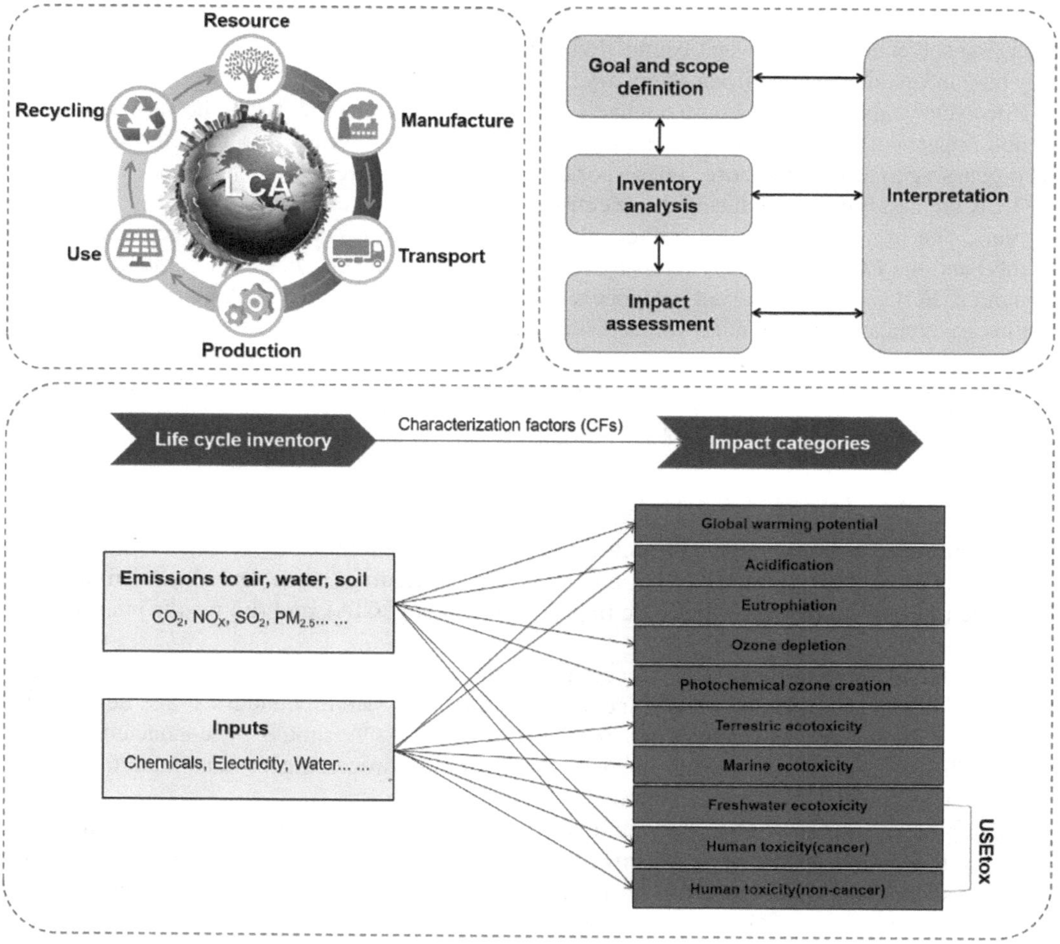

FIGURE 12.2 Mechanism framework and main impact categories in LCA.

(Van Slagmaat et al., 2020); however, other studies conducted LCA based on experimental data without process simulation led to biased results without 'economies of scale' operation condition (Loubet et al., 2017). Moreover, the coverage of life cycle stages is also limited due to the lack of data in particular phases. For example, for bio-based solvents, the 'cradle' phase of biomass feedstock cultivation is usually excluded from the system boundary. However, the cultivation phase represents the more stressful step for the environment due to the land occupation and relative agricultural practices of irrigation, insecticide, and fertilization (Cespi et al., 2015). Thus, land footprint should be considered as a part of sustainability assessment in the comparative LCA studies of green solvents (Khoo et al., 2015). EoL phases play a crucial role in solvents' LCA, but in some cases, are excluded from the system boundary. The choice of solvent EoLs depends on their properties and the waste compositions. Recycling eliminates impacts from solvent generation but introduces extensive energy demand for distillation. However, for those solvents whose environmental impacts associated with production or concentration of waste are high, recovery is still preferable than incineration (Luis et al., 2013).

A number of comparative LCA studies revealed that the asserted green solvents might exhibit higher potential impacts than traditional solvents or display impact shifting among different categories or life cycle stages. Kapanji et al. (2021) conducted life cycle sustainability assessments of biorefineries and indicated the scenario for producing GVL is most unsustainable, because of the manufacture, transportation, and use of n-butyl acetate solvent and hydrogen from natural gas, despite attractive socioeconomic outcomes. The bio-based solvents in current LCA studies are usually placed as intermediate products, thus neglecting the impacts generated by their emissions. Generally, LCA study for bio-based solvents is limited heretofore. Existing LCA studies mainly involved GVL (Choe et al., 2021), Cyrene™ (Garcia-Cruz et al., 2022), limonene (Jahandideh et al., 2017), and 2-MeTHF (Khoo et al., 2015) production; however, studies for DMI and *p*-cymene are still lacking.

For ILs, butylmethylimidazolium ($[Bmim]^+$) cation-based ILs are the largest type analyzed in LCA to date. A series of comparative LCA studies show that impacts from the IL production phase are highest during their life cycle and higher than those of the other traditional solvents (Mehrkesh and Karunanithi, 2013). Solvent recovery is a crucial phase to make use of ILs and reduce environmental impact accordingly; however, such advantage offsets by the less significant contribution from ILs production. The 'leakage' of environmental impacts, referring to the burden shifting to the early life cycle stage, such as ILs synthesis, was observed in various LCA studies for ILs. Additionally, ecotoxicity with precursors and ILs themselves is usually excluded in the assessment. SCFs demonstrate a great performance over the same solvents in traditional conditions. However, it is hard to ensure the achievement of these advantages in the upscale process from current lab-scale or pilot-scale studies, especially in these energy-sensitive cases. Therefore, sensitivity analysis and uncertainty analysis should be further conducted and discussed to support the robustness of LCA results. In addition, energy inputs for synthesis conditions of critical temperatures and pressures are strongly correlated with the solvent types. The critical temperatures of different SCFs are orders of magnitude (water: 374°C, Wagner and Pruss, 2002; CO_2: 31°C, Span and Wagner, 1996); meanwhile, energy inputs for raising temperature are significantly higher than increasing pressure. Thus, the environmental performance of SCFs depends on the selected solvent.

Overall, the LCA study should cover a comprehensive set of environmental impacts related to the entire life cycle stages for identifying the trade-offs as well as avoiding burden shifting into different impact categories rather than focusing on a single category, for example, global warming potential, or CED, etc. Despite the industrial-scale data shortage in this emerging field, lab-scale data can still be used to conduct prospective LCA to reveal the technical feasibility via the process simulation tool. Additionally, the scarcity of CFs for these emerging solvents hindered the inclusion of human toxicity and ecotoxicity in LCIA comprehensively.

12.4.3 (Eco)toxicity of Emerging Solvents

In general, (eco)toxicity as toxicity on the ecosystem and human health should give extra consideration. It is not only because the majority of green chemistry principles are related to safety and toxicity, but it should also fulfill the demands of workplace protocols and regulatory requirements.

Bio-based solvents, which are reviewed above to be used in biomass pretreatment, enable closed-loop biorefineries (Azadi et al., 2012). However, it should be noted that many bio-based solvents could be problematic on their (eco)toxicity. Bottoms et al. (2018) elucidated that GVL exerts toxicity to yeast, and the toxicity may be magnified in combination with ethanol produced or other inhibitors. On the contrary, Kerkel et al. (2021a) demonstrate the low toxicity of GVL for causing acute or sub-chronic adverse effects compared to N-methyl-2-pyrrolidone (NMP), dimethylacetamide (DMAc), and dimethylformamide (DMF) by series of ecotoxicology assessments. Nevertheless, due to the vapor pressure being 32 Pa and higher than the threshold of 10 Pa at 20°C, GVL is classified as a VOC (Kerkel et al., 2021b). Cyrene™ is slightly more toxic to bacteria compared to dimethyl sulfoxide (Camp et al., 2020). Furfural, a precursor for 2-MeTH and GVL (Clarke et al., 2018), exhibits toxicity in the environment which is similar to toluene and benzene (Zuriaga et al., 2018).

Many pieces of evidences show that ILs emitted in freshwater may cause damage to various organisms, such as bacteria (Ghanem et al., 2015), algae (Ventura et al., 2013), fish (Ruokonen et al., 2016), and even mammalian cell (Mikkola et al., 2015). The (eco)toxic concerns of ILs are associated with the combination of volatile chemicals or unsustainable raw materials and, subsequently, disposal and transformation (Amde et al., 2015). ILs containing BF_4^- and PF_6^- anions will release corrosive and toxic hydrofluoric acid when hydrolyzed in water (Wang et al., 2017). The quantitative structure–activity relationship (QSAR) method has been used to predict the toxicity data of 153 ILs using machine learning, which provided a complementary method to access the toxicity of vast ILs (Yan et al., 2021).

DESs are normally regarded as nontoxic, environmentally friendly, and biodegradable solvents. However, recent researches on DESs toxicity were contradictory. Several studies have shown cytotoxicity on fish and human cell lines of [ChCl]-based DES (Radosevic et al., 2015). The hydrogen bonding between HBA and HBD was suspected to be the candidate for causing toxicity (Hayyan et al., 2013). The toxicological behavior of DES is different from the 'sum' of the toxicological behaviors of the individual components, indicating that DESs' toxicity should be studied as a simple compound rather than a mixture (De Morais et al., 2015).

In addition, existing researches on (eco)toxicity of presumed emerging solvents still focus on the toxicology of solvents themselves. Few studies considered the (eco)toxic impacts throughout their upstream production to downstream use. Besides, toxicology assessments are incomplete without degradation studies (Radosevic et al., 2015). (Eco)toxicity of solvents during their transition, degradation, and exposure after their release to the environment is of interest for a holistic assessment. Very limited number of LCA studies has considered the (eco)toxicity impacts of emerging solvents due to the lack of (eco)toxicity CFs. To fill this gap, USEtox was developed by United Nations Environmental Program as a supplementary LCIA model for developing (eco)toxicity CFs. USEtox treats the environmental compartments as well-mixed boxes containing and exchanging contaminants. The model simulates and quantifies the entire process from a contaminant entering the environment till generating (eco)toxicity, including three steps as contaminants fate and distribution (fate), aquatic ecosystem and human exposure (exposure), and associated toxicity effects (effect) quantification (Fantke et al., 2017; Figure 12.3). For example, a life cycle (eco)toxicity study via USEtox model assessed the toxicities of five conventional ILs after their release to the freshwater (Mehrkesh and Karunanithi, 2016). Since ILs are a huge class of green solvents containing thousands of forms potentially, QSAR can be used to predict the toxicity data of ILs as mentioned above. However, to the best of our knowledge, there are no studies using QSAR to supplement the CFs of (eco)toxicity for other emerging solvents.

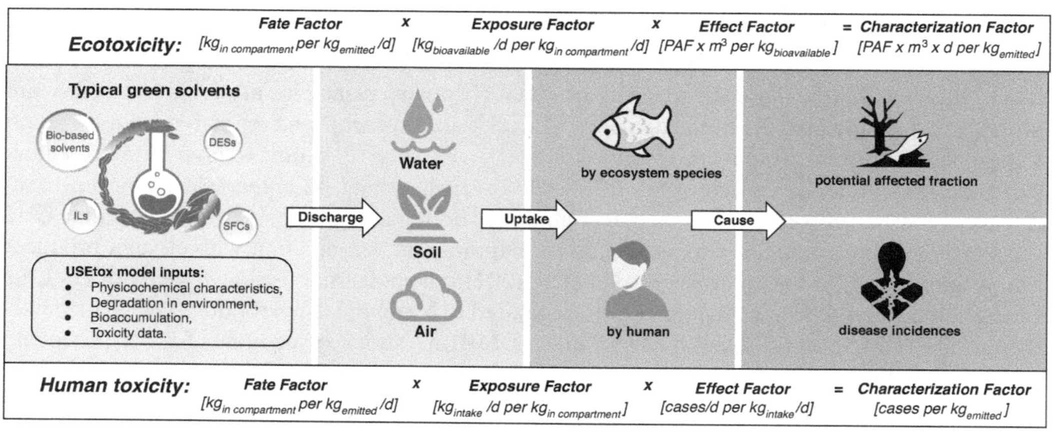

FIGURE 12.3 Mechanism framework of USEtox model for assessing freshwater ecotoxicity and human toxicity effects.

Source: Modified from Zhang et al. (2022).

12.5 CONCLUDING REMARKS

Solvents play an important role in lignocellulosic biomass pretreatment. Based on the proposed green chemistry concept, emerging solvents have been applied to replace the traditional organic solvent with the aim to improve the environmental performance of their applications. Researches have declared emerging solvents including bio-based ones, ILs, DES, and SCFs as green solvents but, most of the time, without comprehensive assessments. The incomplete assessment either focuses on solvent derivation or solvent use may lead to shifting on environmental burdens. Therefore, it is necessary to conduct a more comprehensive assessment considering the life cycle thinking concept and multi-criteria of environment, (eco)toxicity, and economic to verify the sustainability of the proposed green solvents.

12.5.1 Comprehensive Assessment Framework for Green Solvents

Since green solvents were proposed as a new concept, the evaluation framework to comprehensively reveal the potential impacts of green solvents is rare. Given that the chemical industry is now moving beyond cradle-to-grave and considering a circular economy as cradle-to-cradle, it is essential to reform the conventional linear economy to the circular economy to advance sustainable production. Recovery, regeneration, and recirculation should be considered as design elements in all future chemical processes to minimize the impacts on ecology and health. With regards to impact categories in the assessment, (eco)toxicity still remains as gap in the characterization of impacts from these solvents release to the environment which requires attention from researchers with multi-disciplinary backgrounds.

12.5.2 Application of USEtox in LCA

To the best of our knowledge, there is still no LCA study using USEtox to characterize (eco)toxicity for biomass pretreatment involving the use of emerging solvents. However, the USEtox model has been applied to measure the (eco)toxicity CFs of a variety of organic and inorganic chemicals in other fields, such as organic solvents in perovskite solar cell production (Vidal et al., 2021), decamethylcyclopentasiloxane solvents for replacing water in nonaqueous dyeing (Zhang et al., 2022), regionalized (eco)toxicity footprints development in Sweden (Nordborg et al., 2017), and common ILs (Mehrkesh and Karunanithi, 2016). However, key parameters such as (eco)toxicity data, especially chronic data of EC_{50} and ED_{50}, which are essential to generate CFs of (eco)toxicities categories, are usually lacking. The research trends in this area include using machine learning in the field of artificial intelligence to assist the prediction of their cytotoxicity (Yan et al., 2021); conducting LCA prediction with deep neural networks (Zhu et al., 2020), and calibration (eco)toxicity results with epidemic data (Holmquist et al., 2020). With regards to the data selection, USEtox guidelines recommend a tiered approach for selecting (eco)toxicity data, that is, human test prior to animal test, chronic data prior to acute data, etc. (Fantke et al., 2017). Nevertheless, besides the uncertainty of the predicting results, the extrapolation method of acute–chronic and animal–human also adds uncertainties to the results (Igos et al., 2014), indicating the necessity of conducting uncertainty analysis for life cycle toxicity assessment results calculated by the USEtox model. In addition, given that the variety of data sources may be used in the single study, sensitivity and uncertainty analyses are crucial to improve the underlining impacts to conclusions (Zhang et al., 2022).

REFERENCES

Abbott, A. P., D. Boothby, G. Capper, D. L. Davies and R. K. Rasheed. 2004. Deep eutectic solvents formed between choline chloride and carboxylic acids: Versatile alternatives to ionic liquids. *Journal of the American Chemical Society* 126:9142–9147.

Abbott, A. P., G. Capper and D. L. Davies, et al. 2001. Preparation of novel, moisture-stable, Lewis-acidic ionic liquids containing quaternary ammonium salts with functional side chains electronic supplementary information (ESI) available: Plot of conductivity vs. temperature for the ionic liquid formed from zinc chloride and choline chloride (2: 1). *Chemical Communications* 2001:2010–2011.

Abbott, A. P., G. Capper, D. L. Davies, K. J. McKenzie and S. U. Obi. 2006. Solubility of metal oxides in deep eutectic solvents based on choline chloride. *Journal of Chemical & Engineering Data* 51:1280–1282.

Abbott, A. P., G. Capper, D. L. Davies, R. K. Rasheed and V. Tambyrajah. 2003. Novel solvent properties of choline chloride/urea mixtures. *Chemical Communications* 2003:70–71.

Abbott, A. P., J. C. Barron, K. S. Ryder and D. Wilson. 2007. Eutectic-based ionic liquids with metal-containing anions and cations. *Chemistry – A European Journal* 13:6495–6501.

Amde, M., J. F. Liu and L. Pang. 2015. Environmental application, fate, effects, and concerns of ionic liquids: A review. *Environmental Science & Technology* 49:12611–12627.

Amini, E., C. Valls and M. B. Roncero. 2021. Ionic liquid-assisted bioconversion of lignocellulosic biomass for the development of value-added products. *Journal of Cleaner Production* 326:129275.

Anastas, P. T. and J. B. Zimmerman. 2019. The periodic table of the elements of green and sustainable chemistry. *Green Chemistry* 21:6545–6566.

Ashcroft, C. P., P. J. Dunn, J. D. Hayler and A. S. Wells. 2015. Survey of solvent usage in papers published in organic process research & development 1997–2012. *Organic Process Research & Development* 19:740–747.

Azadi, P., R. Carrasquillo-Flores and Y. J. Pagan-Torres, et al. 2012. Catalytic conversion of biomass using solvents derived from lignin. *Green Chemistry* 14:1573–1576.

Benvenutti, L., A. A.F. Zielinski and S. R. S. Ferreira. 2019. Which is the best food emerging solvent: IL, DES or NADES? *Trends in Food Science & Technology* 90:133–146.

Biddy, M. J., C. Scarlata and C. Kinchin. 2016. *Chemicals from Biomass: A Market Assessment of Bioproducts with Near-Term Potential.* Golden, CO: National Renewable Energy Laboratory.

Borra, E. F., O. Seddiki and R. Angel, et al. 2007. Deposition of metal films on an ionic liquid as a basis for a lunar telescope. *Nature* 447:979–981.

Bottoms, S., Q. Dickinson and M. McGee. 2018. Chemical genomic guided engineering of gamma-valerolactone tolerant yeast. *Microbial Cell Factories* 17:5.

Byrne, F. P., S. Jin and G. Paggiola, et al. 2016. Tools and techniques for solvent selection: Green solvent selection guides. *Sustainable Chemical Processes* 4:1–24.

Camp, J. E., S. B. Nyamini and F. J. Scott. 2020. Cyrene(tm) is a green alternative to DMSO as a solvent for antibacterial drug discovery against ESKAPE pathogens. *RSC Medicinal Chemistry* 11:111–117.

Capello, C., U. Fischer and K. Hungerbühler. 2007. What is a green solvent? A comprehensive framework for the environmental assessment of solvents. *Green Chemistry* 9:927–934.

Cespi, D., F. Passarini and G. Mastragostino, et al. 2015. Glycerol as feedstock in the synthesis of chemicals: A life cycle analysis for acrolein production. *Green Chemistry* 17:343–355.

Chatel, G., E. Naffrechoux and M. Draye. 2017. Avoid the PCB mistakes: A more sustainable future for ionic liquids. *Journal of Hazardous Materials* 324:773–780.

Chemat, F., N. Rombaut and A. Meullemiestre, et al. 2017. Review of green food processing techniques. Preservation, transformation, and extraction. *Innovative Food Science & Emerging Technologies* 41:357–377.

Chen, Y. and T. Mu. 2019. Application of deep eutectic solvents in biomass pretreatment and conversion. *Green Energy & Environment* 4:95–115.

Choe, B., S. Lee and W. Won. 2021. Coproduction of butene oligomers and adipic acid from lignocellulosic biomass: Process design and evaluation. *Energy* 235:121278.

Choi, Y. H., J. van Spronsen and Y. Dai, et al. 2011. Are natural deep eutectic solvents the missing link in understanding cellular metabolism and physiology? *Plant Physiology* 156:1701–1705.

Clark, J. H., T. J. Farmer, A. J. Hunt and J. Sherwood. 2015. Opportunities for bio-based solvents created as petrochemical and fuel products transition towards renewable resources. *International Journal of Molecular Sciences* 16:17101–17159.

Clarke, C. J., W. C. Tu, O. Levers, A. Brohl and J. P. Hallett. 2018. Green and sustainable solvents in chemical processes. *Chemical Reviews* 118:747–800.

Dahadha, S., Z. Amin, A. A. Bazyar Lakeh and E. Elbeshbishy. 2017. Evaluation of different pretreatment processes of lignocellulosic biomass for enhanced biomethane production. *Energy & Fuels* 31:10335–10347.

Dai, Y., J. van Spronsen, G. J. Witkamp, R. Verpoorte and Y. H. Choi. 2013. Natural deep eutectic solvents as new potential media for green technology. *Analytica Chimica Acta* 766:61–68.

De Morais, P., F. Goncalves, J. A. P. Coutinho and S. P. M. Ventura. 2015. Ecotoxicity of cholinium-based deep eutectic solvents. *ACS Sustainable Chemistry & Engineering* 3:3398–3404.

Diorazio, L. J., D.R. J. Hose and N. K. Adlington. 2016. Toward a more holistic framework for solvent selection. *Organic Process Research & Development* 20:760–773.

Escobar, E. L. N., T. A. da Silva, C. L. Pirich, M. L. Corazza and L. P. Ramos. 2020. Supercritical fluids: A promising technique for biomass pretreatment and fractionation. *Frontiers in Bioengineering and Biotechnology* 8:252.

Fadel, C. and K. Tarabieh. 2019. Development of an industrial environmental index to assess the sustainability of industrial solvent-based processes. *Resources* 8:115.

Fantke, P., M. Bijster and C. Guignard, et al. 2017. *USEtox 2.0: Documentation (Version 1).* USEtox International Center.

Gao, F., R. Bai, and F. Ferlin. 2020. Replacement strategies for non-green dipolar aprotic solvents. *Green Chemistry* 22:6240–6257.

Garcia-Cruz, A., L. Diaz-Jimenez, A. Ilyina and S. Carlos-Hernandez. 2022. Prospective life cycle assessment of a based orange wax fungicide. *Industrial Crops and Products* 180:114769.

Ghanem, O. B., M. A. Mutalib and M. El-Harbawi, et al. 2015. Effect of imidazolium-based ionic liquids on bacterial growth inhibition investigated via experimental and QSAR modelling studies. *Journal of Hazardous Materials* 297:198–206.

Gu, Y. and F. Jerome. 2013. Bio-based solvents: An emerging generation of fluids for the design of eco-efficient processes in catalysis and organic chemistry. *Chemical Society Reviews* 42:9550–9570.

Hansen, B. B., S. Spittle and B. Chen, et al. 2020. Deep eutectic solvents: A review of fundamentals and applications. *Chemical Reviews* 121:1232–1285.

Hayyan, M., M. A. Hashim and M. A. Al-Saadi, et al. 2013. Assessment of cytotoxicity and toxicity for phosphonium-based deep eutectic solvents. *Chemosphere* 93:455–459.

Henderson, R. K., C. Jimenez-Gonzalez and D. J. Constable, et al. 2011. Expanding GSK's solvent selection guide-embedding sustainability into solvent selection starting at medicinal chemistry. *Green Chemistry* 13:854–862.

Hessel, V., N. N. Tran and M. R. Asrami, et al. 2022. Sustainability of green solvents-review and perspective. *Green Chemistry* 24:410–437.

Holmquist, H., P. Fantke and I. T. Cousins, et al. 2020. An (eco)toxicity life cycle impact assessment framework for per- and polyfluoroalkyl substances. *Environmental Science & Technology* 54:6224–6234.

Ibanez, J. A. Mendiola and M. Castro-Puyana. 2016. Supercritical fluid extraction. In *Encyclopedia of Food and Health*, Eds. B. Caballero, P. M. Finglas and F. Toldra, pp. 227–233. London: Academic Press.

Igos, E., R. Moeller and e. Benetto. 2014. Development of USEtox characterisation factors for dishwasher detergents using data made available under REACH. *Chemosphere* 100:160–166.

ISO. 2006a. *ISO 14040 – Environmental Management – Life Cycle Assessment – Principals and Framework. International Standard.* Geneva: International Organization for Standardization.

ISO. 2006b. *ISO 14044 – Environmental Management – Life Cycle Assessment – Requirements and Guidelines.* Geneva: International Organization for Standardization.

Jahandideh, A., T. J. Johnson and N. Esmaeili, et al. 2017. Life cycle analysis of a large-scale limonene production facility utilizing filamentous N_2-fixing cyanobacteria. *Algal Research* 23:1–11.

Jin, S., F. Byrne and C. R. McElroy, et al. 2017. Challenges in the development of bio-based solvents: A case study on methyl (2, 2-dimethyl-1, 3-dioxolan-4-yl) methyl carbonate as an alternative aprotic solvent. *Faraday Discussions* 202:157–173.

Jindal, R. and A. Sablok. 2015. Preparation and applications of room temperature ionic liquids in organic synthesis: A review on recent efforts. *Current Green Chemistry* 2:135–155.

Kapanji, K. K., S. Farzad and J. F. Gorgens. 2021. Life cycle and sustainability assessments of biorefineries producing glucaric acid, sorbitol or levulinic acid annexed to a sugar mill. *Journal of Cleaner Production* 295:126339.

Kerkel F, D Brock, D Touraud and W Kunz. 2021b. Stabilisation of biofuels with hydrophilic, natural antioxidants solubilised by glycerol derivatives. *Fuel* 284:119055

Kerkel, F., M. Markiewicz, S. Stolte, E. Muller and W. Kunz. 2021a. The green platform molecule gamma-valerolactone-ecotoxicity, biodegradability, solvent properties, and potential applications. *Green Chemistry* 23:2962–2976.

Khoo, H. H., L. L. Wong, J. Tan, V. Isoni and P. Sharratt. 2015. Synthesis of 2-methyl tetrahydrofuran from various lignocellulosic feedstocks: Sustainability assessment via LCA. *Resources, Conservation and Recycling* 95:174–182.

Kim, D. E. and X. Pan. 2010. Preliminary study on converting hybrid poplar to high-value chemicals and lignin using organosolv ethanol process. *Industrial & Engineering Chemistry Research* 49:12156–12163.

Lei, Z., B. Chen, Y. M. Koo and D. R. MacFarlane. 2017. Introduction: Ionic liquids. *Chemical Reviews* 117:6633–6635.

Li, S. X., M. F. Li and J. Bian, et al. 2017. Biphasic 2-methyltetrahydrofuran/oxalic acid/water pretreatment to enhance cellulose enzymatic hydrolysis and lignin valorization. *Bioresource Technology* 243:1105–1111.

Lima, F., L. C. Branco, A. J. D. Silvestre and I. M. Marrucho. 2021. Deep desulfurization of fuels: Are deep eutectic solvents the alternative for ionic liquids? *Fuel* 293:120297.

Loubet, P., M. Tsang, E. Gemechu, A. Foulet and G. Sonnemann. 2017. Life cycle assessment for green solvents. In *Bio-based Solvents*, Eds. F. Jerome and R. Luque, pp. 131–148. Hoboken, NJ: John Wiley & Sons

Luis, P., A. Amelio, S. Vreysen, V. Calabro and B. van der Bruggen. 2013. Life cycle assessment of alternatives for waste-solvent valorization: Batch and continuous distillation vs incineration. *International Journal of Life Cycle Assessment* 18:1048–1061.

Luterbacher, J. S., A. Azarpira and a. H. Motagamwala, et al. 2015. Lignin monomer production integrated into the γ-valerolactone sugar platform. *Energy & Environmental Science* 8:2657–2663.

Martínez, C. M., D. A. Cantero and M. J. Cocero. 2018. Production of saccharides from sugar beet pulp by ultrafast hydrolysis in supercritical water. *Journal of Cleaner Production* 204:888–895.

Mbous, Y. P., M. Hayyan and A. Hayyan, et al. 2017. Applications of deep eutectic solvents in biotechnology and bioengineering – Promises and challenges. *Biotechnology Advances* 35:105–134.

Mehrkesh, A. and A. T. Karunanithi. 2013. Energetic ionic materials: How green are they? A comparative life cycle assessment study. *ACS Sustainable Chemistry & Engineering* 1:448–455.

Mehrkesh, A. and A. T. Karunanithi. 2016. Life-cycle perspectives on aquatic ecotoxicity of common ionic liquids. *Environmental Science & Technology* 50:6814–6821.

Meng, X., Y. Pu, M. Li and A. J. Ragauskas. 2020. A biomass pretreatment using cellulose-derived solvent Cyrene. *Green Chemistry* 22:2862–2872.

Mikkola, S. K., A. Robciuc and J. Lokajova, et al. 2015. Impact of amphiphilic biomass-dissolving ionic liquids on biological cells and liposomes. *Environmental Science & Technology* 49:1870–1878.

Nordborg, M., R. Arvidsson and G. Finnveden, et al. 2017. Updated indicators of Swedish national human toxicity and ecotoxicity footprints using USEtox 2.01. *Environmental Impact Assessment Review* 62:110–114.

Ogmundarson, O., M. J. Herrgard, J. Forster, M. Z. Hauschild and P. Fantke. 2020. Addressing environmental sustainability of biochemicals. *Nature Sustainability* 3:167–174.

Ong, V. Z. and T. Y. Wu. 2020. An application of ultrasonication in lignocellulosic biomass valorisation into bio-energy and bio-based products. *Renewable and Sustainable Energy Reviews* 132:109924.

Parkerton, T. F., D. J. Letinski and E. J. Febbo, et al. 2021. Assessing toxicity of hydrophobic aliphatic and monoaromatic hydrocarbons at the solubility limit using novel dosing methods. *Chemosphere* 265:129174.

Parvatker, A. G. and M. J. Eckelman. 2018. Comparative evaluation of chemical life cycle inventory generation methods and implications for life cycle assessment results. *ACS Sustainable Chemistry & Engineering* 7:350–367.

Radosevic, K., M. J. Bubalo and V. G. Srcek, et al. 2015. Evaluation of toxicity and biodegradability of choline chloride based deep eutectic solvents. *Ecotoxicology and Environmental Safety* 112:46–53.

Rastogi, M. and S. Shrivastava. 2017. Recent advances in second generation bioethanol production: An insight to pretreatment, saccharification and fermentation processes. *Renewable and Sustainable Energy Reviews* 80:330–340.

Rogers, R. D. and K. R. Seddon. 2003. Ionic liquids – Solvents of the future? *Science* 302:792–793.

Roy, S. and S. P. Chundawat. 2022. Ionic liquid-based pretreatment of lignocellulosic biomass for bioconversion: A critical review. *BioEnergy Research*. https://doi.org/10.1007/s12155-022-10425-1.

Ruokonen, S. K., C. Sanwald and M. Sundvik, et al. 2016. Effect of ionic liquids on zebrafish (*Danio rerio*) viability, behavior, and histology; correlation between toxicity and ionic liquid aggregation. *Environmental Science & Technology* 50:7116–7125.

Shamshina, J. L., O. Zavgorodnya and R. D. Rogers. 2019. Ionic liquids. In *Encyclopedia of Analytical Science (Third Edition)*, Eds. P. Worsfold, C. Poole, A. Townshend and M. Miro, pp. 218–225. London: Academic Press.

Sheldon, R. A. 1992. Organic synthesis-past, present and future. *Chemistry and Industry* 1992:903–906.

Smith, E. L., A. P. Abbott and K. S. Ryder. 2014. Deep eutectic solvents (DESs) and their applications. *Chemical Reviews* 114:11060–11082.

Span, R. and W. Wagner. 1996. A new equation of state for carbon dioxide covering the fluid region from the triple-point temperature to 1100 K at pressures up to 800 MPa. *Journal of Physical and Chemical Reference Data* 25:1509–1596.

Trost, B. M. 1991. The atom economy – A search for synthetic efficiency. *Science* 254:1471–1477.

Usmani, Z., M. Sharma and P. Gupta, et al. 2020. Ionic liquid based pretreatment of lignocellulosic biomass for enhanced bioconversion. *Bioresource Technology* 304:123003.

Van Slagmaat, C. A., M. A. Delgove and J. Stouten, et al. 2020. Solvent-free hydrogenation of levulinic acid to γ-valerolactone using a Shvo catalyst precursor: Optimization, thermodynamic insights, and life cycle assessment. *Green Chemistry* 22:2443–2458.

Ventura, S. N. P., M. Gurbisz and M. Ghavre, et al. 2013. Imidazolium and pyridinium ionic liquids from mandelic acid derivatives: Synthesis and bacteria and algae toxicity evaluation. *ACS Sustainable Chemistry & Engineering* 1:393–402.

Vidal, R., J. A. Alberola-Borras and S. Habisreutinger, et al. 2021. Assessing health and environmental impacts of solvents for producing perovskite solar cells. *Nature Sustainability* 4:1–9.

Vovers, J., K. H. Smith and G. W. Stevens. 2017. Bio-based molecular solvents. In *The Application of Green Solvents in Separation Processes*, Eds. F. Pena-Pereira and M. Tobiswevski, pp. 91–110. Amsterdam: Elsevier.

Wagner, W. and A. Pruss. 2002. The IAPWS formulation 1995 for the thermodynamic properties of ordinary water substance for general and scientific use. *Journal of Physical and Chemical Reference Data* 31:387–535.

Walden, P. 1914. Ueber die molekulargrosse und elektrische leitfahigkeit einiger geschmolzenen salze [On the molecular size and electrical conductivity of some molten salts]. *News of the Russian Academy of Sciences. Mathematical series* 8:405–422.

Wang, H., X. Meng, G. Zhao and S. Zhang. 2017. Isobutane/butene alkylation catalyzed by ionic liquids: A more sustainable process for clean oil production. *Green Chemistry* 19:1462–1489.

Winterton, N. 2021. The green solvent: A critical perspective. *Clean Technologies and Environmental Policy*, 23:2499–2522.

Yan, J., X. Yan, S. Hu, H. Zhu and B. Yan. 2021. Comprehensive interrogation on acetylcholinesterase inhibition by ionic liquids using machine learning and molecular modeling. *Environmental Science & Technology* 55:14720–14731.

Yang, G. H., T. Y. Wu, Y. L. Loow and Y. A. Lin. 2016. Application of deep eutectic solvents in biomass valorization. In *The 4th International Conference on Sustainable Solid Waste Management (CYPRUS)*. Malaysia.

Yang, S., X. Yang, X. Meng and L. Wang. 2022. Efficient pretreatment using dimethyl isosorbide as a biobased solvent for potential complete biomass valorization. *Green Chemistry* 24:4082–4094.

Yuan, Z., H. Liu, W. F. Yong, Q. She and J. Esteban. 2022. Status and advances of deep eutectic solvents for metal separation and recovery. *Green Chemistry* 24:1895–1929.

Zhang, Y., S. Guo, Y. Gong and L. Wang, L. 2022. Potential trade-off between water consumption and water quality: Life cycle assessment of nonaqueous solvent dyeing. *Water Research* 215:118222.

Zhao, M. J., Q. Q. Xu and G. M. Li, et al. 2019. Pretreatment of agricultural residues by supercritical CO_2 at 50-80° C to enhance enzymatic hydrolysis. *Journal of Energy Chemistry* 31:39–45.

Zhao, X., K. Cheng and D. Liu. 2009. Organosolv pretreatment of lignocellulosic biomass for enzymatic hydrolysis. *Applied Microbiology and Biotechnology* 82:815–827.

Zhu, X., C. H. Ho and X. Wang. 2020. Application of life cycle assessment and machine learning for high-throughput screening of green chemical substitutes. *ACS Sustainable Chemistry & Engineering* 8:11141–11151.

Zuriaga, E., B. Giner, M. P. Ribate, C. B. Garcia and L. Lomba. 2018. Exploring the usefulness of key green physicochemical properties: Quantitative structure-activity relationship for solvents from biomass. *Environmental Toxicology and Chemistry* 37:1014–1023.

Hydrothermal Biomass Pretreatments for Bioethanol Production

13 Hydrothermal Pretreatment of the Biomass

Scientometric Study

Ozcan Konur
(Formerly) Ankara Yildirim Beyazit University

13.1 INTRODUCTION

The crude oil-based gasoline fuels (Ma et al., 2002; Newman and Kenworthy, 1989) have been widely used in the transportation sector since the 1920s. However, there have been great public concerns over the adverse environmental and human impact of these fuels (Hill et al., 2006, 2009). Hence, biomass-based bioethanol fuels (Alvira et al., 2010; Hill et al., 2006; Konur, 2012e, 2015, 2019, 2020a; Sun and Cheng, 2002) have increasingly been used in blending gasoline fuels (Hsieh et al., 2002; Najafi et al., 2009), in the fuel cells (Antolini, 2007, 2009), and in the biochemical production (Angelici et al., 2013; Morschbacker, 2009) in a biorefinery context (Fernando et al., 2006; Huang et al., 2008). Bioethanol fuels also play a critical role in maintaining the energy security (Kruyt et al., 2009; Winzer, 2012) in the supply shocks (Kilian, 2008, 2009) related to oil price shocks (Hamilton, 2003, 2009) or wars (Jones, 2012) in the aftermath of the Russian invasion of Ukraine starting in February 2022 (Reeves, 2014).

The research in the fields of hydrothermal pretreatments of the biomass has also intensified in recent years. Steam explosion (Jacquet et al., 2015; Pu et al., 2013) and liquid hot water (LHW)-based hydrothermal treatments such as LHW (Liu, 2010; Pu et al., 2013), hot-compressed water (HCW) (Yu et al., 2008; Yuan et al., 2010), and autohydrolysis (Garrote et al., 1999, 2001) pretreatments of the biomass have been widely researched to increase the sugar and bioethanol yield in this context in recent years.

However, it is essential to develop efficient incentive structures (North, 1991) for the primary stakeholders to enhance the research in this field (Konur, 2000, 2002a–c, 2006a, b, 2007a, b). The scientometric analysis has been used in this context to inform the primary stakeholders about the current state of the research in a selected research field (Garfield, 1955; Konur, 2011, 2012a–i, 2015, 2019, 2020a).

As there have been no scientometric studies on the hydrothermal pretreatments of the biomass as of April 2022, this book chapter presents a scientometric study of the research in hydrothermal pretreatments of the biomass. It examines the scientometric characteristics of both the sample and population data presenting scientometric characteristics of these both datasets in the order of documents, authors, publication years, institutions, funding bodies, source titles, countries, Scopus subject categories, keywords, and research fronts, as well as highly cited 25 papers.

13.2 MATERIALS AND METHODS

The search for this study was carried out using Scopus database (Burnham, 2006) in April 2022.

As a first step for the search of the relevant literature, the keywords were selected using the first most-cited 200 papers. The selected keyword list was optimized to obtain a representative sample of papers for the searched research field. This keyword list was provided in the Appendix for future replication studies.

DOI: 10.1201/9781003226536-18

As a second step, two sets of data were used for this study. First, a population sample of around 1,950 papers was used to examine the scientometric characteristics of the population data. Secondly, a sample of 195 most-cited papers, corresponding to 10% of the population papers, was used to examine the scientometric characteristics of these citation classics.

The scientometric characteristics of these both sample and population datasets were presented in the order of documents, authors, publication years, institutions, funding bodies, source titles, countries, Scopus subject categories, keywords, and research fronts.

Lastly, the key scientometric findings for both datasets were discussed to highlight the research landscape for the hydrothermal pretreatments of the biomass. Additionally, a number of brief conclusions were drawn, and a number of relevant recommendations were made to enhance the future research landscape.

13.3 RESULTS

13.3.1 The Most-Prolific Documents in the Hydrothermal Pretreatments of the Biomass

The information on the types of documents for both datasets is given in Table 13.1. The articles and conference papers together with notes dominate both the sample (96%) and population (98%) papers as they are under-represented in the sample papers. Further, review papers and short surveys have a surplus as they are over-represented in the sample papers by 2%.

It is further notable that 98% of the population papers were published in journals while 2% and 1% of them were published in book series and books, respectively. On the contrary, all of the sample papers were published in the journals.

13.3.2 The Most-Prolific Authors in the Hydrothermal Pretreatments of the Biomass

The information about the most-prolific 13 authors with at least 3.1% of sample papers each is given in Table 13.2.

The most-prolific author is Guido Zacchi of Lund University of Sweden with 12.8% of the sample papers working primarily on the steam explosion pretreatment of the biomass, closely followed by his colleague, Mats Galbe of the same institution with 9.2% of the sample papers. Next, John N.

TABLE 13.1
Documents in the Hydrothermal Pretreatments of the Biomass

Documents	Sample Dataset (%)	Population Dataset (%)	Surplus (%)
Article	85.6	92.9	–7.3
Review	4.1	1.7	2.4
Conference paper	9.7	4.4	5.3
Short survey	0.0	0.1	–0.1
Book chapter	0.0	0.6	–0.6
Book	0.0	0.0	0.0
Editorial	0.0	0.0	0.0
Letter	0.0	0.1	–0.1
Note	0.5	0.3	0.2
Sample size	195	1,948	

Sample dataset: the number of papers (%) in the set of 195 highly cited papers. Population dataset: the number of papers (%) in the set of the 1,948 population papers.

TABLE 13.2
Most-Prolific Authors in the Hydrothermal Pretreatments of the Biomass

No.	Author Name	Author Code	Sample Papers %	Population Papers %	Surplus	Institution	Country	HI	N	Res. Front
1	Zacchi, Guido	7006727748	12.8	2.7	10.1	Lund Univ.	Sweden	67	204	SE
2	Galbe, Mats	7003788758	9.2	2.3	6.9	Lund Univ.	Sweden	50	130	SE
3	Saddler, John N.	7005297559	7.7	2.8	4.9	Univ. British Columbia	Canada	96	403	SE
4	Ballesteros, Mercedes*	7006110611	5.6	2.1	3.5	CIEMAT	Spain	49	134	SE, LHW
5	Ballesteros, Ignacio	6602732963	5.6	1.7	3.9	CIEMAT	Spain	37	70	SE, LHW
6	Parajo, Juan C.	7005594129	5.1	1.2	3.9	Univ. Vigo	Spain	61	300	A
7	Negro, Maria J.*	6701512649	5.1	1.1	4.0	CIEMAT	Spain	37	72	SE, LHW
8	Garrote, Gil	6603849654	4.6	1.2	3.4	Univ. Vigo	Spain	42	98	A
9	Oliva, Jose M.	57194220606	4.6	1.1	3.5	CIEMAT	Spain	34	53	SE, LHW
10	Manzanares, Paloma*	55779406300	4.6	1.0	3.6	CIEMAT	Spain	34	68	SE, LHW
11	Ladisch, Michael R.	7005670397	4.6	0.7	2.9	Purdue Univ.	USA	59	290	LHW
12	Dominguez, Herminia*	57226880096	3.1	0.9	2.2	Univ. Vigo	Spain	56	273	A
13	Kim, Youngmi	55699576800	3.1	0.5	2.6	Univ. Wisconsin River Falls	USA	26	35	LHW

Author code: the unique code given by Scopus to the authors. Sample papers: the number of papers authored in the sample dataset. Population papers: the number of papers authored in the population dataset.
HI, H-index; N, number of papers published by each author; *, female; SE, steam explosion pretreatment; LHW, liquid hot water pretreatment; A, autohydrolysis.

Saddler of the University of the British Columbia of Canada publishes 7.2% of the sample papers while Mercedes Ballesteros and Ignacio Ballesteros of the Center for Energy, Environmental and Technological Research (CIEMAT) publish 6.6% of the sample papers each.

The most influential author is Guido Zacchi with 10% surplus, followed by Mats Galbe with 7% surplus. The other influential authors are Michael R. Ladisch, Paloma Manzanares, Maria J. Negro, Ignacio Ballesteros, and Juan C. Parajo with 4% surplus each.

The most-prolific institution for the sample dataset is the CIEMAT with five authors while the University of Vigo and Lund University house three and two authors, respectively. On the other hand, the most-prolific country for the sample dataset is Spain with seven authors while the Sweden and the USA house two authors each.

The most-prolific research front is the steam explosion pretreatment with nine authors, closely followed by the LHW pretreatment with seven authors. Additionally, three authors study primarily autohydrolysis of the biomass.

On the other hand, there is significant gender deficit (Beaudry and Lariviere, 2016) for the sample dataset as surprisingly only four of these top researchers are female.

Additionally, there are other authors with the relatively low citation impact and with 0.8%–1.7% of the population papers each: Hongzhang Chen, Runcang Sun, Andrey Pranovich, Zhenhong Yuan, Shiro Saka, Qiang Yong, Xinshu Zhuang, Qiang Yu, Luiz P. Ramos, Wei Qi, Stefan Wilfor, Feng Xu, Eulogio Castro, Shahab Sokhansanj, Qiong Wang, Wen Wang, and Charles E. Wyman.

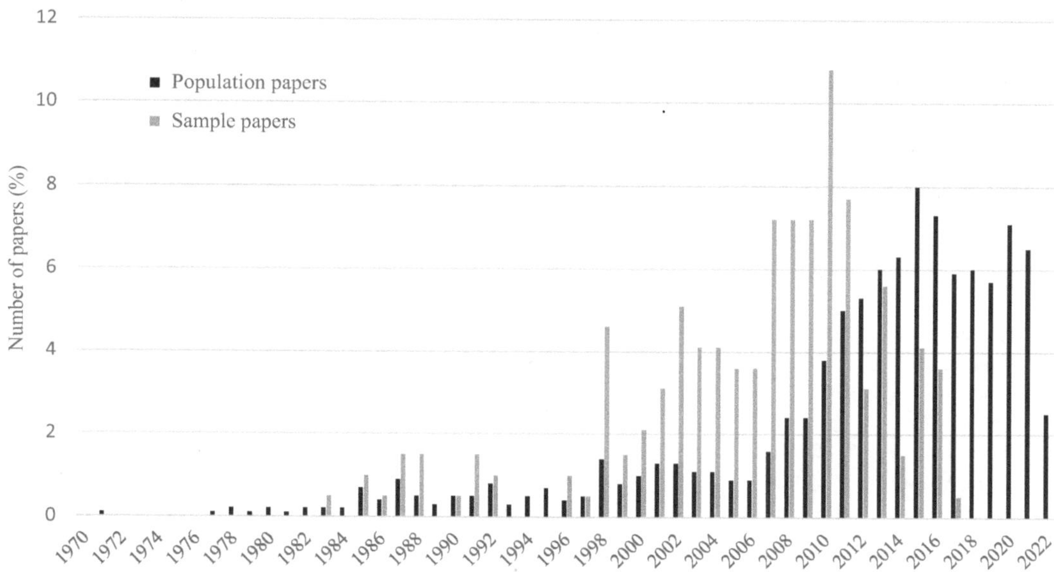

FIGURE 13.1 The research output by years regarding the hydrothermal pretreatments of the biomass.

13.3.3 The Most-Prolific Research Output by Years in Hydrothermal Pretreatments of the Biomass

Information about papers published between 1970 and 2022 is given in Figure 13.1. This figure clearly shows that the bulk of the research papers in the population dataset were published primarily in the 2010s with 59% of the population dataset. The publication rates for the 2020s, 2000s, 1990s, 1980s, and 1970s were 14%, 14%, 6%, 4%, and 1% respectively. Additionally, 0.4% of the population papers were published between 1955 and 1969.

Similarly, the bulk of the research papers in the sample dataset were published in the 2000s and 2010s with 47% and 37% of the sample dataset, respectively. The publication rates for the 1990s, 1980s, and 1970s were 11%, 4%, and 2% of the sample papers, respectively.

The most-prolific publication year for the population dataset was 2015 with 8% while 73% of the population papers were published between 2010 and 2021. Similarly, 79% of the sample papers were published between 2002 and 2016 while the most-prolific publication year was 2010 with 11% of the sample papers.

13.3.4 The Most-Prolific Institutions in the Hydrothermal Pretreatments of the Biomass

Information about the most-prolific 16 institutions publishing papers on the hydrothermal pretreatments of the biomass with at least 2.1% of the sample papers each is given in Table 13.3.

The most-prolific institution is the Lund University with over 13% of the sample papers each while CIEMAT and University of British Columbia are the other prolific institutions with 9% and 6% of the sample papers, respectively. Further, Chinese Academy of Sciences and University of Vigo publish over 5% of the sample papers each.

The top country for these most-prolific institutions is the USA with four institutions, closely followed by Spain with three institutions while Canada, China, and Sweden house two institutions each. In total, eight countries house these top institutions.

On the other hand, the institution with the most citation impact is the Lund University with 10% surplus, closely followed by the CIEMAT with 6% surplus. The other prolific institutions are

TABLE 13.3
The Most-Prolific Institutions in Hydrothermal Pretreatments of the Biomass

No.	Institutions	Country	Sample Papers (%)	Population Papers (%)	Surplus (%)
1	Lund Univ.	Sweden	13.3	3.7	9.6
2	CIEMAT	Spain	8.7	2.7	6.0
3	Univ. British Columbia	Canada	6.2	2.7	3.5
4	Chin. Acad. Sci.	China	5.1	5.0	0.1
5	Univ. Vigo	Spain	5.1	1.8	3.3
6	Georgia Inst. Technol.	USA	4.1	1.2	2.9
7	Purdue Univ.	USA	4.1	0.7	3.4
8	Univ. Hawaii Manoa	USA	4.1	0.7	3.4
9	Dartmouth Coll.	USA	3.1	0.6	2.5
10	Univ. Jaen	Spain	2.6	1.1	1.5
11	S. China Univ. Technol.	China	2.1	2.9	–0.8
12	Tech. Univ. Denmark	Denmark	2.1	1.2	0.9
13	Royal Inst. Technol.	Sweden	2.1	1.1	1
14	VTT Tech. Res. Ctr.	Finland	2.1	1.1	1
15	Forintek Inc.	Canada	2.1	0.7	1.4
16	Budapest Univ. Technol. Econ.	Hungary	2.1	0.6	1.5

University of British Columbia, Purdue University, University of Hawaii Manoa, and University of Vigo with 3% surplus each. Similarly, the institution with the least impact is the South China University of Technology with 1% deficit.

Additionally, there are other institutions with the relatively low citation impact and with 1.1%–2.9% of the population papers each: Beijing Forestry University, Nanjing Forestry University, Kyoto University, University of Sao Paulo, Tianjin University of Science & Technology, Aalto University, Abo Akademi University, State University of New York, Oak Ridge National Laboratory, Federal University of Parana, and Chalmers University of Technology.

13.3.5 The Most-Prolific Funding Bodies in the Hydrothermal Pretreatments of the Biomass

Information about the most-prolific nine funding bodies funding at least 2.1% of the sample papers each is given in Table 13.4. Only 42% and 55% of the sample and population papers were funded.

The most-prolific funding body is the European Commission with 7.7% of the sample papers, closely followed by the Ministry of Science and Technology of China and National Natural Science Foundation of China with 6.7% and 5.1% of the sample papers, respectively.

The most-prolific country for these top funding bodies is China with four funding bodies while the European Union (EU) has two funding bodies. In total, four countries and the EU house these top funding bodies.

The funding body with the most citation impact is the European Commission with 4.1% surplus. The other prolific funding bodies are the Ministry of Science and Technology of China and National Basic Research Program of China with 2.5% and 1.5% surplus, respectively. Similarly, the funding body with the least citation impact is National Natural Science Foundation of China with 5.4% deficit.

TABLE 13.4
The Most-Prolific Funding Bodies in Hydrothermal Pretreatments of the Biomass

No.	Funding Bodies	Country	Sample Paper No. (%)	Population Paper No. (%)	Surplus (%)
1	Eur. Commis.	EU	7.7	3.6	4.1
2	Minist Sci. Technol. China	China	6.7	4.2	2.5
3	Natl. Natr. Sci. Found. China	China	5.1	10.5	–5.4
4	Natl. Key Res. Devnt. Prog. China	China	3.6	3.0	0.6
5	Eur. Reg. Devnt. Fund	EU	3.1	2.4	0.7
6	Natl. Basic Res. Prog. China	China	3.1	1.6	1.5
7	Natl. Counc. Sci. Technol. Devnt.	Brazil	2.6	3.0	–0.4
8	US Dept. Ener.	USA	2.6	2.0	0.6
9	Govnt. Canada	Canada	2.1	1.4	0.7

The other funding bodies with the relatively low citation impact and with 1.0%–1.9% of the population papers each are Ministry of Education, Culture, Sports, Science and Technology of Japan, Japan Society for the Promotion of Science, Higher Education Personnel Improvement Coordination of Brazil, Ministry of Education of China, Fundamental Research Funds for the Central Universities of China, Natural Sciences and Engineering Research Council of Canada, Chinese Academy of Sciences, Ministry of Science, Technology and Innovation of Brazil, Swedish Energy Agency, Sao Paulo Research Foundation of Brazil, Ministry of Finance of Japan, Priority Academic Program Development of Jiangsu Higher Education Institutions of China, Ministry of Economy and Competitiveness of Spain, National High-tech Research and Development Program of China, and the U.S. Department of Agriculture.

13.3.6 The Most-Prolific Source Titles in the Hydrothermal Pretreatments of the Biomass

Information about the most-prolific 12 source titles publishing at least 2.1% of the sample papers each in hydrothermal pretreatments of the biomass is given in Table 13.5.

The most-prolific source title is Bioresource Technology with over 24% of the sample papers. The other prolific journal is Biotechnology and Bioengineering with 9% of the sample papers while Biomass and Bioenergy, Applied Biochemistry and Biotechnology Part A, and Enzyme and Microbial Technology publish over 6% of the sample papers each.

On the other hand, the source title with the most impact is the Bioresource Technology with over 9% surplus, closely followed by Biotechnology and Bioengineering with over 7% surplus. Enzyme and Microbial Technology and Applied Biochemistry and Biotechnology Part A are the other prolific journals with 5% and 4% surplus, respectively. Similarly, the source title with the least impact is the Industrial Crops and Products with 1% deficit.

The other source titles with the relatively low citation impact with 1.5%–4.3% of the population papers each are Applied Biochemistry and Biotechnology, Carbohydrate Polymers, Biomass Conversion and Biorefinery, Wood Science and Technology, Journal of Wood Chemistry and Technology, Energy and Fuels, Fuel, Journal of Chemical Technology and Biotechnology, Process Biochemistry, Cellulose, Bioenergy Research, Renewable Energy, ACS Sustainable Chemistry

TABLE 13.5
The Most-Prolific Source Titles in Hydrothermal Pretreatments of the Biomass

No.	Source Titles	Sample Papers (%)	Population Papers (%)	Surplus (%)
1	Bioresource Technology	24.1	14.8	9.3
2	Biotechnology and Bioengineering	8.7	1.5	7.2
3	Biomass and Bioenergy	6.2	3.6	2.6
4	Applied Biochemistry and Biotechnology Part A	6.2	2.0	4.2
5	Enzyme and Microbial Technology	6.2	1.1	5.1
6	Industrial And Engineering Chemistry Research	5.1	1.8	3.3
7	Biotechnology for Biofuels	4.6	2.9	1.7
8	Process Biochemistry	4.6	1.2	3.4
9	Applied Biochemistry and Biotechnology	3.6	2.4	1.2
10	Industrial Crops and Products	3.1	4.5	−1.4
11	Biotechnology Progress	2.6	0.8	1.8
12	Holzforschung	2.1	2.5	−0.4

and Engineering, Energies, Journal of Wood Science, Waste and Biomass Valorization, and Biotechnology Progress.

13.3.7 The Most-Prolific Countries in the Hydrothermal Pretreatments of the Biomass

Information about the most-prolific 11 countries publishing at least 3.1% of sample papers each in hydrothermal pretreatments of the biomass is given in Table 13.6.

The most-prolific country is the USA with 20% of the sample papers, followed by Spain, Sweden, China, and Canada with 12%–16% of the sample papers each. On the other hand, six European countries listed in Table 13.6 produce 46% and 25% of the sample and population papers, respectively, with 21% surplus.

On the other hand, the country with the most citation impact is Sweden with 9% surplus, closely followed by Spain with 8% surplus. The other influential countries are Canada, the USA, and Hungary with 3%–5% surplus each. Similarly, the country with the least citation impact is China with 11% deficit. Brazil, Japan, and Finland are the other countries with 1%–2% deficit each.

Additionally, there are other countries with relatively low citation impact and with 0.9%–2.9% of the sample papers each: South Korea, Italy, Thailand, France, the UK, Australia, Portugal, India, Malaysia, Austria, Indonesia, Taiwan, Norway, South Africa, Netherlands, Turkey, Mexico, and Switzerland.

13.3.8 The Most-Prolific Scopus Subject Categories in the Hydrothermal Pretreatments of the Biomass

Information about the most-prolific nine Scopus subject categories indexing at least 8% of the sample papers each is given in Table 13.7.

The most-prolific Scopus subject category in the hydrothermal pretreatments of the biomass is Chemical Engineering with 72% of sample papers. The other prolific subject categories are Environmental Science, Biochemistry, Genetics and Molecular Biology, Energy, and Immunology and Microbiology with 36%–42% of the sample papers each.

TABLE 13.6
The Most-Prolific Countries in the Hydrothermal Pretreatments of the Biomass

No.	Countries	Sample Papers (%)	Population Papers (%)	Surplus (%)
1	USA	19.5	15.2	4.3
2	Spain	16.4	8.5	7.9
3	Sweden	15.4	6.8	8.6
4	China	14.9	25.7	−10.8
5	Canada	11.8	7.0	4.8
6	Japan	5.6	6.9	−1.3
7	Finland	4.6	5.2	−0.6
8	Brazil	4.1	5.8	−1.7
9	Hungary	3.6	0.9	2.7
10	Denmark	3.1	2.2	0.9
11	Germany	3.1	1.6	1.5

TABLE 13.7
The Most-Prolific Scopus Subject Categories in the Hydrothermal Pretreatments of the Biomass

No.	Scopus Subject Categories	Sample Papers (%)	Population Papers (%)	Surplus (%)
1	Chemical Engineering	72.3	48.3	24.0
2	Environmental Science	41.5	37.2	4.3
3	Biochemistry, Genetics and Molecular Biology	41.5	20.7	20.8
4	Energy	41.0	39.6	1.4
5	Immunology and Microbiology	35.9	14.7	21.2
6	Chemistry	14.9	17.6	−2.7
7	Agricultural and Biological Sciences	12.3	20.3	−8.0
8	Engineering	8.7	13.5	−4.8
9	Materials Science	8.2	17.0	−8.8

On the other hand, the Scopus subject category with the most citation impact is the Chemical Engineering with 24% surplus, closely followed by Immunology and Microbiology and Biochemistry, Genetics and Molecular Biology with 21% surplus each.

Similarly, the Scopus subject category with the least citation impact is Materials Science with 9% deficit, closely followed by Agricultural and Biological Sciences. The other such categories are Engineering and Chemistry with 5% and 3% deficit, respectively.

13.3.9 The Most-Prolific Keywords in the Hydrothermal Pretreatments of the Biomass

Information about the keywords used in at least 8% or 5% of the sample or population papers, respectively, is given in Table 13.8. For this purpose, keywords related to the keyword set given in the appendix are selected from a list of the most-prolific keyword set provided by Scopus database.

These keywords are grouped under the four headings: biomass, hydrothermal pretreatments, other processes, and products of the hydrothermal pretreatments.

TABLE 13.8
The Most-Prolific Keywords in Hydrothermal Pretreatments of the Biomass

No.	Keywords	Sample Papers (%)	Population Papers (%)	Surplus (%)
1	**Biomass**			
	Cellulose	55.4	41.5	13.9
	Lignin	52.8	38.8	14.0
	Biomass	38.5	24.6	13.9
	Zea mays (corn, maize)	27.7	14.2	13.5
	Wood	25.6	14.7	10.9
	Hemicellulose	25.1	14.4	10.7
	Lignocellulose	20.5	10.1	10.4
	Triticum aestivum (wheat)	15.4	12.9	2.5
	Straw	12.3	9.3	3.0
	Xylan	10.3	6.0	4.3
	Wheat straw	9.7	9.5	0.2
	Corn stover	9.2	4.3	4.9
	Lignocellulosic biomass	8.7	8.2	0.5
	Glucan	7.7	3.5	4.2
	Bagasse	6.7	7.3	–0.6
2	**Hydrothermal Pretreatment**			
	Steam	39.0	26.3	12.7
	Water vapor	29.2	13.6	15.6
	Water	25.1	23.6	1.5
	Steam explosion	20.5	21.1	–0.6
	Temperature	17.4	9.5	7.9
	Explosion	12.8	15.6	–2.8
	Steam pretreatment	11.3	3.7	7.6
	Autohydrolysis	10.8	9.1	1.7
	Hot water	10.3	7.1	3.2
	Hot temperature	7.2	5.5	1.7
	Heat	7.2	4.9	2.3
	Hydrothermal pretreatment	5.1	10.1	–5.0
	Water treatment	4.1	5.7	–1.6
3	**Other Processes**			
	Hydrolysis	66.2	38.6	27.6
	Enzymatic hydrolysis	36.9	27.5	9.4
	Fermentation	29.7	18.2	11.5
	Enzyme activity	28.7	19.0	9.7
	Enzymes	26.7	12.2	14.5
	Cellulase	21.0	9.5	11.5
	Pretreatment	18.5	13.1	5.4
	Pre-treatment	17.9	15.7	2.2
	Saccharification	15.9	14.1	1.8
	Saccharomyces cerevisiae	14.4	5.2	9.2

(*Continued*)

TABLE 13.8 (*Continued*)
The Most-Prolific Keywords in Hydrothermal Pretreatments of the Biomass

No.	Keywords	Sample Papers (%)	Population Papers (%)	Surplus (%)
	Degradation	11.8	7.4	4.4
	Ethanol production	10.8	5.0	5.8
	Sulfuric acid	10.8	4.3	6.5
	Delignification	9.7	7.6	2.1
	Alcohol production	9.7	4.1	5.6
	Fractionation	8.7	4.9	3.8
	Acetic acid	7.7	5.4	2.3
	Yeast	7.7	5.3	2.4
	Ph	7.7	5.2	2.5
	Enzymolysis	7.2	5.8	1.4
	Extraction	6.7	11.3	–4.6
4	**Products**			
	Ethanol	36.4	20.9	15.5
	Sugar	34.4	19.0	15.4
	Glucose	29.2	19.7	9.5
	Alcohol	22.6	8.6	14
	Xylose	17.9	8.7	9.2
	Biofuel	16.9	13.1	3.8
	Polysaccharides	13.3	10.2	3.1
	Bioethanol	11.3	9.4	1.9
	Furfural	5.6	4.5	1.1

There are 15 keywords selected related to the biomass and biomass constituents: The prolific keywords are cellulose, lignin, biomass, *Zea mays* (corn, maize), wood, hemicellulose, and lignocellulose with 21%–55% of the sample papers each.

The prolific keywords related to the hydrothermal pretreatments of the biomass are steam, water vapor, water, steam explosion, and temperature with 17%–39% of the sample papers each while those related to the other processes are hydrolysis, enzymatic hydrolysis, fermentation, enzyme activity, enzymes, and cellulase with 21%–66% of the sample papers each.

The prolific keywords related to the products of the hydrothermal pretreatments are ethanol, sugar, glucose, alcohol, xylose, biofuel, and polysaccharides with 13%–36% of the sample papers each.

13.3.10 The Most-Prolific Research Fronts in Hydrothermal Pretreatments of the Biomass

Information about the research fronts for the sample papers in hydrothermal pretreatments of the biomass is given in Table 13.9. As Table 13.9 shows, there are two primary research fronts for this field: steam explosion pretreatment and LHW-based hydrothermal pretreatments with 58% and 48% of the 195 highly cited papers.

The prolific research fronts in the latter research front are LHW pretreatment, autohydrolysis pretreatment, HCW pretreatment, and hydrothermal pretreatment in general with 7%–13% of the sample papers each.

Information about the research fronts for the sample papers in hydrothermal pretreatments of the biomass with regard to the biomass used in these pretreatments is given in Table 13.10. As

TABLE 13.9
The Most-Prolific Research Fronts for the Hydrothermal Pretreatments of the Biomass

No.	Research Fronts	N Paper % Sample
1	Steam explosion pretreatment	57.9
2	**Liquid Water-based Hydrothermal Pretreatments**	**39.5**
	LHW pretreatment	12.8
	Autohydrolysis pretreatment	11.3
	HCW pretreatment	8.7
	Hydrothermal pretreatment in general	6.7

N paper (%) sample: the number of papers in the population sample of 195 papers.

TABLE 13.10
The Most-Prolific Research Fronts for the Biomass for the Hydrothermal Pretreatments of the Biomass

No.	Research Fronts	N Paper % Sample
1	**Biomass Constituents**	**29.7**
	Lignin	13.8
	Cellulose	9.7
	Hemicellulose	4.1
	Other constituents	2.1
2	**Wood**	**39.5**
	Wood	5.1
	Softwood	14.9
	Spruce	3.6
	Pine	2.6
	Hardwood	19.5
	Eucalyptus	5.6
	Poplar	4.1
3	Grass	3.1
4	**Agricultural Residues**	**42.1**
	Other residues	12.8
	Corn stover	11.8
	Wheat straw	8.7
	Sugarcane bagasse	6.2
	Corncobs	2.6
5	**Biomass in General**	**14.4**
	Biomass	8.7
	Lignocellulosic biomass	5.6

N paper (%) sample: the number of papers in the population sample of 195 papers.

Table 13.10 shows, there are four primary research fronts for this field: agricultural residues, wood, biomass constituents, and biomass in general with 42%, 40%, 30%, and 14% of the 195 highly cited papers, respectively. The other research front is grass with 3% of the sample papers.

The most-prolific agricultural residues are corn stover, wheat straw, and sugarcane bagasse with 12%, 9%, and 6% of the sample papers, respectively. The most-prolific wood types are softwood and hardwood with 15% and 20% of the sample papers, respectively, while spruce, pine, eucalyptus, and poplar are the most-prolific wood species. Further, lignin and cellulose are the most-studied biomass constituents with 14% and 10% of the sample papers. Finally, there are two research fronts for the field of biomass in general: biomass and lignocellulosic biomass with 9% and 6% of the sample papers, respectively.

13.4 DISCUSSION

13.4.1 Introduction

The crude oil-based gasoline fuels have been widely used in the transportation sector since the 1920s. However, there have been great public concerns over the adverse environmental and human impact of these fuels. Hence, biomass-based bioethanol fuels have increasingly been used in blending gasoline fuels in the fuel cells, and in the biochemical production in a biorefinery context.

The research in the field of hydrothermal pretreatments of the biomass such as steam explosion and LHW-based hydrothermal pretreatments to increase the sugar and bioethanol yield has also intensified in recent years.

However, it is essential to develop efficient incentive structures for the primary stakeholders to enhance the research in this field. This is especially important to maintain energy security in the cases of supply shocks such as oil shocks or war-related chocks as in the case of Russian invasion of Ukraine.

The scientometric analysis has been used in this context to inform the primary stakeholders about the current state of the research in a selected research field.

As there have been no scientometric studies on the hydrothermal pretreatments of the biomass, this book chapter presents a scientometric study of the research in the hydrothermal pretreatments of the biomass. It examines the scientometric characteristics of both the sample and population data presenting scientometric characteristics of these both datasets in the order of documents, authors, publication years, institutions, funding bodies, source titles, countries, Scopus subject categories, keywords, and research fronts.

As a first step for the search of the relevant literature, the keywords were selected using the first most-cited 200 papers. The selected keyword list was optimized to obtain a representative sample of papers for the searched research field. This keyword list was provided in the Appendix for future replication studies. Further, a selected list of the keywords is presented in Table 13.8.

As a second step, two sets of data were used for this study. First, a population sample of over 1,948 papers was used to examine the scientometric characteristics of the population data. Secondly, a sample of 195 most-cited papers, corresponding to 10% of the population dataset, was used to examine the scientometric characteristics of these citation classics.

The scientometric characteristics of these both sample and population datasets were presented in the order of documents, authors, publication years, institutions, funding bodies, source titles, countries, Scopus subject categories, keywords, and research fronts.

Lastly, the key scientometric findings for both datasets were discussed to highlight the research landscape for the hydrothermal pretreatments of the biomass. Additionally, a number of brief conclusions were drawn, and a number of relevant recommendations were made to enhance the future research landscape.

13.4.2 The Most-Prolific Documents in the Hydrothermal Pretreatments of the Biomass

Articles (together with conference papers) dominate the sample and population papers (Table 13.1). Further, review papers and articles have a surplus and deficit, respectively.

Scopus differs from the Web of Science database in differentiating and showing articles (86%) and conference papers (10%) published in the journals separately. However, it should be noted that these conference papers are also published in journals as articles, compared to those published only in the conference proceedings. Similarly, Scopus differs from Web of Science database in introducing short surveys (0%). Hence, the total number of articles and review papers in the sample dataset are 98% and 4%, respectively.

It is observed during the search process that there has been inconsistency in the classification of the documents in Scopus as well as in other databases such as Web of Science. This is especially relevant for the classification of papers as reviews or articles as the papers not involving a literature review may be erroneously classified as a review paper. There is also a case of review papers being classified as articles.

In this context, it would be helpful to provide a classification note for the published papers in the books and journals at the first instance. It would also be helpful to use the document types listed in Table 13.1 for this purpose. Book chapters may also be classified as articles or reviews as an additional classification to differentiate review chapters from the experimental chapters as it is done by the Web of Science. It would be further helpful to additionally classify the conference papers as articles or review papers as well as it is done in the Web of Science database.

13.4.3 The Most-Prolific Authors in the Hydrothermal Pretreatments of the Biomass

There have been most-prolific 13 authors with at least 3.1% of the sample papers each as given in Table 13.2. These authors have shaped the development of the research in this field.

The most-prolific authors are Guido Zacchi and Mats Galbe of Lund University of Sweden, John N. Saddler of the University of the British Columbia of Canada, Mercedes Ballesteros and Ignacio Ballesteros of the CIEMAT working primarily on steam explosion pretreatments of the biomass.

It is important to note the inconsistencies in indexing of the author names in Scopus and other databases. It is especially an issue for the names with more than two components such as 'Blake Sam de Hyun Zacchi'. The probable outcomes are 'Zacchi, B.S.D.H.', 'de Hyun Zacchi, B.S.', or 'Hyun Zacchi, B.S.D.'. The first choice is the gold standard of the publishing sector as the last word in the name is taken as the last name. In most of the academic databases such as PUBMED and EBSCO, this version is used predominantly. The second choice is a strong alternative while the last choice is an undesired outcome as two last words are taken as the last name. It is good practice to combine the words of the last name by a hyphen: 'Hyun-Zacchi, B.S.D.'. It is notable that inconsistent indexing of the author names may cause substantial inefficiencies in the search process for the papers as well as allocating credit to the authors as there are different author entries for each outcome in the databases.

There are also inconsistencies in the shortening Chinese names. For example, 'Yuoying Zhang' is often shortened as 'Zhang, Y.', 'Zhang, Y.-Y.', and 'Zhang Y.Y' as it is done in the Web of Science database as well. However, the gold standard in this case is 'Zhang Y' where the last word is taken as the last name and the first word is taken as a single forename. In most of the academic databases such as PUBMED and EBSCO, this first version is used predominantly. However, there is an increasing trend to use the second option: 'Zhang, Y.Y.' to differentiate Chinese names efficiently. Therefore, there have been difficulties to locate papers for the Chinese authors. In such cases, the use of the unique author codes provided for each author by the Scopus database has been helpful.

There is also a difficulty in allowing credit for the authors especially for the authors with common names such as 'Zhang, X', or 'Huang, X', or 'Zhu, X' in conducting scientometric studies. These difficulties strongly influence the efficiency of the scientometric studies as well as allocating credit to the authors as there are the same author entries for different authors with the same name, for example, 'Zhang X' in the databases.

In this context, the coding of authors in Scopus database is a welcome innovation compared to the other databases such as Web of Science. In this process, Scopus allocates a unique number to each author in the database (Aman, 2018). However, there might still be substantial inefficiencies in this coding system especially for common names. For example, some of the papers for a certain author may be allocated to another researcher with a different author code. It is possible that Scopus uses a number of software programs to differentiate the author names and the program may not be false-proof (D'Angelo and van Eck, 2020).

In this context, it does not help that author names are not given in full in some journals and books. This makes difficult to differentiate authors with common names and makes the scientometric studies further difficult in the author domain. Therefore, the author names should be given in all books and journals at the first instance. There is also a cultural issue where some authors do not use their full names in their papers. Instead, they use initials for their forenames: 'Zacchi, H.J.' or just 'Zacchi' instead of 'Zacchi, Hyun Jae'.

There are also inconsistencies in naming of the authors with more than two components by the authors themselves in journal papers and book chapters. For example, 'Zacchi, A.P.C.', 'Galbe, C.E.', and 'Negro, S.J.' might be given as 'Zacchi, A', 'Galbe, C', or 'Negro, S' in the journals and books, respectively. This also makes the scientometric studies difficult in the author domain. Hence, contributing authors should use their name consistently in their publications.

The other critical issue regarding the author names is the inconsistencies in the spelling of the author names in the national spellings (e.g., Gonçalves, Üşengeç) rather than in the English spellings (e.g., Goncalves, Usengec) in Scopus database. Scopus differs from the Web of Science database and many other databases in this respect where the author names are given only in the English spellings. It is observed that national spellings of the author names do not help in conducting scientometric studies as well as in allocating credits to the authors as sometimes there are the different author entries for the English and National spellings in the Scopus database.

The most-prolific institutions for the sample dataset are the CIEMAT, University of Vigo, and Lund University. The most-prolific countries for the sample dataset are Spain, Sweden, the USA, and the UK.

The most-prolific research fronts are the steam explosion and LHW pretreatment of the biomass. The other front is the autohydrolysis pretreatment of the biomass.

It is also notable that there is significant gender deficit for the sample dataset as surprisingly only four of these top researchers are female. This finding is the most thought-provoking with strong public policy implications. Hence, institutions, funding bodies, and policy-makers should take efficient measures to reduce the gender deficit in this field as well as other scientific fields with strong gender deficit. In this context, it is worth to note the level of representation of the researchers from the minority groups in science on the basis of race, sexuality, age, and disability, besides the gender (Blankenship, 1993; Dirth and Branscombe, 2017; Konur, 2000, 2002a–c, 2006a, b, 2007a, b).

13.4.4 The Most-Prolific Research Output by Years in the Hydrothermal Pretreatments of the Biomass

The research output observed between 1970 and 2021 is illustrated in Figure 13.1. This figure clearly shows that the bulk of the research papers in the population dataset were published primarily in the 2010s. Similarly, the bulk of the research papers in the sample dataset were published in the last two decades.

These findings suggest that the most-prolific sample and population papers were primarily published in the last two decades. These are the thought-provoking findings as there has been no significant research in this field in the pre-2000s, but there has been significant research boom in the last two decades. In this context, the increasing public concerns about climate change (Change, 2007), greenhouse gas emissions (Carlson et al., 2017), and global warming (Kerr, 2007) have been certainly behind the boom in the research in this field in the last two decades.

Based on these findings, the size of the population papers likely to more than double in the current decade provided that the public concerns about climate change, greenhouse gas emissions, and global warming are translated efficiently to the research funding in this field.

13.4.5 The Most-Prolific Institutions in the Hydrothermal Pretreatments of the Biomass

The most-prolific 16 institutions publishing papers on the hydrothermal pretreatments of the biomass with at least 2.1% of the sample papers each given in Table 13.3 have shaped the development of the research in this field.

The most-prolific institutions are the Lund University, CSIC, University of British Columbia, Chinese Academy of Sciences, and University of Vigo.

The top countries for these most-prolific institutions are the USA, Spain, Canada, China, and Sweden. In total, only eight countries house these top institutions.

On the other hand, the institutions with the most and least impact are Lund University and South China University, respectively. These findings confirm the dominance of the US, Canadian, Chinese, and European institutions in this research field.

13.4.6 The Most-Prolific Funding Bodies in the Hydrothermal Pretreatments of the Biomass

The most-prolific nine funding bodies funding at least 2.1% of the sample papers each are given in Table 13.4. It is notable that only 42% and 55% of the sample and population papers were funded.

The most-prolific funding bodies are the European Commission, Ministry of Science and Technology of China, and National Natural Science Foundation of China.

The most-prolific countries for these top funding bodies are China and the EU. In total, four countries and the EU house these top funding bodies. The funding bodies with the most and least citation impact are European Commission and the National Natural Science Foundation of China, respectively.

These findings on the funding of the research in this field suggest that the level of the funding, mostly in the last two decades, is highly intensive, and it has been largely instrumental in enhancing the research in this field (Ebadi and Schiffauerova, 2016) in light of North's institutional framework (North, 1991). However, considering the relatively low levels of funding especially for the sample papers, there is ample room to enhance funding in this field.

13.4.7 The Most-Prolific Source Titles in Hydrothermal Pretreatments of the Biomass

The most-prolific 12 source titles publishing at least 2.1% of the sample papers each in hydrothermal pretreatments of the biomass have shaped the development of the research in this field (Table 13.5).

The most-prolific source titles are Bioresource Technology, Biotechnology and Bioengineering, Biomass and Bioenergy, Applied Biochemistry and Biotechnology Part A, and Enzyme and Microbial Technology.

It is notable that these top source titles are primarily related to the biomass, bioresources, biotechnology, and bioenergy. This finding suggests that the journals in this field have significantly shaped the development of the research in this field as they focus on the hydrothermal pretreatment aspects of bioethanol fuels.

13.4.8 The Most-Prolific Countries in Hydrothermal Pretreatments of the Biomass

The most-prolific 11 countries publishing at least 3.1% of the sample papers each have significantly shaped the development of the research in this field (Table 13.6).

The most-prolific countries are the USA, Spain, Sweden, China, and Canada. On the other hand, the countries with the most and least citation impact are Sweden and China, respectively. Additionally, it is notable that Japan, Brazil, and Finland were under-represented in the sample papers while Sweden, Spain, Canada, and the USA were over-represented. On the other hand, six

European countries listed in Table 13.6 produce 46% and 25% of the sample and population papers, respectively, with 21% surplus.

The close examination of these findings suggests that Europe, China, and Canada, the USA are the major producers of the research in this field. It is a fact that the USA has been a major player in science (Leydesdorff and Wagner, 2009; Leydesdorff et al., 2014). The USA has further developed a strong research infrastructure to support its corn- and grass-based bioethanol industry (Vadas et al., 2008).

However, China has been a rising star in scientific research in competition with the USA and Europe (Leydesdorff and Zhou, 2005). China is also a major player in this field as a major producer of bioethanol (Li and Chan-Halbrendt, 2009).

Next, Europe has been a persistent player in the scientific research in competition with both the USA and China (Leydesdorff, 2000). Europe has also been a persistent producer of bioethanol along with the USA and Brazil (Gnansounou, 2010).

Additionally, Brazil has also been a persistent player in scientific research at a moderate level (Leta and de Meis, 1996). Brazil has also developed a strong research infrastructure to support its biomass-based bioethanol industry (Macedo et al., 2008).

13.4.9 The Most-Prolific Scopus Subject Categories in Hydrothermal Pretreatments of the Biomass

The most-prolific nine Scopus subject categories indexing at least 8% of the sample papers each, respectively, given in Table 13.7 have shaped the development of the research in this field.

The most-prolific Scopus subject categories in the hydrothermal pretreatments of the biomass are Chemical Engineering, Environmental Science, Biochemistry, Genetics and Molecular Biology, Energy, and Immunology and Microbiology.

These findings are thought-provoking suggesting that the primary subject categories are Chemical Engineering and Environmental Science. The other key finding is that social sciences are not well represented in both the sample and population papers, as in the most fields in bioethanol fuels. These findings are not surprising as the key research fronts in this field relate to the development and application of hydrothermal pretreatments of the biomass.

13.4.10 The Most-Prolific Keywords in Hydrothermal Pretreatments of the Biomass

A limited number of keywords have shaped the development of the research in this field as shown in Table 13.8 and the Appendix.

These keywords are grouped under the four headings: biomass and biomass constituents, hydrothermal pretreatments, other processes, and products of the hydrothermal pretreatments.

These findings suggest that it is necessary to determine the keyword set carefully to locate the relevant research in each of these research fronts. Additionally, the size of the samples for each keyword highlights the intensity of the research in the relevant research areas. The relevant keywords are presented in Table 13.7 and the Appendix.

13.4.11 The Most-Prolific Research Fronts in Hydrothermal Pretreatments of the Biomass

As Table 13.9 shows, there are two primary research fronts for this research field: steam explosion pretreatment and HLW-based hydrothermal pretreatments. The prolific research fronts in the latter research front are LHW pretreatment, autohydrolysis pretreatment, HCW pretreatment, and hydrothermal pretreatment in general.

Information about the research fronts for the sample papers in hydrothermal pretreatments of the biomass with regard to the biomass or biomass constituents used in these pretreatments is given in

Table 13.10. As Table 13.10 shows, there are four primary research fronts for this field: agricultural residues, wood, biomass constituents, and biomass in general.

The most-prolific agricultural residues are corn stover, wheat straw, and sugarcane bagasse while the most-prolific wood types are softwood and hardwood where spruce, pine, eucalyptus, and poplar are the most-prolific wood species. Further, lignin and cellulose are the most-studied biomass constituents. Finally, there are two research fronts for the field of biomass in general: biomass in general and lignocellulosic biomass.

These findings are thought-provoking in seeking ways to increase bioethanol yield through the hydrothermal pretreatments of the biomass at the global scale. It is clear that all of these research fronts with regard to both pretreatment types and biomass types have public importance and merit substantial funding and other incentives.

In the end, these most-cited papers in this field hint that the efficiency of bioethanol fuels and their derivatives could be optimized using the structure, processing, and property relationships of these hydrothermal pretreatments (Formela et al., 2016; Konur, 2018a, 2020b, 2021a–d; Konur and Matthews, 1989).

13.5 CONCLUSION AND FUTURE RESEARCH

The research on the hydrothermal pretreatments of the biomass has been mapped through a scientometric study of both sample (195 papers) and population (1,948 papers) datasets.

The critical issue in this study has been to obtain a representative sample of the research as in any other scientometric study. Therefore, the keyword set has been carefully devised and optimized after a number of runs in the Scopus database. It is a representative sample of the wider population studies. This keyword set was provided in the Appendix, and the relevant keywords are presented in Table 13.8.

The other issue has been the selection of a multidisciplinary database to carry out the scientometric study of the research in this field. For this purpose, Scopus database has been selected. The journal coverage of this database has been notably wider than that of Web of Science and other multi-subject databases.

The key scientometric properties of the research in this field have been determined and discussed in this book chapter. It is evident that a limited number of documents, authors, institutions, publication periods, institutions, funding bodies, source titles, countries, Scopus subject categories, Scopus keywords, and research fronts have shaped the development of the research in this field.

There is ample scope to increase the efficiency of the scientometric studies in this field in the author and document domains by developing consistent policies and practices in both domains across all the academic databases. In this respect, it seems that authors, journals, and academic databases have a lot to do. Furthermore, the significant gender deficit as in most scientific fields emerges as a public policy issue. The potential deficits on the basis of age, race, disability, and sexuality need also to be explored in this field as in other scientific fields.

The research in this field has boomed in the last two decades possibly promoted by the public concerns on global warming, greenhouse gas emissions, and climate change. The institutions from the USA, Canada, Europe, and China have mostly shaped the research in this field.

The relatively well funding rate of 55% for the population papers suggests that funding in this field significantly enhanced the research in this field primarily in the last two decades, possibly more than doubling in the current decade. However, it is evident that there is ample room for more funding and other incentives to enhance the research in this field further as only 42% of the sample papers declared any funding.

The most-prolific source titles have been mostly indexed by the subject categories of Chemical Engineering and Environmental Science as the focus of the sample papers has been on the development and utilization of the hydrothermal pretreatments of the biomass to increase the sugar and bioethanol yield at large.

Europe, the USA, China, and Canada have been the major producers of the research in this field as the major producers and users of bioethanol fuels from different types of biomass such as corn, sugarcane, and grass as well as other types of biomass. It is evident that these countries have well-developed research infrastructure in bioethanol fuels and their derivatives.

The primary Scopus subject categories have been Chemical Engineering and Environmental Science as the focus of the research has been on the development and utilization of the hydrothermal pretreatments of the biomass to increase the sugar and bioethanol yield at large. Further, social sciences are not well represented in both the sample and population papers, as in the most fields in bioethanol fuels. These findings are not surprising.

It emerges that ethanol is more popular than bioethanol as a keyword with strong implications for the search strategy. In other words, the search strategy using only bioethanol keyword would not be much helpful. The keywords are grouped under the four headings: biomass and biomass constituents, hydrothermal pretreatment, other processes, and products of the hydrothermal pretreatments. These groups of keywords highlight the potential primary research fronts for these fields.

There are two primary research fronts with regard to the hydrothermal pretreatment of the steam explosion and HLW-based hydrothermal pretreatments. The prolific research fronts in the latter research front are LHW pretreatment, autohydrolysis pretreatment, HCW pretreatment, and hydrothermal pretreatment in general.

Similarly, there are four primary research fronts for the biomass or biomass constituents used for the hydrothermal pretreatments of the biomass: agricultural residues, wood, biomass constituents, and biomass in general. Further, the most-prolific specific biomass in these four classes of biomass are corn stover, wheat straw, sugarcane bagasse, softwood, hardwood, spruce, pine, eucalyptus, poplar, lignin, cellulose, biomass, and lignocellulosic biomass.

These findings are thought-provoking. The focus of these most-cited 195 papers as well as 1,948 population papers is the development and utilization of hydrothermal pretreatments to increase the sugar and bioethanol yield. These studies highlight strong structure–processing–property relationships for hydrothermal pretreatments for bioethanol fuels and their derivatives.

Thus, the scientometric analysis has a great potential to gain valuable insights into the evolution of the research in this field as in other scientific fields especially in the aftermath of the significant global supply shocks such as the Russian invasion of Ukraine.

It is recommended that further scientometric studies are carried out for the primary types of the hydrothermal pretreatments and the biomass used as well as about the other types of pretreatments. It is further recommended that reviews of the most-cited papers are carried out for each research front to complement these scientometric studies. Next, the scientometric studies of the hot papers in these primary fields are carried out.

ACKNOWLEDGMENTS

The contribution of the highly cited researchers in the field of the hydrothermal pretreatments of the biomass has been gratefully acknowledged.

APPENDIX: THE KEYWORD SET FOR HYDROTHERMAL PRETREATMENTS OF THE BIOMASS

TITLE ('steam explosion' OR 'steam pretreat*' OR 'steaming pretreat*' OR 'steam treat*' OR 'steam explod*' OR 'liquid hot water' OR 'hot water' OR 'hot-compressed water' OR 'hot compressed liquid water' OR 'compressed hot water' OR 'hydrothermal pretreat*' OR 'hydrothermal pre-treat*' OR 'low steam' OR 'steam assisted' OR autohydrolysis OR 'wet torrefaction' OR 'hydrothermally pretreat*' OR autoclaving OR 'pressurized hot water' OR 'steam fractionat*' OR aquasolve OR lhw) AND TITLE (*mannan OR waste* OR *grass OR grass* OR biomass* OR lignocellul* OR bagasse* OR cellulos* OR *wood OR woody OR *woods OR *algae OR *algal OR microalga OR olive OR chlamydomonas OR straw OR stover* OR eucalyptus OR miscanthus OR pine OR chlorella OR lignin OR cellulolytic OR hemicellulose* OR xylan OR porphyridium OR nannochloropsis OR poplar* OR cedar OR pulp* OR 'corn cob*' OR spruce OR beech OR oak OR pinus OR residue* OR willow* OR cypress OR sawdust OR prosopis OR 'rice hull*' OR '*bean hull*' OR husk* OR birch OR cynara OR '*cane tops' OR 'cotton stalk*' OR 'corn stalk*' OR alfalfa OR 'maize stem*' OR cornstalk* OR hyacinth OR salix OR corncob* OR 'fruit bunch*' OR populus OR sage OR 'orange peel*' OR '*weed stem*' OR bamboo* OR birch OR '*flower stalk*' OR 'barley hull*' OR 'maize stem*') AND NOT (TITLE (sludge OR anaerobic OR nano* OR soil* OR fiber* OR biogas OR methane OR lauan OR electricity OR carbonization OR liquefaction OR dialdehyde OR hydrogen OR composite* OR gasification OR supercapacitor* OR ester* OR char OR epoxy OR combustion OR milk OR biomethane OR levulinic OR antioxidant* OR aldehyde OR hydrochar OR polyols OR vapours OR 'bio-oil') OR SUBJAREA (phar OR medi OR vete OR nurs OR heal)) AND (LIMIT-TO (SRCTYPE, 'j') OR LIMIT-TO (SRCTYPE, 'k') OR LIMIT-TO (SRCTYPE, 'b')) AND (LIMIT-TO (DOCTYPE, 'ar') OR LIMIT-TO (DOCTYPE, 'cp') OR LIMIT-TO (DOCTYPE, 're') OR LIMIT-TO (DOCTYPE, 'ch') OR LIMIT-TO (DOCTYPE, 'no') OR LIMIT-TO (DOCTYPE, 'le') OR LIMIT-TO (DOCTYPE, 'sh')) AND (LIMIT-TO (LANGUAGE, 'English'))

REFERENCES

Alvira, P., E. Tomas-Pejo, M. Ballesteros and M. J. Negro. 2010. Pretreatment technologies for an efficient bioethanol production process based on enzymatic hydrolysis: A review. *Bioresource Technology* 101:4851–4861.

Aman, V. 2018. Does the Scopus author ID suffice to track scientific international mobility? A case study based on Leibniz laureates. *Scientometrics* 117:705–720.

Angelici, C., B. M. Weckhuysen and P. C. A. Bruijnincx. 2013. Chemocatalytic conversion of ethanol into butadiene and other bulk chemicals. *ChemSusChem* 6:1595–1614.

Antolini, E. 2007. Catalysts for direct ethanol fuel cells. *Journal of Power Sources* 170:1–12.

Antolini, E. 2009. Palladium in fuel cell catalysis. *Energy and Environmental Science* 2:915–931.

Beaudry, C. and V. Lariviere. 2016. Which gender gap? Factors affecting researchers' scientific impact in science and medicine. *Research Policy* 45:1790–1817.

Blankenship, K. M. 1993. Bringing gender and race in: US employment discrimination policy. *Gender & Society* 7:204–226.

Burnham, J. F. 2006. Scopus database: A review. *Biomedical Digital Libraries* 3:1–8.

Carlson, K. M., J. S. Gerber and D. Mueller, et al. 2017. Greenhouse gas emissions intensity of global croplands. *Nature Climate Change* 7:63–68.

Change, C. 2007. Climate change impacts, adaptation and vulnerability. *Science of the Total Environment* 326:95–112.

D'Angelo, C. A. and N. J. van Eck. 2020. Collecting large-scale publication data at the level of individual researchers: A practical proposal for author name disambiguation. *Scientometrics* 123:883–907.

Dirth, T. P. and N. R. Branscombe. 2017. Disability models affect disability policy support through awareness of structural discrimination. *Journal of Social Issues* 73:413–442.

Ebadi, A. and A. Schiffauerova. 2016. How to boost scientific production? A statistical analysis of research funding and other influencing factors. *Scientometrics* 106:1093–1116.

Fernando, S., S. Adhikari, C. Chandrapal and M. Murali. 2006. Biorefineries: Current status, challenges, and future direction. *Energy & Fuels* 20:1727–1737.

Formela, K., A. Hejna, L. Piszczyk, M. R. Saeb and X. Colom. 2016. Processing and structure-property relationships of natural rubber/wheat bran biocomposites. *Cellulose* 23:3157–3175.

Garfield, E. 1955. Citation indexes for science. *Science* 122:108–111.

Garrote, G., H. Dominguez and J. C. Parajo. 1999. Mild autohydrolysis: An environmentally friendly technology for xylooligosaccharide production from wood. *Journal of Chemical Technology and Biotechnology* 74:1101–1109.

Garrote, G., H. Dominguez and J. C. Parajo. 2001. Kinetic modelling of corncob autohydrolysis. *Process Biochemistry* 36:571–578.

Gnansounou, E. 2010. Production and use of lignocellulosic bioethanol in Europe: Current situation and perspectives. *Bioresource Technology* 101:4842–4850.

Hamilton, J. D. 2003. What is an oil shock? *Journal of Econometrics* 113:363–398.

Hamilton, J. D. 2009. Causes and consequences of the oil shock of 2007–08. *Brookings Papers on Economic Activity* 2009:215–261.

Hill, J., E. Nelson, D. Tilman, S. Polasky and D. Tiffany. 2006. Environmental, economic, and energetic costs and benefits of biodiesel and ethanol biofuels. *Proceedings of the National Academy of Sciences of the United States of America* 103:11206–11210.

Hill, J., S. Polasky and E. Nelson, et al. 2009. Climate change and health costs of air emissions from biofuels and gasoline. *Proceedings of the National Academy of Sciences of the United States of America* 106:2077–2082.

Hsieh, W. D., R. H. Chen, T. L. Wu and T. H. Lin. 2002. Engine performance and pollutant emission of an SI engine using ethanol-gasoline blended fuels. *Atmospheric Environment* 36:403–410.

Huang, H. J., S. Ramaswamy, U. W. Tschirner and B. V. Ramarao. 2008. A review of separation technologies in current and future biorefineries. *Separation and Purification Technology* 62:1–21.

Jacquet, N., G. Maniet, C. Vanderghem, F. Delvigne and A. Richel. 2015. Application of steam explosion as pretreatment on lignocellulosic material: A review. *Industrial and Engineering Chemistry Research* 54:2593–2598.

Jones, T. C. 2012. America, oil, and war in the Middle East. *Journal of American History* 99:208–218.

Kerr, R. A. 2007. Global warming is changing the world. *Science* 316:188–190.

Kilian, L. 2008. Exogenous oil supply shocks: How big are they and how much do they matter for the US economy? *Review of Economics and Statistics* 90:216–240.

Kilian, L. 2009. Not all oil price shocks are alike: Disentangling demand and supply shocks in the crude oil market. *American Economic Review* 99:1053–1069.

Konur, O. 2000. Creating enforceable civil rights for disabled students in higher education: An institutional theory perspective. *Disability & Society* 15:1041–1063.

Konur, O. 2002a. Access to nursing education by disabled students: Rights and duties of nursing programs. *Nurse Education Today* 22:364–374.

Konur, O. 2002b. Assessment of disabled students in higher education: Current public policy issues. *Assessment and Evaluation in Higher Education* 27:131–152.

Konur, O. 2002c. Access to employment by disabled people in the UK: Is the Disability Discrimination Act working? *International Journal of Discrimination and the Law* 5:247–279.

Konur, O. 2006a. Participation of children with dyslexia in compulsory education: Current public policy issues. *Dyslexia* 12:51–67.

Konur, O. 2006b. Teaching disabled students in Higher Education. *Teaching in Higher Education* 11:351–363.

Konur, O. 2007a. A judicial outcome analysis of the Disability Discrimination Act: A windfall for the employers? *Disability & Society* 22:187–204.

Konur, O. 2007b. Computer-assisted teaching and assessment of disabled students in higher education: The interface between academic standards and disability rights. *Journal of Computer Assisted Learning* 23:207–219.

Konur, O. 2011. The scientometric evaluation of the research on the algae and bio-energy. *Applied Energy* 88:3532–3540.

Konur, O. 2012a. Prof. Dr. Ayhan Demirbas' scientometric biography. *Energy Education Science and Technology Part A: Energy Science and Research* 28:727–738.

Konur, O. 2012b. The evaluation of the biogas research: A scientometric approach. *Energy Education Science and Technology Part A: Energy Science and Research* 29:1277–1292.

Konur, O. 2012c. The evaluation of the global energy and fuels research: A scientometric approach. *Energy Education Science and Technology Part A: Energy Science and Research* 30:613–628.

Konur, O. 2012d. The evaluation of the research on the biodiesel: A scientometric approach. *Energy Education Science and Technology Part A: Energy Science and Research* 28:1003–1014.

Konur, O. 2012e. The evaluation of the research on the bioethanol: A scientometric approach. *Energy Education Science and Technology Part A: Energy Science and Research* 28:1051–1064.

Konur, O. 2012f. The evaluation of the research on the biofuels: A scientometric approach. *Energy Education Science and Technology Part A: Energy Science and Research* 28:903–916.

Konur, O. 2012g. The evaluation of the research on the biohydrogen: A scientometric approach. *Energy Education Science and Technology Part A: Energy Science and Research* 29:323–338.

Konur, O. 2012h. The evaluation of the research on the microbial fuel cells: A scientometric approach. *Energy Education Science and Technology Part A: Energy Science and Research* 29:309–322.

Konur, O. 2012i. The scientometric evaluation of the research on the production of bioenergy from biomass. *Biomass and Bioenergy* 47:504–515.

Konur, O. 2015. Current state of research on algal bioethanol. In *Marine Bioenergy: Trends and Developments*, Eds. S. K. Kim and C. G. Lee, pp. 217–244. Boca Raton, FL: CRC Press.

Konur, O. 2019. Cyanobacterial bioenergy and biofuels science and technology: A scientometric overview. In *Cyanobacteria: From Basic Science to Applications*, Eds. A. K. Mishra, D. N. Tiwari and A. N. Rai, pp. 419–442. Amsterdam: Elsevier.

Konur, O. 2020a. The scientometric analysis of the research on the bioethanol production from green macroalgae. In *Handbook of Algal Science, Technology and Medicine*, Ed. O. Konur, pp. 385–401. London: Academic Press.

Konur, O., Ed. 2018a. *Bioenergy and Biofuels*. Boca Raton, FL: CRC Press.

Konur, O., Ed. 2020b. *Handbook of Algal Science, Technology and Medicine*. London: Academic Press.

Konur, O., Ed. 2021a. *Handbook of Biodiesel and Petrodiesel Fuels: Science, Technology, Health, and Environment*. Boca Raton, FL: CRC Press.

Konur, O., Ed. 2021b. *Handbook of Biodiesel and Petrodiesel Fuels: Science, Technology, Health, and Environment. Volume 1. Biodiesel Fuels: Science, Technology, Health, and Environment*. Boca Raton, FL: CRC Press.

Konur, O., Ed. 2021c. *Handbook of Biodiesel and Petrodiesel Fuels: Science, Technology, Health, and Environment. Volume 2. Biodiesel Fuels Based on the Edible and Nonedible Feedstocks, Wastes, and Algae: Science, Technology, Health, and Environment*. Boca Raton, FL: CRC Press.

Konur, O., Ed. 2021d. *Handbook of Biodiesel and Petrodiesel Fuels: Science, Technology, Health, and Environment. Volume 3. Petrodiesel Fuels: Science, Technology, Health, and Environment*. Boca Raton, FL: CRC Press.

Konur, O. and F. L. Matthews. 1989. Effect of the properties of the constituents on the fatigue performance of composites: A review. *Composites* 20:317–328.

Kruyt, B., D. P. van Vuuren, H. J. de Vries and H. Groenenberg. 2009. Indicators for energy security. *Energy Policy* 37:2166–2181.

Leta, J. and L. de Meis. 1996. A profile of science in Brazil. *Scientometrics* 35:33–44.

Leydesdorff, L. 2000. Is the European Union becoming a single publication system? *Scientometrics* 47:265–280.

Leydesdorff, L. and C. Wagner. 2009. Is the United States losing ground in science? A global perspective on the world science system. *Scientometrics* 78:23–36.

Leydesdorff, L., C. S. Wagner and L. Bornmann. 2014. The European Union, China, and the United States in the top-1% and top-10% layers of most-frequently cited publications: Competition and collaborations. *Journal of Informetrics* 8:606–617.

Leydesdorff, L. and P. Zhou. 2005. Are the contributions of China and Korea upsetting the world system of science? *Scientometrics* 63:617–630.

Li, S. Z. and C. Chan-Halbrendt. 2009. Ethanol production in (the) People's Republic of China: Potential and technologies. *Applied Energy* 86:S162–S169.

Liu, S. 2010. Woody biomass: Niche position as a source of sustainable renewable chemicals and energy and kinetics of hot-water extraction/hydrolysis. *Biotechnology Advances* 28:563–582.

Ma, X., L. Sun and C. Song. 2002. A new approach to deep desulfurization of gasoline, diesel fuel and jet fuel by selective adsorption for ultra-clean fuels and for fuel cell applications. *Catalysis Today* 77:107–116.

Macedo, I. C., J. E. A. Seabra and J. E. A. R. Silva. 2008. Greenhouse gases emissions in the production and use of ethanol from sugarcane in Brazil: The 2005/2006 averages and a prediction for 2020. *Biomass and Bioenergy* 32:582–595.

Morschbacker, A. 2009. Bio-ethanol based ethylene. *Polymer Reviews* 49:79–84.

Najafi, G., B. Ghobadian and T. Tavakoli, et al. 2009. Performance and exhaust emissions of a gasoline engine with ethanol blended gasoline fuels using artificial neural network. *Applied Energy* 86:630–639.

Newman, P. W. G. and J. R. Kenworthy. 1989. Gasoline consumption and cities: A comparison of U.S. cities with a global survey. *Journal of the American Planning Association* 55:24–37.

North, D. C. 1991. Institutions. *Journal of Economic Perspectives* 5:97–112.

Pu, Y., F. Hu, F. Huang, B. H. Davison and A. J. Ragauskas. 2013. Assessing the molecular structure basis for biomass recalcitrance during dilute acid and hydrothermal pretreatments. *Biotechnology for Biofuels* 6:15.

Reeves, S. 2014. To Russia with love: How moral arguments for a humanitarian intervention in Syria opened the door for an invasion of the Ukraine. *Michigan State University International Law Review* 23:199+.

Sun, Y. and J. Cheng. 2002. Hydrolysis of lignocellulosic materials for ethanol production: A review. *Bioresource Technology* 83:1–11.

Vadas, P. A., K. H. Barnett and D. J. Undersander 2008. Economics and energy of ethanol production from alfalfa, corn, and switchgrass in the Upper Midwest, USA. *Bioenergy Research* 1:44–55.

Winzer, C. 2012. Conceptualizing energy security. *Energy Policy* 46:36–48.

Yu, Y., X. Lou and H. Wu, H. 2008. Some recent advances in hydrolysis of biomass in hot-compressed water and its comparisons with other hydrolysis methods. *Energy and Fuels* 22:46–60.

Yuan, Z., S. Cheng, M. Leitch and C. C. Xu. 2010. Hydrolytic degradation of alkaline lignin in hot-compressed water and ethanol. *Bioresource Technology* 101:9308–9313.

14 Hydrothermal Pretreatment of the Biomass

Review

Ozcan Konur
(Formerly) Ankara Yildirim Beyazit University

14.1 INTRODUCTION

The crude oil-based gasoline fuels (Ma et al., 2002; Newman and Kenworthy, 1989) have been widely used in the transportation sector since the 1920s. However, there have been great public concerns over the adverse environmental and human impact of these fuels (Hill et al., 2006, 2009). Hence, biomass-based bioethanol fuels (Hill et al., 2006; Konur, 2012, 2015, 2020) have increasingly been used in blending gasoline fuels (Hsieh et al., 2002; Najafi et al., 2009), in the fuel cells (Antolini, 2007, 2009), and in the biochemical production (Angelici et al., 2013; Morschbacker, 2009) in a biorefinery context (Fernando et al., 2006; Huang et al., 2008).

However, it is necessary to pretreat the biomass (Alvira et al., 2010; Taherzadeh and Karimi, 2008) to enhance the yield of the bioethanol (Hahn-Hagerdal et al., 2006; Sanchez and Cardona, 2008) prior to the bioethanol production through the hydrolysis (Alvira et al., 2010; Taherzadeh and Karimi, 2007) and fermentation (Lin and Tanaka, 2006; Olsson and Hahn-Hagerdal, 1996) of the biomass.

The research in the fields of hydrothermal pretreatments of the biomass has also intensified in recent years. Steam explosion (Jacquet et al., 2015; Pu et al., 2013) and liquid hot water (LHW)-based hydrothermal treatments such as LHW (Liu, 2010; Zhuang et al., 2016), hot compressed water (HCW) (Bach and Skreiberg, 2016; Yu et al., 2008), and autohydrolysis (Garrote et al., 1999, 2001) pretreatments of the biomass have been widely researched to increase the sugar and bioethanol yield in this context.

However, it is essential to develop efficient incentive structures (North, 1991) for the primary stakeholders to enhance the research in this field (Konur, 2000, 2002a–c, 2006a, b, 2007a, b). Although there have been a number of review papers on the hydrothermal pretreatment of the biomass (Bach and Skreiberg, 2016; Jacquet et al., 2015; Pu et al., 2013), there has been no review of the most-cited 25 articles in this field.

Thus, this book chapter presents a review of the most-cited 25 articles in the field of the hydrothermal pretreatments of the biomass. Then, it discusses the key findings of these highly influential papers and comments on the future research priorities in this field.

14.2 MATERIALS AND METHODS

The search for this study was carried out using Scopus database (Burnham, 2006) in April 2022.

As a first step for the search of the relevant literature, the keywords were selected using all of the population papers. The selected keyword list was then optimized to obtain a representative sample of papers for the searched research field. This final keyword set was provided in the Appendix of Konur (2023) for future replication studies.

As a second step, a sample dataset was used for this study. The first 25 articles with at least 409 citations each were selected for the review study. Key findings from each paper were taken

DOI: 10.1201/9781003226536-19

from the abstracts of these papers and were discussed. Additionally, a number of brief conclusions were drawn, and a number of relevant recommendations were made to enhance the future research landscape.

14.3 RESULTS

A brief information about 25 most-cited papers with at least 409 citations each on the hydrothermal pretreatments of the biomass is given below. The primary research fronts are steam explosion and HLW-based hydrothermal pretreatments of the biomass with 14 and 12 papers, respectively.

14.3.1 The Steam Explosion Pretreatment of the Biomass

Li et al. (2007) studied lignin depolymerization and repolymerization and its critical role for delignification of aspen wood by steam explosion pretreatment in a paper with 640 citations. They observed the competition between lignin depolymerization and repolymerization and identified the conditions required for these two types of reaction. Addition of a reactive phenol, 2-naphthol, inhibited the repolymerization reaction strongly, resulting in a highly improved delignification by subsequent solvent extraction and an extracted lignin of uniform structure.

Sun et al. (2005) investigated the characteristics of degraded cellulose obtained from steam-exploded wheat straw in a paper with 525 citations. They steamed straw at 200°C, 15 bar for 10 and 33 min, and 220°C, 22 bar for 3, 5, and 8 min with a solid-to-liquid ratio of 2:1 (w/w) and 220°C, 22 bar for 5 min with a solid-to-liquid ratio of 10:1, respectively. They obtained 61.3%, 60.2%, 66.2%, 63.1%, 60.3%, and 61.3% of the straw residue, respectively. They delignified and bleached by 2% alkaline hydrogen peroxide (H_2O_2) at 50°C for 5 h under pH 11.5, which yielded 34.9%, 32.6%, 40.0%, 36.9%, 30.9%, and 36.1% (% dry wheat straw) of the cellulose preparation, respectively. They obtained the optimum cellulose yield (40.0%) when the steam explosion pretreatment was performed at 220°C, 22 bar for 3 min with a solid-to-liquid ratio of 2:1, in which the cellulose fraction obtained had a viscosity average degree of polymerization of 587 and contained 14.6% hemicelluloses and 1.2% klason lignin. The steam explosion pretreatment led to a significant loss in hemicelluloses and H_2O_2 post-treatment resulted in substantial dissolution of lignin and an increase in cellulose crystallinity.

Laser et al. (2002) compared the liquid hot water (LHW) and steam explosion pretreatments of the sugarcane bagasse in a paper with 470 citations. Solid concentration ranged from 1% to 8% for LHW pretreatment and was ≥50% for steam pretreatment. Reaction temperature and time ranged from 170°C to 230°C and 1–46 min, respectively. In four cases, LHW pretreatment achieved ≥80% conversion by simultaneous saccharification and fermentation (SSF), ≥80% xylan recovery, and no hydrolysate inhibition of glucose fermentation yield. Combined effectiveness was not as good for steam pretreatment due to low xylan recovery. SSF conversion increased and xylan recovery decreased as xylan dissolution increased for both modes. SSF conversion, xylan dissolution, hydrolysate furfural concentration, and hydrolysate inhibition increased, while xylan recovery and hydrolysate pH decreased, as a function of increasing LHW pretreatment solid concentration (1%–8%). They asserted that autohydrolysis played an important, if not exclusive, role in batch hydrothermal pretreatment.

Ohgren et al. (2007a) studied the effect of hemicellulose and lignin removal on enzymatic hydrolysis of steam-pretreated corn stover in a paper with 466 citations. They observed that a near-theoretical glucose yield (96%–104%) from acid-catalyzed steam-pretreated corn stover could be obtained if xylanases were used to supplement cellulases during hydrolysis. Xylanases hydrolyzed residual hemicellulose, thereby improving the access of enzymes to cellulose. Under these conditions, xylose yields reached 70%–74%. When pretreatment severity was reduced by using autocatalysis instead of acid-catalyzed steam pretreatment, xylose yields were increased to 80%–86%. The overall glucose yield increased slightly due to delignification, but the overall xylose yield decreased due to hemicellulose loss in the delignification step.

Palonen et al. (2004) studied the adsorption of *Trichoderma reesei* enzymes on steam-pretreated softwood and isolated lignin in a paper with 399 citations. They compared the adsorption of two purified cellulases from *T. reesei*, CBH I (Cel7A) and EG II (Cel5A), and their catalytic domains on steam pretreated softwood (SPS) and lignin using tritium-labeled enzymes. They observed that both CBH I and its catalytic domain exhibited a higher affinity to SPS than EG II or its catalytic domain. Removal of cellulose-binding domain markedly decreased the binding efficiency. Significant amounts of CBH I and EG II also bound to isolated lignin. The catalytic domains of the two enzymes of *T. reesei* differed essentially in the adsorption to isolated lignin. The catalytic domain of EG II adsorbed to alkaline-isolated lignin with a high affinity, whereas the catalytic domain of CBH I did not adsorb to any of the lignins tested. The cellulose-binding domain had a significant role in the unspecific binding of cellulases to lignin.

Kumar et al. (2012) investigated lignin present in steam-pretreated Douglas-fir softwood binding enzymes and limiting cellulose accessibility in a paper with 295 citations. They observed that the lignin component significantly influenced the swelling and accessibility of cellulose, as at low protein loadings (5 FPU/g cellulose), only 16% of the cellulose present in the steam pretreated softwood was hydrolyzed while almost complete hydrolysis was achieved with the delignified substrate. When lignin was added back in the same proportions, it was originally found to the highly accessible and swollen, delignified steam pretreated softwood, and to a cellulose control such as Avicel, the hydrolysis yields decreased by 9% and 46%, respectively. However, when higher enzyme loadings were employed, the greater availability of the enzyme could overcome the limitations imposed by both the lignin's restrictions on cellulose accessibility and direct binding of the enzymes, resulting in a near-complete hydrolysis of the cellulose.

Ohgren et al. (2007b) compared SSF and separate hydrolysis and fermentation (SHF) using steam-pretreated corn stover in a paper with 270 citations. The enzymatic loading in these experiments was 10 FPU/g water-insoluble solids (WIS), and the yeast concentration in SSF was 1 g/L (dry weight) of a *Saccharomyces cerevisiae* strain. When the whole slurry from the pretreatment stage was used as it was, they observed that diluted to 8% WIS with water and pH adjusted, SSF gave a 13% higher overall ethanol yield than SHF (72.4% versus 59.1% of the theoretical). The impact of the inhibitory compounds in the liquid fraction of the pretreated slurry affected SSF and SHF in different ways. The overall ethanol yield decreased when SSF was run in the absence of inhibitors compared to SSF with inhibitors present. On the contrary, the presence of inhibitors decreased the overall ethanol yield in the case of SHF. However, the SHF yield achieved in the absence of inhibitors was still lower than the SSF yield achieved with inhibitors present.

Ballesteros et al. (2006) investigated ethanol production from steam explosion-pretreated wheat straw in a paper with 264 citations. They employed a SSF process, using diluted acid [H_2SO_4 0.9% (w/w)] and water as preimpregnation agents. They steam-exploded acid- or water-impregnated biomass at different temperatures (160°C–200°C) and residence times (5, 10, and 20 min). They observed that the best pretreatment conditions to obtain high conversion yield to ethanol (approx. 80% of theoretical) of cellulose-rich residue after steam-explosion were 190°C and 10 min or 200°C and 5 min, in acid-impregnated straw. However, 180°C for 10 min in acid-impregnated biomass provided the highest ethanol yield referred to raw material (140 L/t wheat straw), and sugar recovery yield in the filtrate (300 g/kg wheat straw).

Heitz et al. (1991) investigated the fractionation of aspen wood (*Populus tremuloides*) at the pilot plant scale by optimizing steam pretreatment conditions in a paper with 251 citations. They observed that the maximum recovery of hemicelluloses was 65% of the potential at the pretreatment severity $\log_{10}R_o=3.8$. Under these conditions, lignin recovery by caustic extraction was about 80% of total lignin. Cellulose was completely recovered at this severity but needed to be bleached to achieve its natural coloration. At the pretreatment severities both greater and smaller than $\log_{10}R_o=3.8$, hemicellulose recovery decreased. The cellulose derived from the steam fractionation process was rapidly depolymerized as severity increased, while its enzymatic digestibility was close to theoretical at severities beyond $\log_{10}R_o=3.2$.

Hongzhang and Liying (2007) investigated the unpolluted fractionation of wheat straw by steam explosion and ethanol extraction in a paper with 246 citations. They steam-exploded wheat straw for 4.5 min with a moisture of 34.01% and a pressure of 1.5 MPa without acid or alkali. They observed that the ratio of monosaccharides to oligosaccharides was 1:9, and the main component, xylose, was 85.9% in content. The total recovery rate of hemicellulose was 80%. The optimum extraction conditions were 40% ethanol, fiber/liquor ratio 1:50 (w/v), the pretreatment severity $\log(R)=3.7$ (180°C for 20 min), 0.1% NaOH. The lignin yield was 75% by acid precipitation and 85% ethanol solvent was recovered. The purity rate of lignin reached 85.3%. The cellulose recovery rate was 94%, and the hemicellulose and lignin content decreased after steam explosion and ethanol extraction.

Cara et al. (2006) studied the enhanced enzymatic hydrolysis of olive tree wood by steam explosion and alkaline hydrogen peroxide delignification in a paper with 227 citations. They carried out the pretreatment at 190°C, 210°C, 230°C, and 240°C for 5 min using a cellulolytic complex supplemented with β-glucosidase at 10% (w/v) pretreated material concentration. They observed that delignification enhanced enzymatic hydrolysis yields of steam-pretreated olive tree wood. Up to 80% of the lignin in the original wood was solubilized, leaving a cellulose-rich residue that led to a concentrated glucose solution (51.3 g/L after 72 h enzymatic hydrolysis in the best case). The maximum overall process yield, taking into account both sugars present in the liquid from steam pretreatment and glucose from the steamed, delignified and hydrolyzed solid was obtained at the lowest steam pretreatment temperature assayed. As a result, from 100 g of raw material, 28.8 g of sugars were recovered of 54.7 g available (52.6%).

Grous et al. (1986) studied the effect of steam-explosion pretreatment on pore size and enzymatic hydrolysis of poplar wood in a paper with 225 citations. They observed that the glucose yield after 24 h of enzymatic hydrolysis using enzymes from *Trichoderma reesei* and *Aspergillus niger* was in excess of 90% of the potential, whereas the corresponding yield from unpretreated substrate was only 15%. This was primarily due to the increase in pore surface area accessible to enzyme molecules with a considerable increase in pore volume available to 5–9 nm solute probes. Pretreated wood that was subsequently oven-dried hydrolyzed poorly and showed a reduction in available pore volume after drying. Xylans were readily hydrolyzed to xylose during pretreatment, and owing to decomposition, the amount of xylose in solution after steam pretreatment decreased as the severity of the reaction conditions increased; the converse was true for glucose.

Cantarella et al. (2004) studied the effect of inhibitors released during steam explosion pretreatment of poplar wood on subsequent enzymatic hydrolysis and SSF in a paper with 223 citations. They used a blend of Celluclast and Novozym cellulase complexes in the presence of the inhibiting compounds produced during the preceding steam explosion pretreatment process. The steam explosion temperature and time conditions were 214°C and 6 min, resulting in a $\log R_0$ of 4.13. In enzymatic hydrolysis tests at 45°C, the biomass loading in the bioreactor was 100 gDW/L (dry weight) and the enzyme-to-biomass ratio 0.06 g/gDW. The enzyme activities for endoglucanase, exoglucanase, and β-glucosidase were 5.76, 0.55, and 5.98 U/mg, respectively. Acetic acid (2 g/L), furfural, 5-HMF, syringaldehyde, 4-hydroxybenzaldehyde, and vanillin (0.5 g/L) did not significantly influence the enzyme activity, whereas formic acid (11.5 g/L) inactivated the enzymes and levulinic acid (29.0 g/L) partially affected the cellulase. Untreated SE biomass during the enzymatic attack gave rise to a non-fermentable hydrolysate, which became fermentable when rinsed SE biomass was used. The presence of acetic acid, vanillin, and 5-HMF (0.5 g/L) in SSF of 100 gDw/L biomass gave rise to ethanol yields of 84.0%, 73.5%, and 91.0%, respectively, with respective lag phases of 42, 39, and 58 h.

Rocha et al. (2012) studied the steam explosion pretreatment reproduction and alkaline delignification reactions performed on a pilot scale with sugarcane bagasse for bioethanol production in a paper with 212 citations. They used different varieties of natural sugarcane bagasse, pretreated bagasse, and delignified pretreated bagasse (cellulosic pulp) in the 20 processes of pretreatment and delignification on the pilot scale. They observed that the 20 samples of the most diverse varieties and origins of natural sugarcane bagasse considering planting soils, planting periods, and weather showed no significant chemical differences. The pilot process of steam explosion pretreatment and the alkaline delignification process of pretreated material showed a good capacity of reproduction, as the standard differences were below 2.7. The average allowed in the pretreatment and alkaline delignification processes were 66.1 and 51.5, respectively, ensuring an excellent reproduction capacity of the processes obtained through chemical characterizations (Table 14.1).

14.3.2 The Liquid Hot Water Pretreatment of the Biomass

There are four, six, and four highly cited papers (HCPs) for the research fronts pf the LHW, HCW, and autohydrolysis pretreatments, respectively.

14.3.2.1 The LHW Pretreatment of the Biomass

Laser et al. (2002) compared LHW and steam explosion pretreatments of sugarcane bagasse for bioconversion to ethanol in a paper with 470 citations. Solid concentration ranged from 1% to 8% for LHW pretreatment and was ≥50% for steam pretreatment. Reaction temperature and time ranged from 170°C to 230°C and 1–46 min, respectively. In four cases, LHW pretreatment achieved ≥80% ethanol conversion by SSF, ≥80% xylan recovery, and no hydrolysate inhibition of glucose fermentation yield. Combined effectiveness was not as good for steam pretreatment due to low xylan recovery. SSF conversion increased and xylan recovery decreased as xylan dissolution increased for both modes. SSF conversion, xylan dissolution, hydrolysate furfural concentration, and hydrolysate inhibition increased, while xylan recovery and hydrolysate pH decreased, as a function of increasing LHW pretreatment solids concentration (1%–8%). They asserted that autohydrolysis played an important, if not exclusive, role in batch hydrothermal pretreatment.

Mosier et al. (2005) optimized pH controlled LHW pretreatment of corn stover without any chemical pretreatment in a paper with 452 citations. They observed maximization of the solubilization of the hemicellulose fraction as liquid soluble oligosaccharides while minimizing the formation of monomeric sugars. The optimized condition for controlled pH, LHW pretreatment of a 16% slurry of corn stover in water was 190°C for 15 min. At the optimal condition, 90% of the cellulose was hydrolyzed to glucose by 15 filter paper cellulase activity (FPU) of cellulase per gram of glucan. When the resulting pretreated slurry, in undiluted form, was hydrolyzed by 11 FPU of cellulase per gram of glucan, a hydrolysate containing 32.5 g/L glucose and 18 g/L xylose was formed. Both the xylose and the glucose in this undiluted hydrolysate were fermented by recombinant yeast 424A(LNH-ST) to ethanol at 88% of theoretical yield.

Ko et al. (2015) studied the effect of LHW pretreatment severity on properties of hardwood lignin and enzymatic hydrolysis of cellulose in a paper with 239 citations. They observed that hardwood pretreated with LHW at severities ranging from severity factor $\log R_o = 8.3–12.5$ resulted in 80%–90% recovery of the initial lignin in the residual solids. The ratio of acid-insoluble lignin (AIL) to acid-soluble lignin (ASL) increased, and the formation of spherical lignin droplets on the cell wall surface was observed. When lignins were isolated from hardwoods pretreated at increasing severities, the glass transition temperature (T_g) of isolated lignins increased from 171°C to 180°C as the severity increased from $\log R_o = 10.4–12.5$. The condensation reactions of lignin molecules occurred during pretreatment and altered the lignin structure. Lignins derived from more severely pretreated hardwoods had higher T_g values and showed more pronounced inhibition of enzymatic hydrolysis.

Perez et al. (2008) optimized LHW pretreatment conditions (temperature and residence time) to enhance sugar recovery from wheat straw for bioethanol production in a paper with 209 citations. They observed that optimal conditions were 188°C and 40 min, leading to hemicellulose-derived sugars (HDS) recovery yield of 43.6% in raw material and enzymatic hydrolysis yield of 79.8% of theoretical. However, when response variables were analyzed separately, 71.2% of HDS recovery was obtained at 184°C and 24 min, whereas conditions of 214°C and 2.7 min led to a maximum hydrolysis yield of 90.6% of theoretical. They asserted that a two-step pretreatment would be the most adequate process configuration to get a maximum recovery of fermentable sugars.

TABLE 14.1
The Steam Explosion Pretreatment of the Biomass

No.	Papers	Biomass	Prt.	Other Prts.	Parameters	Keywords	Lead Author	Affil.	Cits
1	Li et al. (2007)	Wood, lignin	Steam explosion	Na	Lignin degradation	Lignin, wood, steam explosion	Li, Jiebing 7410069979	Royal Inst. Technol. Sweden	640
2	Sun et al. (2005)	Wheat straw, cellulose	Steam explosion	Alkaline H_2O_2	Cellulose degradation, cellulose yield	Wheat straw, steam exploded	Sun, Runcang 55661525600	Dalian Polytech. Univ. China	525
3	Laser et al. (2002)	Sugarcane bagasse, xylan	LHW, steam explosion	Na	Pretreatment type, solid concentration, reaction temperature and time, ethanol conversion, xylan recovery	Sugar cane bagasse, LHW, steam	Lynd, Lee R. 35586183800	Dartmouth Coll. USA	470
4	Ohgren et al. (2007a)	Corn stover, hemicellulose, lignin	Steam explosion	Acid, enzymes	Lignin and hemicellulose degradation, glucose yield	Corn stover, hemicellulose, lignin, steam pretreated	Zacchi, Guido 7006727748	Lund Univ. Sweden	466
5	Palonen et al. (2004)	Softwood, lignin	Steam explosion	Enzymes	Enzyme type, enzymatic activity, lignin degradation	Softwood, lignin, steam	Palonen, Hetti 56575011300	VTT Tech. Res. Ctr. Finland	399
6	Kumar et al. (2012)	Softwood, cellulose, lignin	Steam explosion	Enzymes	Cellulose degradation, cellulose accessibility, sugar yields	Steam pretreated, lignin, cellulose, softwood	Arantes, Valdeir 56344492500	Univ. Sao Paulo Brazil	295
7	Ohgren et al. (2007b)	Corn stover	Steam explosion	Enzymes	Ethanol yield, fermentation type	Corn stover, steam pretreated	Zacchi, Guido 7006727748	Lund Univ. Sweden	270
8	Ballesteros et al. (2006)	Wheat straw	Steam explosion	Acid, water	Sugar and ethanol yield, temperature, time	Wheat straw, steam explosion	Manzanares, Paloma 55779406300	CIEMAT Spain	264
9	Heitz et al. (1991)	Aspen wood, lignin, cellulose	Steam explosion	Na	Lignin, cellulose, hemicellulose degradation	Populus, steam	Chornet, Estaban 7005128090	CRB Innov. Canada	251
10	Hongzhang and Liying (2007)	Wheat straw, hemicellulose, lignin, cellulose	Steam explosion	Acid, alkali	hemicellulose, lignin, cellulose degradation, sugar and ethanol yield	Wheat straw, steam explosion	Hongzhang, Chen 7501614171	Chinese Acad. Sci. China	246
11	Cara et al. (2006)	Olive tree wood, lignin	Steam explosion	Alkaline H_2O_2	Lignin degradation, sugar yield	Wood, steam explosion	Castro, Eulogio 7102441948	Univ. Jaen Spain	227

(Continued)

TABLE 14.1 (*Continued*)
The Steam Explosion Pretreatment of the Biomass

No.	Papers	Biomass	Prt.	Other Prts.	Parameters	Keywords	Lead Author	Affil.	Cits
12	Grous et al. (1986)	Poplar wood, xylan	Steam explosion	Enzymes	Sugar yield,	Poplar, steam explosion	Converse, Alvin O. 7003591261	Dartmouth Coll. USA	225
13	Cantarella et al. (2004)	Poplar wood	Steam explosion	Enzymes	Sugar yield, ethanol yield, enzymes, inhibitors	Wood, steam explosion	Cantarella, Maria 7003630895	Univ. L'Aquila Italy	223
14	Rocha et al. (2012)	Sugarcane bagasse	Steam explosion	Alkali	Bagasse type, composition, reproduction, delignification	Sugarcane bagasse, steam explosion	Rocha, George J. M. 14012497200	Brazil. Biorenew. Natl. Lab. Brazil	212

Prt, hydrothermal pretreatments; other prts., other pretreatments; na, nonavailable; cits., number of citations received for each paper.

14.3.2.2 The HCW Pretreatment of the Biomass

Mok and Antal (1992) studied the uncatalyzed solvolysis of whole biomass hemicellulose by HCW in a paper with 384 citations. They used woody and four herbaceous biomass species for 0–15 min at 200°C–230°C. They observed that between 40% and 60% of the sample mass was solubilized. In all cases, 100% of the hemicellulose was solubilized, of which 90% (on average) was recoverable as monomeric sugar. Concurrently, between 4% and 22% of the cellulose and 35% and 60% of the lignin were also solubilized. The results were independent of temperature and time within the range studied. They observed variations occurred primarily as a function of species. High lignin solubilization apparently reduced the ability to recover hemicellulose sugar.

Minowa et al. (1998) studied the cellulose decomposition in HCW with alkaline or nickel catalyst in a paper with 330 citations. They employed different reaction temperatures from 200°C to 350°C. Hydrolysis could play an important role in forming glucose/oligomer, and the obtained glucose/oligomer could decompose quickly to nonglucose aqueous products, oil, char, and gases. Under the catalyst-free condition, the interpretation of the observation led to a simplified reaction scheme. With regard to the alkaline catalyst, the observation suggested a role of the alkaline catalyst in inhibiting the formation of char from oil (stabilization of oil), resulting in oil production. On the other hand, the nickel catalyst could catalyze the steam reforming reaction of aqueous products as intermediates and the methanation reaction.

Liu and Wyman (2003) studied the effect of flow rate of HCW on xylan, lignin, and total mass removal from corn stover in a paper with 265 citations. They employed a small tubular flow-through reactor at 180°C, 200°C, and 220°C. They observed that solubilization of hemicellulose increased with flow, especially at high temperatures. The dissolved xylan in the hydrolysate was mostly oligomers over this temperature range, and the fraction as oligomers increased with flow rate. Further, lignin removal increased from less than about 30% for batch reactors to about 75% at high flow rates and was nearly linearly related to hemicellulose release for the flow-through reactor. They asserted that mass transfer or other physical factors impacted hemicellulose hydrolysis. In addition, lignin was released throughout hydrolysis, but its fate might be governed by subsequent precipitation reactions unless it was removed first.

Liu and Wyman (2005) studied the partial flow of HCW through corn stover to enhance hemicellulose sugar recovery and enzymatic digestibility of cellulose in a paper with 264 citations. They applied flow with HCW at selected intervals at 200°C. They observed that partial flow reduced water consumption by 60% compared with continuous flow-through operation but still achieved higher xylose sugar yields (84%–89%) compared to batch pretreatment (46.6%). In addition, corn stover cellulose pretreated by partial flow had higher enzymatic digestibility (88%–90%) than batch operations (~85%) at otherwise identical conditions, apparently due to much higher lignin removal for the former (40%–45% vs 10%–12%). Partial flow also reduced degradation, with the recovery of xylose and glucose in the solids and hydrolysate increased to 90%–92% vs only about 76% for batch operation.

Nitsos et al. (2013) optimized the HCW pretreatment of beech wood in the bioethanol production process in a paper with 232 citations. They carried out hydrothermal pretreatment experiments in a batch-mode, high-pressure reactor under autogenous pressure at varying temperature (130°C–220°C) and time (15–180 min) regimes, and at a liquid-to-solid ratio (LSR) of 15. They observed that the major changes induced in the solid biomass were the dissolution or removal of hemicellulose to the process liquid and the partial removal and relocation of lignin on the external surface of biomass particles in the form of recondensed droplets. These structural changes led to a 2.5-fold increase in surface area and total pore volume of the pretreated biomass solids. The enzymatic hydrolysis of cellulose to glucose increased from less than 7 wt% for the parent biomass to as high as 70 wt% for the treated solids. Maximum xylan recovery (60 wt%) in the hydrothermal process liquid was observed at about 80 wt% hemicellulose removal; this was accomplished by moderate treatment severities ($\log R_o = 3.8–4.1$). At higher severities ($\log R_o = 4.7$), xylose degradation products were the predominant chemicals formed.

Yuan et al. (2010) studied the hydrolytic degradation of alkaline lignin in HCW and ethanol in a paper with 212 citations. They degraded alkaline lignin of a very high molecular weight into oligomers in a HCW-ethanol medium with NaOH as the catalyst and phenol as the capping agent at 220°C–300°C. Under the optimal reaction conditions, i.e., 260°C, 1 h, with the lignin/phenol ratio of 1:1 (w/w), almost complete degradation was achieved, producing less than 1% solid residue and negligible gas products. The obtained degraded lignin had a number-average molecular weight Mn and weight-average molecular weight Mw of 450 and 1,000 g/mol respectively, significantly lower than the Mn and Mw of 10,000 and 60,000 g/mol of the original lignin. A higher temperature and a longer reaction time favored phenol combination but increased the formation of solid residue due to the condensation reactions of the degradation intermediates/products. The degraded lignin products were soluble in organic solvents.

14.3.2.3 The Autohydrolysis Pretreatment of the Biomass

Garrote et al. (1999) studied the mild autohydrolysis as an environmental-friendly technology for xylooligosaccharide production from wood in a paper with 374 citations. They subjected *Eucalyptus globulus* wood samples to hydrothermal treatments under mild operational conditions (145°C–190°C, liquid-to-solid ratio 6–10 g/g, reaction times up to 7.5 h). They observed that negligible effects were caused by hydrothermal treatments on both cellulose and lignin, and developed kinetic models which described the hydrolysis of hemicelluloses. They found that xylan degradation, xylooligosaccharide and xylose generation, and xylose dehydration to furfural were accurately described by models.

Garrote et al. (2002) studied the autohydrolysis of corncob as a nonisothermal operation for xylooligosaccharide production in a paper with 232 citations. They pretreated corncob samples with water under nonisothermal conditions (reaction temperature up to 216°C) using a liquid-to-solid ratio (SLR) of 8:1 kg/kg. They observed that most of the cellulose was retained in the solid phase, whereas partial delignification (up to 26% lignin removal) was achieved. Up to 94% of xylan was removed, producing xylooligomers and other reaction byproducts (including xylose and furfural). They modeled the decomposition of xylan into xylooligomers, with further generation of xylose, furfural, and degradation products by a series of parallel and consecutive reactions, which were assumed to be first order with coefficients showing an Arrhenius-type dependence on temperature (Table 14.2).

14.4 DISCUSSION

14.4.1 Introduction

The crude oil-based gasoline fuels have been widely used in the transportation sector since the 1920s. However, there have been great public concerns over the adverse environmental and human impact of these fuels. Hence, biomass-based bioethanol fuels have increasingly been used in blending gasoline and petrodiesel fuels, in the fuel cells, and in the biochemical production in a biorefinery context.

The research in the field of hydrothermal pretreatments of the biomass such as steam explosion and LHW-based hydrothermal pretreatments to increase the sugar and bioethanol yield has also intensified in recent years.

However, it is essential to develop efficient incentive structures for the primary stakeholders to enhance the research in this field. Although there have been a number of review papers for this field, there has been no review of the most-cited 25 articles in this field.

Thus, this book chapter presents a review of the most-cited 25 articles in this field. Then, it discusses the key findings of these highly influential papers and comments on the future research priorities in this field.

TABLE 14.2
The LHW Pretreatment of the Biomass

No.	Papers	Biomass	Prt.	Other Prts.	Parameters	Keywords	Lead Author	Affil.	Cits
1	Laser et al. (2002)	Sugarcane bagasse	LHW, steam explosion	Na	Pretreatment type, solid concentration, reaction temperature and time, ethanol conversion, xylan recovery	Sugar cane bagasse, LHW, steam	Lynd, Lee R. 35586183800	Dartmouth Coll. USA	470
2	Mosier et al. (2005)	Corn stover, hemicellulose	LHW	Enzymes	Hemicellulose degradation, sugar and ethanol yield, hydrolysis, fermentation	Corn stover, LHW	Ladisch, Michael R. 7005670397	Purdue Univ. USA	452
3	Mok and Antal (1992)	Biomass, hemicellulose	HCW	Na	Hemicellulose degradation, temperature, time	Biomass, hemicellulose, hot compressed liquid water	Antal, Michael, J. 7103384857	Univ. Hawaii Manoa	384
4	Garrote et al. (1999)	Wood, hemicellulose	Autohydrolysis	Na	Hemicellulose degradation, hydrolysis modeling	Wood, autohydrolysis	Parajo, Juan C. 7005594129	Univ. Vigo Spain	374
5	Minowa et al. (1998)	Cellulose	HCW	Alkali	Cellulose degradation mechanisms	Cellulose, HCW	Minowa, Tomoaki 7006698355	Natl. Inst. Adv. Ind. Sci. Technol. Japan	330
6	Liu and Wyman (2003)	Corn stover, xylan, hemicellulose, lignin	HCW	Flowthorugh	Xylan, lignin, and hemicellulose degradation	Corn stover, xylan, lignin, compressed hot water	Wyman, Charles E. 7004396809	Dartmouth Coll. USA	265
7	Liu and Wyman (2005)	Corn stover, hemicellulose, cellulose, lignin	HCW	Flowthrough	Hemicellulose, cellulose, lignin degradation, sugar yield, water consumption	Compressed hot water, corn stover, hemicellulose, cellulose	Wyman, Charles E. 7004396809	Dartmouth Coll. USA	264
8	Ko et al. (2015)	Hardwood, lignin, cellulose	LHW	Na	Lignin, cellulose degradation, pretreatment severity	Hardwood, lignin, cellulose, LHW	Ladisch, Michael R. 7005670397	Purdue Univ. USA	239

(Continued)

TABLE 14.2 (*Continued*)
The LHW Pretreatment of the Biomass

No.	Papers	Biomass	Prt.	Other Prts.	Parameters	Keywords	Lead Author	Affil.	Cits
9	Nitsos et al. (2013)	Beech wood, Lignin, cellulose hemicellulose, xylan	HCW	Na	Lignin, cellulose hemicellulose, xylan degradation, pretreatment severity	Lignocellulosic, hydrothermal, biomass	Triantafyllidis, Konstantinos S. 8406363900	Aristotle Univ. Thessaloniki Greece	232
10	Garrote et al. (2001)	Corncob, xylan, lignin	Autohydrolysis	Na	Xylan and lignin degradation	Corncob, autohydrolysis	Parajo, Juan C. 7005594129	Univ. Vigo Spain	232
11	Yuan et al. (2010)	Lignin	HCW	Alkali, ethanol	Lignin degradation	Lignin, HCW	Xu, Chunbao C. 35207737400	Lakehead Univ. Canada	212
12	Perez et al. (2008)	Wheat straw	LHW	Na	Temperature, residence time, hemicellulose degradation, sugar yield	Wheat straw, LHW,	Manzanares, Paloma 55779406300	CIEMAT Spain	209

Prt, hydrothermal pretreatments; other prts., other pretreatments; na, nonavailable; HCW, hot compressed water; LHW, hot liquid water.

As a first step for the search of the relevant literature, the keywords were selected using the whole set of papers. The selected keyword list was optimized to obtain a representative sample of papers for the searched research field. This keyword list was provided in the Appendix of Konur (2023) for future replication studies.

As a second step, a sample dataset was used for this study. The first 25 articles with at least 409 citations each were selected for the review study. Key findings from each paper were taken from the abstracts of these papers and were discussed. Additionally, a number of brief conclusions were drawn, and a number of relevant recommendations were made to enhance the future research landscape.

There are two primary research fronts for these most-cited 25 papers: steam explosion and LHW-based hydrothermal pretreatments with 56% and 48% of the 25 HCPs, respectively (Table 14.3). However, LHW-based pretreatments are over-represented in the reviewed papers compared to the 195 HCPs with 8% surplus. On the other hand, there is no representation of hydrothermal pretreatments in general in the reviewed papers.

Furthermore, HCW and autohydrolysis pretreatments are under-represented with 15% and 3% deficit, respectively, while LHW pretreatment is over-represented with a 3% surplus.

Table 14.4 provides information on these biomass used in these 25 HCPs. There are four primary types of the biomass: biomass constituents, wood, agricultural residues, and biomass in general with 80%, 44%, 48%, and 12% of the 25 HCPs, respectively. Further, lignin, cellulose, and hemicellulose are the primary constituents with 28%, 24%, and 24% of the 25 HCPs, respectively. Further, corn stover and wheat straw are the key agricultural residues with 20% and 16% of the 25 HCPs, respectively, while wood in general is the most prolific keyword with 20% of the 25 HCPs.

It is notable that the papers related to biomass constituents are over-represented with 50% surplus where lignin, cellulose, and hemicellulose are over-represented with 14%, 14%, and 10% surplus, respectively. Further, papers related to wood in general are over-represented with 20% surplus.

14.4.2 The Steam Explosion Pretreatment of the Biomass

Li et al. (2007) studied lignin depolymerization and repolymerization and its critical role for delignification of aspen wood by steam explosion and observed the competition between lignin depolymerization and repolymerization. On the other hand, Sun et al. (2005) investigated the characteristics of degraded cellulose obtained from steam-exploded wheat straw and observed that steam explosion pretreatment led to a significant loss in hemicelluloses.

Laser et al. (2002) compared the LHW and steam explosion pretreatments of the sugarcane bagasse and observed that the LHW pretreatment achieved ≥80% conversion by SSF, ≥80% xylan

TABLE 14.3
The Most-Prolific Research Fronts for the Biomass for the Hydrothermal Pretreatments of the Biomass

No.	Research Fronts	N Paper % Review	N Paper % Sample	Surplus
1.	Steam explosion-based hydrothermal pretreatments	56	58	−2
2.	**LHW-based Hydrothermal Pretreatments**	**48**	**40**	**8**
	HCW pretreatment	24	9	15
	LHW pretreatment	16	13	3
	Autohydrolysis	8	11	−3
3.	Hydrothermal pretreatments in general	0	7	−7
	Sample	25	195	

N paper (%) sample: the number of papers in the population sample of 195 papers. N paper (%) review: the number of papers in the sample of 25 most-cited papers.

TABLE 14.4
The Most-Prolific Research Fronts for the Biomass for the Hydrothermal Pretreatments of the Biomass

No.	Research Fronts	N Paper % Review	N Paper % Sample	Surplus %
1	**Biomass Constituents**	**80**	**30**	**50**
	Lignin	28	14	14
	Cellulose	24	10	14
	Hemicellulose	24	4	20
	Other constituents	4	2	2
2	**Wood**	**44**	**40**	**5**
	Wood	24	5	19
	Softwood	8	15	−7
	Spruce	0	4	−4
	Pine	0	3	−3
	Hardwood	12	20	−8
	Eucalyptus	0	6	−6
	Poplar	4	4	0
3	Grass	0	3	−3
4	**Agricultural Residues**	**48**	**42**	**6**
	Other residues	0	13	−13
	Corn stover	20	12	8
	Wheat straw	16	9	7
	Sugarcane bagasse	8	6	2
	Corncobs	4	3	1
5	**Biomass in General**	**12**	**14**	**−2**
	Biomass	8	9	−1
	Lignocellulosic biomass	4	6	−2

N paper (%) sample: the number of papers in the population sample of 195 papers. N paper (%) review: the number of papers in the sample of 25 most cited papers.

recovery, and no hydrolysate inhibition of glucose fermentation yield. On the other hand, Ohgren et al. (2007a) studied the effect of hemicellulose and lignin removal on enzymatic hydrolysis of steam-pretreated corn stover and observed that a near-theoretical glucose yield from acid-catalyzed steam-pretreated corn stover could be obtained if xylanases were used to supplement cellulases during hydrolysis.

Palonen et al. (2004) studied the adsorption of *Trichoderma reesei* enzymes on steam-pretreated softwood and isolated lignin and observed that both CBH I and its catalytic domain exhibited a higher affinity to SPS than EG II or its catalytic domain. On the other hand, Kumar et al. (2012) investigated lignin present in steam-pretreated Douglas-fir softwood binding enzymes and limiting cellulose accessibility and observed that the lignin component significantly influenced the swelling and accessibility of cellulose as at low protein loadings.

Ohgren et al. (2007b) compared SSF and SHF using steam-pretreated corn stover and observed that SSF gave a 13% higher overall ethanol yield than SHF (72.4% versus 59.1% of the theoretical). On the other hand, Ballesteros et al. (2006) investigated ethanol production from steam-explosion-pretreated wheat straw and observed that the best pretreatment conditions to obtain high conversion yield to ethanol of cellulose-rich residue after steam-explosion were 190°C and 10 min or 200°C and 5 min, in acid-impregnated straw.

Heitz et al. (1991) investigated the fractionation of aspen wood at the pilot plant scale by optimizing steam pretreatment conditions and observed that the maximum recovery of hemicelluloses was 65% of the potential at the pretreatment severity $\log_{10}R_o=3.8$. On the other hand, Hongzhang and Liying (2007) investigated the unpolluted fractionation of wheat straw by steam explosion and ethanol extraction and observed that the ratio of monosaccharides to oligosaccharides was 1:9 and the main component, xylose, was 85.9% in content.

Cara et al. (2006) studied the enhanced enzymatic hydrolysis of olive tree wood by steam explosion and alkaline hydrogen peroxide delignification and observed that delignification enhanced enzymatic hydrolysis yields of steam pretreated the biomass. On the other hand, Grous et al. (1986) studied the effect of steam explosion pretreatment on pore size and enzymatic hydrolysis of poplar wood and observed that the glucose yield after 24h of enzymatic hydrolysis using enzymes from *Trichoderma reesei* and *Aspergillus niger* was in excess of 90% of the potential.

Cantarella et al. (2004) studied the effect of inhibitors released during steam explosion pretreatment of poplar wood on subsequent enzymatic hydrolysis and SSF and observed that acetic acid, furfural, 5-HMF, syringaldehyde, 4-hydroxybenzaldehyde, and vanillin did not significantly influence the enzyme activity. On the other hand, Rocha et al. (2012) studied the steam explosion pretreatment reproduction and alkaline delignification reactions performed on a pilot scale with sugarcane bagasse for bioethanol production and observed that the 20 samples of most diverse varieties and origins of natural sugarcane bagasse considering planting soils, planting periods, and weather showed no significant chemical differences.

Overall, these HCPs clearly show that these steam explosion pretreatments degrade the biomass to improve the hydrolysis of the biomass and fermentation of the resulting hydrolysates with an improved sugar and bioethanol yield.

14.4.3 The Liquid Hot Water Pretreatment of the Biomass

There are four, six, and four HCPs for the research fronts pf the LHW, HCW, and autohydrolysis pretreatments, respectively.

14.4.3.1 The LHW Pretreatment of the Biomass

Ko et al. (2015) studied the effect of LHW pretreatment severity on properties of hardwood lignin and enzymatic hydrolysis of cellulose and observed that hardwood pretreated with LHW at severities ranging from severity factor $\log R_o=8.3$–12.5 resulted in 80%–90% recovery of the initial lignin in the residual solids. On the other hand, Perez et al. (2008) optimized LHW pretreatment conditions to enhance sugar recovery from wheat straw for bioethanol production and observed that optimal conditions were 188°C and 40min, leading to hemicellulose-derived sugars (HDS) recovery yield of 43.6% in raw material and enzymatic hydrolysis yield of 79.8% of theoretical.

Overall, these HCPs clearly show that these LHW pretreatments degrade the biomass to improve the hydrolysis of the biomass and fermentation of the resulting hydrolysates with an improved sugar and bioethanol yield.

14.4.3.2 The HCW Pretreatment of the Biomass

Mok and Antal (1992) studied the uncatalyzed solvolysis of whole biomass hemicellulose by HCW and observed that between 40% and 60% of the sample mass was solubilized. On the other hand, Minowa et al. (1998) studied the cellulose decomposition in HCW with alkaline or nickel catalyst and observed that hydrolysis could play an important role in forming glucose/oligomer, and the obtained glucose/oligomer could decompose quickly to nonglucose aqueous products, oil, char, and gases.

Liu and Wyman (2003) studied the effect of flow rate of HCW on xylan, lignin, and total mass removal from corn stover and observed that solubilization of hemicellulose increased with flow, especially at high temperatures. On the other hand, Liu and Wyman (2005) studied the partial flow of HCW through corn stover to enhance hemicellulose sugar recovery and enzymatic digestibility of cellulose and observed that partial flow reduced water consumption by 60% compared with continuous flow-through operation but still achieved higher xylose sugar yields compared to batch pretreatment.

Nitsos et al. (2013) optimized the HCW pretreatment of beech wood in the bioethanol production process and observed that the major changes induced in the solid biomass were the dissolution or removal of hemicellulose to the process liquid and the partial removal and relocation of lignin on the external surface of biomass particles. On the other hand, Yuan et al. (2010) studied the hydrolytic degradation of alkaline lignin in HCW and ethanol and observed that under the optimal reaction conditions almost complete degradation was achieved, producing less than 1% solid residue and negligible gas products.

Overall, these HCPs clearly show that these HCW pretreatments degrade the biomass to improve the hydrolysis of the biomass and fermentation of the resulting hydrolysates with an improved sugar and bioethanol yield.

14.4.3.3 The Autohydrolysis Pretreatment of the Biomass

Garrote et al. (1999) studied the mild autohydrolysis as an environmental-friendly technology for xylooligosaccharide production from eucalyptus wood and observed that negligible effects were caused by hydrothermal treatments on both cellulose and lignin and developed kinetic models which described the hydrolysis of hemicelluloses. Further, Garrote et al. (2002) studied the autohydrolysis of corncob as a nonisothermal operation for xylooligosaccharide production and observed that most of the cellulose was retained in the solid phase, whereas partial delignification was achieved.

Overall, these HCPs clearly show that these autohydrolysis explosion pretreatments degrade the biomass to improve the hydrolysis of the biomass and fermentation of the resulting hydrolysates with an improved sugar and bioethanol yield.

14.5 CONCLUSION AND FUTURE RESEARCH

The brief information about the key research fronts covered by the 25 most-cited papers with at least 409 citations each is given under two primary headings: steam explosion and LHW-based hydrothermal pretreatments where HCW, LHW, and autohydrolysis are the prolific research fronts.

The usual characteristics of these HCPs are that these pretreatments are often used in combination of other pretreatments for the hydrolysis of the biomass as seen in Tables 14.1 and 14.2. In this way, these hydrothermal pretreatments are effective in disrupting the biomass constituents such as cellulose, lignin, and hemicellulose of these feedstocks resulting in improved sugar and bioethanol yield.

The key findings on these research fronts should be read in the light of the increasing public concerns about climate change, GHG emissions, and global warming as these concerns have been certainly behind the boom in the research on the bioethanol production as an alternative to crude oil-based gasoline and diesel fuels in the last decades.

As Table 14.3 shows, there are two primary research fronts for these most-cited 25 papers: steam explosion and LHW-based hydrothermal pretreatments. Further, Table 14.4 provides information on the biomass used in these 25 HCPs. There are four primary types of the biomass: biomass constituents, wood, agricultural residues, and biomass in general. Further, lignin, hemicellulose, and cellulose are the primary constituents, while corn stover and wheat straw are the key agricultural residues.

These studies emphasize the importance of proper incentive structures for the efficient development and application of the hydrothermal pretreatments to enhance sugar and bioethanol yield of the biomass after the hydrolysis of the biomass and the following fermentation of the resulting hydrolysates in the light of North's institutional framework (North, 1991). In this context, the major producers and users of bioethanol fuels such as the USA, Canada, China, and Europe had developed strong incentive structures for the effective development and application of hydrothermal pretreatments for bioethanol and sugar production.

It is recommended that such review studies are performed for the primary research fronts of hydrothermal pretreatments as well as the biomass constituents, wood, grass, and agricultural residues. Further, such studies should also be carried out for the other pretreatments of the biomass.

ACKNOWLEDGMENTS

The contribution of the highly cited researchers in the field of the hydrothermal pretreatment of the biomass has been gratefully acknowledged.

REFERENCES

Alvira, P., E. Tomas-Pejo, M. Ballesteros and M. J. Negro. 2010. Pretreatment technologies for an efficient bioethanol production process based on enzymatic hydrolysis: A review. *Bioresource Technology* 101:4851–4861.

Angelici, C., B. M. Weckhuysen and P. C. A. Bruijnincx. 2013. Chemocatalytic conversion of ethanol into butadiene and other bulk chemicals. *ChemSusChem* 6:1595–1614.

Antolini, E. 2007. Catalysts for direct ethanol fuel cells. *Journal of Power Sources* 170:1–12.

Antolini, E. 2009. Palladium in fuel cell catalysis. *Energy and Environmental Science* 2:915–931.

Bach, Q. V. and O. Skreiberg. 2016. Upgrading biomass fuels via wet torrefaction: A review and comparison with dry torrefaction. *Renewable and Sustainable Energy Reviews* 54:665–677.

Ballesteros, I., M. J. Negro and J. M. Oliva, et al. 2006. Ethanol production from steam-explosion pretreated wheat straw. *Applied Biochemistry and Biotechnology* 130:496–508.

Burnham, J. F. 2006. Scopus database: A review. *Biomedical Digital Libraries* 3:1–8.

Cantarella, M., L. Cantarella, A. Gallifuoco, A. Spera and F. Alfani, F. 2004. Effect of inhibitors released during steam-explosion treatment of poplar wood on subsequent enzymatic hydrolysis and SSF. *Biotechnology Progress* 20:200–206.

Cara, C., E. Ruiz, I. Ballesteros, M. J. Negro and E. Castro. 2006. Enhanced enzymatic hydrolysis of olive tree wood by steam explosion and alkaline peroxide delignification. *Process Biochemistry* 41:423–429.

Fernando, S., S. Adhikari, C. Chandrapal and M. Murali. 2006. Biorefineries: Current status, challenges, and future direction. *Energy & Fuels* 20:1727–1737.

Garrote, G., H. Dominguez and J. C. Parajo. 1999. Mild autohydrolysis: An environmentally friendly technology for xylooligosaccharide production from wood. *Journal of Chemical Technology and Biotechnology* 74:1101–1109.

Garrote, G., H. Dominguez and J. C. Parajo. 2001. Kinetic modelling of corncob autohydrolysis. *Process Biochemistry* 36:571–578.

Garrote, G., H. Dominguez and J. C. Parajo. 2002. Autohydrolysis of corncob: Study of non-isothermal operation for xylooligosaccharide production. *Journal of Food Engineering* 52:211–218.

Grous, W. R., A. O. Converse and H. E. Grethlein. 1986. Effect of steam explosion pretreatment on pore size and enzymatic hydrolysis of poplar. *Enzyme and Microbial Technology* 8:274–280.

Hahn-Hagerdal, B., M. Galbe, M. F. Gorwa-Grauslund, G. Liden and G. Zacchi. 2006. Bio-ethanol – The fuel of tomorrow from the residues of today. *Trends in Biotechnology* 24:549–556.

Heitz, M., E. Capek-Menard and P. G. Koeberle et al. 1991. Fractionation of *Populus tremuloides* at the pilot plant scale: Optimization of steam pretreatment conditions using the STAKE II technology. *Bioresource Technology* 35:23–32.

Hill, J., E. Nelson, D. Tilman, S. Polasky and D. Tiffany. 2006. Environmental, economic, and energetic costs and benefits of biodiesel and ethanol biofuels. *Proceedings of the National Academy of Sciences of the United States of America* 103:11206–11210.

Hill, J., S. Polasky and E. Nelson, et al. 2009. Climate change and health costs of air emissions from biofuels and gasoline. *Proceedings of the National Academy of Sciences of the United States of America* 106:2077–2082.

Hongzhang, C. and L. Liying. 2007. Unpolluted fractionation of wheat straw by steam explosion and ethanol extraction. *Bioresource Technology* 98:666–676.

Hsieh, W. D., R. H. Chen, T. L. Wu and T. H. Lin. 2002. Engine performance and pollutant emission of an SI engine using ethanol-gasoline blended fuels. *Atmospheric Environment* 36:403–410.

Huang, H. J., S. Ramaswamy, U. W. Tschirner and B. V. Ramarao. 2008. A review of separation technologies in current and future biorefineries. *Separation and Purification Technology* 62:1–21.

Jacquet, N., G. Maniet, C. Vanderghem, F. Delvigne and A. Richel. 2015. Application of steam explosion as pretreatment on lignocellulosic material: A review. *Industrial and Engineering Chemistry Research* 54:2593–2598.

Ko, J. K., Y. Kim, E. Ximenes and M. R. Ladisch. 2015. Effect of liquid hot water pretreatment severity on properties of hardwood lignin and enzymatic hydrolysis of cellulose. *Biotechnology and Bioengineering* 112:252–262.

Konur, O. 2000. Creating enforceable civil rights for disabled students in higher education: An institutional theory perspective. *Disability & Society* 15:1041–1063.

Konur, O. 2002a. Access to nursing education by disabled students: Rights and duties of nursing programs. *Nurse Education Today* 22:364–374.

Konur, O. 2002b. Assessment of disabled students in higher education: Current public policy issues. *Assessment and Evaluation in Higher Education* 27:131–52.

Konur, O. 2002c. Access to employment by disabled people in the UK: Is the Disability Discrimination Act working? *International Journal of Discrimination and the Law* 5:247–279.

Konur, O. 2006a. Participation of children with dyslexia in compulsory education: Current public policy issues. *Dyslexia* 12:51–67.

Konur, O. 2006b. Teaching disabled students in Higher Education. *Teaching in Higher Education* 11:351–363.

Konur, O. 2007a. A judicial outcome analysis of the Disability Discrimination Act: A windfall for the employers? *Disability & Society* 22:187–204.

Konur, O. 2007b. Computer-assisted teaching and assessment of disabled students in higher education: The interface between academic standards and disability rights. *Journal of Computer Assisted Learning* 23:207–219.

Konur, O. 2012. The evaluation of the research on the bioethanol: A scientometric approach. *Energy Education Science and Technology Part A: Energy Science and Research* 28:1051–1064.

Konur, O. 2015. Current state of research on algal bioethanol. In *Marine Bioenergy: Trends and Developments*, Eds. S. K. Kim and C. G. Lee, pp. 217–244. Boca Raton, FL: CRC Press.

Konur, O. 2020. The scientometric analysis of the research on the bioethanol production from green macroalgae. In *Handbook of Algal Science, Technology and Medicine*, Ed. O. Konur, pp. 385–401. London: Academic Press.

Konur, O. 2023. Hydrothermal pretreatment of the biomass: Scientometric study. In *Bioethanol Fuel Production Processes. I: Biomass Pretreatments. Handbook of Bioethanol Fuels Volume 1*, Ed. O. Konur. Boca Raton, FL: CRC Press.

Kumar, L., V. Arantes, R. Chandra and J. Saddler. 2012. The lignin present in steam pretreated softwood binds enzymes and limits cellulose accessibility. *Bioresource Technology* 103:201–208.

Laser, M., D. Schulman and S. G. Allen, et al. 2002. A comparison of liquid hot water and steam pretreatments of sugar cane bagasse for bioconversion to ethanol. *Bioresource Technology* 81:33–44.

Li, J., G. Henriksson and G. Gellerstedt. 2007. Lignin depolymerization/repolymerization and its critical role for delignification of aspen wood by steam explosion. *Bioresource Technology* 98:3061–3068.

Lin, Y. and S. Tanaka. 2006. Ethanol fermentation from biomass resources: Current state and prospects. *Applied Microbiology and Biotechnology* 69:627–642.

Liu, C. and C. E. Wyman. 2003. The effect of flow rate of compressed hot water on xylan, lignin, and total mass removal from corn stover. *Industrial and Engineering Chemistry Research* 42:5409–5416.

Liu, C. and C. E. Wyman. 2005. Partial flow of compressed-hot water through corn stover to enhance hemicellulose sugar recovery and enzymatic digestibility of cellulose. *Bioresource Technology* 96:1978–1985.

Liu, S. 2010. Woody biomass: Niche position as a source of sustainable renewable chemicals and energy and kinetics of hot-water extraction/hydrolysis. *Biotechnology Advances* 28:563–582.

Ma, X., L. Sun and C. Song. 2002. A new approach to deep desulfurization of gasoline, diesel fuel and jet fuel by selective adsorption for ultra-clean fuels and for fuel cell applications. *Catalysis Today* 77:107–116.

Minowa, T., F. Zhen and T. Ogi. 1998. Cellulose decomposition in hot-compressed water with alkali or nickel catalyst. *Journal of Supercritical Fluids* 13:253–259.

Mok, W. S. L. and M. J. Antal. 1992. Uncatalyzed solvolysis of whole biomass hemicellulose by hot compressed liquid water. *Industrial and Engineering Chemistry Research* 31:1157–1161.

Morschbacker, A. 2009. Bio-ethanol based ethylene. *Polymer Reviews* 49:79–84.

Mosier, N., R. Hendrickson, N. Ho, M. Sedlak and M. R. Ladisch. 2005. Optimization of pH controlled liquid hot water pretreatment of corn stover. *Bioresource Technology* 96:1986–1993.

Najafi, G., B. Ghobadian and T. Tavakoli, et al. 2009. Performance and exhaust emissions of a gasoline engine with ethanol blended gasoline fuels using artificial neural network. *Applied Energy* 86:630–639.

Newman, P. W. G. and J. R. Kenworthy. 1989. Gasoline consumption and cities: A comparison of U.S. cities with a global survey. *Journal of the American Planning Association* 55:24–37.

Nitsos, C. K., K. A. Matis and K. S. Triantafyllidis. 2013. Optimization of hydrothermal pretreatment of lignocellulosic biomass in the bioethanol production process. *ChemSusChem* 6:110–122.

North, D. C. 1991. Institutions. *Journal of Economic Perspectives* 5:97–112.

Ohgren, K., R. Bura, G. Lesnicki, J. Saddler and G. Zacchi. 2007b. A comparison between simultaneous saccharification and fermentation and separate hydrolysis and fermentation using steam-pretreated corn stover. *Process Biochemistry* 42:834–839.

Ohgren, K., R. Bura, J. Saddler and G. Zacchi. 2007a. Effect of hemicellulose and lignin removal on enzymatic hydrolysis of steam pretreated corn stover. *Bioresource Technology* 98:2503–2510.

Olsson, L. and B. Hahn-Hagerdal. 1996. Fermentation of lignocellulosic hydrolysates for ethanol production. *Enzyme and Microbial Technology* 18:312–331.

Palonen, H., F. Tjerneld, G. Zacchi and M. Tenkanen. 2004. Adsorption of *Trichoderma reesei* CBH I and EG II and their catalytic domains on steam pretreated softwood and isolated lignin. *Journal of Biotechnology* 107:65–72.

Perez, J. A., I. Ballesteros and M. Ballesteros. 2008. Optimizing liquid hot water pretreatment conditions to enhance sugar recovery from wheat straw for fuel-ethanol production. *Fuel* 87:3640–3647.

Pu, Y., F. Hu, F. Huang, B. H. Davison and A. J. Ragauskas. 2013. Assessing the molecular structure basis for biomass recalcitrance during dilute acid and hydrothermal pretreatments. *Biotechnology for Biofuels* 6:15.

Rocha, G. J. M., A. R. Goncalves, B. R. Oliveira, E. G. Olivares and C. E. V. Rossell. 2012. Steam explosion pretreatment reproduction and alkaline delignification reactions performed on a pilot scale with sugarcane bagasse for bioethanol production. *Industrial Crops and Products* 35:274–279.

Sanchez, O. J. and C. A. Cardona. 2008. Trends in biotechnological production of fuel ethanol from different feedstocks. *Bioresource Technology* 99:5270–5295.

Sun, X. F., F. Xu, R. C. Sun, P. Fowler and M. S. Baird. 2005. Characteristics of degraded cellulose obtained from steam-exploded wheat straw. *Carbohydrate Research* 340:97–106.

Taherzadeh, M. J. and K. Karimi. 2007. Enzyme-based hydrolysis processes for ethanol from lignocellulosic materials: A review. *Bioresources* 2:707–738.

Taherzadeh, M. J. and K. Karimi. 2008. Pretreatment of lignocellulosic wastes to improve ethanol and biogas production: A review. *International Journal of Molecular Sciences* 9:1621–1651.

Yu, Y., X. Lou and H. Wu, H. 2008. Some recent advances in hydrolysis of biomass in hot-compressed water and its comparisons with other hydrolysis methods. *Energy and Fuels* 22:46–60.

Yuan, Z., S. Cheng, M. Leitch and C. C. Xu. 2010. Hydrolytic degradation of alkaline lignin in hot-compressed water and ethanol. *Bioresource Technology* 101:9308–9313.

Zhuang, X., W. Wang and Q. Yu, et al. 2016. Liquid hot water pretreatment of lignocellulosic biomass for bioethanol production accompanying with high valuable products. *Bioresource Technology* 199:68–75.

Mechanical Biomass Pretreatments

15 Mechanical Pretreatment of the Biomass

Scientometric Study

Ozcan Konur
(Formerly) Ankara Yildirim Beyazit University

15.1 INTRODUCTION

The crude oil-based gasoline fuels (Ma et al., 2002; Newman and Kenworthy, 1989) have been widely used in the transportation sector since the 1920s. However, there have been great public concerns over the adverse environmental and human impact of these fuels (Hill et al., 2006, 2009). Hence, biomass-based bioethanol fuels (Alvira et al., 2010; Hill et al., 2006; Konur, 2012e, 2015, 2020a; Sun and Cheng, 2002) have increasingly been used in blending gasoline fuels (Hsieh et al., 2002; Najafi et al., 2009), in the fuel cells (Antolini, 2007, 2009), and in the biochemical production (Angelici et al., 2013; Morschbacker, 2009) in a biorefinery context (Fernando et al., 2006; Huang et al., 2008).

The research in the fields of mechanical pretreatments of the biomass has also intensified in recent years. Milling (Balaz, 2018; Mayer-Laigle et al., 2018; Zhu, 2011), microwave (Kostas et al., 2017; Li et al., 2016; Saleem et al., 2015), and ultrasonic (Bussemaker and Zhang, 2013; Luo et al., 2014; Subhedar and Gogate, 2013) pretreatments of the biomass have been widely researched to increase the bioethanol yield in this context in recent years.

However, it is essential to develop efficient incentive structures (North, 1991) for the primary stakeholders to enhance the research in this field (Konur, 2000, 2002a–c, 2006a, b, 2007a, b). The scientometric analysis has been used in this context to inform the primary stakeholders about the current state of the research in a selected research field (Garfield, 1955; Konur, 2011, 2012a–i, 2015, 2018b, 2019, 2020a).

As there have been no scientometric studies on the mechanical pretreatments of the biomass, this book chapter presents a scientometric study of the research in mechanical pretreatments of the biomass. It examines the scientometric characteristics of both the sample and population data presenting scientometric characteristics of these both datasets in the order of documents, authors, publication years, institutions, funding bodies, source titles, countries, Scopus subject categories, keywords, and research fronts, as well as highly cited 20 papers.

15.2 MATERIALS AND METHODS

The search for this study was carried out using Scopus database (Burnham, 2006) in February 2022.

As a first step for the search of the relevant literature, the keywords were selected using the first most-cited 200 papers for each research front. The selected keyword list was optimized to obtain a representative sample of papers for the searched research field. This keyword list was provided in the Appendix for future replication studies.

As a second step, two sets of data were used for this study. First, a population sample of over 1,800 papers was used to examine the scientometric characteristics of the population data. Second,

DOI: 10.1201/9781003226536-21

a sample of 181 most-cited papers, corresponding to 10% of the population papers, was used to examine the scientometric characteristics of these citation classics.

The scientometric characteristics of these both sample and population datasets were presented in the order of documents, authors, publication years, institutions, funding bodies, source titles, countries, Scopus subject categories, keywords, and research fronts as well as the 20 highly cited papers (HCPs).

Lastly, the key scientometric findings for both datasets were discussed to highlight the research landscape for the mechanical pretreatments of the biomass. Additionally, a number of brief conclusions were drawn, and a number of relevant recommendations were made to enhance the future research landscape.

15.3 RESULTS

15.3.1 The Most-Prolific Documents in the Mechanical Pretreatments of the Biomass

The information on the types of documents for both datasets is given in Table 15.1. The articles and conference papers dominate both the sample and population papers as they are under-represented slightly in the sample papers. Further, review papers have a surplus.

It is further notable that 95% of the population papers were published in journals, while 4% and 1% of them were published in book series and books, respectively. On the contrary, all of the sample papers were published in the journals.

15.3.2 The Most-Prolific Authors in the Mechanical Pretreatments of the Biomass

The information about the most-prolific ten authors with at least 2.2% of sample papers each is given in Table 15.2.

The most-prolific author is Runcang Sun with 5.5% of the sample papers. Takashi Endo, Yuanxin Wu, Ziniu Yu, and Shengdong Zhu follow this top author with 3.3%–3.9% of the sample papers each.

The most-prolific institution for the sample dataset is the Wuhan Institute of Chemical Technology with five authors. The National Institute of Advanced Industrial Science and Technology (AIST) follows this top institution with two authors. In total, five institutions house these prolific authors.

TABLE 15.1
Documents in the Mechanical Pretreatments of the Biomass

Documents	Sample Dataset (%)	Population Dataset (%)	Surplus (%)
Article	90.1	91	–0.9
Review	7.7	3	4.7
Conference paper	2.2	5	–2.8
Book chapter	0.0	1	–1
Book	0.0	0	0
Editorial	0.0	0	0
Letter	0.0	0	0
Note	0.0	0	0
Short survey	0.0	0	0
Sample size	181	1,811	

Sample dataset: the number of papers (%) in the set of 181 HCPs. Population dataset: the number of papers (%) in the set of the 1,811 population papers.

TABLE 15.2
The Most-Prolific Authors in the Mechanical Pretreatments of the Biomass

No.	Author Name	Author Code	Sample Papers %	Population Papers %	Institution	Country	HI	N	Res. Front
1	Sun, Runcang	55661525600	5.5	1.3	Dalian Polytech Univ.	China	93 13	985 57	Ultrasonics
2	Endo, Takashi	7003538734	3.9	0.7	AIST	Japan	39	113	Milling
3	Wu, Yuanxin	8919301100	3.3	0.4	Wuhan Inst. Chem. Tech.	China	27	201	Microwave
4	Yu, Ziniu	8919301100	3.3	0.4	Huazhong Agr. Univ.	China	50	296	Microwave
5	Zhu, Shengdong	7404346720	3.3	0.4	Wuhan Inst. Chem. Tech.	China	19	49	Microwave
6	Gogate, Parag R.	7003942509	2.8	0.7	Inst. Chem. Technol.	India	74	396	Ultrasonics
7	Inoue, Hiroyuki	55733159900	2.8	0.3	AIST	Japan	29	84	Milling
8	Xu, Feng	56420960200	2.8	0.7	Beijing Forestry Univ.	China	67	433	Milling
9	Wang, Cunwen	8628270200	2.2	0.2	Wuhan Inst. Chem. Technol.	China	30	139	Microwave
10	Yu, Faquan	8960054800	2.2	0.2	Wuhan Inst. Chem. Technol.	China	36	142	Microwave

Author code: the unique code given by Scopus to the authors. Sample papers: the number of papers authored in the sample dataset. Population papers: the number of papers authored in the population dataset.
HI, H-inde15; N, number of papers published by each author.

The most-prolific country for the sample dataset is China with seven authors and Japan has two authors. In total, three countries house these authors.

The most-prolific research front is the microwave pretreatment of the biomass with five authors. The other fronts are milling and ultrasonic pretreatment of the biomass with three and two authors, respectively.

On the other hand, there is significant gender deficit (Beaudry and Lariviere, 2016) for the sample dataset as surprisingly all of these top researchers are male.

15.3.3 The Most-Prolific Research Output by Years in Mechanical Pretreatments of the Biomass

Information about papers published between 1970 and 2022 is given in Figure 15.1. This figure clearly shows that the bulk of the research papers in the population dataset were published primarily in the 2010s with 56% of the population dataset. The publication rates for the 2020s, 2000s, 1990s, 1980s, and 1970s were 18%, 1%, 5%, 3%, and 2% respectively.

Similarly, the bulk of the research papers in the sample dataset were published in the 2010s and 2000s with 61% and 30% of the sample dataset, respectively. The publication rates for the 1990s, 1980s, and 1970s were 6%, 4%, and 2% of the sample papers, respectively.

The most-prolific publication year for the population dataset was 2021 with over 10%, while 72% of the population papers were published between 2011 and 2021. Similarly, 76% of the sample papers were published between 2006 and 2018, while the most-prolific publication years were 2011 and 2013 with 8.8% and 8.3% of the sample papers, respectively.

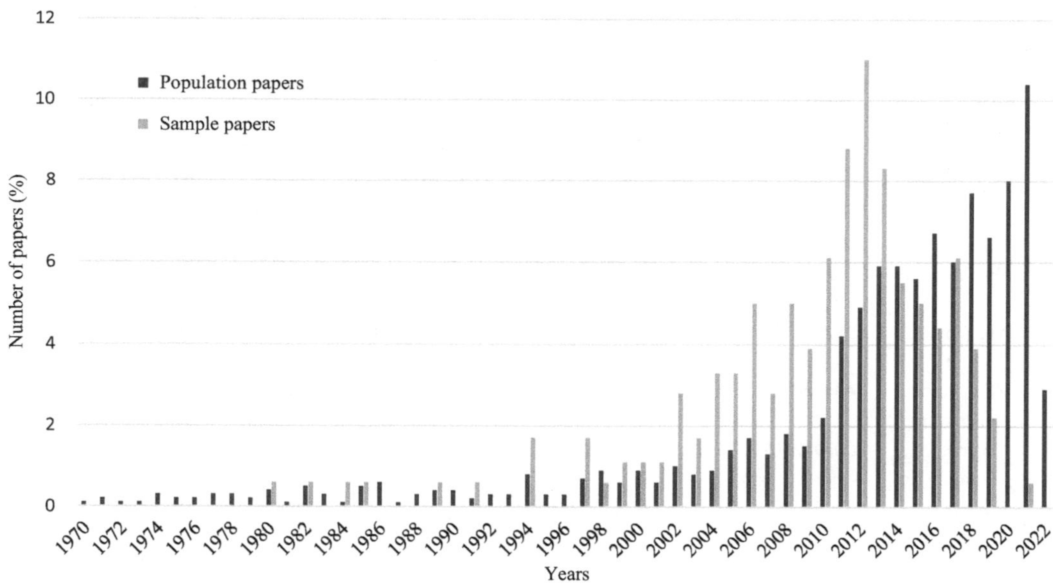

FIGURE 15.1 The research output by years regarding the mechanical pretreatments of the biomass.

15.3.4 The Most-Prolific Institutions in the Mechanical Pretreatments of the Biomass

Information about the most-prolific 11 institutions publishing papers on the mechanical pretreatments of the biomass with at least 2% of the sample papers is given in Table 15.3.

The most-prolific institution is the South China University of Technology with 7.2% of the sample papers. Beijing Forestry University and Chinese Academy of Sciences follow this top institution with 5.5% of the sample papers each.

The top country for these most-prolific institutions is China with five institutions. Next, Brazil follows these countries with three institutions. In total, seven countries house these top institutions.

On the other hand, the institution with the most citation impact is the South China University of Technology with 4.5% surplus. Similarly, the institution with the least impact is the Institute of Chemical Technology with 1.5% surplus.

15.3.5 The Most-Prolific Funding Bodies in the Mechanical Pretreatments of the Biomass

Information about the most-prolific nine funding bodies funding at least 2.2% of the sample papers is given in Table 15.4.

The most-prolific funding body is the National Natural Science Foundation of China with 14.4% of the sample papers. The Ministry of Science and Technology of India, Ministry of Education, Culture, Sports, Science and Technology of Japan, and Ministry of Science and Technology of China are the other prolific funding bodies with 3.3% of sample papers each.

It is notable that 49% and 43% of the sample and population papers are funded, respectively.

The most-prolific country for these top funding bodies is China with six funding bodies, and Japan has two funding bodies. In total, three countries house these top funding bodies.

The funding body with the most citation impact is the National Natural Science Foundation of China with 5.1% surplus. Similarly, the funding body with the least citation impact is The Chinese Academy of Sciences with at least 0.6% surplus.

TABLE 15.3
The Most-Prolific Institutions in Mechanical Pretreatments of the Biomass

No.	Institutions	Country	Sample Papers (%)	Population Papers (%)	Surplus (%)
1	South China Univ. Technol.	China	7.2	2.7	4.5
2	Beijing Forestry Univ.	China	5.5	1.8	3.7
3	Chin. Acad. Sci.	China	5.5	2.2	3.3
4	AIST	Japan	5.0	1.1	3.9
5	Wuhan Inst. Chem. Technol.	China	3.9	0.7	3.2
6	Huazhong Agric. Univ.	China	3.9	0.6	3.3
7	North Carolina State Univ.	USA	2.8	0.7	2.1
8	Inst. Chem. Technol.	India	2.8	1.3	1.5
9	Slovak Acad. Sci.	Slovakia	2.2	0.4	1.8
10	Univ. Saskatchewan	Canada	2.2	0.6	1.6
11	Univ. Wales	UK	2.2	0.5	1.7

TABLE 15.4
The Most-Prolific Funding Bodies in Mechanical Pretreatments of the Biomass

No.	Funding Bodies	Country	Sample Paper No. (%)	Population Paper No. (%)	Surplus (%)
1	Natl. Natr. Sci. Found. China	China	14.4	9.3	5.1
2	Minist. Educ. Cult. Sci. Technol.	Japan	3.3	1.3	2.0
3	Minist. Sci. Technol.	China	3.3	1.2	2.1
4	Minist. Sci. Technol.	India	3.3	1.0	2.3
5	Chin. Acad. Sci.	China	2.8	2.2	0.6
6	Japan Sci. Technol. Agcy.	Japan	2.8	0.6	2.2
7	Natl. Sci. Found. Quangdong	China	2.8	0.6	2.2
8	Minist. Educ.	China	2.2	1.4	0.8
9	Natl. Hi-Tech. Res. Devnt. Prog.	China	2.2	0.4	1.8

15.3.6 The Most-Prolific Source Titles in the Mechanical Pretreatments of the Biomass

Information about the most-prolific 11 source titles publishing at least 2.2% of the sample papers each in mechanical pretreatments of the biomass is given in Table 15.5.

The most-prolific source title is Bioresource Technology with 22% of the sample papers, followed by Ultrasonics Sonochemistry with 12% of the sample papers.

On the other hand, the source title with the most citation impact is the Bioresource Technology with 15% surplus, followed by Ultrasonics Sonochemistry with 8% of the sample papers. Similarly, the source title with the least impact is the Industrial Crops and Products with 1% surplus.

15.3.7 The Most-Prolific Countries in the Mechanical Pretreatments of the Biomass

Information about the most-prolific 11 countries publishing at least 3.3% of sample papers each in mechanical pretreatments of the biomass is given in Table 15.6.

The most-prolific countries are China with 27% of the sample papers, followed by the USA and Japan with 17% and 15% of the sample papers, respectively. The other prolific countries are India, the UK, and Spain with 9%, 8%, and 6% of the sample papers, respectively. On the other hand, three European countries listed in Table 15.6 produce 18% and 12% of the sample papers as a whole.

TABLE 15.5
The Most-Prolific Source Titles in Mechanical Pretreatments of the Biomass

No.	Source Titles	Sample Papers (%)	Population Papers (%)	Surplus (%)
1	Bioresource Technology	22.1	6.9	15.2
2	Ultrasonics Sonochemistry	11.6	3.2	8.4
3	Carbohydrate Polymers	5.5	1.4	4.1
4	Biomass and Bioenergy	3.3	1.0	2.3
5	Industrial Crops and Products	2.8	2.0	0.8
6	Green Chemistry	2.8	0.9	1.9
7	Biosystems Engineering	2.8	0.4	2.4
8	Applied Biochemistry and Biotechnology	2.2	0.8	1.4
9	Energy Conversion and Management	2.2	0.6	1.6
10	Biochemical Engineering Journal	2.2	0.4	1.8
11	Renewable and Sustainable Energy Reviews	2.2	0.3	1.9

TABLE 15.6
The Most-Prolific Countries in the Mechanical Pretreatments of the Biomass

No.	Countries	Sample Papers (%)	Population Papers (%)	Surplus (%)
1	China	26.5	20.2	6.3
2	USA	17.1	13.0	4.1
3	Japan	14.9	7.5	7.4
4	India	8.8	7.4	1.4
5	UK	7.7	5.5	2.2
6	Spain	5.5	2.7	2.8
7	France	5.0	3.5	1.5
8	Canada	5.0	3.0	2.0
9	S. Korea	3.9	3.6	0.3
10	Australia	3.9	2.9	1.0
11	Brazil	3.3	5.4	–2.1

On the other hand, the country with the most citation impact is Japan with 7% surplus, followed by China with 6% surplus. Similarly, the country with the least citation impact is Brazil with 2% deficit.

15.3.8 The Most-Prolific Scopus Subject Categories in the Mechanical Pretreatments of the Biomass

Information about the most-prolific 11 Scopus subject categories indexing at least 9% of the sample papers is given in Table 15.7.

The most-prolific Scopus subject category in the mechanical pretreatments of the biomass is Chemical Engineering with 56% of sample papers, followed by Environmental Science with 54% of the sample papers.

On the other hand, the Scopus subject categories with the most citation impact are the Environmental Science and Chemical Engineering with 24% surplus each. Similarly, the Scopus subject category with the least citation impact is Materials Science with 10% deficit, followed by Engineering and Agricultural and Biological Sciences with 7% and 5% deficit, respectively.

TABLE 15.7
The Most-Prolific Scopus Subject Categories in the Mechanical Pretreatments of the Biomass

No.	Scopus Subject Categories	Sample Papers (%)	Population Papers (%)	Surplus (%)
1	Chemical Engineering	56.4	32.1	24.3
2	Environmental Science	53.6	29.3	24.3
3	Energy	37.0	22.4	14.6
4	Chemistry	28.7	22.4	6.3
5	Biochemistry, Genetics, and Molecular Biology	18.8	9.2	9.6
6	Agricultural and Biological Sciences	16.0	20.8	–4.8
7	Engineering	13.8	20.6	–6.8
8	Physics and Astronomy	13.8	9.6	4.2
9	Medicine	13.3	6.5	6.8
10	Materials Science	11.0	21.0	–10.0
11	Immunology and Microbiology	8.8	3.6	5.2

15.3.9 The Most-Prolific Keywords in the Mechanical Pretreatments of the Biomass

Information about the keywords used in at least 11% and 4% of the sample or population papers, respectively, is given in Table 15.8. For this purpose, keywords related to the keyword set given in the Appendix are selected from a list of the most-prolific keyword set provided by Scopus database.

These keywords are grouped under the four headings: biomass, mechanical pretreatments, other processes, and products of the mechanical pretreatments.

There are nine keywords selected related to the biomass: The prolific keywords are cellulose, biomass, and lignin with 47%, 30%, and 30% of the sample papers, respectively.

The prolific keywords related to the mechanical pretreatments are ultrasonics, ultrasound, and microwaves with 29%, 29%, and 24% of the sample papers, respectively, while those related to the other processes are hydrolysis, enzymatic hydrolysis, and pretreatment 37%, 27%, and 23% of the sample papers, respectively.

The prolific keywords related to the products of the mechanical pretreatments are ethanol, glucose, sugar, and bioethanol with 20%, 18%, 25%, and 9% of the sample papers, respectively.

15.3.10 The Most-Prolific Research Fronts in Mechanical Pretreatments of the Biomass

Information about the most-prolific research fronts for the sample papers in mechanical pretreatments of the biomass is given in Table 15.9. As Table 15.9 shows, there are three primary research fronts for this field: milling, microwave, and ultrasonic pretreatments of the biomass with 28%, 34%, and 38% of the 181 HCPs and 37%, 33%, and 41% of the population papers, respectively.

15.3.11 The Highly Cited Papers of the Research on the Mechanical Pretreatments of the Biomass

Table 15.10 presents the brief data on the 20 HCPs for the mechanical pretreatments of the biomass. Eleven and nine HCPs are related to the milling and microwave pretreatments of the biomass, respectively, while there is no HCP for the ultrasonic pretreatment of the biomass.

TABLE 15.8
The Most-Prolific Keywords in Mechanical Pretreatments of the Biomass

No.	Keywords	Sample Papers (%)	Population Papers (%)	Surplus (%)
1	**Biomass**			
	Cellulose	47.0	21.9	25.1
	Biomass	30.4	16.2	14.2
	Lignin	29.8	13.7	16.1
	Lignocellulose	15.5	5.2	10.3
	Straw	15.5	4.3	11.2
	Lignocellulosic biomass	13.3	6.3	7
	Hemicellulose	13.3	3.5	9.8
	Bagasse	12.2	4.8	7.4
	Wood	10.5	12.5	–2
2	**Mechanical Pretreatment**			
	Ultrasonics	28.7	17.8	10.9
	Ultrasound	28.7	14.5	14.2
	Microwaves	23.8	18.5	5.3
	Milling	21.0	20.7	0.3
	Microwave radiation	14.9	7.6	7.3
	Ball milling	11.6	6.8	4.8
	Ultrasonic frequencies	9.4	4.9	4.5
	Microwave irradiation	8.8	5.8	3
	Grinding	6.1	8.9	–2.8
	Microwave heating	5.5	4.6	0.9
	Sonication	5.5	4.5	1
3	**Other Processes**			
	Hydrolysis	36.5	15.2	21.3
	Enzymatic hydrolysis	26.5	10.5	16
	Pretreatment	22.7	7.2	15.5
	Enzyme activity	18.2	6.1	12.1
	Saccharification	15.5	6.1	9.4
	Pretreatment	14.9	6.7	8.2
	Fermentation	14.9	5.9	9
	Cellulase	14.4	3.6	10.8
	Irradiation	12.2	5.5	6.7
	Particle size	11.0	8.0	3
	Delignification	6.6	4.0	2.6
	Crystallinity	6.1	4.1	2
4	**Products**			
	Sugar	24.9	9.6	15.3
	Ethanol	20.4	7.0	13.4
	Glucose	18.2	6.7	11.5
	Bioethanol	9.4	4.9	4.5

TABLE 15.9
The Most-Prolific Research Fronts for the Mechanical Pretreatments of the Biomass

No.	Research Fronts	N Paper % Sample	N Paper % Population	Surplus %
1	Milling pretreatment	28.2	36.7	−8.5
2	Microwave pretreatment	33.7	33.2	0.5
3	Ultrasonic pretreatment	37.6	40.5	−2.9

N paper (%) sample: the number of papers in the population sample of 181 papers. N paper (%) population: the number of papers in the population sample of 1,811 papers.

15.4 DISCUSSION

15.4.1 Introduction

The crude oil-based gasoline fuels have been widely used in the transportation sector since the 1920s. However, there have been great public concerns over the adverse environmental and human impact of these fuels. Hence, biomass-based bioethanol fuels have increasingly been used in blending gasoline fuels, in the fuel cells, and in the biochemical production in a biorefinery context.

The research in the field of mechanical pretreatments of the biomass such as milling, microwave, and ultrasonic pretreatments to increase the bioethanol yield has also intensified in recent years.

However, it is essential to develop efficient incentive structures for the primary stakeholders to enhance the research in this field. The scientometric analysis has been used in this context to inform the primary stakeholders about the current state of the research in a selected research field.

As there have been no scientometric studies on the mechanical pretreatments of the biomass, this book chapter presents a scientometric study of the research in mechanical pretreatments of the biomass. It examines the scientometric characteristics of both the sample and population data presenting scientometric characteristics of these both datasets in the order of documents, authors, publication years, institutions, funding bodies, source titles, countries, Scopus subject categories, keywords, and research fronts, as well as the 20 HCPs.

As a first step for the search of the relevant literature, the keywords were selected using the first most-cited 200 papers for each research front. The selected keyword list was optimized to obtain a representative sample of papers for the searched research field. This keyword list was provided in the Appendix for future replication studies.

As a second step, two sets of data were used for this study. First, a population sample of over 1,811 papers was used to examine the scientometric characteristics of the population data. Second, a sample of 181 most-cited papers, corresponding to 10% of the population dataset, was used to examine the scientometric characteristics of these citation classics.

The scientometric characteristics of these both sample and population datasets were presented in the order of documents, authors, publication years, institutions, funding bodies, source titles, countries, Scopus subject categories, keywords, and research fronts as well as the 20 HCPs.

Lastly, the key scientometric findings for both datasets were discussed to highlight the research landscape for the mechanical pretreatments of the biomass. Additionally, a number of brief conclusions were drawn, and a number of relevant recommendations were made to enhance the future research landscape.

15.4.2 The Most-Prolific Documents in the Mechanical Pretreatments of the Biomass

Both articles (together with conference papers) and review papers dominate the sample and population papers (Table 15.1). Further, review papers and articles have a surplus and deficit, respectively.

TABLE 15.10
The 20 Highly Cited Papers for the Mechanical Pretreatments of the Biomass

No.	Papers	Biomass	Milling	Other Pts.	Parameters	Keywords	Lead Author	Affil.	Cits
1	Schwanninger et al. (2004)	Wood, cellulose	Vibratory ball milling,	None	Milling, temperature, particle size, oxygen	Milling, wood, cellulose	Schwanninger, Manfred 6602877236	Univ. Natr. Sci. App. Sci. Austria	870
2	Mani et al. (2004)	Wheat and barley straws, corn stover, switchgras	Hammer milling	None	Biomass type, energy consumption, density	Grinding, straw, stover, switchgrass	Tabil, Lope G 6701349307	Univ. Saskatchewan Canada	507
3	Hu and Wen (2008)	Switchgrass	Microwave	Alkali, enzymes	Alkali loading, temperature, solid content, treatment time	Digestibility, switchgrass, microwave, pretreatment, alkali	Wen, Zhiyou 14012373200	Virg. Polytech. Inst. State Univ. USA	339
4	Binod et al. (2012)	Sugarcane bagasse	Microwave	Alkali (NaOH), acid (H2SO4), enzymes	Alkali and acid loadings, energy loading, treatment time	Microwave, pretreatment, enzymatic, saccharification, fermentable, sugar, sugarcane bagasse	Pandey, Ashok 7201771319	CSIR India	266
5	Da Silva et al. (2010)	Sugarcane bagasse and straw	Ball milling, wet disk milling	Enyzmes	Milling type, sugar yield, bioethanol yield	Bagasse, straw, hydrolysis, enzymatic, ethanol, fermentation, milling	Bon, Elba P S 7007036976	Fed. Univ. Rio de Janeiro Brazil	264
6	Zhao et al. (2006)	Cellulose	Ball milling	Acid (H_2SO_4)	Crystallinity	Hydrolysis, cellulose, milling	Holladay, Johnathan E 7006800080	Pacific NW Natl. Lab. USA	246
7	Zhu et al. (2005)	Rice straw	Microwave	Alkali, enzymes	Alkali loading, energy loading, treatment time, solid content, enzyme loading	Pretreatment, microwave, alkali, rice straw, enzymic, hydrolysis	Zhu, Shendong 8919301300	Wuhan Inst. Chem. Technol. China	230
8	Chen et al. (2011)	Sugarcane bagasse	Microwave	Acid (H2SO4)	Treatment time, temperature	Disruption, sugarcane bagasse, lignocellulosic, acid, pretreatment, microwave	Chen, Wei- Hsin 57200873137	Natl. Univ. Tainan Taiwan	228
9	Ma et al. (2009)	Rice straw	Microwave	Enzymes	Energy loading, solid content, and treatment time	Enzymatic, saccharification, rice straw, microwave, pretreatment	Wu, Yue-Jin 55531536300	Chin. Acad. Sci. China	224

(*Continued*)

TABLE 15.10 (*Continued*)
The 20 Highly Cited Papers for the Mechanical Pretreatments of the Biomass

No.	Papers	Biomass	Milling	Other Pts.	Parameters	Keywords	Lead Author	Affil.	Cits
10	Wu et al. (2010)	Cellulose	Microwave	Acid (SO3H, H2SO4)	Acid type	Microwave, hydrolysis, cellulose, biomass, acid	Fu, Zaihui 7403346468	Hunan Normal Univ., China	208
11	Zhu et al. (2009)	Wood	Disk milling, hammer milling	Enzymes, sulfite	Glucose yield, energy consumption	Milling, wood, saccharification	Zhu, Junyong 7405692678	US Forest Serv. USA	199
12	Hideno et al. (2009)	Rice straw	Wet disk milling, ball milling	Enzymes, hot-compressed water	Sugar yield, energy consumption	Milling, hydrolysis, straw	Inoue, Hiroyuki 55733159900	AIST Japan	181
13	Zhang and Zhao (2009)	Cellulose	Microwave	Acid (H-form ZSM-5 zeolites, sulfated ion- exchanging resin NKC-9, ionic liquids	Acid type	Acid, microwave, hydrolysis, cellulose, ionic liquid	Zhao, Zongbao K 35197609300	Chin. Acad. Sci. China	179
14	Silva et al. (2012)	Wheat straw	Ball milling, jet milling, sieve-based grindings	Enzymes, steam explosion	Sugar yield, Milling type, carbohydrate yield, particle size	Straw, grinding, enzymatic	Rouau, Xavier 55973961500	INRA France	178
15	Chen et al. (2012)	Sugarcane bagasse	Microwave	Acid (H2SO4)	Acid content	Hydrolysis, sugarcane bagasse, pretreated, acid, microwave	Chen, Wei- Hsin 57200873137	Natl. Univ. Tainan Taiwan	162
16	Ling et al. (2019)	Cellulose	Ball milling	None	crystallinity	Milling, cellulose	French, Alfred D 7202095026	Agr. Res. Serv. USA	158
17	Fan et al. (2013)	Cellulose	Microwave	None	Temperature	Microwave, depolymerization, cellulose	Clark, James H 57218227445	Univ. York UK	157
18	Inoue et al. (2008)	Eucalyptus	Ball milling	Enzymes, hot compressed water	Temperature, treatment time, enzyme content, sugar yield	Milling, eucalyptus, hydrolysis, enzymatic	Inoue, Hiroyuki 55733159900	AIST Japan	155
19	Avolio et al. (2012)	Cellulose	Dry ball milling	None	Treatment time, crystallinity index, crystallite size	Milling, cellulose	Gentile, Gennaro 7102027136	CNR Italy	148
20	Zhu et al. (2010)	Pine wood chips	Disk milling	Enzymes, hot-water, dilute acid (H_2SO_4), sulfite ($NaHSO_3$)	Energy consumption, solid content, sugar yield, disk plate gap	Pine, enzymatic, saccharification, milling	Zhu, Junyong 7405692678	US Forest Serv. USA	146

Scopus differs from the Web of Science database in differentiating and showing articles and conference papers published in the journals separately. Similarly, Scopus differs from Web of science database in introducing short surveys. Hence, the total number of articles and review papers in the sample dataset are 92% and 3%, respectively.

It is observed during the search process that there has been inconsistency in the classification of the documents in Scopus as well as in other databases such as Web of Science. This is especially relevant for the classification of papers as reviews or articles as the papers not involving a literature review may be erroneously classified as a review paper. There is also a case of review papers being classified as articles.

In this context, it would be helpful to provide a classification note for the published papers in the books and journals at the first instance. It would also be helpful to use the document types listed in Table 15.1 for this purpose. Book chapters may also be classified as articles or reviews as an additional classification to differentiate review chapters from the experimental chapters as it is done by the Web of Science. It would be further helpful to additionally classify the conference papers as articles or review papers as well as it is done in the Web of Science database.

15.4.3 The Most-Prolific Authors in the Mechanical Pretreatments of the Biomass

There have been most-prolific ten authors with at least 2.2% of the sample papers each as given in Table 15.2. These authors have shaped the development of the research in this field.

The most-prolific authors are Runcang Sun, Takashi Endo, Yuanxin Wu, Ziniu Yu, and Shengdong Zhu, mostly from China.

It is important to note the inconsistencies in indexing of the author names in Scopus and other databases. It is especially an issue for the names with more than two components such as 'Judge Sam de Alonso Fatty'. The probable outcomes are 'Fatty, J.S.D.A.', 'de Alonso Fatty, J.S.' or 'Alonso Fatty, J.S.'. The first choice is the gold standard of the publishing sector as the last word in the name is taken as the last name. In most of the academic databases such as Pubmed and EBSCO databases, this version is used predominantly. The second choice is a strong alternative, while the last choice is an undesired outcome as two last words are taken as the last name. It is a good practice to combine the words of the last name by a hyphen: 'Alonso-Fatty, J.S.D.'. It is notable that inconsistent indexing of the author names may cause substantial inefficiencies in the search process for the papers as well as allocating credit to the authors as there are different author entries for each outcome in the databases.

There are also inconsistencies in the shortening Chinese names. For example, 'Yuoying Wang' is often shortened as 'Wang, Y.', 'Wang, Y.-Y.', and 'Wang Y.Y.', as it is done in the Web of Science database as well. However, the gold standard in this case is 'Wang Y', where the last word is taken as the last name and the first word is taken as a single forename. In most of the academic databases such as Pubmed and EBSCO, this first version is used predominantly. However, it makes sense to use the third option to differentiate Chinese names efficiently: 'Wang Y.Y.'. Therefore, there have been difficulties to locate papers for the Chinese authors. In such cases, the use of the unique author codes provided for each author by the Scopus database has been helpful.

There is also a difficulty in allowing credit for the authors especially for the authors with common names such as 'Wang, Y', or 'Huang, Y' or 'Zhu, Y' in conducting scientometric studies. These difficulties strongly influence the efficiency of the scientometric studies as well as allocating credit to the authors as there are the same author entries for different authors with the same name, e.g., 'Wang Y' in the databases.

In this context, the coding of authors in Scopus database is a welcome innovation compared to the other databases such as Web of Science. In this process, Scopus allocates a unique number to each author in the database (Aman, 2018). However, there might still be substantial inefficiencies in this coding system especially for common names. For example, some of the papers for a certain author maybe allocated to another researcher with a different author code. It is possible that Scopus uses a number of software programs to differentiate the author names, and the program may not be false-proof (D'Angelo and van Eck, 2020).

In this context, it does not help that author names are not given in full in some journals and books. This makes difficult to differentiate authors with common names and makes the scientometric studies further difficult in the author domain. Therefore, the author names should be given in all books and journals at the first instance. There is also a cultural issue where some authors do not use their full names in their papers. Instead, they use initials for their forenames: 'Coutancy, A.P.' or just 'Coutancy' instead of 'Coutancy, Alas Padras'.

There are also inconsistencies in naming of the authors with more than two components by the authors themselves in journal papers and book chapters. For example, 'Alaspanda, A.P.C.', 'Sakoura, C.E.', and 'Mentaslo, S.J.' might be given as 'Alaspanda, A', 'Sakoura, C', or 'Mentaslo, S' in the journals and books, respectively. This also makes the scientometric studies difficult in the author domain. Hence, contributing authors should use their name consistently in their publications.

The other critical issue regarding the author names is the inconsistencies in the spelling of the author names in the national spellings (e.g., Gonçalves, Üzeyir) rather than in the English spellings (e.g., Goncalves, Uzeyir) in Scopus database. Scopus differs from the Web of science database and many other databases in this respect, where the author names are given only in the English spellings. It is observed that national spellings of the author names do not help in conducting scientometric studies as well in allocating credits to the authors as sometimes there are the different author entries for the English and National spellings in the Scopus database.

The most-prolific institutions for the sample dataset are the Wuhan Institute of Chemical Technology and National Institute of Advanced Industrial Science and Technology (AIST). In total, five institutions house these prolific authors. The most-prolific countries for the sample dataset are China and Japan. In total, three countries house these authors.

The most-prolific research front is the microwave pretreatment of the biomass, followed by the milling and ultrasonic pretreatments.

It is also notable that there is a significant gender deficit for the sample dataset as surprisingly all of these top researchers are male. This finding is the most thought-provoking with strong public policy implications. Hence, institutions, funding bodies, and policymakers should take efficient measures to reduce the gender deficit in this field as well as other scientific fields with strong gender deficit. In this context, it is worth to note the level of representation of the researchers from the minority groups in science on the basis of race, sexuality, age, and disability, besides the gender (Blankenship, 1993; Dirth and Branscombe, 2017; Konur, 2000, 2002a–c, 2006a, b, 2007a, b).

15.4.4 The Most-Prolific Research Output by Years in the Mechanical Pretreatments of the Biomass

The research output observed between 1970 and 2021 is illustrated in Figure 15.1. This figure clearly shows that the bulk of the research papers in the population dataset were published primarily in the 2010s and to a lesser extent 2020s. Similarly, the bulk of the research papers in the sample dataset were published in the 2010s and to a lesser extent 2000s.

These data suggest that the most-cited sample and population papers were primarily published in the 2000s and 2010s, respectively. These are the thought-provoking findings as there has been no significant research in this field in the pre-2000s, but there has been significant research boom in the last two decades. In this context, the increasing public concerns about climate change (Change, 2007), greenhouse gas emissions (Carlson et al., 2017), and global warming (Kerr, 2007) have been certainly behind the boom in the research in this field in the last two decades.

Based on these findings, the size of the population papers is likely to more than double in the current decade, provided that the public concerns about climate change, greenhouse gas emissions, and global warming are translated efficiently to the research funding in this field.

15.4.5 The Most-Prolific Institutions in the Mechanical Pretreatments of the Biomass

The most-prolific 11 institutions publishing papers on the mechanical pretreatments of the biomass with at least 2% of the sample papers each given in Table 15.3 have shaped the development of the research in this field.

The most-prolific institutions are the South China University of Technology, Beijing Forestry University, and Chinese Academy of Sciences, all from China.

The top countries for these most-prolific institutions are China and Brazil. In total, seven countries house these top institutions.

On the other hand, the institutions with the most and least impact are South China University of Technology and the Institute of Chemical Technology, respectively.

These findings confirm the dominance of Chinese institutions this research field.

15.4.6 The Most-Prolific Funding Bodies in the Mechanical Pretreatments of the Biomass

The most-prolific six funding bodies funding at least 2.2% of the sample papers each are given in Table 15.4.

The most-prolific funding bodies are the National Natural Science Foundation of China, Ministry of Science and Technology of India, Ministry of Education, Culture, Sports, Science and Technology of Japan and Ministry of Science and Technology of China. It is notable that 49% and 43% of the sample and population papers are funded, respectively.

The most-prolific countries for these top funding bodies are China and Japan. In total, three countries house these top funding bodies. The funding bodies with the most and least citation impact are National Natural Science Foundation of China and Chinese Academy of Sciences, respectively.

These findings on the funding of the research in this field suggest that the level of the funding, mostly in the last two decades, is highly intensive and it has been largely instrumental in enhancing the research in this field (Ebadi and Schiffauerova, 2016) in the light of North's institutional framework (North, 1991). However, considering the relatively low levels of funding at the macroscale, there is ample room to enhance funding in this field.

15.4.7 The Most-Prolific Source Titles in Mechanical Pretreatments of the Biomass

The most-prolific 11 source titles publishing at least 2.2% of the sample papers each in mechanical pretreatments of the biomass have shaped the development of the research in this field (Table 15.5).

The most-prolific source titles are Bioresource Technology and Ultrasonics Sonochemistry. On the other hand, the source titles with the most and least impact are Bioresource Technology and Industrial Crops and Products, respectively.

It is notable that these top source titles are related to the biomass, ultrasonics, and bioenergy. This finding suggests that the journals in this field have significantly shaped the development of the research in this field as they focus on the pretreatment aspects of bioethanol fuels.

15.4.8 The Most-Prolific Countries in Mechanical Pretreatments of the Biomass

The most-prolific 11 countries publishing at least 3.3% of the sample papers each have significantly shaped the development of the research in this field (Table 15.6).

The most-prolific countries are the China, the USA, and Japan, while the other prolific countries are India, the UK, and Spain. On the other hand, the countries with the most and least citation impact are the Japan and Brazil, respectively. Further, three European countries listed in Table 15.6 produce 18% and 12% of the sample papers with 6% surplus as a whole.

The close examination of these findings suggests that China, the USA, and Japan as well as Europe are the major producers of the research in this field. It is a fact that the USA has been a major player in science (Leydesdorff and Wagner, 2009; Leydesdorff et al., 2014). The USA has further developed a strong research infrastructure to support its corn and grass-based bioethanol industry (Vadas et al., 2008).

However, China has been a rising star in scientific research in competition with the USA and Europe (Leydesdorff and Zhou, 2005). China is also a major player in this field as a major producer of bioethanol (Li and Chan-Halbrendt, 2009).

Next, Europe has been a persistent player in the scientific research in competition with both the USA and China (Leydesdorff, 2000). Europe has also been a persistent producer of bioethanol along with the USA and Brazil (Gnansounou, 2010).

Additionally, Brazil has also been a persistent player in scientific research at a moderate level (Glanzel et al., 2006). Brazil has also developed a strong research infrastructure to support its biomass-based bioethanol industry (Macedo et al., 2008).

15.4.9 The Most-Prolific Scopus Subject Categories in Mechanical Pretreatments of the Biomass

The most-prolific 11 Scopus subject categories indexing at least 9% of the sample papers, respectively, given in Table 15.7 have shaped the development of the research in this field.

The most-prolific Scopus subject categories in the mechanical pretreatments of the biomass are Chemical Engineering and Environmental Science.

On the other hand, the Scopus subject categories with the most citation impact are the Environmental Science and Chemical Engineering. Similarly, the Scopus subject categories with the least citation impact are Materials Science and Engineering and Agricultural and Biological Sciences.

These findings are thought-provoking, suggesting that the primary subject categories are Chemical Engineering and Environmental Science. The other key finding is that social sciences are not well represented in both the sample and population papers, unlike most fields in bioethanol fuels. These findings are not surprising as the key research fronts in this field relate to the pretreatments of thee biomass.

15.4.10 The Most-Prolific Keywords in Mechanical Pretreatments of the Biomass

A limited number of keywords have shaped the development of the research in this field as shown in Table 15.8 and the Appendix.

These keywords are grouped under the four headings: biomass, mechanical pretreatment, other processes, and products of the mechanical pretreatments.

There are nine keywords selected related to the biomass: The prolific keywords are cellulose, biomass, and lignin. The prolific keywords related to the mechanical pretreatments are ultrasonics, ultrasound, and microwaves, while those related to the other processes are hydrolysis, enzymatic hydrolysis, and pretreatment. Finally, the prolific keywords related to the products of the mechanical pretreatments are ethanol, glucose, sugar, and bioethanol.

On the other hand, the keywords with the most citation impact are cellulose, hydrolysis, lignin, enzymatic hydrolysis, pretreatments, sugar, biomass, and ultrasound.

These findings suggest that it is necessary to determine the keyword set carefully to locate the relevant research in each of these research fronts. Additionally, the size of the samples for each keyword highlights the intensity of the research in the relevant research areas.

15.4.11 The Most-Prolific Research Fronts in Mechanical Pretreatments of the Biomass

As Table 15.9 shows, there are three primary research fronts for this research field: mMilling, microwave, and ultrasonic pretreatments of the biomass. It is evident that the research for milling and ultrasonic pretreatments of the biomass is under-represented in the sample papers.

These findings are thought-provoking in seeking ways to increase bioethanol yield through the mechanical pretreatments of the biomass at the global scale. It is clear that all of these three research fronts have public importance and merit substantial funding and other incentives.

15.4.12 The Highly Cited Papers of the Research on the Mechanical Pretreatments of the Biomass

Table 15.10 presents the brief data on the 20 HCPs for the mechanical pretreatments of the biomass. Eleven and nine HCPs are related to the milling and microwave pretreatments of the biomass, respectively, while there is no HCP for the ultrasonic pretreatment of the biomass.

In the end, these most-cited papers in this field hint that the efficiency of bioethanol fuels and their derivatives could be optimized using the structure, processing, and property relationships of these three pretreatments (Formela et al., 2016; Konur, 2018a, 2020b, 2021a–d; Konur and Matthews, 1989).

15.5 CONCLUSION AND FUTURE RESEARCH

The research on the mechanical pretreatments of the biomass has been mapped through a scientometric study of both sample and population datasets.

The critical issue in this study has been to obtain a representative sample of the research as in any other scientometric study. Therefore, the keyword set has been carefully devised and optimized after a number of runs in the Scopus database. It is a representative sample of the wider population studies.

The other issue has been the selection of a multidisciplinary database to carry out the scientometric study of the research in this field. For this purpose, Scopus database has been selected. The journal coverage of this database has been notably wider than that of Web of Science.

The key scientometric properties of the research in this field have been determined and discussed in this book chapter. It is evident that a limited number of documents, authors, institutions, publication periods, institutions, funding bodies, source titles, countries, Scopus subject categories, keywords, and research fronts as well as HCPs have shaped the development of the research in this field.

There is ample scope to increase the efficiency of the scientometric studies in this field in the author and document domains by developing consistent policies and practices in both domains across all the academic databases. In this respect, it seems that authors, journals, and academic databases have a lot to do. Furthermore, the significant gender deficit as in most scientific fields emerges as a public policy issue. The potential deficits on the basis of age, race, disability, and sexuality need also to be explored in this field as in other scientific fields.

The research in this field has boomed in the 2000s and 2010s possibly promoted by the public concerns on global warming, greenhouse gas emissions, and climate change. The institutions from China and Brazil have mostly shaped the research in this field. Further, the sharp rise in the research output in 2020 and 2021 might be related to the supply shocks caused by the COVID-19 pandemics.

The relatively well funding rate of 49% and 43% for sample and population papers, respectively, suggests that funding in this field significantly enhanced the research in this field primarily in the 2010s and the early 2020s, possibly more than doubling in the current decade. However, it is evident that there is ample room for more funding and other incentives to enhance the research in this field further.

The most-prolific source titles have been mostly indexed by the subject categories of chemical engineering, energy, and environmental science as the focus of the sample papers has been on the development and utilization of the milling, microwave, and ultrasonic pretreatments of the biomass to increase the bioethanol yield at large.

China, the USA, Japan, and Europe have been the major producers of the research in this field as the major producers and users of bioethanol fuels from different types of biomass such as corn, sugarcane, and grass as well as other types of biomass. It is evident that these countries have well-developed research infrastructure in bioethanol fuels and their derivatives.

The primary subject categories have been Chemical Engineering and Environmental Science as the focus of the research has been on the development and utilization of the milling, microwave, and ultrasonic pretreatments of the biomass to increase the bioethanol yield at large. Further, social sciences are not well represented in both the sample and population papers, as in the most fields in bioethanol fuels. These findings are not surprising.

It emerges that ethanol is more popular than bioethanol as a keyword with strong implications for the search strategy. In other words, the search strategy using only bioethanol keyword would not be much helpful. These keywords are grouped under the four headings: biomass, mechanical pretreatments, other processes, and products of the mechanical pretreatments. These groups of keywords highlight the potential primary research fronts for these fields.

These findings are thought-provoking. The focus of these most-cited 181 papers as well as 1,811 population papers is the development and utilization of mechanical pretreatments to increase the bioethanol yield.

The 20 HCPs are related primarily to the milling and microwave pretreatments at the expense of the ultrasonic pretreatments. These studies highlight strong structure-processing-property relationships for mechanical pretreatments for bioethanol fuels and their derivatives.

Thus, the scientometric analysis has a great potential to gain valuable insights into the evolution of the research in this field as in other scientific fields.

It is recommended that further scientometric studies are carried out about the other types of pretreatments. It is further recommended that reviews of the most-cited papers are carried out for each research front to complement these scientometric studies. Next, the scientometric studies of the hot papers in these primary fields are carried out.

ACKNOWLEDGMENTS

The contribution of the highly cited researchers in the field of the mechanical pretreatments of the biomass has been gratefully acknowledged.

APPENDIX: THE KEYWORD SET FOR MECHANICAL PRETREATMENTS OF THE BIOMASS

((title (microwave* or "micro wave*") and title (wastes or waste or *grass or biomass* or lignocellulos* or bagasse* or cellulos* or hemicellulos* or *wood or stover* or miscanthus or husk* or straw or straws or hyacinth or sorghum or sawdust or residue* or corncob* or stem or "wheat bran" or pseudostem* or bluestem* or *stalk or reed or arundo or cellobiose or bamboo* or eucalyptus or cypress or stillage* or xylan or pinus or sawmill* or "corn cob*" or ulva or *algae or seaweed* or biowaste* or shoots or "fruit bunch*" or *algal or foliage or poplar* or rhizome or agave or peel* or coffee* or molass* or spruce or prosopis or birch or triticale or arborera or saccharum or bioresource* or eucheuma or laminaria or bode or woody or taro or willow or szarvasi or eichhornia or sago)) and not title (*furfural or *hydrogen or sludge or carbonization or digestion or digester* or *gas or cell or nano* or pyroly* or *char or methane or torrefaction or "activated carbon*" or ferul* or earth or arsen* or spectro* or pet or extract* or ions or h2 or chelat* or ester* or polycarbonate or gas* or protein* or sensor* or phenolic* or *ethylene or chromat* or synthesis or absorber* or reinforced or combustion or acrylamide* or aggregate* or aluminium or surfactant* or moisture or lipid* or hmf or olive or popping or lactic or "bio-oil*" or *phenolic or dye* or acetylation or soil* or *butanol or wool or dechlor* or drying or dried or modificat* or sheep or papaya or pectin or transester* or ash* or impregn* or radioact* or tissue* or *viral or *diesel or properties or nuclear or chitosan or inciner* or "essential oil*" or analysis or fabricat* or dxn* or butyric or fire* or belt or tar or decanter or leach* or syringol or biochem* or isomer* or pellet* or dietary or antioxidant* or furan* or magnet* or derived or *porous or levunic or absorption or recycl* or harvest* or anaerobic or wave* or liquefaction or neur* or levog* or composite* or fiber* or rubber* or veget* or adsor* or *coal or *chars or dielectric or graft or polysaccharide* or "oil prod*" or proteom* or *stability or emiss* or carbon or "oil recov*" or imaging or flax or fruit* or phosphor* or comput* or carbam* or scan* or starch or testing or cook* or knot* or oxid* or foam* or tobac* or caco3 or carp or spore* or polar* or levul* or tomogh* or termite* or mycel* or carboxymethyl* or algor* or cnt* or "value add*" or wound* or alcohol* or measur* or resonat* or fame* or petrol* or beam or beetle* or steril* or epoxy or processing or cation* or xylanase*) and not srctitle (food* or agric* or hazard* or nano* or cement* or carbohydr* or lwt*)) or (((title (ultrasound or ultrason* or sonicat* or sono* or cavitation) or srctitle (ultrasound or ultrason*)) and (title (waste* or *grass or grass* or biomass* or lignocellul* or bagasse* or cellulos* or *wood or woody or *algae or *algal or straw* or stover* or eucalyptus or miscanthus or pine or chlorella or lignin or *cellulose or porphyridium or nannochloropsis or palm or poplar or cedar or corncob* or "corn cob*" or spruce or beech or oak or bunch* or pinus or residue* or willow* or cypress or sawdust or prosopis or sugarcane or "rice hull*" or husk* or *cane or birch or cynara or kenaf or "*cane tops") or srctitle (cellulose or biomass or sugar))) and not (title (removal or concrete or *diesel or *gas or digestion or drug* or pozzolanic or *composites or composite or cyclodext* or sorbitol or uranium or lactic or polysaccharide* or wastewater* or ore or ester* or medic* or reforming or separation or ruminal or recycl* or hematite or mine or graphite or cement* or diet* or coal* or *sorption or calcined or circuit* or inciner* or feldspar or food* or binder* or gypsum or batter* or pyroly* or aluminum or magnet* or *furfural or hydrogen* or metal* or methane or steel or hmf or lactose or *ethylene or alloy* or seed* or h2 or feed or zirconium or cereal* or salt or reinforced or flour or synthes* or ferul* or leach* or torref* or pigment* or vitrif* or electrode* or blend* or ash or stone or nano* or microfibril* or anaerobic or polyphenol* or lipid* or *harvest or fibril* or *methane or *hydrogen or carotenoid* or phenolic or rubber or dehydration or cancer or carrot* or sonogash* or antioxid* or "waste water*" or properties or alcohol or *proteins or protein* or cook* or "waste gas*" or elastic* or flavon* or transester* or anthocyan* or drying or pectin* or "added value" or liquefaction or fiber* or sonora* or moisture or sonoma or sludge or microcapsule* or construction or aeru* or veget* or emuls* or "active carbon" or crystal* or edta or fusion or calcium or suspension* or compounds or chlorogenic or dye* or freez* or alkalo* or betul* or chromat* or stiff* or dog* or milk* or hydrogel* or wine or decay or retro* or kraft or salvia or "microalgae oil" or velocity or biochemical* or defect* or sward or imaging or conductivity or sensor* or effluent* or desulfur or copper or vocal or pellet* or stress or lutein or bladder or pesticide* or color* or refract* or transduc* or hazard* or neur* or taurine or coagu* or vacuum or radical* or caco3 or cyanide or ceil* or xylem or microcyst* or ether*) or srctitle (food* or dairy or phar* or medi* or cereal* or nutrit* or feed* or brew* or lwt or animal* or biological or agric* or zoo* or genetic* or lipid* or soil* or composite* or postharvest))) or ((((title (milling or grinding)) or (title (milled or milling or grinding or "mill pretreatment*" or grinded) and title (ethanol or bioethanol or pretreat* or "pre-treat*" or *hydrolysis or saccharification or ferment* or hydrolytic or digestibility or disruption or fractionation or decomposition or dissolution or delignification or depolymerization or sugar or sugars or *oligosaccharides or glucose or deconstruction or xylose or disintegration or "size reduction"))) and (title (waste* or *grass or grass* or biomass* or lignocellul* or bagasse* or cellulos* or *wood or woody or *algae or *algal or straw* or stover* or eucalyptus or miscanthus or pine or chlorella or scrap* or lignin or *cellulose or porphyridium or nannochloropsis or starch or palm or poplar or cedar or corncob* or "corn cob*" or spruce or beech or fir or oak or bunch* or pinus or stalk* or residue* or willow* or cypress or stem* or sawdust or prosopis or sugarcane or hull* or husk* or shell* or *cane or birch or cynara) or srctitle (cellulose or wood* or biomass or sugar))) and not (title (removal or concrete or *diesel or *gas or digestion or drug* or pozzolanic or *composites or composite or cyclodext* or sorbitol or uranium or lactic or polysaccharide* or wastewater* or ore or ester* or medic* or "milling waste*" or reforming or separation or ruminal or "wood milling" or recycl* or hematite or mine or graphite or cement* or diet* or "rice milling" or coal* or *sorption or calcined or circuit* or inciner* or feldspar or food* or binder* or gypsum or batter* or pyroly* or aluminum or magnet* or *furfural or hydrogen or metal* or methane or steel or hmf or lactose or *ethylene or alloy* or reforming or seed* or h2 or feed or zirconium or cereal* or salt or reinforced or flour or synthes* or ferul* or leach* or torref* or pigment* or vitrif* or electrode* or blend* or ash or stone or nano* or microfibril*) or srctitle (food* or dairy or phar* or medi* or cereal* or nutrit* or feed* or brew* or lwt or animal* or biological or agric* or zoo* or genetic* or lipid* or soil* or nano*))) and (limit-to (srctype, "j") or limit-to (srctype, "k") or limit-to (srctype, "b")) and (limit-to (language, "english")) and (limit-to (doctype, "ar") or limit-to (doctype, "cp") or limit-to (doctype, "re") or limit-to (doctype, "ch") or limit-to (doctype, "no") or limit-to (doctype, "le") or limit-to (doctype, "sh") or limit-to (doctype, "bk") or limit-to (doctype, "ed"))

REFERENCES

Alvira, P., E. Tomas-Pejo, M. Ballesteros and M. J. Negro. 2010. Pretreatment technologies for an efficient bioethanol production process based on enzymatic hydrolysis: A review. *Bioresource Technology* 101:4851–4861.

Aman, V. 2018. Does the Scopus author ID suffice to track scientific international mobility? A case study based on Leibniz laureates. *Scientometrics* 117:705–720.

Angelici, C., B. M. Weckhuysen and P. C. A. Bruijnincx. 2013. Chemocatalytic conversion of ethanol into butadiene and other bulk chemicals. *ChemSusChem* 6:1595–1614.

Antolini, E. 2007. Catalysts for direct ethanol fuel cells. *Journal of Power Sources* 170:1–12.

Antolini, E. 2009. Palladium in fuel cell catalysis. *Energy and Environmental Science* 2:915–931.

Avolio, R., I. Bonadies, and D. Capitani, et al. 2012. A multitechnique approach to assess the effect of ball milling on cellulose. *Carbohydrate Polymers* 87:265–273.

Balaz, M. 2018. Ball milling of eggshell waste as a green and sustainable approach: A review. *Advances in Colloid and Interface Science* 256:256–275.

Beaudry, C. and V. Lariviere. 2016. Which gender gap? Factors affecting researchers' scientific impact in science and medicine. *Research Policy* 45:1790–1817.

Binod, P., K. Satyanagalakshmi and R. Sindhu, et al. 2012. Short duration microwave assisted pretreatment enhances the enzymatic saccharification and fermentable sugar yield from sugarcane bagasse. *Renewable Energy* 37:109–116.

Blankenship, K. M. 1993. Bringing gender and race in: US employment discrimination policy. *Gender & Society* 7:204–226.

Burnham, J. F. 2006. Scopus database: A review. *Biomedical Digital Libraries* 3:1–8.

Bussemaker, M. J. and D. Zhang. 2013. Effect of ultrasound on lignocellulosic biomass as a pretreatment for biorefinery and biofuel applications. *Industrial and Engineering Chemistry Research* 52:3563–3580.

Carlson, K. M., J. S. Gerber and D. Mueller, et al. 2017. Greenhouse gas emissions intensity of global croplands. *Nature Climate Change* 7:63–68.

Change, C. 2007. Climate change impacts, adaptation and vulnerability. *Science of the Total Environment* 326:95–112.

Chen, W. H., S. C. Ye and H. K. Sheen. 2012. Hydrolysis characteristics of sugarcane bagasse pretreated by dilute acid solution in a microwave irradiation environment. *Applied Energy* 93:237–244.

Chen, W. H., Y. J. Tu and H. K. Sheen. 2011. Disruption of sugarcane bagasse lignocellulosic structure by means of dilute sulfuric acid pretreatment with microwave- assisted heating. *Applied Energy* 88:2726–2734.

D'Angelo, C. A. and N. J. van Eck. 2020. Collecting large-scale publication data at the level of individual researchers: A practical proposal for author name disambiguation. *Scientometrics* 123:883–907.

Da Silva, A.S., H. Inoue, T. Endo, S. Yano and E. P. S. Bon. 2010. Milling pretreatment of sugarcane bagasse and straw for enzymatic hydrolysis and ethanol fermentation. *Bioresource Technology* 101:7402V7409.

Dirth, T. P. and N. R. Branscombe. 2017. Disability models affect disability policy support through awareness of structural discrimination. *Journal of Social Issues* 73:413–442.

Ebadi, A. and A. Schiffauerova. 2016. How to boost scientific production? A statistical analysis of research funding and other influencing factors. *Scientometrics* 106:1093–1116.

Fan, J., M. de Bruyn and V. L. Budarin, et al. 2013. Direct microwave-assisted hydrothermal depolymerization of cellulose. *Journal of the American Chemical Society* 135:12728–12731.

Fernando, S., S. Adhikari, C. Chandrapal and M. Murali. 2006. Biorefineries: Current status, challenges, and future direction. *Energy & Fuels* 20:1727–1737.

Formela, K., A. Hejna, L. Piszczyk, M. R. Saeb and X. Colom. 2016. Processing and structure-property relationships of natural rubber/wheat bran biocomposites. *Cellulose* 23:3157–3175.

Garfield, E. 1955. Citation indexes for science. *Science* 122:108–111.

Glanzel, W., J. Leta and B. Thijs. 2006. Science in Brazil. Part 1: A macro-level comparative study. *Scientometrics* 67:67–86.

Gnansounou, E. 2010. Production and use of lignocellulosic bioethanol in Europe: Current situation and perspectives. *Bioresource Technology* 101:4842–4850.

Hideno, A., H. Inoue and K. Tsukahara, et al. 2009. Wet disk milling pretreatment without sulfuric acid for enzymatic hydrolysis of rice straw. *Bioresource Technology* 100:2706–2711.

Hill, J., E. Nelson, D. Tilman, S. Polasky and D. Tiffany. 2006. Environmental, economic, and energetic costs and benefits of biodiesel and ethanol biofuels. *Proceedings of the National Academy of Sciences of the United States of America* 103:11206–11210.

Hill, J., S. Polasky and E. Nelson, et al. 2009. Climate change and health costs of air emissions from biofuels and gasoline. *Proceedings of the National Academy of Sciences of the United States of America* 106:2077–2082.

Hsieh, W. D., R. H. Chen, T. L. Wu and T. H. Lin. 2002. Engine performance and pollutant emission of an SI engine using ethanol-gasoline blended fuels. *Atmospheric Environment* 36:403–410.

Hu, Z. and Z. Wen. 2008. Enhancing enzymatic digestibility of switchgrass by microwave-assisted alkali prtreatment. *Biochemical Engineering Journal* 38:369–378.

Huang, H. J., S. Ramaswamy, U. W. Tschirner and B. V. Ramarao. 2008. A review of separation technologies in current and future biorefineries. *Separation and Purification Technology* 62:1–21.

Inoue, H., S. Yano, T. Endo, T. Sakaki and S. Sawayama. 2008. Combining hot-compressed water and ball milling pretreatments to improve the efficiency of the enzymatic hydrolysis of eucalyptus. *Biotechnology for Biofuels* 1:2.

Kerr, R. A. 2007. Global warming is changing the world. *Science* 316:188–190.

Konur, O. 2000. Creating enforceable civil rights for disabled students in higher education: An institutional theory perspective. *Disability & Society* 15:1041–1063.

Konur, O. 2002a. Access to nursing education by disabled students: Rights and duties of nursing programs. *Nurse Education Today* 22:364–374.

Konur, O. 2002b. Assessment of disabled students in higher education: Current public policy issues. *Assessment and Evaluation in Higher Education* 27:131–52.

Konur, O. 2002c. Access to employment by disabled people in the UK: Is the Disability Discrimination Act working? *International Journal of Discrimination and the Law* 5:247–279.

Konur, O. 2006a. Participation of children with dyslexia in compulsory education: Current public policy issues. *Dyslexia* 12:51–67.

Konur, O. 2006b. Teaching disabled students in Higher Education. *Teaching in Higher Education* 11:351–363.

Konur, O. 2007a. A judicial outcome analysis of the Disability Discrimination Act: A windfall for the employers? *Disability & Society* 22:187–204.

Konur, O. 2007b. Computer-assisted teaching and assessment of disabled students in higher education: The interface between academic standards and disability rights. *Journal of Computer Assisted Learning* 23:207–219.

Konur, O. 2011. The scientometric evaluation of the research on the algae and bio-energy. *Applied Energy* 88:3532–3540.

Konur, O. 2012a. Prof. Dr. Ayhan Demirbas' scientometric biography. *Energy Education Science and Technology Part A: Energy Science and Research* 28:727–738.

Konur, O. 2012b. The evaluation of the biogas research: A scientometric approach. *Energy Education Science and Technology Part A: Energy Science and Research* 29:1277–1292.

Konur, O. 2012c. The evaluation of the global energy and fuels research: A scientometric approach. *Energy Education Science and Technology Part A: Energy Science and Research* 30:613–628.

Konur, O. 2012d. The evaluation of the research on the biodiesel: A scientometric approach. *Energy Education Science and Technology Part A: Energy Science and Research* 28:1003–1014.

Konur, O. 2012e. The evaluation of the research on the bioethanol: A scientometric approach. *Energy Education Science and Technology Part A: Energy Science and Research* 28:1051–1064.

Konur, O. 2012f. The evaluation of the research on the biofuels: A scientometric approach. *Energy Education Science and Technology Part A: Energy Science and Research* 28:903–916.

Konur, O. 2012g. The evaluation of the research on the biohydrogen: A scientometric approach. *Energy Education Science and Technology Part A: Energy Science and Research* 29:323–338.

Konur, O. 2012h. The evaluation of the research on the microbial fuel cells: A scientometric approach. *Energy Education Science and Technology Part A: Energy Science and Research* 29:309–322.

Konur, O. 2012i. The scientometric evaluation of the research on the production of bioenergy from biomass. *Biomass and Bioenergy* 47:504–515.

Konur, O. 2015. Current state of research on algal bioethanol. In *Marine Bioenergy: Trends and Developments*, Eds. S. K. Kim and C. G. Lee, pp. 217–244. Boca Raton, FL: CRC Press.

Konur, O. 2018b. Bioenergy and biofuels science and technology: Scientometric overview and citation classics. In *Bioenergy and Biofuels*, Ed. O. Konur, pp. 3–63. Boca Raton: CRC Press.

Konur, O. 2019. Cyanobacterial bioenergy and biofuels science and technology: A scientometric overview. In *Cyanobacteria: From Basic Science to Applications*, Eds. A. K. Mishra, D. N. Tiwari and A. N. Rai, pp. 419–442. Amsterdam: Elsevier.

Konur, O. 2020a. The scientometric analysis of the research on the bioethanol production from green macroalgae. In *Handbook of Algal Science, Technology and Medicine*, Ed. O. Konur, pp. 385–401. London: Academic Press.

Konur, O., Ed. 2018a. *Bioenergy and Biofuels*. Boca Raton, FL: CRC Press.

Konur, O., Ed. 2020b. *Handbook of Algal Science, Technology and Medicine*. London: Academic Press.

Konur, O., Ed. 2021a. *Handbook of Biodiesel and Petrodiesel Fuels: Science, Technology, Health, and Environment*. Boca Raton, FL: CRC Press.

Konur, O., Ed. 2021b. *Handbook of Biodiesel and Petrodiesel Fuels: Science, Technology, Health, and Environment. Volume 1. Biodiesel Fuels: Science, Technology, Health, and Environment*. Boca Raton, FL: CRC Press.

Konur, O., Ed. 2021c. *Handbook of Biodiesel and Petrodiesel Fuels: Science, Technology, Health, and Environment. Volume 2. Biodiesel Fuels based on the Edible and Nonedible Feedstocks, Wastes, and Algae: Science, Technology, Health, and Environment*. Boca Raton, FL: CRC Press.

Konur, O., Ed. 2021d. *Handbook of Biodiesel and Petrodiesel Fuels: Science, Technology, Health, and Environment. Volume 3. Petrodiesel Fuels: Science, Technology, Health, and Environment*. Boca Raton, FL: CRC Press.

Konur, O. and F. L. Matthews. 1989. Effect of the properties of the constituents on the fatigue performance of composites: A review. *Composites* 20:317–328.

Kostas, E. T., D. Beneroso and J. P. Robinson. 2017. The application of microwave heating in bioenergy: A review on the microwave pre-treatment and upgrading technologies for biomass. *Renewable and Sustainable Energy Reviews* 77:12–27.

Leydesdorff, L. 2000. Is the European Union becoming a single publication system? *Scientometrics* 47:265–280.

Leydesdorff, L. and C. Wagner. 2009. Is the United States losing ground in science? A global perspective on the world science system. *Scientometrics* 78:23–36.

Leydesdorff, L., C. S. Wagner and L. Bornmann. 2014. The European Union, China, and the United States in the top-1% and top-10% layers of most-frequently cited publications: Competition and collaborations. *Journal of Informetrics* 8:606–617.

Leydesdorff, L. and P. Zhou. 2005. Are the contributions of China and Korea upsetting the world system of science? *Scientometrics* 63:617–630.

Li, H., Y. Qu, Y. Yang, S. Chang and J. Xu. 2016. Microwave irradiation – A green and efficient way to pretreat biomass. *Bioresource Technology* 199:34–41.

Li, S. Z. and C. Chan-Halbrendt. 2009. Ethanol production in (the) People's Republic of China: Potential and technologies. *Applied Energy* 86:S162–S169.

Ling, Z., T. Wang and M. Makarem, et al. 2019. Effects of ball milling on the structure of cotton cellulose. *Cellulose* 26:305–328.

Luo, J., Z. Fang and R. L. Smith. 2014. Ultrasound-enhanced conversion of biomass to biofuels. *Progress in Energy and Combustion Science* 41:56–93.

Ma, H., W. W. Liu, X. Chen, Y. J. Wu and Z. L. Yu. 2009. Enhanced enzymatic saccharification of rice straw by microwave pretreatment. *Bioresource Technology* 100:1279–1284.

Ma, X., L. Sun and C. Song. 2002. A new approach to deep desulfurization of gasoline, diesel fuel and jet fuel by selective adsorption for ultra-clean fuels and for fuel cell applications. *Catalysis Today* 77:107–116.

Macedo, I. C., J. E. A. Seabra and J. E. A. R. Silva. 2008. Green house gases emissions in the production and use of ethanol from sugarcane in Brazil: The 2005/2006 averages and a prediction for 2020. *Biomass and Bioenergy* 32:582–595.

Mani, S., L. G. Tabil and S. Sokhansanj. 2004. Grinding performance and physical properties of wheat and barley straws, corn stover and switchgrass. *Biomass and Bioenergy* 27:339–352.

Mayer-Laigle, C., N. Blanc, R. K. Rajaonarivony and X. Rouau. 2018. Comminution of dry lignocellulosic biomass, a review: Part I. From fundamental mechanisms to milling behavior. *Bioengineering* 5:41.

Morschbacker, A. 2009. Bio-ethanol based ethylene. *Polymer Reviews* 49:79–84.

Najafi, G., B. Ghobadian and T. Tavakoli, et al. 2009. Performance and exhaust emissions of a gasoline engine with ethanol blended gasoline fuels using artificial neural network. *Applied Energy* 86:630–639.

Newman, P. W. G. and J. R. Kenworthy. 1989. Gasoline consumption and cities: A comparison of U.S. cities with a global survey. *Journal of the American Planning Association* 55:24–37.

North, D. C. 1991. Institutions. *Journal of Economic Perspectives* 5:97–112.

Saleem, M. E., R. Omar, S. M. M. Kamal and D. R. A. Biak. 2015. Microwave-assisted pretreatment of lignocellulosic biomass: A review. *Journal of Engineering Science and Technology* 10:97–109.

Schwanninger, M., J. C. Rodrigues, H. Pereira and B. Hinterstoisser. 2004. Effects of short-time vibratory ball milling on the shape of FT-IR spectra of wood and cellulose. *Vibrational Spectroscopy* 36:23–40.

Silva, G. G. D., M. Couturier, J. G. Berrin, A. Buleon and X. Rouau. 2012. Effects of grinding processes on enzymatic degradation of wheat straw. *Bioresource Technology* 103:192–200.

Subhedar, P. B. and P. R. Gogate. 2013. Intensification of enzymatic hydrolysis of lignocellulose using ultrasound for efficient bioethanol production: A review. *Industrial and Engineering Chemistry Research* 52:11816–11828.

Sun, Y. and J. Cheng. 2002. Hydrolysis of lignocellulosic materials for ethanol production: A review. *Bioresource Technology* 83:1–11.

Vadas, P. A., K. H. Barnett and D. J. Undersander. 2008. Economics and energy of ethanol production from alfalfa, corn, and switchgrass in the Upper Midwest, USA. *Bioenergy Research* 1:44–55.

Wu, Y., Z. Fu and D. Yin, et al. 2010. Microwave-assisted hydrolysis of crystalline cellulose catalyzed by biomass char sulfonic acids. *Green Chemistry* 12:696–970.

Zhang, Z. and Z. K. Zhao. 2009. Solid acid and microwave-assisted hydrolysis of cellulose in ionic liquid. *Carbohydrate Research* 344:2069–2072.

Zhao, H., J. H. Kwak and Y. Wang, et al. 2006. Effects of crystallinity on dilute acid hydrolysis of cellulose by cellulose ball-milling study. *Energy and Fuels* 20:807–811.

Zhu, J. Y. 2011. Physical pretreatment – Woody biomass size reduction – For forest biorefinery. *ACS Symposium Series* 1067:89–107.

Zhu, J. Y., G. S. Wang, X. J. Pan and R. Gleisner. 2009. Specific surface to evaluate the efficiencies of milling and pretreatment of wood for enzymatic saccharification. *Chemical Engineering Science* 64:474–485.

Zhu, S., Y. Wu, Z. Yu, J. Liao and Y. Zhang. 2005. Pretreatment by microwave/alkali of rice straw and its enzymic hydrolysis. *Process Biochemistry* 40:3082–3086.

Zhu, W., J. Y. Zhu, R. Gleisner and X. J. Pan. 2010. On energy consumption for size-reduction and yields from subsequent enzymatic saccharification of pretreated lodgepole pine. *Bioresource Technology* 101:2782–2792.

16 Milling Pretreatment of the Biomass

Review

Ozcan Konur
(Formerly) Ankara Yildirim Beyazit University

16.1 INTRODUCTION

Crude oil-based gasoline fuels (Ma et al., 2002; Newman and Kenworthy, 1989) have been widely used in the transportation sector since the 1920s. However, there have been great public concerns over the adverse environmental and human impact of these fuels (Hill et al., 2006, 2009). Hence, biomass-based bioethanol fuels (Hill et al., 2006; Konur, 2012, 2015, 2020a) have increasingly been used in blending gasoline fuels (Hsieh et al., 2002; Najafi et al., 2009), in fuel cells (Antolini, 2007, 2009), and in biochemical production (Angelici et al., 2013; Morschbacker, 2009) in a biorefinery context (Fernando et al., 2006; Huang et al., 2008).

However, it is necessary to pretreat biomass (Taherzadeh and Karimi, 2008; Yang and Wyman, 2008) to enhance the yield of the bioethanol (Hahn-Hagerdal et al., 2006; Sanchez and Cardona, 2008) before the bioethanol production through the hydrolysis (Sun and Cheng, 2002; Taherzadeh and Karimi, 2007) and fermentation (Lin and Tanaka, 2006; Olsson and Hahn-Hagerdal, 1996) of biomass. The milling pretreatment as a mechanical pretreatment is one of these pretreatment methods.

However, it is essential to develop efficient incentive structures (North, 1991) for the primary stakeholders to enhance research in this field (Konur, 2000, 2002a–c, 2006a, b, 2007a, b). Although there have been several review papers on the microwave pretreatment of biomass (Balaz, 2018; Karinkanta et al., 2018; Piras et al., 2019), there has been no review of the 25 most-cited articles in this field.

Thus, this book chapter presents a review of the 25 most-cited articles in this field. Then, it discusses the key findings of these highly influential papers and comments on future research priorities in this field.

16.2 MATERIALS AND METHODS

The search for this study was carried out using the Scopus database (Burnham, 2006) in March 2022.

As a first step for the search of the relevant literature, the keywords were selected using the first 200 highly cited papers (HCPs). The selected keyword list was optimized to obtain a representative sample of papers for this research field. This final keyword set was provided in the Appendix for future replication studies.

As a second step, a sample dataset was used for this study. The first 25 articles with at least 78 citations each were selected for the review study. Key findings from each paper were taken from the abstracts of these papers and were discussed briefly. Additionally, many brief conclusions were drawn and several relevant recommendations were made to enhance future research landscape.

DOI: 10.1201/9781003226536-22

16.3 RESULTS

The brief information about the 25 most-cited papers with at least 78 citations each on the milling pretreatment of biomass is given as follows.

16.3.1 Milling Pretreatment of the Wood

Schwanninger et al. (2004) studied the effect of short-time vibratory ball milling (BM) on the chemical and structural changes in wood and cellulose in a paper with 869 citations. They observed that the milling itself was the main influencing factor, and these changes were mainly associated with a decrease in the degree of crystallinity and/or a decrease in the degree of polymerization of the cellulose.

Zhu et al. (2009) characterized the specific surface of wood to evaluate the efficiencies of milling and sulfite pretreatment of wood for enzymatic saccharification in a paper with 199 citations. They observed that the developed methodology was effective in differentiating various size reduction and chemical pretreatment processes in terms of cellulose-to-glucose conversion efficiency and size-reduction energy consumption. Thermomechanical disk milling (DM-I), exposing cellulose, was more effective than high-pressure thermomechanical disk milling (DM-II), exposing lignin, in subsequent enzymatic hydrolysis. However, DM-I was more energy-intensive than DM-II. Both DMs produced fibers, which were more efficient in enzymatic hydrolysis than hammer milling (HM) that produced fiber bundles. Sulfite pretreatment increased cellulose conversion and reduced mechanical milling energy consumption. The sulfite pretreatment was the most efficient pretreatment in terms of glucose yield and milling energy consumption.

Inoue et al. (2008) enhanced the enzymatic digestibility of eucalyptus with combined pretreatments of hot-compressed water (HCW) and BM in a paper with 155 citations. They observed that the BM treatment simultaneously improved the digestibility of both glucan and xylan and was effective in lowering the enzyme loading compared with the HCW treatment. The combined HCW and BM treatment reduced the BM time. The eucalyptus pretreated with HCW (160°C, 30 min) followed by BM (20 min) had an approximately 70% yield of total sugar with a cellulase loading. This yield was comparable to the yields from samples treated with HCW (200°C, 30 min) or BM (40 min) hydrolyzed with 40 filter paper unit (FPU)/g substrate. They concluded that the combined HCW-BM treatment could save energy and enzyme loading.

Zhu et al. (2010) enhanced the enzymatic digestibility of pine wood chips by the combined pretreatment of disk milling (DM) and chemical pretreatments in a paper with 146 citations. They pretreated chips with hot water, dilute acid (H_2SO_4), and two sulfite (SPORL) pretreatments. They then milled these chips under various solid loadings and disk-plate gaps. They observed that DM could decrease size-reduction energy consumption by 20%–80% depending on the pretreatment applied under 20% solid loading and a disk-plate gap of 0.76 mm in milling. Sulfite pretreatment to overcome recalcitrance of lignocellulose (SPORL) with a sodium bisulfite ($NaHSO_3$) charge of 8% and H_2SO_4 charge of 2.21% on wood was the most effective in decreasing size-reduction energy consumption. Solid loading had the most significant effect on DM energy. When solid loading was reduced from 30% to 3%, DM energy could be decreased by more than a factor of 10 for wood chips pretreated by both SPORL and H_2SO_4 at an acid charge of 2.21%. The glucose yields from biomass produced by all pretreatments were independent of the solid loading in DM, indicating that these energy savings in size reduction could be realized without affecting the yield. When biomass was pretreated by SPORL with 2.21% acid charge, size-reduction energy consumption was decreased at a practical solid loading of approximately 10%–20% with excellent yield. Similar effects on size-reduction energy savings and excellent yield were also achieved when large disk-plate gaps (up to 1.52 mm studied) were applied in DM of biomass pretreated by SPORL with acid.

Schell and Harwood (1994) studied the energy requirements and size distribution for milling wood in a paper with 136 citations. They observed that HM used less energy than DM, but produced particles with a larger-size distribution. They also studied shredding paper and switchgrass as biomass.

Teramoto et al. (2008) studied the milling of eucalyptus wood chips for enzymatic saccharification using ethanol/water/acetic acid (CH_3COOH) mixed solvent in a paper with 120 citations. They observed that this mixture activated the original wood and induced a pore formation by the removal of part of the lignin and hemicellulose fractions. Furthermore, glucose was obtained from cellulosic components completely under optimal conditions and the scale affected the improvement of enzymatic digestibility.

Kleine et al. (2013) degraded lignin and wood by solvent-free grinding in a reactive medium in a paper with 113 citations. They targeted the cleavage of β-O-4 linkages in lignin. This method was transition metal- and solvent-free, requiring only inexpensive reagents.

Fujimoto et al. (2005) studied the effect of milling on the wood lignin structure in a paper with 113 citations. They milled coarse wood meal of sweetgum (*Liquidambar styraciflua*) for 150–1,200 min with the use of a planetary BM under six different conditions. They observed that β-O-4 structure degraded by 25% when milling was continued until 40% of lignin in wood became extractable and preferential erythro-degradation proceeded under all milling conditions.

Mais et al. (2002) enhanced the enzymatic hydrolysis of α-cellulose and SO_2-impregnated steam-exploded wood chips by simultaneous BM in a paper with 102 citations. They further pretreated wood chips with water and hot alkaline hydrogen peroxide (H_2O_2) to remove more than 90% of the original lignin. They observed that the optimal conditions for the enzymatic hydrolysis of α-cellulose were a higher number of beads, while the presence of air–liquid interface did not affect the rate of saccharification. Similarly, when employing the chips, they achieved up to 100% hydrolysis with a minimum enzyme loading, at lower feedstock concentrations, and with a greater number of reaction beads during milling (Table 16.1).

16.3.2 Milling Pretreatment of the Agricultural Wastes

Mani et al. (2004) studied the milling behavior of wheat and barley straws, corn stover, and switchgrass in a paper with 507 citations. They milled this feedstock using a HM with three different screen sizes. They observed that switchgrass and corn stover had the highest and lowest specific energy consumption at 3.2 mm screen size, respectively. They modeled the bulk and particle densities of grinds to a geometric mean diameter within the range of 0.18–1.43 mm. Finally, they found that switchgrass had the highest calorific value and the lowest ash content among biomass species tested.

Da Silva et al. (2010) studied the milling behavior of sugarcane bagasse and straw using BM and wet DM (WDM) in a paper with 264 citations. They observed that glucose and xylose hydrolysis yields at optimal conditions for BM-treated bagasse and straw were 78.7% and 72.1% and 77.6% and 56.8%, respectively. Maximum glucose and xylose yields for bagasse and straw using WDM were 49.3% and 36.7% and 68.0% and 44.9%, respectively. BM improved the enzymatic hydrolysis by decreasing the crystallinity, while the defibrillation effect observed for WDM samples favored enzymatic conversion. Bioethanol yields from total fermentable sugars using a C_6 fermenting strain reached 89.8% and 91.8% for bagasse and straw hydrolysates, respectively, and 82% and 78% when using a C_6/C_5 fermenting strain.

Hideno et al. (2009) studied the WDM behavior of rice straw for enzymatic hydrolysis, compared with BM and HCW pretreatment in a paper with 181 citations. They observed that glucose and xylose yields by WDM, BM, and HCW treatments were 78.5% and 41.5%, 89.4% and 54.3%, and 70.3% and 88.6%, respectively. Both WDM and HCW pretreatments increased sugar yields without decreasing their crystallinity. The feature size of the WDMed rice straw was similar to that of the HCW-treated rice straw. The energy consumption of WDM was lower than that of other pretreatments. They concluded that WDM was an economical and practical pretreatment for the enzymatic hydrolysis of straw.

Silva et al. (2012) studied the milling behavior of wheat straw in a paper with 167 citations. They produced a large range of wheat straw powders from coarse to fine particles by sieve-based

TABLE 16.1
Milling Pretreatment of the Wood

No.	Papers	Biomass	Milling	Other Pts.	Parameters	Keywords	Lead Author	Affil.	Cits
1	Schwanninger et al. (2004)	Wood, cellulose	Vibratory ball milling	None	Milling, temperature, particle size, oxygen	Milling, wood, cellulose	Schwanninger, Manfred 6602877236	Univ. Natr. Sci. App. Sci. Austria	869
2	Zhu et al. (2009)	Wood	Disk milling, hammer milling	Enzymes, sulfite	Glucose yield, energy consumption	Milling, wood, saccharification	Zhu, Junyong 7405692678	US Forest Serv. USA	199
3	Inoue et al. (2008)	Eucalyptus	Ball milling	Enzymes, hot-compressed water	Temperature, treatment time, enzyme content, sugar yield	Milling, eucalyptus, hydrolysis, enzymatic	Inoue, Hiroyuki 55733159900	AIST Japan	155
4	Zhu et al. (2010)	Pine wood chips	Disk milling	Enzymes, hot water, dilute acid (H_2SO_4), sulfite ($NaHSO_3$)	Energy consumption, solid content, sugar yield, disk-plate gap	Pine, enzymatic, saccharification, milling	Zhu, Junyong 7405692678	US Forest Serv. USA	146
5	Schell and Harwood (1994)	Wood	Hammer milling, disk milling	None	Energy consumption, particle size	Milling, lignocellulosic, biomass	None	NREL USA	136
6	Teramoto et al. (2008)	Eucalyptus wood chips	Ball milling	Enzymes, CH_3COOH	Glucose yield, particle size	Eucalyptus wood chips, saccharification, milling	Teramoto, Yoshikuni 7006223067	AIST Japan	120
7	Kleine et al. (2013)	Wood lignin	Milling	None	Degradation	Grinding, wood	Bolm, Carsten 35565611000	RWTH Univ. Germany	113
8	Fujimoto et al. (2005)	Wood lignin	Planetary ball milling	None	Degradation	Milling, lignin, wood	Fujimoto, Aya 10244510600	Univ. Tokyo Japan	113
9	Mais et al. (2002)	α-cellulose, wood chips	Ball milling	Enzymes, H_2O_2, steam explosion	Solid content, enzyme loading, reaction volumes, number of ball beads	Milling, enzymatic, hydrolysis	Mansfield, Shawn D 7006421766	Univ. British Columbia Canada	102

Other pts., other pretreatments.

grinding and then ultrafine particles by jet milling (JM) and by BM. They observed that the wheat straw degradability was enhanced by the decrease in particle size until a limit of 100 μm with up to 36% total carbohydrate and 40% glucose hydrolysis yields. BM samples had up to 46% total carbohydrate and 72% glucose yields as a consequence of cellulose crystallinity reduction (from 22% to 13%). They concluded that BM was an effective pretreatment with similar glucose yield and superior carbohydrate yield compared with steam explosion pretreatment.

Adapa et al. (2011) studied the milling performance of barley, canola, oat, and wheat straw in a paper with 122 citations. They used a forage chopper and a HM. They observed that the bulk density of native straw was significantly higher than the bulk density of steam-exploded straw. For native straw, the particle density of canola and oat straw significantly increased with a decrease in HM screen size from 30 to 1.6 mm. On the contrary, the particle density of steam-exploded barley and oat straw was significantly higher than native straw, except for barley at 6.4 mm HM screen size. However, the particle density of steam-exploded canola straw was not statistically different from native straw. The chopper consumed the highest and lowest specific energy to chop barley and canola straw, respectively. The highest and lowest specific energy was consumed by wheat at 1.6 mm and canola straws ground using 30 mm HM screen size, respectively. For steam-exploded straw, the highest and lowest specific energy was consumed by oat at 1.6 mm and canola straws ground using 6.4 mm HM screen size, respectively. The specific energy required by HM to grind non-treated and steam-exploded barley, canola, oat, and wheat straw showed a negative power correlation with HM screen sizes.

Lin et al. (2010) milled corn stover for enhancing the efficiency of enzymatic hydrolysis using BM in a paper with 104 citations. They used acid (H_2SO_4, HCl, HNO_3, CH_3COOH, HCOOH, and H_3PO_4) and alkaline pretreatments (NaOH, KOH, $Ca(OH)_2$, and NH_3H_2O) for this purpose. They obtained the optimal enzymatic hydrolysis efficiencies under the conditions of BM in the alkali medium due to delignification where BM pretreatment was a robust process. The particle size of material was decreased, and the fiber structure was more loosely organized. The treatment effect of WBM was better than that of dry BM (DBM). The optimal parameters for the milling process were a ball speed of 350 r/min, solid/liquid ratio of 1:10, raw material particle size of 0.5 mm, number of balls of 20, and grinding for 30 min. In comparison with the water milling process, alkaline milling pretreatment could increase the enzymatic hydrolysis efficiency of corn stover by 110%; through the digestion process with the combination of xylanase and cellulase mixture, the hydrolysis efficiency could increase by 160%.

Jin and Chen (2006) milled steam-exploded rice straw (SERS) for its enzymatic hydrolysis by fluidized bed opposed JM in a paper with 99 citations. They obtained 78% superfine ground SERS products with a mean fiber length of 60 μm where the particle yield was 179% higher than that from the native rice straw. Enzymatic hydrolysis, chemical composition, fiber characteristics, composed cells proportion of the superfine ground SERS product, and the ground residues all showed great differences (Table 16.2).

16.3.3 Milling Pretreatment of the Cellulose

Zhao et al. (2006) studied the effects of crystallinity on dilute acid (0.05 M H_2SO_4) hydrolysis of ball-milled (BM)ed cellulose with varying fractions of crystalline (α-form) and amorphous forms at 175°C in a paper with 244 citations. They modeled mechanical disruption of crystallinity by breaking hydrogen bonds in α-cellulose, opening up the structure, and making more β-1,4 glycosidic bonds readily accessible to the dilute acid. However, there were reactions of amorphous cellulose that formed solid degradation products.

Ling et al. (2019) studied the effects of BM on the structure of cotton cellulose in a paper with 157 citations. They used 13 methods to follow the changes in cellulose after BM for 15, 45, and 120 min. They observed profound changes as the cellulose was decrystallized with changes in particle size and chemical changes as specific surface areas and pore sizes increased while there was a progressive increase in initial moisture content and some loss in stability.

TABLE 16.2
Milling Pretreatment of the Agricultural Wastes

No.	Papers	Biomass	Milling	Other Pts.	Parameters	Keywords	Lead Author	Affil.	Cits
1	Mani et al. (2004)	Wheat and barley straws, corn stover, switchgrass	Hammer milling	None	Biomass type, energy consumption, density	Grinding, straw, stover, switchgrass	Tabil, Lope G 6701349307	Univ. Saskatchewan Canada	507
2	Da Silva et al. (2010)	Sugarcane bagasse and straw	Ball milling, wet disk milling	Enzymes	Milling type, sugar yield, bioethanol yield	Bagasse, straw, hydrolysis, enzymatic, ethanol, fermentation, milling	Bon, Elba P S 7007036976	Fed. Univ. Rio de Janeiro Brazil	264
3	Hideno et al. (2009)	Rice straw	Wet disk milling, ball milling	Enzymes, hot-compressed water	Sugar yield, energy consumption	Milling, hydrolysis, straw	Inoue, Hiroyuki 55733159900	AIST Japan	181
4	Silva et al. (2012)	Wheat straw	Ball milling, jet milling, sieve-based grinding	Enzymes, steam explosion	Sugar yield, milling type, carbohydrate yield, particle size	Straw, grinding, enzymatic	Rouau, Xavier 55973961500	INRA France	167
5	Adapa et al. (2011)	Barley, canola, oat, and wheat straw	Hammer milling	Steam explosion	Energy consumption, particle size, distribution, density	Grinding, straw	Tabil, Lope G 6701349307	Univ. Saskatchewan Canada	122
6	Lin et al. (2010)	Corn stover	Wet ball milling, dry ball milling	Enzymes, acids (H_2SO_4, HCl, HNO_3, CH_3COOH, HCOOH, and H_3PO_4), alkali (NaOH, KOH, $Ca(OH)_2$, and NH_3H_2O)	Particle size, acid/alkali type, wet/dry milling, ball speed, solid content, number of balls, treatment time	Milling, stover, enzymatic, hydrolysis	Huang, He 55386071100	Nanjing Univ. Technol. China	104
7	Jin and Chen (2006)	Rice straw	Jet milling	Steam explosion	Particle size, particle yield	Grinding, enzymatic, hydrolysis	Chen, Hongzhang 7501614171	Chin. Acad. Sci. China	99

Other pts., other pretreatments.

Avolio et al. (2012) studied the effect of DBM on cellulose structure, morphology, and properties in a paper with 148 citations. They observed that both the crystallinity index and the mean size of the cellulose crystallite domains had a progressive decrease as a function of the milling time. In particular, the crystallinity index decreased from 0.53 to 0.15 after 60 min DBM, whereas the mean size of the cellulose crystallite domains showed a reduction from the original value of about 4.0 nm up to about 3.4 nm after 30 min. Furthermore, the original fibrous structure almost disappeared and was modified to a quasi-circular shape due to the DBM process.

Ago et al. (2004) studied the crystalline transformation of native cellulose from cellulose I to cellulose II polymorph by a BM method in a paper with 112 citations. They observed that the extent of transformation into cellulose ID increased with milling time. The cellulose molecules were most mobile when the water content ID was around 30 wt%.

Ryu et al. (1982) studied the effect of compression milling (CM) on cellulose structure in a paper with 111 citations. They proposed a two-phase model, which was based on the degradation of cellulose by simultaneous actions of the cellulase complex on the crystalline and amorphous phases. They compared the relationships between cellulose accessibility and the kinetic parameters with those predicted by the model and found that there was a good agreement.

Zhang et al. (2007) studied the effect of pan milling (PM) on the morphological and structural development of hardwood cellulose in a paper with 99 citations. They observed that compared with BM, the PM showed a much higher efficiency in mechanochemical pretreatment of this feedstock as the average particle size reduced and the specific surface area increased after 40 milling cycles. The PM also led to the collapse of hydrogen bonds and the reduction in crystallinity. The crystallinity index of feedstock decreased after milling for 40 cycles. PMed cellulose had lower thermal stability and higher solubility in aqueous alkali.

Yu and Wu (2011) studied the effect of BM on the hydrolysis of microcrystalline cellulose in HCW in a paper with 72 citations. They observed that BM led to a considerable reduction in cellulose particle size and crystallinity, as well as a significant increase in the specific reactivity of cellulose during hydrolysis in HCW. BM also significantly influenced the distribution of glucose oligomers in the primary liquid products of cellulose hydrolysis as it increased the selectivity of glucose oligomers at low conversions. At high conversions, the reduction in chain length played an important role in glucose oligomer formation as cellulose samples become more crystalline. They further observed that an extensive BM completely converted the crystalline cellulose into amorphous cellulose, substantially enhancing the formation of glucose oligomers with high degrees of polymerization (Table 16.3).

16.3.4 Milling Pretreatment of the Algae

Postma et al. (2015) milled *Chlorella vulgaris* using bead milling (BM) in a paper with 116 citations. They observed that biomass was successfully disintegrated at different biomass concentrations over a range of agitator speeds with over 97% of cell disintegration. A clear optimum rate of disintegration and protein release took place at an agitator speed of 9–10 m/s regardless of biomass concentration. Selective extraction of water-soluble proteins took place as proteins released sooner than cell disintegration took place. Proteins could be released at 85% lower energy input than for cell disintegration.

Postma et al. (2017) milled *Chlorella vulgaris*, *Neochloris oleoabundans*, and *Tetraselmis suecica* using BM at different bead sizes (0.3–1 mm) in a paper with 83 citations. They observed that cell disintegration, proteins, and carbohydrates released into the water phase followed a first-order kinetics. The process was selective toward proteins over carbohydrates during the early stages of BM. In general, smaller beads led to higher kinetic rates, with a minimum specific energy consumption for 0.3 mm beads. The optimal disintegration and energy usage for all strains took place in the 0.3–0.4 mm range (Table 16.4).

TABLE 16.3
Milling Pretreatment of the Cellulose

No.	Papers	Biomass	Milling	Other Pts.	Parameters	Keywords	Lead Author	Affil.	Cits
1	Zhao et al. (2006)	Cellulose	Ball milling	Acid (H_2SO_4)	Crystallinity	Hydrolysis, cellulose, milling	Holladay, Johnathan E 7006800080	Pacific NW Natl. Lab. USA	244
2	Ling et al. (2019)	Cellulose	Ball milling	None	Crystallinity	Milling, cellulose	French, Alfred D 7202095026	Agr. Res. Serv. USA	157
3	Avolio et al. (2012)	Cellulose	Dry ball milling	None	Treatment time, crystallinity index, crystallite size	Milling, cellulose	Gentile, Gennaro 7102027136	CNR Italy	148
4	Ago et al. (2004)	Cellulose	Ball milling	None	Treatment time	Milling, cellulose	Ago, Mariko 8871694300	AIST Japan	112
5	Ryu et al. (1982)	Cellulose	Compression milling	Enzymes	Crystallinity, accessibility, specific surface area, degree of polymerization	Milling, enzymatic, hydrolysis	Ryu, Dewey DY 35081182300	Korea Adv. Inst. Sci. S. Korea	111
6	Mais et al. (2002)	α-cellulose, wood chips	Ball milling	Enzymes, H_2O_2, steam explosion	Solid content, enzyme loading, reaction volumes, number of ball beads	Milling, enzymatic, hydrolysis	Mansfield, Shawn D 7006421766	Univ. British Columbia Canada	102
7	Zhang et al. (2007)	Hardwood cellulose	Pan milling, ball milling	None	Particle size, surface area, crystallinity index	Milling, cellulose, hardwood	Lu, Canhui 23009599600	Sichuan Univ. China	94
8	Yu and Wu (2011)	Cellulose	Ball milling	Hot compressed water	Particle size and crystallinity	Milling, cellulose, hydrolysis	Wu, Hongwei 15043364100	Curtin Univ. Technol. Australia	72

Other pts., other pretreatments.

TABLE 16.4
Milling Pretreatment of the Algae

No.	Papers	Biomass	Milling	Parameters	Keywords	Lead Author	Affil.	Cits
15	Postma et al. (2015)	*Chlorella vulgaris*	Bead milling	Biomass content, agitator speed, energy consumption	Microalgae, chlorella, milling	Postma, P Richard 56402885700	Wageningen Univ. Netherlands	116
24	Postma et al. (2017)	*Neochloris oleoabundans, Tetraselmis suecica*	Bead milling	Bead size	Milling, microalgae	Postma, P Richard 56402885700	Wageningen Univ. Netherlands	83

16.3.5 Milling Treatment of the Grass

Mani et al. (2004) studied the grinding behavior of wheat and barley straws, corn stover, and switchgrass in a paper with 507 citations. They found that switchgrass had the highest calorific value and the lowest ash content among biomass species tested (Table 16.5).

16.4 DISCUSSION

16.4.1 Introduction

Crude oil-based gasoline fuels have been widely used in the transportation sector since the 1920s. However, there have been great public concerns over the adverse environmental and human impact of these fuels. Hence, biomass-based bioethanol fuels have increasingly been used in blending gasoline and petrodiesel fuels, in fuel cells, and in biochemical production in a biorefinery context. Research in the field of milling pretreatment of biomass intensified in recent years.

However, it is essential to develop efficient incentive structures for the primary stakeholders to enhance research in this field. Although there have been several review papers on the milling pretreatment of biomass, there has been no review of the 25 most-cited articles in this field.

Thus, this book chapter presents a review of the 25 most-cited articles in this field. Then, it discusses the key findings of these highly influential papers and comments on future research priorities in this field.

As a first step for the search of the relevant literature, the keywords were selected using the first 200 HCPs. The selected keyword list was optimized to obtain a representative sample of papers for the searched research field. This keyword list was provided in the Appendix for future replication studies.

As a second step, a sample dataset was used for this study. The first 25 articles with at least 78 citations each were selected for the review study. Key findings from each paper were taken from the abstracts of these papers and were briefly discussed. Additionally, many brief conclusions were drawn and several relevant recommendations were made to enhance future research landscape.

As Table 16.6 shows, there are three primary research fronts for these 25 most-cited papers: agricultural wastes, cellulose, and wood with seven, eight, and nine reviewed articles, respectively. The other research fronts are the algae and grass with two and one reviewed papers.

However, agricultural wastes, woody biomass, and cellulose are overrepresented in the reviewed papers compared with the population papers. On the contrary, kitchen wastes, plants, grass, and biomass in general are underrepresented or not represented at all among the reviewed papers.

Straw and corn stover are overrepresented in the set of reviewed papers by 15% and 3%, respectively. It is notable the papers related to bagasse, other agricultural biomass, and corncobs are not represented in the set of reviewed papers at all.

TABLE 16.5
Milling Pretreatment of the Grass

No.	Papers	Biomass	Milling	Parameters	Keywords	Lead Author	Affil.	Cits
2	Mani et al. (2004)	Wheat and barley straws, corn stover, switchgrass	Hammer milling	Biomass type, energy consumption, density	Grinding, straw, stover, switchgrass	Tabil, Lope G 6701349307	Univ. Saskatchewan Canada	507

TABLE 16.6
Milling Pretreatment of the Biomass: Research Fronts

No.	Research Fronts	Papers (%)	Reviewed Papers (%)
1	**Agricultural Wastes**	**25.2**	**28**
	Straw	9.3	24
	Corn stover	5.3	8
	Bagasse	4.9	0
	Other agricultural biomass	4.4	0
	Corncobs	1.3	0
2	**Woody Biomass**	**26.2**	36
	Wood	20.9	24
	Wood chips	5.3	12
3	Cellulose	26.7	32
4	Kitchen wastes	1.8	0
5	**Plants**	**10.7**	8
	Corn	4.0	0
	Algae	2.7	8
	Others	2.7	0
	Sugarcane	1.3	0
6	Grass	4.0	4
7	Biomass in general	1.8	0
	Sample size	225	25

Reviewed papers: 25 papers reviewed in this study. Papers: the sample of population papers (225 papers).

The research front for woody biomass includes papers related to wood and wood wastes. Sugarcane, corn, and other plants are the plants studied in the population papers, while only algae are represented in the set of reviewed articles.

16.4.2 Milling Pretreatment of the Wood

There were nine HCPs related to the milling pretreatment of the wood biomass. Four HCPs studied the milling pretreatment as a stand-alone mechanical pretreatment (Fujimoto et al., 2005; Kleine et al., 2013; Schell and Harwood, 1994; Schwanninger et al., 2004), while five HCPs studied milling pretreatments with other pretreatments such as enzymatic pretreatment (Inoue et al., 2008; Mais et al., 2002; Teramoto et al., 2008; Zhu et al., 2009, 2010), sulfite/SPORL pretreatment (Zhu et al., 2009,

2010), HCW/steam explosion pretreatment (Inoue et al., 2008; Mais et al., 2002; Zhu et al., 2010), acid pretreatment (Teramoto et al., 2008; Zhu et al., 2010), and H_2O_2 pretreatment (Teramoto et al., 2008).

The most prolific milling methods are BM (Fujimoto et al., 2005; Inoue et al., 2008; Mais et al., 2002; Schwanninger et al., 2004; Teramoto et al., 2008), DM (Schell and Harwood, 1994; Zhu et al., 2009, 2010), and HM (Schell and Harwood, 1994; Zhu et al., 2009) with five, three, and two HCPs, respectively.

The wood biomass used in these studies is the wood biomass in general (Schell and Harwood, 1994; Schwanninger et al., 2004; Zhu et al., 2009), wood chips (Mais et al., 2002; Teramoto et al., 2008; Zhu et al., 2010), eucalyptus (Inoue et al., 2008; Teramoto et al., 2008), wood lignin (Fujimoto et al., 2005; Kleine et al., 2013), and pine (Zhu et al., 2010).

A number of variables are studied in these studies such as sugar yield (Inoue et al., 2008; Teramoto et al., 2008; Zhu et al., 2009, 2010), particle size (Schell and Harwood, 2004; Schwanninger et al., 2004; Teramoto et al., 2008), energy consumption (Schell and Harwood, 1994; Zhu et al., 2009, 2010), milling temperature (Inoue et al., 2008; Schwanninger et al., 2004), and enzyme content (Inoue et al., 2008; Mais et al., 2002).

Junyong Zhu is the most prolific lead author with two HCPs, while Association for Iron & Steel Technology (AIST) and the US Forest Service are the most prolific institutions with two institutions each.

A number of keywords are used in the titles of these studies. The most prolific keyword is milling with eight HCPs, closely followed by wood with five HCPs. The other prolific keywords are enzymatic, saccharification, cellulose, and lignin with three HCPs each, while two studies each use hydrolysis and biomass in their titles.

The key finding from these papers is that the milling pretreatment in combination with these other pretreatments or stand-alone was effective in enhancing the digestibility of wood. Thus, milling pretreatment increased the bioethanol yield from wood (Gregg and Sadler, 1996; Hu et al., 2008).

16.4.3 Milling Pretreatment of the Cellulose

There were eight HCPs related to the milling pretreatment of the cellulose. Four HCPs studied the milling pretreatment as a stand-alone mechanical pretreatment (Ago et al., 2004; Avolio et al., 2012; Ling et al., 2019; Zhang et al., 2007), while four HCPs studied milling pretreatments with other pretreatments such as enzymatic pretreatments (Mais et al., 2002; Ryu et al., 1982), acid pretreatment (Zhao et al., 2006), HCW/steam explosion (Mais et al., 2002; Yu and Wu, 2011), and H_2O_2 (Mais et al., 2002).

The most prolific milling method is BM with seven HCPs (Ago et al., 2004; Avolio et al., 2012; Ling et al., 2019; Mais et al., 2002; Yu and Wu, 2011; Zhang et al., 2007; Zhao et al., 2006). The other milling methods are PM (Zhang et al., 2007) and CM (Ryu et al., 1982).

A number of variables are studied in these studies such as crystallinity (Ago et al., 2004; Avolio et al., 2012; Ling et al., 2019; Ryu et al., 1982; Yu and Wu, 2011; Zhang et al., 2007; Zhao et al., 2006), particle size (Avolio et al., 2012; Yu and Wu, 2011; Zhang et al., 2007), treatment time (Ago et al., 2004; Avolio et al., 2012), and surface area (Ryu et al., 1982; Zhang et al., 2007).

A number of keywords are used in the titles of these studies. The most prolific keyword is milling and cellulose with eight HCPs each. The other prolific keywords are enzymatic and saccharification with two HCPs each.

The key finding from these papers is that the milling pretreatment in combination with these other pretreatments or stand-alone was effective in enhancing the digestibility of cellulose. Thus, milling pretreatment increased the bioethanol yield from cellulose (Beguin and Aubert, 1994; Gregg and Saddler, 1996).

16.4.4 Milling Pretreatment of the Agricultural Wastes

There were seven HCPs related to the milling pretreatment of agricultural wastes. Only one HCP studied the milling pretreatment as a stand-alone mechanical pretreatment (Mani et al., 2004), while six HCPs studied milling pretreatments with other pretreatments such as enzymatic pretreatments (Da Silva et al., 2010; Hideno et al., 2009; Lin et al., 2010; Silva et al., 2012), acid pretreatment (Lin et al., 2010), HCW/steam explosion (Adapa et al., 2011; Hideno et al., 2009; Jin and Chen, 2006; Silva et al., 2012), and alkali (Lin et al., 2010).

The most prolific milling method is BM with four HCPs (Da Silva et al., 2010; Hideno et al., 2009; Lin et al., 2010; Silva et al., 2012). The other milling methods are DM (Da Silva et al., 2010; Hideno et al., 2009), JM (Jin and Chen, 2006; Silva et al., 2012), and DM (Da Silva et al., 2010; Hideno et al., 2009) with two HCPs each.

A number of variables are studied in these studies such as particle size (Adapa et al, 2011; Jin and Chen, 2006; Lin et al., 2010; Silva et al., 2012), energy consumption (Adapa et al., 2011; Hideno et al., 2009; Mani et al., 2004), sugar yield (Da Silva et al., 2010; Hideno et al., 2009; Silva et al., 2012), milling type (Da Silva et al., 2010; Silva et al., 2012), and density (Adapa et al., 2011; Mani et al., 2004).

The most prolific agricultural biomass used in these studies is straw from wheat (Adapa et al., 2011; Mani et al., 2004), barley (Adapa et al., 2011; Mani et al., 2004), rice (Hideno et al., 2009; Jin and Chen, 2006), and canola and oat (Adapa et al., 2011). The other prolific agricultural biomass is corn stover (Lin et al., 2010; Mani et al., 2004).

A number of keywords are used in the titles of these studies. The most prolific keyword is straw with five HCPs, closely followed by enzymatic and hydrolysis with four HCPs each. The other prolific keywords are milling, stover, and ethanol with at least two HCPs each.

The key finding from these papers is that the milling pretreatment in combination with these other pretreatments or stand-alone was effective in enhancing the digestibility of agricultural wastes, notably straw and corn stover. Thus, milling pretreatment increased the bioethanol yield from agricultural residues (Binod et al., 2010; Cardona et al., 2010).

16.4.5 Milling Pretreatment of the Algae

There are only two HCPs related to the milling pretreatment of the algae (Postma et al., 2015, 2017). They use only BM for the algae and employ variables of biomass content, energy consumption, and bead size. The key finding from these papers is that milling pretreatment enhanced the digestibility of the algae for sugar and bioethanol production. Thus, this pretreatment increased the bioethanol yield from algae (Ho et al., 2013; John et al., 2011).

16.4.6 Milling Pretreatment of the Grass

There is only one HCP related to the milling pretreatment of the grass (Mani et al., 2014). It uses HM for the switchgrass besides wheat and barley straws and corn stover and employs variables of energy consumption and density. The key finding from this paper is that switchgrass had the highest specific energy consumption and highest calorific value and the lowest ash content compared with the other biomass. Thus, this pretreatment increased the bioethanol yield from switchgrass (McLaughlin and Kszos, 2005; Parrish and Fike, 2005).

16.5 CONCLUSION AND FUTURE RESEARCH

The brief information about the key research fronts covered by the 25 most-cited papers with at least 78 citations each is given under four headings: grass, cellulose, agricultural wastes, and wood.

The usual characteristics of these HCPs are that they are often used in combination with other pretreatments such as acids, alkali, enzymes, HCW/steam explosion, sulfite, and H_2O_2. In this way, milling pretreatment was effective in disrupting the structures of these feedstocks.

The key findings on these research fronts should be read in light of the increasing public concerns about climate change, greenhouse gas (GHG) emissions, and global warming as these concerns have been certainly behind the boom in research on bioethanol production as an alternative to crude oil-based gasoline and diesel fuels in the last decades.

The data in Table 16.6 show that agricultural wastes, woody biomass, cellulose, and plants are the most prolific feedstocks to study the effect of the milling pretreatment stand-alone or in combination with other pretreatments. This is in line with research on bioethanol production and utilization focusing more on wastes and algae as nonfood feedstocks compared to the food feedstocks.

These studies emphasize the importance of proper incentive structures for the efficient development and application of the milling pretreatment to enhance the digestibility of bioethanol feedstocks to increase the sugar yield and bioethanol yield after hydrolysis and fermentation of these feedstocks in light of North's institutional framework (North, 1991). In this context, the major producers and users of bioethanol fuels such as China, the USA, Europe, and Japan, and to a lesser extent Canada, Austria, Italy, the Netherlands, and Norway had developed strong incentive structures for the effective development of and application of microwave pretreatment for bioethanol and sugar production. However, the relatively low funding rate is a major barrier to the further development of research in this field.

It is recommended that such review studies are performed for the other research fronts related to the other pretreatments, including acid, enzymatic, alkali, and HCW/steam explosion pretreatments, which are the most prolific pretreatments used in combination with milling pretreatment in this book chapter, complementing this book chapter.

ACKNOWLEDGMENTS

The contribution of the highly cited researchers in the field of the milling pretreatment of biomass has been gratefully acknowledged.

APPENDIX: THE KEYWORD SET

((((TITLE (milling OR grinding)) OR (TITLE (milled OR milling OR grinding OR "mill pre-treatment*" OR grinded) AND TITLE (ethanol OR bioethanol OR pretreat* OR "pre-treat*" OR *hydrolysis OR saccharification OR ferment* OR hydrolytic OR digestibility OR disruption OR fractionation OR decomposition OR dissolution OR delignification OR depolymerization OR sugar OR sugars OR *oligosaccharides OR glucose OR deconstruction OR xylose OR disintegration OR "size reduction"))) AND (TITLE (waste* OR *grass OR grass* OR biomass* OR lignocellul* OR bagasse* OR cellulos* OR *wood OR woody OR *algae OR *algal OR straw* OR stover* OR eucalyptus OR miscanthus OR pine OR chlorella OR scrap* OR lignin OR *cellulose OR porphyridium OR nannochloropsis OR starch OR palm OR poplar OR cedar OR corncob* OR "corn cob*" OR spruce OR beech OR fir OR oak OR bunch* OR pinus OR stalk* OR residue* OR willow* OR cypress OR stem* OR sawdust OR prosopis OR sugarcane OR hull* OR husk* OR shell* OR *cane OR birch OR cynara) OR SRCTITLE (cellulose OR wood* OR biomass OR sugar))) AND NOT (TITLE (removal OR concrete OR *diesel OR *gas OR digestion OR drug* OR pozzolanic OR *composites OR composite OR cyclodext* OR sorbitol OR uranium OR lactic OR polysaccharide* OR wastewater* OR ore OR ester* OR medic* OR "milling waste*" OR reforming OR separation OR ruminal OR "wood milling" OR recycl* OR hematite OR mine OR graphite OR cement* OR diet* OR "rice milling" OR coal* OR *sorption OR calcined OR circuit* OR inciner* OR feldspar OR food* OR binder* OR gypsum OR batter* OR pyroly* OR aluminum OR magnet* OR *furfural OR hydrogen OR metal* OR methane OR steel OR hmf OR lactose OR *ethylene OR alloy* OR reforming OR seed* OR h2 OR feed OR zirconium OR cereal* OR salt OR reinforced OR flour OR synthes* OR ferul* OR leach* OR torref* OR pigment* OR vitrif* OR electrode* OR blend* OR ash OR stone OR nano* OR microfibril*) OR SRCTITLE (food* OR dairy OR phar* OR medi* OR cereal* OR nutrit* OR feed* OR brew* OR lwt OR animal* OR biological OR agric* OR zoo* OR genetic* OR lipid* OR soil* OR nano*)) AND (LIMIT-TO (DOCTYPE, "ar") OR LIMIT-TO

(DOCTYPE, "cp") OR LIMIT-TO (DOCTYPE, "re") OR LIMIT-TO (DOCTYPE, "ch") OR LIMIT-TO (DOCTYPE, "sh") OR LIMIT-TO (DOCTYPE, "bk") OR LIMIT-TO (DOCTYPE, "no")) AND (LIMIT-TO (LANGUAGE, "English")) AND (LIMIT-TO (SRCTYPE, "j") OR LIMIT-TO (SRCTYPE, "k") OR LIMIT-TO (SRCTYPE, "b"))

REFERENCES

Adapa, P., L. Tabil and G. Schoenau. 2011. Grinding performance and physical properties of non-treated and steam exploded barley, canola, oat and wheat straw. *Biomass and Bioenergy* 35:549–561.

Ago, M., T. Endo and T. Hirotsu. 2004. Crystalline transformation of native cellulose from cellulose I to cellulose II polymorph by a ball-milling method with a specific amount of water. *Cellulose* 11:163–167.

Angelici, C., B. M. Weckhuysen and P. C. A. Bruijnincx. 2013. Chemocatalytic conversion of ethanol into butadiene and other bulk chemicals. *ChemSusChem* 6:1595–1614.

Antolini, E. 2007. Catalysts for direct ethanol fuel cells. *Journal of Power Sources* 170:1–12.

Antolini, E. 2009. Palladium in fuel cell catalysis. *Energy and Environmental Science* 2:915–931.

Avolio, R., I. Bonadies, and D. Capitani, et al. 2012. A multitechnique approach to assess the effect of ball milling on cellulose. *Carbohydrate Polymers* 87:265–273.

Balaz, M. 2018. Ball milling of eggshell waste as a green and sustainable approach: A review. *Advances in Colloid and Interface Science* 256:256–275.

Beguin, P. and J. P. Aubert. 1994. The biological degradation of cellulose. *FEMS Microbiology Reviews* 13:25–58.

Binod, P., R. Sindhu and R. R. Singhania, et al. 2010. Bioethanol production from rice straw: An overview. *Bioresource Technology* 101:4767–4774.

Burnham, J. F. 2006. Scopus database: A review. *Biomedical Digital Libraries* 3:1–8.

Cardona, C. A., J. A. Quintero and I. C. Paz. 2010. Production of bioethanol from sugarcane bagasse: Status and perspectives. *Bioresource Technology* 101:4754–4766.

Da Silva, A.S., H. Inoue, T. Endo, S. Yano and E. P. S. Bon. 2010. Milling pretreatment of sugarcane bagasse and straw for enzymatic hydrolysis and ethanol fermentation. *Bioresource Technology* 101:7402–7409.

Fernando, S., S. Adhikari, C. Chandrapal and M. Murali. 2006. Biorefineries: Current status, challenges, and future direction. *Energy & Fuels* 20:1727–1737.

Fujimoto, A., Y. Matsumoto, H. M. Chang and G. Meshitsuka. 2005. Quantitative evaluation of milling effects on lignin structure during the isolation process of milled wood lignin. *Journal of Wood Science* 51:89–91.

Gregg, D. J. and J. N. Saddler. 1996. Factors affecting cellulose hydrolysis and the potential of enzyme recycle to enhance the efficiency of an integrated wood to ethanol process. *Biotechnology and Bioengineering* 51:375–383.

Hahn-Hagerdal, B., M. Galbe, M. F. Gorwa-Grauslund, G. Liden and G. Zacchi. 2006. Bio-ethanol – The fuel of tomorrow from the residues of today. *Trends in Biotechnology* 24:549–556.

Hideno, A., H. Inoue and K. Tsukahara, et al. 2009. Wet disk milling pretreatment without sulfuric acid for enzymatic hydrolysis of rice straw. *Bioresource Technology* 100:2706–2711.

Hill, J., E. Nelson, D. Tilman, S. Polasky and D. Tiffany. 2006. Environmental, economic, and energetic costs and benefits of biodiesel and ethanol biofuels. *Proceedings of the National Academy of Sciences of the United States of America* 103:11206–11210.

Hill, J., S. Polasky and E. Nelson, et al. 2009. Climate change and health costs of air emissions from biofuels and gasoline. *Proceedings of the National Academy of Sciences of the United States of America* 106:2077–2082.

Ho, S. H., S. W. Huang and C. Y. Chen, et al. 2013. Bioethanol production, using carbohydrate-rich microalgae biomass as feedstock. *Bioresource Technology* 135:191–198.

Hsieh, W. D., R. H. Chen, T. L. Wu and T. H. Lin. 2002. Engine performance and pollutant emission of an SI engine using ethanol-gasoline blended fuels. *Atmospheric Environment* 36:403–410.

Hu, G., J. A. Heitmann and O. J. Rojas. 2008. Feedstock pretreatment strategies for producing ethanol from wood, bark, and forest residues. *BioResources* 3:270–294.

Huang, H. J., S. Ramaswamy, U. W. Tschirner and B. V. Ramarao. 2008. A review of separation technologies in current and future biorefineries. *Separation and Purification Technology* 62:1–21.

Inoue, H., S. Yano, T. Endo, T. Sakaki and S. Sawayama. 2008. Combining hot-compressed water and ball milling pretreatments to improve the efficiency of the enzymatic hydrolysis of eucalyptus. *Biotechnology for Biofuels* 1:2.

Jin, S. and H. Chen. 2006. Superfine grinding of steam-exploded rice straw and its enzymatic hydrolysis. *Biochemical Engineering Journal* 30:225–230.

John, R. P., G. S. Anisha, K. M. Nampoothiri and A. Pandey. 2011. Micro and macroalgal biomass: A renewable source for bioethanol. *Bioresource Technology* 102:186–193.

Karinkanta, P., A. Ammala, M. Illikainen and J. Niinimaki. 2018. Fine grinding of wood – Overview from wood breakage to applications. *Biomass and Bioenergy* 113:31–44.

Kleine, T., J. Buendia and C. Bolm. 2013. Mechanochemical degradation of lignin and wood by solvent-free grinding in a reactive medium. *Green Chemistry* 15:160–166.

Konur, O. 2000. Creating enforceable civil rights for disabled students in higher education: An institutional theory perspective. *Disability & Society* 15:1041–1063.

Konur, O. 2002a. Access to nursing education by disabled students: Rights and duties of nursing programs. *Nurse Education Today* 22:364–374.

Konur, O. 2002b. Assessment of disabled students in higher education: Current public policy issues. *Assessment and Evaluation in Higher Education* 27:131–152.

Konur, O. 2002c. Access to employment by disabled people in the UK: Is the Disability Discrimination Act working? *International Journal of Discrimination and the Law* 5:247–279.

Konur, O. 2006a. Participation of children with dyslexia in compulsory education: Current public policy issues. *Dyslexia* 12:51–67.

Konur, O. 2006b. Teaching disabled students in Higher Education. *Teaching in Higher Education* 11:351–363.

Konur, O. 2007a. A judicial outcome analysis of the Disability Discrimination Act: A windfall for the employers? *Disability & Society* 22:187–204.

Konur, O. 2007b. Computer-assisted teaching and assessment of disabled students in higher education: The interface between academic standards and disability rights. *Journal of Computer Assisted Learning* 23:207–219.

Konur, O. 2012. The evaluation of the research on the bioethanol: A scientometric approach. *Energy Education Science and Technology Part A: Energy Science and Research* 28:1051–1064.

Konur, O. 2015. Current state of research on algal bioethanol. In *Marine Bioenergy: Trends and Developments*, Eds. S. K. Kim and C. G. Lee, pp. 217–244. Boca Raton, FL: CRC Press.

Konur, O. 2020a. The scientometric analysis of the research on the bioethanol production from green macroalgae. In *Handbook of Algal Science, Technology and Medicine*, Ed. O. Konur, pp. 385–401. London: Academic Press.

Lin, Y. and S. Tanaka. 2006. Ethanol fermentation from biomass resources: Current state and prospects. *Applied Microbiology and Biotechnology* 69:627–642.

Lin, Z., H. Huang, and H. Zhang, et al. 2010. Ball milling pretreatment of corn stover for enhancing the efficiency of enzymatic hydrolysis. *Applied Biochemistry and Biotechnology* 162:1872–1880.

Ling, Z., T. Wang and M. Makarem, et al. 2019. Effects of ball milling on the structure of cotton cellulose. *Cellulose* 26:305–328.

Ma, X., L. Sun and C. Song. 2002. A new approach to deep desulfurization of gasoline, diesel fuel and jet fuel by selective adsorption for ultra-clean fuels and for fuel cell applications. *Catalysis Today* 77:107–116.

Mais, U., A. R. Esteghlalian, J. N. Saddler and S. D. Mansfield. 2002. Enhancing the enzymatic hydrolysis of cellulosic materials using simultaneous ball milling. *Applied Biochemistry and Biotechnology – Part A Enzyme Engineering and Biotechnology* 98–100:815–832.

Mani, S., L. G. Tabil and S. Sokhansanj. 2004. Grinding performance and physical properties of wheat and barley straws, corn stover and switchgrass. *Biomass and Bioenergy* 27:339–352.

McLaughlin, S. B. and L. A. Kszos. 2005. Development of switchgrass (*Panicum virgatum*) as a bioenergy feedstock in the United States. *Biomass and Bioenergy* 28:515–535.

Morschbacker, A. 2009. Bio-ethanol based ethylene. *Polymer Reviews* 49:79–84.

Najafi, G., B. Ghobadian and T. Tavakoli, et al. 2009. Performance and exhaust emissions of a gasoline engine with ethanol blended gasoline fuels using artificial neural network. *Applied Energy* 86:630–639.

Newman, P. W. G. and J. R. Kenworthy. 1989. Gasoline consumption and cities: A comparison of U.S. cities with a global survey. *Journal of the American Planning Association* 55:24–37.

North, D. C. 1991. Institutions. *Journal of Economic Perspectives* 5:97–112.

Olsson, L. and B. Hahn-Hagerdal. 1996. Fermentation of lignocellulosic hydrolysates for ethanol production. *Enzyme and Microbial Technology* 18:312–331.

Parrish, D. J. and J. H. Fike. 2005. The biology and agronomy of switchgrass for biofuels. *Critical Reviews in Plant Sciences* 24:423–459.

Piras, C. C., S. Fernandez-Prieto and W. M. de Borggraeve. 2019. Ball milling: A green technology for the preparation and functionalisation of nanocellulose derivatives. *Nanoscale Advances* 1:937–947.

Postma, P. R., E. Suarez-Garcia and K. Yonathan, et al. 2017. Energy efficient bead milling of microalgae: Effect of bead size on disintegration and release of proteins and carbohydrates. *Bioresource Technology* 224:670–679.

Postma, P. R., T. L. Miron and G. Olivieri, et al. 2015. Mild disintegration of the green microalgae *Chlorella vulgaris* using bead milling. *Bioresource Technology* 184:297–304.

Ryu, D. D. Y., S. B. Lee, T. Tassinari and C. Macy. 1982. Effect of compression milling on cellulose structure and on enzymatic hydrolysis kinetics. *Biotechnology and Bioengineering* 24:1047–1067.

Sanchez, O. J. and C. A. Cardona. 2008. Trends in biotechnological production of fuel ethanol from different feedstocks. *Bioresource Technology* 99:5270–5295.

Schell, D. J. and C. Harwood. 1994. Milling of lignocellulosic biomass – Results of pilot-scale testing. *Applied Biochemistry and Biotechnology* 45–46:159–168.

Schwanninger, M., J. C. Rodrigues, H. Pereira and B. Hinterstoisser. 2004. Effects of short-time vibratory ball milling on the shape of FT-IR spectra of wood and cellulose. *Vibrational Spectroscopy* 36:23–40.

Silva, G. G. D., M. Couturier, J. G. Berrin, A. Buleon and X. Rouau. 2012. Effects of grinding processes on enzymatic degradation of wheat straw. *Bioresource Technology* 103:192–200.

Sun, Y. and J. Cheng. 2002. Hydrolysis of lignocellulosic materials for ethanol production: A review. *Bioresource Technology* 83:1–11.

Taherzadeh, M. J. and K. Karimi. 2007. Enzyme-based hydrolysis processes for ethanol from lignocellulosic materials: A review. *Bioresources* 2:707–738

Taherzadeh, M. J. and K. Karimi. 2008. Pretreatment of lignocellulosic wastes to improve ethanol and biogas production: A review. *International Journal of Molecular Sciences* 9:1621–1651.

Teramoto, Y., N. Tanaka, S. H. Lee and T. Endo. 2008. Pretreatment of eucalyptus wood chips for enzymatic saccharification using combined sulfuric acid-free ethanol cooking and ball milling. *Biotechnology and Bioengineering* 99:75–85.

Yang, B. and C. E. Wyman. 2008. Pretreatment: The key to unlocking low-cost cellulosic ethanol. *Biofuels Bioproducts & Biorefining-BIOFPR* 2:26–40.

Yu, Y. and H. Wu. 2011. Effect of ball milling on the hydrolysis of microcrystalline cellulose in hot-compressed water. *AIChE Journal* 57:793–800.

Zhang, W., M. Liang and C. Lu. 2007. Morphological and structural development of hardwood cellulose during mechanochemical pretreatment in solid state through pan-milling. *Cellulose* 14:447–456.

Zhao, H., J. H. Kwak and Y. Wang, et al. 2006. Effects of crystallinity on dilute acid hydrolysis of cellulose by cellulose ball-milling study. *Energy and Fuels* 20:807–811.

Zhu, J. Y., G. S. Wang, X. J. Pan and R. Gleisner. 2009. Specific surface to evaluate the efficiencies of milling and pretreatment of wood for enzymatic saccharification. *Chemical Engineering Science* 64:474–485.

Zhu, W., J. Y. Zhu, R. Gleisner and X. J. Pan. 2010. On energy consumption for size-reduction and yields from subsequent enzymatic saccharification of pretreated lodgepole pine. *Bioresource Technology* 101:2782–2792.

17 Microwave Pretreatment of the Biomass
Review

Anh Tuan Hoang
HUTECH University

Ozcan Konur
(Formerly) Ankara Yildirim Beyazit University

17.1 INTRODUCTION

Crude oil–based gasoline fuels (Ma et al., 2002; Newman and Kenworthy, 1989) have been widely used in the transportation sector since the 1920s. However, there have been great public concerns over the adverse environmental and human impact of these fuels (Hill et al., 2006, 2009). Hence, biomass-based bioethanol fuels (Hahn-Hagerdal et al., 2006; Hill et al., 2006; Konur, 2012, 2015, 2020a) have increasingly been used in blending gasoline fuels (Hsieh et al., 2002; Najafi et al., 2009), in fuel cells (Antolini, 2007, 2009), and in biochemical production (Angelici et al., 2013; Morschbacker, 2009) in a biorefinery context (Fernando et al., 2006; Huang et al., 2008).

However, it is necessary to pretreat the biomass (Taherzadeh and Karimi, 2008; Yang and Wyman, 2008) to enhance the yield of the bioethanol (Hahn-Hagerdal et al., 2006; Sanchez and Cardona, 2008) prior to bioethanol's production through hydrolysis (Sun and Cheng, 2002; Taherzadeh and Karimi, 2007) and fermentation (Lin and Tanaka, 2006; Olsson and Hahn-Hagerdal, 1996) of the biomass. Microwave pretreatment is one of these pretreatment methods.

However, it is essential to develop efficient incentive structures (North, 1991) for the primary stakeholders to enhance the research in this field (Konur, 2000, 2002a–c, 2006a, b, 2007a, b). Although there have been a number of review papers on microwave pretreatment of biomass (Kostas et al., 2017; Li et al., 2016; Puligundla et al., 2016), there has been no review of the most-cited 25 articles in this field.

Thus, this chapter presents a review of the most-cited 25 articles in these fields. Then, it discusses the key findings of these highly influential papers and comments on future research priorities in this field.

17.2 MATERIALS AND METHODS

The search for this study was carried out using the Scopus database (Burnham, 2006) in March 2022.

As a first step for the search of the relevant literature, keywords were selected using all population papers. The selected keyword list was optimized to obtain a representative sample of papers for the searched research field. This final keyword set was provided in the Appendix for future replication studies.

As a second step, a sample dataset was used for this study. The first 25 articles with at least 88 citations each were selected for the review study. Key findings from each paper were taken from the abstracts of these papers and were discussed. Additionally, a number of brief conclusions were drawn and a number of relevant recommendations were made to enhance the future research landscape.

DOI: 10.1201/9781003226536-23

17.3 RESULTS

Brief information about the 25 most-cited papers with at least 88 citations each on microwave pretreatment of the biomass is given below.

17.3.1 Microwave Pretreatment of the Grass

There are three Highly cited papers (HCPs) for microwave pretreatment of the grass.

Hu and Wen (2008) enhanced the enzymatic digestibility of switchgrass by microwave-assisted alkaline pretreatment in a paper with 339 citations. They soaked the switchgrass in alkaline solutions (0.05–0.3 g alkali/g biomass), treated it using microwave, and then hydrolyzed it using cellulase enzymes. They observed that the alkali–microwave combined pretreatment and hydrolysis resulted in a higher sugar yield compared to conventional heating while they achieved the highest yield of 90% of maximum potential sugars at 0.1 g/g of alkali loading due to the disruption of recalcitrant structures of the switchgrass. At optimal conditions of 0.1 g/g of alkali loading, 190°C, 50 g/L solid content, and 30 min treatment time (TT), the sugar yield from the combined pretreatment and hydrolysis was 58.7 g/100 g of biomass, with 99% of potential maximum sugars. They concluded that combined microwave–alkaline pretreatment enhanced the reduction of sugar yield significantly.

Chen and Wan (2018) enhanced the digestibility of the switchgrass, miscanthus, and corn stover by microwave-assisted deep eutectic solvent (DES) pretreatment in a paper with 134 citations. They facilitated this pretreatment for 45 s with microwave irradiation at 800 W using a DES composed of choline chloride and lactic acid. They observed that this pretreatment was highly effective in removing lignin and xylan while retaining most of the cellulose in the pretreated solids. This ultrafast fractionation increased the cellulose digestibility by 2–5 folds. Switchgrass gave a similar glucose yield to corn stover but experienced less cellulose loss during pretreatment. Miscanthus was also substantially fractionated but showed lower digestibility than switchgrass and corn stover. Lignin was readily recoverable with relatively high purity from all the pretreatment liquor. They concluded that this combined pretreatment was effective for biomass fractionation and cellulose digestibility improvement as well as lignin extraction.

Orozco et al. (2007) enhanced the digestibility of cellulose and grass by microwave-assisted dilute acid (phosphoric acid, H_3PO_4) pretreatment in a paper with 104 citations. They observed a rapid hydrolysis reaction at moderate temperatures. The optimum conditions for grass hydrolysis were 2.5% H_3PO_4 at a temperature of 175°C. Sugar degradation occurred at acid concentrations greater than 2.5% (v/v) and temperatures greater than 175°C. The optimum conditions gave a yield of 90% of glucose. Primary advantages of employing microwave heating were to achieve a high rate constant at moderate temperatures and prevent a hotspot formation within the reactor, which would have caused localized degradation of glucose (Table 17.1).

17.3.2 Microwave Pretreatment of the Cellulose

There are six HCPs for the microwave pretreatment of cellulose. Notes of Orozco et al. (2007) were included in Section 17.3.1.

Wu et al. (2010) enhanced the digestibility of the cellulose in water by microwave-assisted acid pretreatment catalyzed by biomass char sulfonic acids (BC-SO_3H), prepared cheaply from natural bamboo, cotton, and starch, in a paper with 208 citations. They observed that this acid had a much higher turnover for this reaction compared to a dilute H_2SO_4 solution, due to its strong affinity to the β-1,4-glycosidic bonds of cellulose. Further, microwave pretreatment played a key role in activating cellulose molecules and strengthening particle collision, leading to a substantial acceleration effect on this heterogeneous catalytic process.

Zhang and Zhao (2009) enhanced the digestibility of cellulose in water by microwave-assisted solid acid pretreatment in ionic liquid (IL) in a paper with 179 citations. They observed that microwave pretreatment promoted the acid-catalyzed hydrolysis of cellulose in IL where H-form ZSM-5 zeolites with a lower Si/Al molar ratio and a larger surface area showed a relatively higher catalytic activity. These solid catalysts exhibited better performance than the sulfated ion-exchanging resin NKC-9. Microwave irradiation

TABLE 17.1
Microwave Pretreatment of Grass

No.	Papers	Biomass	Other Pts.	Parameters	Keywords	Lead Author	Affil.	Cits
1	Hu and Wen (2008)	Switchgrass	Alkali, enzymes	Alkali loading, temperature, solid content, treatment time	Digestibility, switchgrass, microwave, pretreatment, alkali	Wen, Zhiyou 14012373200	Virg. Polytech. Inst. State Univ. USA	339
2	Chen and Wan (2018)	Miscanthus, switchgrass, corn stover	Deep eutectic solvent-DES (ChCl:LA)	Biomass type	Fractionation, lignocellulosic, biomass, microwave, DES, pretreatment	Wan, Caixia 35772603400	Univ. Missouri USA	134
3	Orozco et al. (2007)	Cellulose, grass	Acid (H_3PO_4)	Acid content, temperature	Acid, hydrolysis, cellulose, cellulosic, biowaste, microwave	Walker, Rooney G 35583992400	Queen's Univ. UK	104

Other pts., other pretreatments.

at an appropriate power significantly reduced the TT and increased the yields of reducing sugars as a typical hydrolysis reaction with Avicel cellulose produced a glucose yield of around 37% within 8 min.

Fan et al. (2013) enhanced the digestibility of cellulose by microwave pretreatment over a broad temperature range (150°C–270°C) in a paper with 155 citations. They proposed a mechanism of cellulose interaction with microwaves and found that the degree of freedom of the cellulose-enclosed CH_2OH groups was crucial. This mechanism explained the high efficiency of microwave treatment, the dependence of the selectivity/yield of glucose on the applied microwave density, the observed high glucose-to-hydroxymethylfurfural (HMF) ratio, and the influence of the degree of cellulose crystallinity on the results of the hydrolysis process. The highest selectivity toward glucose was around 75% while the highest glucose yield obtained was 21%.

Budarin et al. (2010) enhanced the digestibility of cellulose by microwave pretreatment at low temperatures in a paper with 134 citations. They identified 180°C as a key in the process mechanism, as the amorphous region of cellulose softened allowing a microwave-induced rearrangement. Proton transfer was then possible under the microwave field resulting in acid-catalyzed decomposition. This low-temperature process was suitable for scale-up.

Ha et al. (2011) enhanced the enzymatic digestibility of cotton cellulose by microwave-assisted IL pretreatment in a paper with 124 citations. They observed that microwave irradiation not only enhanced the solubility of cellulose in ILs but also significantly decreased the degree of polymerization of regenerated cellulose after IL dissolution pretreatment, resulting in a significant improvement in cellulose hydrolysis. The rate of enzymatic hydrolysis of cotton cellulose was increased by at least 12-fold after IL dissolution pretreatment at 110°C and by 50-fold after IL dissolution pretreatment with microwave irradiation. They concluded that cellulose pretreatment with ILs and microwave irradiation was an effective method for the pretreatment of cellulose (Table 17.2).

17.3.3 Microwave Pretreatment of Agricultural Wastes

17.3.3.1 Microwave Pretreatment of Bagasse

There are six HCPs for microwave pretreatment of bagasse.

Binod et al. (2012) enhanced the enzymatic digestibility of sugarcane bagasse by microwave-assisted alkaline and/or acid pretreatment in a paper with 266 citations. They used microwave–acid, microwave–alkaline, and combined microwave-alkaline-acid pretreatments. They observed that microwave treatment of sugarcane bagasse with 1% sodium hydroxide (NaOH) at 600 W for 4 min followed by enzymatic hydrolysis gave a reducing sugar yield of 66.5 g/100 g of dry biomass, while combined microwave-alkaline-acid treatment with 1% of NaOH followed by 1% of sulfuric acid (H_2SO_4), the reducing sugar yield increased to 83 g/100 g of dry biomass. They concluded that combined microwave-alkaline-acid pretreatment for a short duration enhanced the reduction of sugar yield significantly.

Chen et al. (2011) enhanced the digestibility of sugarcane bagasse by microwave-assisted dilute acid pretreatment in a paper with 228 citations. They employed temperatures of 130°C, 160°C, and 190°C with two TTs of 5 and 10 min. They observed that an increase in reaction temperature destroyed the lignocellulosic structure of bagasse significantly. When the reaction temperature was as high as 190°C, the specific surface area of the pretreated bagasse grew substantially. Further, almost all hemicellulose was removed from bagasse and the crystalline structure of cellulose disappeared while the feature of lignin was retained clearly. However, the influence of TT on the lignocellulosic structure was not significant as the pretreatment for 5 min was sufficiently long.

Chen et al. (2012) enhanced the digestibility of sugarcane bagasse by microwave-assisted dilute acid (H_2SO_4) pretreatment in a paper with 162 citations. They pretreated bagasse in an acid solution at 180°C for 30 min, with the concentration ranging from 0 to 0.02 M. A higher concentration destroyed bagasse although they also observed the buffering capacity of the biomass in pretreatment. Around 40–44 wt% of bagasse was degraded from the pretreatment where around 80%–98% of hemicellulose was hydrolyzed. In contrast, crystalline cellulose and lignin were hardly affected by pretreatment. The maximum yields of sugars with minimum furfural selectivity occurred at an acid concentration of 0.005 M. They recommended this optimal concentration for acid concentration.

Ooshima et al. (1984) enhanced the digestibility of rice straw and bagasse by microwave pretreatment in a paper with 129 citations. They pretreated these feedstocks with a microwave (2,450 MHz) in sealed glass vessels. They observed that this treatment enhanced markedly the accessibility of these feedstocks for enzymatic hydrolysis: 1.6 times in the rice straw by the microwave treatment at 170°C for 5 min and 3.2 times in the bagasse by the treatment at 200°C for 5 min, compared to the untreated feedstocks.

Intanakul et al. (2003) enhanced the enzymatic digestibility of rice straw and sugarcane bagasse by microwave pretreatment in a paper with 104 citations. They immersed these feedstocks in a glycerol medium with small amounts of water and treated them with 240 W of microwave irradiation for 10 min at atmospheric pressure where a temperature of about 200°C could be reached in this medium without a high pressure buildup. They observed that more than twice the amount of reducing sugars was produced from enzymatic saccharification utilizing microwave pretreatment, compared with no pretreatment. They concluded that the microwave pretreatment which achieved high temperatures at atmospheric pressure provided some advantages over a steam explosion process which required high pressure and a subsequent sudden pressure drop.

De Souza Moretti et al. (2014) enhanced the enzymatic digestibility of sugarcane bagasse by microwave-assisted acid pretreatment associated with glycerol in a paper with 100 citations. They subjected this feedstock to microwave irradiation for 5 min with distilled water, phosphoric acid (H_3PO_4) (pH 3.0), and glycerol (100%) before being enzymatically hydrolyzed using the cellulase enzyme extract *Myceliophthora thermophila* M.7.7. and the commercial enzyme cocktail Celluclast 1.5 L. Infrared spectra of untreated and treated bagasse in microwave irradiation and glycerol showed significant differences in the regions 1,635, 1,600, and 1,510 related to vibration of the aromatic ring, and the band at 1,100 cm^{-1} was attributed to an overlap of C–O–H elongation of primary and secondary alcohols and at 980 cm^{-1} to stretching of glycosidic linkages C–O–C. Bagasse treated in microwave irradiation and glycerol had higher thermal stability compared to the untreated bagasse.

Around 5.4 and 11.3% w/w of lignin and xylan fractions, respectively, were degraded after pretreatment of bagasse in microwave heating with glycerol. The highest yields of hydrolysis of

TABLE 17.2
Microwave Pretreatment of Cellulose

No.	Papers	Biomass	Other Pts.	Parameters	Keywords	Lead Author	Affil.	Cits
1	Wu et al. (2010)	Cellulose	Acid (SO_3H, H_2SO_4)	Acid type	Microwave, hydrolysis, cellulose, biomass, acid	Fu, Zaihui 7403346468	Hunan Normal Univ., China	208
2	Zhang and Zhao (2009)	Cellulose	Acid (H-form ZSM-5 zeolites, sulfated ion-exchanging resin NKC-9, ionic liquids	Acid type	Acid, microwave, hydrolysis, cellulose, ionic liquid	Zhao, Zongbao K 35197609300	Chin. Acad. Sci. China	179
3	Fan et al. (2013)	Cellulose	NA	Temperature	Microwave, depolymerization, cellulose	Clark, James H 57218227445	Univ. York UK	155
4	Budarin et al. (2010)	Cellulose	NA	Temperature	Microwave, decomposition, cellulose, biomass	Clark, James H 57218227445	Univ. York UK	134
5	Ha et al. (2011)	Cellulose	Ionic liquids, enzymes	Combined effect	Microwave, pretreatment, cellulose, ionic liquid, hydrolysis, enzymatic	Koo, Yoon- Moo 57193813907	Inha Univ. S. Korea	124
6	Orozco et al. (2007)	Cellulose, grass	Acid (H_3PO_4)	Acid content, temperature	Acid, hydrolysis, cellulose, cellulosic, biowaste, microwave	Walker, Rooney G 35583992400	Queen's Univ. UK	104

NA, non-available; Other pts., other pretreatments.

hemicellulose (22.4%) and cellulose (40.2%) w/w were obtained in the reaction mixture containing the enzyme and Celluclast commercial cane bagasse treated in microwave irradiation and glycerol after 24 h of incubation (Table 17.3).

17.3.3.2 Microwave Pretreatment of Straw

There are 12 HCPs for microwave pretreatment of straw. Notes of Ooshima et al. (1984) and Intanakul et al. (2003) were included in Section 17.3.3.1.

Zhu et al. (2005) enhanced the enzymatic digestibility of rice straw by microwave-assisted alkaline pretreatment in a paper with 230 citations. They observed that the rice straw had a weight loss of 44.6% and composition of cellulose 69.2%, lignin 4.9%, and hemicellulose 10.2% after 30 min microwave/alkaline pretreatment at 700 W while it only had a weight loss of 41.5% and composition of cellulose 65.4%, lignin 6.0%, and hemicellulose 14.3% after 70 min alkali-alone pretreatment. They reasoned that microwave/alkali combined pretreatment could remove more lignin and hemicellulose from rice straw with a shorter pretreatment time than the alkali-alone one. Finally, for the pretreated rice straw (solid content 50 g/L, enzyme loading 20 mg/g substrate), they observed that rice straw pretreated by microwave/alkali had a higher hydrolysis rate and sugar content in the hydrolysate compared with the one by alkali-alone. They concluded that combined microwave–alkaline pretreatment enhanced the reduction of sugar yield significantly.

Ma et al. (2009) enhanced the enzymatic digestibility of rice straw by microwave-assisted dilute acid pretreatment in a paper with 224 citations. They optimized this combined pretreatment. They observed that energy loading, TT, and solid content were the key variables directing the enzymatic saccharification of rice straw. Maximal efficiencies of cellulose, hemicellulose, and total saccharification were respectively increased by 30.6%, 43.3%, and 30.3% under the optimal conditions of 680 W, 24 min and 75 g/L. They confirmed that microwave pretreatment could disrupt the silicified waxy surface, break down the lignin–hemicellulose complex, and partially remove silicon and lignin.

Zhu et al. (2006d) enhanced the enzymatic digestibility of rice straw by microwave-assisted alkaline, acid/alkaline, acid/alkaline/H_2O_2 pretreatments in a paper with 138 citations. They observed that the rice straw pretreated by microwave/acid/alkaline/H_2O_2 had the highest weight loss and cellulose content while xylose could not be recovered during the microwave/alkaline pretreatment process, but it could be recovered during the microwave/acid/alkaline and microwave/acid/alkaline/H_2O_2 pretreatment processes. Finally, they found that the rice straw pretreated by microwave/acid/alkaline/H_2O_2 had the highest hydrolysis rate and glucose content in the hydrolysate for optimal conditions at a substrate concentration of 50 g/L, pH 4·8, and enzyme loading of 20 mg/g of the substrate.

Singh et al. (2011) enhanced the enzymatic digestibility of rice straw and hulls by microwave-assisted alkaline pretreatment in a paper with 129 citations. They observed that alkaline concentration (AC), TT, and solid content (SC) were the main factors governing the saccharification of rice straw and hulls. Optimum conditions of pretreatment were AC 2.75%, TT 22.5 min, and SC 30 g/L.

Zhu et al. (2006a) enhanced the enzymatic digestibility of wheat straw by microwave-assisted alkaline pretreatment and optimized simultaneous saccharification and fermentation (SSF) of these pretreated feedstocks in a paper with 108 citations. They used cellulases of *Trichoderma reesei* and *Saccharomyces cerevisiae* YC-097 cells. Optimal SSF for the pretreated straw were 100 g/L substrate, 40°C, 15 mg [cellulase]/g [substrate], initial pH 5.3, and 72 h. Under these optimal conditions, the bioethanol concentration reached 34.3 g/L and the ethanol yield was 69.3%. The optimal SSF for the conventional alkali-pretreated wheat straw were 100 g/L substrate, 40°C, 20 mg [cellulase]/g [substrate], initial pH 5.3, and 96 h. Under these optimal conditions, the ethanol concentration and yield were 31.1 g/L and 64.8%, respectively. They concluded that the production of bioethanol from microwave-assisted pretreated wheat straw had lower enzyme loading and shorter reaction time and they could achieve a higher bioethanol concentration and yield compared to the conventional alkali-pretreated wheat straw.

TABLE 17.3
Microwave Pretreatment of Bagasse

No.	Papers	Biomass	Other Pts.	Parameters	Keywords	Lead Author	Affil.	Cits
1	Binod et al. (2012)	Sugarcane bagasse	Alkali (NaOH), acid (H_2SO_4), enzymes	Alkali and acid loadings, energy loading, treatment time	Microwave, pretreatment, enzymatic, saccharification, fermentable, sugar, sugarcane bagasse	Pandey, Ashok 7201771319	CSIR India	266
2	Chen et al. (2011)	Sugarcane bagasse	Acid (H_2SO_4)	Treatment time, temperature	Disruption, sugarcane bagasse, lignocellulosic, acid, pretreatment, microwave	Chen, Wei- Hsin 57200873137	Natl. Univ. Tainan Taiwan	228
3	Chen et al. (2012)	Sugarcane bagasse	Acid (H_2SO_4)	Acid content	Hydrolysis, sugarcane bagasse, pretreated, acid, microwave	Chen, Wei- Hsin 57200873137	Natl. Univ. Tainan Taiwan	162
4	Ooshima et al. (1984)	Rice straw and bagasse	NA	Biomass, type, temperature, treatment time	Microwave, cellulosic, enzymatic, hydrolysis	None	Osaka City Univ. Japan	129
5	Intanakul et al. (2003)	Rice straw, sugarcane bagasse	Enzymes, glycerol	Irradiation temperature and time	Enzymatic, hydrolysis, lignocellulosic, wastes, microwave, pretreatment	Kitchaiya, Prakob 6508093303	King Mongkut's Inst. Technol. Thailand	104
6	De Souza Moretti et al. (2014)	Sugarcane bagasse	Enzymes, acid (H_3PO_4), glycerol	Treatment time, glycerol content, acid pH	Pretreatment, sugarcane bagasse, microwaves, enzymatic, hydrolysis	De Souza Moretti, Marcia M. 55513072400	Sao Paulo State Univ. Brazil	100

NA, non-available; Other pts., other pretreatments.

Zhu et al. (2006b) enhanced the enzymatic digestibility of wheat straw by microwave-assisted alkaline pretreatment in a paper with 107 citations. They observed that a higher microwave power with a shorter pretreatment time and lower microwave power with a longer pretreatment time had the same effect on the weight loss and composition at the same energy consumption. The wheat straw had a weight loss of 48.4% and a composition of cellulose 79.6%, lignin 5.7%, and hemicellulose 7.8% after 25 min of microwave-assisted alkaline pretreatment at 700 W, compared with a weight loss of 44.7% and a composition of cellulose 73.5%, lignin 7.2%, and hemicellulose 11.2% after 60 min of conventional alkaline pretreatment. The microwave-assisted alkaline pretreatment removed more lignin and hemicellulose from wheat straw with a shorter pretreatment time than the conventional alkaline one. Finally, for the enzymatic hydrolysis of pretreated wheat straw (substrate concentration 50 g/L and enzyme loading 20 mg/g substrate), they found that the microwave-assisted alkaline pretreated wheat straw had a higher hydrolysis rate, reducing sugar concentration and glucose content in the hydrolysate than the conventional alkali-pretreated one.

Zhu et al. (2006c) enhanced the enzymatic digestibility of rice straw by microwave-assisted alkaline pretreatment in a paper with 94 citations. They observed that rice straw pretreated by combining microwave irradiation with alkali could increase the initial hydrolysis rate but the hydrolysis yield remained unchanged. When the enzyme solution was treated by microwave irradiation, the initial hydrolysis rate increased slightly, but the yield was decreased remarkably where the optimal hydrolysis conditions were temperature (45°C), pH (4.8), and enzyme loading (20 mg/g substrate). When intermittent microwave irradiation was used, the initial hydrolysis rate was greatly accelerated but the yield decreased slightly. The optimal hydrolysis conditions were a temperature of 50°C, pH 4.8, and enzyme loading 20 mg/g substrate.

Lu et al. (2011) enhanced the digestibility of rapeseed straw by microwave-assisted dilute acid (H_2SO_4) pretreatment in a paper with 92 citations. They observed the lowest energy consumption when solid loading and energy input were fixed at 50% (w/w) and 54 kJ (900 W for 1 min), respectively, and amounted to 5.5 and 10.9 kJ to produce 1 g of glucose after enzymatic hydrolysis and 1 g ethanol after fermentation, respectively. In general, 1 g of ethanol can produce about 30 kJ of energy, and therefore, the energy input for the pretreatment was only 35% of the energy output. This approach resulted in 92.9% of higher energy savings for producing 1 g of ethanol compared to the results of microwave pretreatments previously reported.

Gong et al. (2010) enhanced the enzymatic digestibility of rice straw by microwave-assisted acid (acetic acid, CH_3COOH; propionic acid, $C_3H_6O_2$) pretreatment in a paper with 91 citations. They observed that the sequence of influential factors was microwave intensity > solid-liquid ratio > acid concentration > microwave TT. Under optimal conditions (25% acid concentration, 1:15 solid–liquid ratio, 230 W microwave intensity, and 5 min TT), the removal ratios of lignin were 46.1% and 51.54%, and the sugar yields were 71.41% and 80.08% when acetic acid and propionic acid were used as solvents, respectively. The sugar yield was only 35.28% in the blank sample. The surface of rice straw changed substantially when microwave-assisted pretreatment was used. They concluded that microwave-assisted acid pretreatment was effective in improving the enzymatic hydrolysis sugar yield of rice straw.

Diaz et al. (2015) enhanced the digestibility of corn straw and rice husk by microwave-assisted pretreatment in alkaline glycerol in a paper with 88 citations. They observed that native and pretreated solids underwent enzyme hydrolysis using the extract obtained from the fermentation of *Myceliophthora heterothallica*. They obtained the highest saccharification yields for both feedstocks when the biomass was pretreated in alkaline glycerol with a more significant modification in the structure of corn straw subjected to this pretreatment. Lignin removals were 30% and 12% for corn straw and rice husk, respectively, and they obtained higher enzymatic hydrolysis yields with the extract of *M. heterothallica* where the enzymatic hydrolysis yields were improved by supplementing with β-glucosidase (Table 17.4).

TABLE 17.4
Microwave Pretreatment of Straw

No.	Papers	Biomass	Other Pts.	Parameters	Keywords	Lead Author	Affil.	Cits
1	Zhu et al. (2005)	Rice straw	Alkali, enzymes	Alkali loading, energy loading, treatment time, solid content, enzyme loading	Pretreatment, microwave, alkali, rice straw, enzymatic, hydrolysis	Zhu, Shendong 8919301300	Wuhan Inst. Chem. Technol. China	230
2	Ma et al. (2009)	Rice straw	Enzymes	Energy loading, solid content, and treatment time	Enzymatic, saccharification, rice straw, microwave, pretreatment	Wu, Yue-Jin 55531536300	Chin. Acad. Sci. China	224
3	Zhu et al. (2006d)	Rice straw	Alkali, acid/alkali, acid/alkali/ H2O2, enzymes	Pretreatment type, solid content, pH, enzyme loading	Microwave, pretreatment, enzymatic, hydrolysis, rice straw	Zhu, Shendong 8919301300	Wuhan Inst. Chem. Technol. China	138
4	Ooshima et al. (1984)	Rice straw and bagasse	NA	Biomass, type, temperature, treatment time	Microwave, cellulosic, enzymatic, hydrolysis	None	Osaka City Univ. Japan	129
5	Singh et al. (2011)	Rice straw and hulls	Alkali, enzymes	Alkali concentration, treatment time, solid content	Saccharification, rice straw, hull, microwave, alkali, pretreatment, lignocellulolytic, enzyme	Singh, Namita 55425779500 Bishnoi, Narsi R 6602910759	Guru Jambheshwar Univ. Sci. Technol. India	120
6	Zhu et al. (2006a)	Wheat straw	Alkali, enzymes	Solid content, temperature, enzyme content, pH, treatment time	Ethanol, microwave, alkali, pretreated, wheat straw	Zhu, Shengdong 8919301300	Wuhan Inst. Chem. Technol. China	108
7	Zhu et al. (2006b)	Wheat straw	Alkali, enzymes	Microwave power, pretreatment time, solid content, enzyme content	Microwave, alkali, pretreatment, wheat straw, enzymatic hydrolysis	Zhu, Shengdong 8919301300	Wuhan Inst. Chem. Technol. China	107
8	Intanakul et al. (2003)	Rice straw, sugarcane bagasse	Enzymes, glycerol	Irradiation temperature and time	Enzymatic, hydrolysis, lignocellulosic, wastes, microwave, pretreatment	Kitchaiya, Prakob 6508093303	King Mongkut's Inst. Technol. Thailand	104
9	Zhu et al. (2006c)	Rice straw	Alkali, enzymes	Temperature, pH, enzyme content	Microwave, enzymatic, hydrolysis, rice straw	Zhu, Shendong 8919301300	Wuhan Inst. Chem. Technol. China	94

(*Continued*)

TABLE 17.4 (*Continued*)
Microwave Pretreatment of Straw

No.	Papers	Biomass	Other Pts.	Parameters	Keywords	Lead Author	Affil.	Cits
10	Lu et al. (2011)	Rapeseed straw	Acid (H_2SO_4)	Energy loading, solid content	Microwave, pretreatment, rape straw, bioethanol production	Lu, Xuebin 14056612200	Tianjin Univ. China	92
11	Gong et al. (2010)	Rice straw	Enzymes, acid (CH_3COOH, $C_3H_6O_2$)	Acid content, solid–liquid ratio, microwave intensity, treatment time, catalyst content	Microwave, acid, pretreatment, enzymatic, hydrolysis, rice straw	Huang, Yudong 7501568993	Harbin Inst. Technol. China	91
12	Diaz et al. (2015)	Corn straw, rice husk	Alkaline, glycerol	Enzyme type, glycerol	Microwave, pretreatment, lignocellulosic, biomass, alkaline, glycerol, fermentable, sugars	Diaz, Ana Belen 15061063400	Univ. Cadiz Spain	88

NA, non-available; Other pts., other pretreatments.

17.3.3.3 Microwave Pretreatment of Other Agricultural Biomasses

There are three HCPs for the microwave pretreatment of other agricultural biomasses. Notes of Singh et al. (2011) and Diaz et al. (2015) were included in Section 17.3.3.2, while notes of Chen and Wan (2018) were included in Section 17.3.1 (Table 17.5).

17.3.4 Microwave Pretreatment of Wood

Liu et al. (2010) enhanced the digestibility of softwood by microwave pretreatment in aqueous glycerol containing a series of organic and inorganic acids (hydrochloric acid-HCl, malonic acid-C3H4O4, and phosphoric acid-H_3PO_4) with different pKa values in a paper with 90 citations. They observed that the pulp obtained by organosolvolysis with 0.1% HCl (pKa −6) at 180°C for 6 min gave the highest sugar yield of 53.1%, based on the weight of the original biomass. The pretreatment efficiency correlated linearly with the pKa of acids, with the exception of $C_3H_4O_4$ and H_3PO_4. Organosolvolysis with 1.0% of H_3PO_4 (pKa 2.15) gave a saccharification yield (50.6%) higher than that what was expected from its pKa, while the catalytic effect of $C_3H_4O_4$ (pKa 2.83) was negligible. They then observed the extensive exposure of crystalline and non-crystalline cellulose by glycerolysis with strong inorganic acids (Table 17.6).

TABLE 17.5
Microwave Pretreatment of Other Agricultural Wastes

No.	Papers	Biomass	Other Pts.	Parameters	Keywords	Lead Author	Affil.	Cits
1	Chen and Wan (2018)	Miscanthus, switchgrass, corn stover	Deep eutectic solvent-DES (ChCl:LA)	Biomass type	Fractionation, lignocellulosic, biomass, microwave, DES	Wan, Caixia 35772603400	Univ. Missouri USA	134
2	Singh et al. (2011)	Rice straw and hulls	Alkali, enzymes	Alkali concentration, treatment time, solid content	Saccharification, rice straw, hull, microwave, alkali, pretreatment, lignocellulolytic, enzyme	Singh, Namita 55425779500 Bishnoi, Narsi R 6602910759	Guru Jambheshwar Univ. Sci. Technol. India	120
3	Diaz et al. (2015)	Corn straw, rice husk	Alkaline, glycerol	Enzyme type, glycerol	Microwave, pretreatment, lignocellulosic, biomass, alkaline, glycerol, fermentable, sugars	Diaz, Ana Belen 15061063400	Univ. Cadiz Spain	88

Other pts., other pretreatments.

TABLE 17.6
Pretreatment of Wood

No.	Papers	Biomass	Other Pts.	Parameters	Keywords	Lead Author	Affil.	Cits
24	Liu et al. (2010)	Softwood	Acids (HCl, $C_3H_4O_4$, H_3PO_4), glycerol	Temperature, treatment time, pKa	Microwave, pretreatment, softwood, glycerol	Watanabe, Takashi 57125173500	Kyoto Univ. Japan	90

Other pts., other pretreatments.

17.4 DISCUSSION

17.4.1 Introduction

Crude oil–based gasoline fuels have been widely used in the transportation sector since the 1920s. However, there have been great public concerns over the adverse environmental and human impact of these fuels. Hence, biomass-based bioethanol fuels have increasingly been used in blending gasoline and petrodiesel fuels, in fuel cells, and in biochemical production in a biorefinery context. Research in the fields of microwave pretreatment of biomass has also intensified in recent years.

However, it is essential to develop efficient incentive structures for the primary stakeholders to enhance the research in this field. Although there have been a number of review papers on microwave pretreatment of biomass for bioethanol production, there has been no review of the most-cited 25 articles in this field.

Thus, this chapter presents a review of the most-cited 25 articles in this field. Then, it discusses the key findings of these highly influential papers and comments on future research priorities in this field.

As a first step for the search of the relevant literature, keywords were selected using the whole set of papers. The selected keyword list was optimized to obtain a representative sample of papers for the searched research field. This keyword list was provided in the Appendix for future replication studies.

As a second step, a sample dataset was used for this study. The first 25 articles with at least 88 citations each were selected for the review study. Key findings from each paper were taken from the abstracts of these papers and were discussed. Additionally, a number of brief conclusions were drawn and a number of relevant recommendations were made to enhance the future research landscape.

As per Table 17.7, there are three primary research fronts for these most-cited 25 papers: agricultural wastes, cellulose, and grass with 68%, 24%, and 12% of the reviewed papers, respectively. Another research front is the woody biomass with 4% of reviewed papers.

However, agricultural wastes, cellulose, and grass are over-represented in the reviewed papers compared to the population papers with 15%, 8%, and 8%, respectively. On the contrary, woody biomass is under-represented by 19%. Furthermore, a number of minor research fronts are not represented at all in the set of reviewed papers: food wastes, plants, biomass in general, starch, and other biomasses.

Straw, bagasse, and other biomasses such as husks and hulls, and corn stover appear in the set of reviewed papers in decreasing order. It is notable that papers related to straw, bagasse, and other agricultural biomasses are over-represented in the set of reviewed papers by 30%, 16%, and 4%, respectively. The research fronts of agricultural biomass in general, bunch, and corncobs are not at all represented in the set of reviewed papers.

The research front for the woody biomass includes papers related to wood, bamboo, sawdust, stem/trunk, and other woody biomasses. Algae, water hyacinth, and corn/sorghum are the plants studied in the population papers.

17.4.2 Microwave Pretreatment of Grass

Three papers related to microwave pretreatment of grass cover alkaline, DESs, acids, and enzymes as joint pretreatments (Chen and Wan, 2018; Hu and Wen, 2008; Orozco et al., 2007). They also cover other feedstocks such as corn stover and cellulose. The common finding from these papers is that microwave pretreatment in combination with other pretreatments was effective in disrupting recalcitrant structures of switchgrass. Thus, these combined pretreatments increased the bioethanol yield from switchgrass (McLaughlin and Kszos, 2005; Parrish and Fike, 2005).

TABLE 17.7
Microwave Pretreatment of Biomass: Research Fronts

No.	Research Fronts	Reviewed Papers (%)	Sample Papers (%)	Surplus
1	**Agricultural Wastes**	**68.0**	**53.1**	**14.9**
	Straw	48.0	17.5	30.5
	Agricultural waste in general	0.0	8.4	–8.4
	Bagasse	24.0	7.8	16.2
	Other agricultural biomasses	12.0	7.8	4.2
	Bunch	0.0	4.4	–4.4
	Corncobs	0.0	4.1	–4.1
	Corn stover	4.0	3.1	0.9
2	**Woody Biomass**	**4.0**	**22.8**	**–18.8**
	Wood	4.0	11.3	–7.3
	Bamboo	0.0	4.4	–4.4
	Sawdust	0.0	2.8	–2.8
	Stem/trunk	0.0	2.2	–2.2
	Other woody biomasses	0.0	2.2	–2.2
3	Cellulose	24.0	15.6	8.4
4	Food wastes	0.0	8.8	–8.8
5	**Plants**	**0.0**	**8.1**	**–8.1**
	Algae	0.0	5.0	–5
	Water hyacinth	0.0	1.3	–1.3
	Corn/sorghum	0.0	1.3	–1.3
6	Grass	12.0	5.6	6.4
7	Biomass in general	0.0	4.1	–4.1
8	Starch	0.0	3.4	14.9
9	Other biomasses	0.0	2.5	30.5
	Sample size	25	320	

Reviewed papers: 25 papers reviewed in this study. Papers: the sample of 320 population papers.

17.4.3 Microwave Pretreatment of Cellulose

Four papers related to microwave pretreatment of cellulose cover acids, ILs, and enzymes as joint pretreatments (Ha et al., 2011; Orozco et al., 2007; Wu et al., 2010; Zhang and Zhao, 2009). They also cover other feedstocks such as grass (Orozco et al., 2007). Additionally, Budarin et al. (2010) and Fan et al. (2013) determined the determinants of enhanced digestibility of cellulose by microwave pretreatment. The common finding from these papers is that microwave pretreatment in combination with other pretreatments or alone was effective in enhancing the digestibility of cellulose. Thus, microwave pretreatment increased the bioethanol yield from cellulose (Beguin and Aubert, 1994; Nechyporchuk et al., 2016).

17.4.4 Microwave Pretreatment of Agricultural Wastes

17.4.4.1 Microwave Pretreatment of Bagasse

Five papers related to microwave pretreatment of sugarcane and rice bagasse cover acids, alkali, and enzymes as joint pretreatments (Binod et al., 2012; Chen et al., 2011, 2012; de Souza Moretti et al., 2014; Intanakul et al., 2003). Additionally, Ooshima et al. (1984) determined the determinants of enhanced digestibility of bagasse by microwave pretreatment. The common finding from these papers is that microwave pretreatment in combination with other pretreatments or alone was effective in enhancing the digestibility of sugarcane and rice bagasse. Thus, microwave pretreatment increased the bioethanol yield from bagasse (Cardona et al., 2010; Pandey et al., 2000).

17.4.4.2 Microwave Pretreatment of Straw

Eleven papers related to microwave pretreatment of straw from rice, wheat, rapeseed, and corn cover the acids, alkali, glycerol, and enzymes as joint pretreatments (Diaz et al., 2015; Gong et al., 2010; Intanakul et al., 2003; Lu et al., 2011; Ma et al., 2009; Singh et al., 2011; Zhu et al., 2005, 2006a–d). Additionally, Ooshima et al. (1984) determined the determinants of enhanced digestibility of straw by microwave pretreatment. The common finding from these papers is that microwave pretreatment in combination with other pretreatments or alone was effective in enhancing the digestibility of straw from rice, wheat, rapeseed, and corn. Thus, microwave pretreatment increased the bioethanol yield from straw (Binod et al., 2010; Kaparaju et al., 2009).

17.4.4.3 Microwave Pretreatment of Other Agricultural Biomasses

Three papers related to microwave pretreatment of corn stover, rice hulls, and rice husks cover the alkali, glycerol, DESs, and enzymes as joint pretreatments (Chen and Wan, 2018; Diaz et al., 2015; Singh et al., 2011). The common finding from these papers is that microwave pretreatment in combination with other pretreatments was effective in enhancing the digestibility of these agricultural wastes. Thus, microwave pretreatment increased the bioethanol yield from these wastes (Copur et al., 2013; Dagnino et al., 2013).

17.4.5 Microwave Pretreatment of Wood

There is only one paper related to microwave pretreatment of the woody biomass covering acids and glycerol as joint pretreatments (Liu et al., 2010). The key finding from this paper is that microwave pretreatment in combination with other pretreatments was effective in enhancing the digestibility of wood. Thus, microwave pretreatment increased the bioethanol yield from wood (Gregg and Sadler, 1996; Hu et al., 2008).

17.5 CONCLUSION AND FUTURE RESEARCH

Brief information about the key research fronts covered by the 25 most-cited papers with at least 88 citations each is given under four headings: Grass, cellulose, agricultural wastes, and wood.

The usual characteristic of these highly cited papers is that microwave pretreatments are often used in combination with other pretreatments such as acids, alkali, enzymes, glycerol, ILs, and DESs. In this way, microwave pretreatment is effective in disrupting the structures of these feedstocks.

The key findings on these research fronts should be read in light of the increasing public concerns about climate change, greenhouse gas emissions, and global warming as these concerns have been certainly behind the boom in the research on bioethanol production as an alternative to crude oil–based gasoline and diesel fuels in the last decades.

As per the data in Table 17.7, agricultural wastes, woody biomass, and cellulose are the most prolific feedstocks to study the effect of microwave pretreatment either stand-alone or in combination with other pretreatments. The other feedstocks are food wastes such as potato peel wastes, plants such as algae, grass, starch, and other biomasses such as lignin and cellobiose. This is in line with the research on bioethanol production and utilization focusing more on wastes and algae as non-food feedstocks.

These studies emphasize the importance of proper incentive structures for the efficient development and application of microwave pretreatment to enhance the digestibility of bioethanol feedstocks to increase the bioethanol yield after the hydrolysis and fermentation of these feedstocks and hydrolysates, respectively, in light of North's institutional framework (North, 1991). In this context, the major producers and users of bioethanol fuels such as China and to a lesser extent the United States, Europe, Brazil, and Japan had developed strong incentive structures for the effective development and application of microwave pretreatment for bioethanol and sugar production.

It is recommended that such review studies be performed for the major research fronts for microwave pretreatment of the biomass as well as other research fronts related to other pretreatments complementing this chapter.

ACKNOWLEDGMENTS

The contribution of highly cited researchers in the field of microwave pretreatment of biomass has been gratefully acknowledged.

APPENDIX: THE KEYWORD SET

[TITLE (microwave* OR "micro wave*") AND TITLE (ethanol OR bioethanol OR pretreat* OR "pre-treat*" OR *hydrolysis OR saccharification OR ferment* OR hydrolytic OR "enzymatic digestibility" OR disruption OR fractionation OR decomposition OR dissolution OR delignification OR depolymerization OR sugar OR sugars OR *oligosaccharides OR glucose OR deconstruction OR xylose) AND TITLE (wastes OR waste OR *grass OR biomass* OR lignocellulos* OR bagasse* OR cellulos* OR hemicellulos* OR *wood OR stover* OR miscanthus OR hull* OR husk* OR straw* OR hyacinth OR sorghum OR starch OR sawdust OR residue* OR trunk* OR corncob* OR stem OR pseudostem* OR bluestem* OR *stalk OR tops OR reed OR arundo OR cellobiose OR bran OR bamboo* OR eucalyptus OR cypress OR corn OR palm OR stillage* OR xylan OR pinus OR sawmill* OR "corn cob*" OR ulva OR *algae OR seaweed* OR biowaste* OR shoots OR bunch* OR *algal OR foliage OR poplar* OR rhizome OR agave OR peel* OR coffee* OR molass* OR spruce OR prosopis OR grain* OR birch OR triticale OR arborera OR saccharum OR bioresource* OR eucheuma OR laminaria OR bode OR woody OR taro OR willow OR szarvasi OR eichhornia OR sago) AND NOT TITLE (*furfural OR *hydrogen OR sludge OR carbonization OR digestion OR digester* OR *gas OR cell OR nano* OR pyroly* OR *char OR methane OR torrefaction OR "activated carbon*" OR ferul* OR earth OR arsen* OR spectromet* OR pet OR extract* OR ions OR h2 OR chelat* OR ester* OR polycarbonate OR gas* OR protein* OR sensor* OR phenolic* OR *ethylene OR chromat* OR synthesis OR absorber* OR reinforced OR combustion OR acrylamide* OR aggregate* OR aluminium OR surfactant* OR moisture OR lipid* OR hmf OR olive OR popping OR lactic OR "bio-oil*" OR *phenolic OR dye* OR acetylation OR soil* OR *butanol OR wool OR dechlor* OR drying OR dried OR modificat* OR sheep OR papaya OR pectin OR transester* OR ash* OR impregn* OR radioact* OR tissue* OR *viral OR *diesel OR properties OR nuclear OR chitosan OR inciner* OR "essential oil*" OR analysis OR fabricat* OR dxn* OR butyric OR fire* OR belt OR tar OR decanter OR leach* OR syringol OR biochem* OR isomer* OR pellet* OR dietary OR antioxidant*) AND [LIMIT-TO (SRCTYPE, "j") OR LIMIT-TO (SRCTYPE, "k") OR LIMIT-TO (SRCTYPE, "b")] AND (LIMIT-TO (DOCTYPE, "ar") OR LIMIT-TO (DOCTYPE, "cp") OR LIMIT-TO (DOCTYPE, "re") OR LIMIT-TO (DOCTYPE, "ch") OR LIMIT-TO (DOCTYPE, "no")] AND (LIMIT-TO (LANGUAGE,"English")]

REFERENCES

Angelici, C., B. M. Weckhuysen and P. C. A. Bruijnincx. 2013. Chemocatalytic conversion of ethanol into butadiene and other bulk chemicals. *ChemSusChem* 6:1595–1614.

Antolini, E. 2007. Catalysts for direct ethanol fuel cells. *Journal of Power Sources* 170:1–12.

Antolini, E. 2009. Palladium in fuel cell catalysis. *Energy and Environmental Science* 2:915–931.

Beguin, P. and J. P. Aubert. 1994. The biological degradation of cellulose. *FEMS Microbiology Reviews* 13:25–58.

Binod, P., K. Satyanagalakshmi and R. Sindhu, et al. 2012. Short duration microwave assisted pretreatment enhances the enzymatic saccharification and fermentable sugar yield from sugarcane bagasse. *Renewable Energy* 37:109–116.

Binod, P., R. Sindhu and R. R. Singhania, et al. 2010. Bioethanol production from rice straw: An overview. *Bioresource Technology* 101:4767–4774.

Budarin, V. L., J. H. Clark, B. A. Lanigan, P. Shuttleworth and D. J. Macquarrie. 2010. Microwave assisted decomposition of cellulose: A new thermochemical route for biomass exploitation. *Bioresource Technology* 101:3776–3779.

Burnham, J. F. 2006. Scopus database: A review. *Biomedical Digital Libraries* 3:1–8.

Cardona, C. A., J. A. Quintero and I. C. Paz. 2010. Production of bioethanol from sugarcane bagasse: Status and perspectives. *Bioresource Technology* 101:4754–4766.

Chen, W. H., S. C. Ye and H. K. Sheen. 2012. Hydrolysis characteristics of sugarcane bagasse pretreated by dilute acid solution in a microwave irradiation environment. *Applied Energy* 93:237–244.

Chen, W. H., Y. J. Tu and H. K. Sheen. 2011. Disruption of sugarcane bagasse lignocellulosic structure by means of dilute sulfuric acid pretreatment with microwave- assisted heating. *Applied Energy* 88:2726–2734.

Chen, Z. and C. Wan. 2018. Ultrafast fractionation of lignocellulosic biomass by microwave- assisted deep eutectic solvent pretreatment. *Bioresource Technology* 250:532–537.

Copur, Y., A. Tozluoglu and M. Ozkan. 2013. Evaluating pretreatment techniques for converting hazelnut husks to bioethanol. *Bioresource Technology* 129:182–190.

Dagnino, E. P., E. R. Chamorro, S. D. Romano, F. E. Felissia and M. C. Area. 2013. Optimization of the acid pretreatment of rice hulls to obtain fermentable sugars for bioethanol production. *Industrial Crops and Products* 42:363–368.

De Souza Moretti, M. M., D. A. Bocchini-Martins and C. da Costa Correria Nunes, et al. 2014. Pretreatment of sugarcane bagasse with microwaves irradiation and its effects on the structure and on enzymatic hydrolysis. *Applied Energy* 122:189–195.

Diaz, A. B., de Souza Moretti, M. M. and C. Bezerra-Bussoli, et al. 2015. Evaluation of microwave-assisted pretreatment of lignocellulosic biomass immersed in alkaline glycerol for fermentable sugars production. *Bioresource Technology* 185:316–323.

Fan, J., M. de Bruyn and V. L. Budarin, et al. 2013. Direct microwave-assisted hydrothermal depolymerization of cellulose. *Journal of the American Chemical Society* 135:12728–12731.

Fernando, S., S. Adhikari, C. Chandrapal and M. Murali. 2006. Biorefineries: Current status, challenges, and future direction. *Energy & Fuels* 20:1727–1737.

Gong, G., D. Liu and Y. Huang. 2010. Microwave-assisted organic acid pretreatment for enzymatic hydrolysis of rice straw. *Biosystems Engineering* 107:67–73.

Gregg, D. J. and J. N. Saddler. 1996. Factors affecting cellulose hydrolysis and the potential of enzyme recycle to enhance the efficiency of an integrated wood to ethanol process. *Biotechnology and Bioengineering* 51:375–383.

Ha, S. H., N. L. Mai, G. An and Y. M. Koo. 2011. Microwave-assisted pretreatment of cellulose in ionic liquid for accelerated enzymatic hydrolysis. *Bioresource Technology* 102:1214–1219.

Hahn-Hagerdal, B., M. Galbe, M. F. Gorwa-Grauslund, G. Liden and G. Zacchi. 2006. Bio- ethanol – The fuel of tomorrow from the residues of today. *Trends in Biotechnology* 24:549–556.

Hill, J., E. Nelson, D. Tilman, S. Polasky and D. Tiffany. 2006. Environmental, economic, and energetic costs and benefits of biodiesel and ethanol biofuels. *Proceedings of the National Academy of Sciences of the United States of America* 103:11206–11210.

Hill, J., S. Polasky and E. Nelson, et al. 2009. Climate change and health costs of air emissions from biofuels and gasoline. *Proceedings of the National Academy of Sciences of the United States of America* 106:2077–2082.

Hsieh, W. D., R. H. Chen, T. L. Wu and T. H. Lin. 2002. Engine performance and pollutant emission of an SI engine using ethanol-gasoline blended fuels. *Atmospheric Environment* 36:403–410.

Hu, G., J. A. Heitmann and O. J. Rojas. 2008. Feedstock pretreatment strategies for producing ethanol from wood, bark, and forest residues. *BioResources* 3:270–294.

Hu, Z. and Z. Wen. 2008. Enhancing enzymatic digestibility of switchgrass by microwave- assisted alkali pretreatment. *Biochemical Engineering Journal* 38:369–378.

Huang, H. J., S. Ramaswamy, U. W. Tschirner and B. V. Ramarao. 2008. A review of separation technologies in current and future biorefineries. *Separation and Purification Technology* 62:1–21.

Intanakul, P., M. Krairiksh and P. Kitchaiya. 2003. Enhancement of enzymatic hydrolysis of lignocellulosic wastes by microwave pretreatment under atmospheric pressure. *Journal of Wood Chemistry and Technology* 23:217–225.

Kaparaju, P., M. Serrano, A. B. Thomsen, P. Kongjan and I. Angelidaki. 2009. Bioethanol, biohydrogen and biogas production from wheat straw in a biorefinery concept. *Bioresource Technology* 100:2562–2568.

Konur, O. 2000. Creating enforceable civil rights for disabled students in higher education: An institutional theory perspective. *Disability & Society* 15:1041–1063.

Konur, O. 2002a. Access to nursing education by disabled students: Rights and duties of nursing programs. *Nurse Education Today* 22:364–374.

Konur, O. 2002b. Assessment of disabled students in higher education: Current public policy issues. *Assessment and Evaluation in Higher Education* 27:131–52.

Konur, O. 2002c. Access to employment by disabled people in the UK: Is the Disability Discrimination Act working? *International Journal of Discrimination and the Law* 5:247–279.

Konur, O. 2006a. Participation of children with dyslexia in compulsory education: Current public policy issues. *Dyslexia* 12:51–67.

Konur, O. 2006b. Teaching disabled students in Higher Education. *Teaching in Higher Education* 11:351–363.

Konur, O. 2007a. A judicial outcome analysis of the Disability Discrimination Act: A windfall for the employers? *Disability & Society* 22:187–204.

Konur, O. 2007b. Computer-assisted teaching and assessment of disabled students in higher education: The interface between academic standards and disability rights. *Journal of Computer Assisted Learning* 23:207–219.

Konur, O. 2012. The evaluation of the research on the bioethanol: A scientometric approach. *Energy Education Science and Technology Part A: Energy Science and Research* 28:1051–1064.

Konur, O. 2015. Current state of research on algal bioethanol. In *Marine Bioenergy: Trends and Developments*, Eds. S. K. Kim and C. G. Lee, pp. 217–244. Boca Raton, FL: CRC Press.

Konur, O. 2020a. The scientometric analysis of the research on the bioethanol production from green macroalgae. In *Handbook of Algal Science, Technology and Medicine*, Ed. O. Konur, pp. 385–401. London: Academic Press.

Kostas, E. T., D. Beneroso and J. P. Robinson, J.P. 2017. The application of microwave heating in bioenergy: A review on the microwave pre-treatment and upgrading technologies for biomass. *Renewable and Sustainable Energy Reviews* 77:12–27.

Li, H., Y. Qu, Y. Yang, S. Chang and J. Xu. 2016. Microwave irradiation – A green and efficient way to pretreat biomass. *Bioresource Technology* 199:34–41.

Lin, Y. and S. Tanaka. 2006. Ethanol fermentation from biomass resources: Current state and prospects. *Applied Microbiology and Biotechnology* 69:627–642.

Liu, J., R. Takada and S. Karita, et al. 2010. Microwave-assisted pretreatment of recalcitrant softwood in aqueous glycerol. *Bioresource Technology* 101:9355–9360.

Lu, X., B. Xi, Y. Zhang and I. Angelidaki. 2011. Microwave pretreatment of rape straw for bioethanol production: Focus on energy efficiency. *Bioresource Technology* 102:7937–7940.

Ma, H., W. W. Liu, X. Chen, Y. J. Wu and Z. L. Yu. 2009. Enhanced enzymatic saccharification of rice straw by microwave pretreatment. *Bioresource Technology* 100: 1279–1284.

Ma, X., L. Sun and C. Song. 2002. A new approach to deep desulfurization of gasoline, diesel fuel and jet fuel by selective adsorption for ultra-clean fuels and for fuel cell applications. *Catalysis Today* 77:107–116.

McLaughlin, S. B. and L. A. Kszos. 2005. Development of switchgrass (*Panicum virgatum*) as a bioenergy feedstock in the United States. *Biomass and Bioenergy* 28:515–535.

Morschbacker, A. 2009. Bio-ethanol based ethylene. *Polymer Reviews* 49:79–84.

Najafi, G., B. Ghobadian and T. Tavakoli, et al. 2009. Performance and exhaust emissions of a gasoline engine with ethanol blended gasoline fuels using artificial neural network. *Applied Energy* 86:630–639.

Nechyporchuk, O., M. N. Belgacem and J. Bras. 2016. Production of cellulose nanofibrils: A review of recent advances. *Industrial Crops and Products* 93:2–25.

Newman, P. W. G. and J. R. Kenworthy. 1989. Gasoline consumption and cities: A comparison of U.S. cities with a global survey. *Journal of the American Planning Association* 55:24–37.

North, D. C. 1991. Institutions. *Journal of Economic Perspectives* 5:97–112.

Olsson, L. and B. Hahn-Hagerdal. 1996. Fermentation of lignocellulosic hydrolysates for ethanol production. *Enzyme and Microbial Technology* 18:312–331.

Ooshima, H., K. Aso, Y. Harano and T. Yamamoto, T. 1984. Microwave treatment of cellulosic materials for their enzymatic hydrolysis. *Biotechnology Letters* 6:289–294.

Orozco, A., M. Ahmad, D. Rooney and G. Walker. 2007. Dilute acid hydrolysis of cellulose and cellulosic bio-waste using a microwave reactor system. *Process Safety and Environmental Protection* 85:446–449.

Pandey, A., C. R. Soccol, P. Nigam and V. T. Soccol. 2000. Biotechnological potential of agro-industrial residues. I: Sugarcane bagasse. *Bioresource Technology* 74:69–80.

Parrish, D. J. and J. H. Fike. 2005. The biology and agronomy of switchgrass for biofuels. *Critical Reviews in Plant Sciences* 24:423–459.

Puligundla, P., S. E. Oh and C. Mok. 2016. Microwave-assisted pretreatment technologies for the conversion of lignocellulosic biomass to sugars and ethanol: A review. *Carbon Letters* 17:1–10.

Sanchez, O. J. and C. A. Cardona. 2008. Trends in biotechnological production of fuel ethanol from different feedstocks. *Bioresource Technology* 99:5270–5295.

Singh, A., S. Tuteja, N. Singh and N. R. Bishnoi. 2011. Enhanced saccharification of rice straw and hull by microwave-alkali pretreatment and lignocellulolytic enzyme production. *Bioresource Technology* 102:1773–1782.

Sun, Y. and J. Cheng. 2002. Hydrolysis of lignocellulosic materials for ethanol production: A review. *Bioresource Technology* 83:1–11.

Taherzadeh, M. J. and K. Karimi. 2007. Enzyme-based hydrolysis processes for ethanol from lignocellulosic materials: A review. *Bioresources* 2:707–738

Taherzadeh, M. J. and K. Karimi. 2008. Pretreatment of lignocellulosic wastes to improve ethanol and biogas production: A review. *International Journal of Molecular Sciences* 9:1621–1651.

Wu, Y., Z. Fu and D. Yin, et al. 2010. Microwave-assisted hydrolysis of crystalline cellulose catalyzed by biomass char sulfonic acids. *Green Chemistry* 12:696–970.

Yang, B. and C. E. Wyman. 2008. Pretreatment: The key to unlocking low-cost cellulosic ethanol. *Biofuels Bioproducts & Biorefining – BIOFPR* 2:26–40.

Zhang, Z. and Z. K. Zhao. 2009. Solid acid and microwave-assisted hydrolysis of cellulose in ionic liquid. *Carbohydrate Research* 344:2069–2072.

Zhu, S., Y. Wu and Z. Yu, et al. 2006a. Production of ethanol from microwave-assisted alkali pretreated wheat straw. *Process Biochemistry* 41:869–873.

Zhu, S., Y. Wu and Z. Yu, et al. 2006c. The effect of microwave irradiation on enzymatic hydrolysis of rice straw. *Bioresource Technology* 97:1964–1968.

Zhu, S., Y. Wu and Z. Yu, et al. 2006d. Comparison of three microwave/chemical pretreatment processes for enzymatic hydrolysis of rice straw. *Biosystems Engineering* 93:279–283.

Zhu, S., Y. Wu, and Z. Yu, et al. 2006b. Microwave-assisted alkali pre-treatment of wheat straw and its enzymatic hydrolysis. *Biosystems Engineering* 94:437–442.

Zhu, S., Y. Wu, Z. Yu, J. Liao and Y. Zhang. 2005. Pretreatment by microwave/alkali of rice straw and its enzymic hydrolysis. *Process Biochemistry* 40:3082–3086.

18 Ultrasound-Assisted Pretreatment of Lignocellulosic Biomass for Bioethanol Production

Dileswar Pradhan
Technological University Dublin

Bahiru Tsegaye
Wolkite University

Sneha S. Mathew
Mahatma Gandhi University

Swarna Jaiswal and Amit K. Jaiswal
Technological University Dublin

18.1 INTRODUCTION

Protection of the environment and energy security are two major challenges in today's world, both of which have increased demand for a renewable and sustainable energy source. Besides, due to the rapid growth in the consumption rate of nonrenewable fuels, it is expected that fossil fuel reserves will be depleted over the next 40–50 years (Zabed et al., 2017). Thus, for combating the current environmental challenges and global energy crisis, bioethanol fuels have emerged as a sustainable and renewable liquid fuels (Aditiya et al., 2016). There are numerous factors that allow its utilization as an alternative fuel: (i) renewable bioresources are utilized for its production instead of nonrenewable petroleum products, (ii) its toxicity is lower than that of other alcoholic fuels, (iii) by-products formed due to incomplete oxidation of ethanol (e.g., acetaldehyde and acetic acid) have lower toxicity than by-products generated during the production of other fuel alcohols (Vohra et al., 2014).

Lignocellulosic biomass (LCB), starch, sugars, and algae are the most common renewable feedstock used for bioethanol production (Zabed et al., 2017). Sugar- and starch-based materials are first generation feedstocks, whereas LCB and algae are second and third generation feedstocks, respectively (Satari and Jaiswal, 2021). However, LCBs are the most utilized renewable raw materials for the production of bioethanol fuels that do not compete with animal feed or food production. Besides, LCBs are generated in large quantities from various inexpensive resources, including agricultural waste, wood, industrial wastes, and municipal wastes (Limayem and Ricke, 2012). Usually, these lignocellulosic materials are converted into bioethanol fuels through three primary processes: pretreatment to obtain individual components of LCB (cellulose, hemicellulose, and lignin), enzymatic hydrolysis to convert cellulose into fermentable sugars, and finally, microbial fermentation to obtain bioethanol fuels (Zhuang et al., 2016). A process flow chart for the production of bioethanol fuels from LCB is depicted in Figure 18.1 (Vasic et al., 2021).

DOI: 10.1201/9781003226536-24

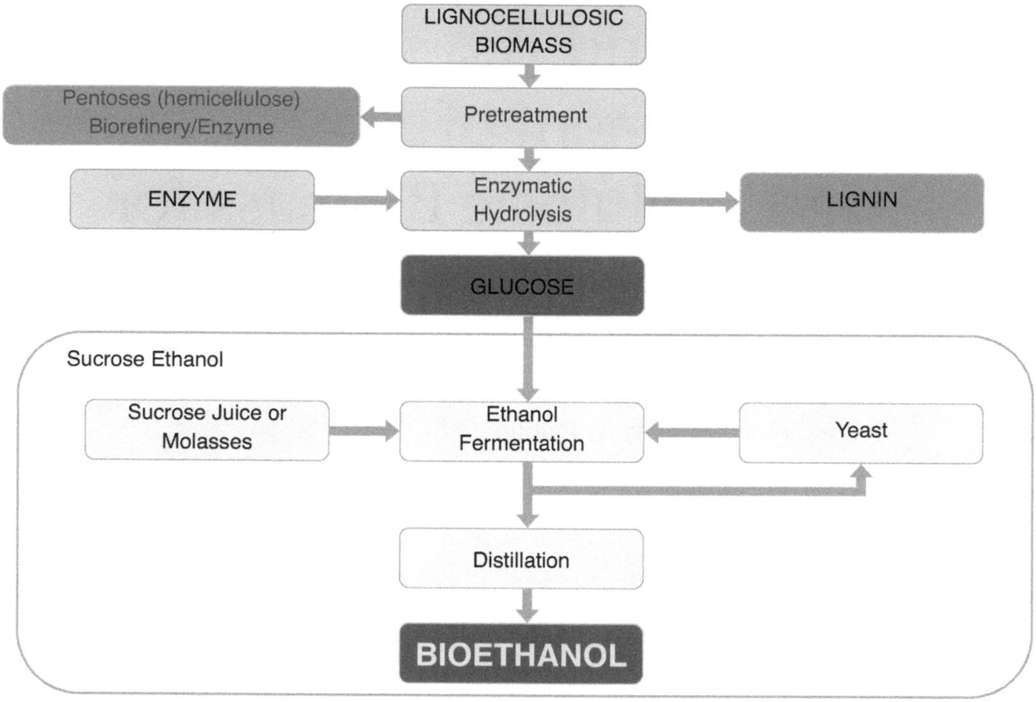

FIGURE 18.1 Process flow chart for production of bioethanol from LCB (Vasic et al., 2021). Publisher: MDPI, Open Access, http://creativecommons.org/licenses/by/4.0/. LCB, lignocellulosic biomass.

The pretreatment of LCB is the most expensive among all the processes involved in bioethanol production, accounting for 20%–48% of the overall operational costs (Ong and Wu, 2020). Nevertheless, biomass pretreatment is a necessary and crucial step in increasing the effectiveness of the following enzymatic hydrolysis process (Zhuang et al., 2016). To date, several chemical, physical, biological, and physicochemical pretreatment technologies have been investigated to improve cellulose accessibility and digestibility and, subsequently, bioethanol production (Bundhoo and Mohee, 2018). However, most of the pretreatment methods are not environmentally friendly and require a significant amount of energy due to the use of traditional heating methodologies and harsh chemicals (Hassan et al., 2018). Nonetheless, among all the pretreatment techniques investigated, ultrasound irradiation or sonication has gained significant interest as a green pretreatment strategy for improving bioethanol production from lignocellulosic feedstock (Bundhoo and Mohee, 2018).

The application of ultrasound technology in biomass valorization includes the pretreatment of biomass, degradation of lignin, conversion of cellulose, and enhancing sugar synthesis via biochemical pathways. When ultrasonic energy is utilized for the pretreatment of biomass, the main observed effects are related to the improvement in convective forces, acoustic cavitation, and the formation of highly reactive radicals (Flores et al., 2021). The cavitation forces result in structural disintegration of LCB, leading to enhanced bioethanol yield (Karimi et al., 2014). Therefore, over the years, ultrasound technology has been extensively utilized as an alternative technique for improving the efficiency of the pretreatment process. Thus, the current chapter intends to provide a comprehensive overview of the application of ultrasound technology in the pretreatment of LCB for bioethanol production. In addition, the application of ultrasonication in enzymatic hydrolysis and fermentation process intensification for bioethanol production has also been discussed.

18.2 PRINCIPLE OF ULTRASONICATION

Based on the frequency range, sound waves can be classified into three categories: ultrasound, human audible sound, and infrasound (Karimi et al., 2014). The application of sound waves with a frequency range greater than the upper limit of human hearing is referred to as ultrasound (Yusaf and Al-Juboori, 2014). Generally, the infrasound frequency is less than 20 Hz, and the frequency of human audible sound is between 20 Hz and 20 kHz. On the other hand, ultrasonic waves typically have a frequency range of 20 kHz–500 MHz (Karimi et al., 2014). By tuning the frequency, ultrasound can be utilized in various processes involved in the valorization of LCB.

Ultrasonic processing can be divided into two types: high-power low-frequency and low-power high-frequency ultrasounds. Ultrasound with low power and high frequency is a non-destructive method that can be utilized for determining various physicochemical parameters, such as physical condition, structure, composition, and flow of a material. The frequency of this form of ultrasound spans from 100 kHz to 1 MHz, and the power is generally less than 1 W/cm^2. On the other hand, ultrasound with high power and low frequency involves the chemical, physical, and mechanical changes of the material through cavitation. This form of ultrasound uses frequencies in the range of 16–100 kHz and power in the range of 10–1,000 W/cm^2 (Soria and Villamiel, 2010). Generally, high-power low-frequency ultrasound has been extensively used in various LCB valorization processes, such as pretreatment, enzyme hydrolysis, and fermentation.

18.3 MECHANISM OF ULTRASONIC ACTION

The application of ultrasound waves in a liquid medium can cause various physical effects such as heating, acoustic streaming, acoustic cavitation, and nebulization, and chemical effects such as the formation of radicals. A schematic depiction of the aforementioned phenomena is illustrated in Figure 18.2. Acoustic cavitation occurs when ultrasonic waves propagate in a liquid media with a specified frequency and energy, causing compression and rarefaction cycles to produce microbubbles containing liquid vapor and gas (Flores et al., 2021). The collapse of cavitation bubbles in liquid media can generate thermal, mechanical, and chemical effects. Mechanical methods involve a number of techniques that include collapse pressure, turbulence, and shear stress. However chemical effects include the generation of free radicals, with the use of oxidizing agents such as chlorine and ozone (Yusaf and Al-Juboori, 2014).

The bubbles produced during ultrasonic treatment are classified into two types (stable cavitation bubbles and internal cavitation bubbles) based on their structure. Stable cavitation bubbles are defined as large bubble clouds with a balanced size formed during pressure cycles. On the other hand, internal cavitation bubbles are smaller in size, which collapse and disintegrate rapidly (Majid et al., 2015). The collapse of cavitation bubbles causes energy buildup in hot areas, which leads to the generation of severe pressures (1,000 atm) and temperatures (5,000 K) in the cavitation zone. Furthermore, the high pressure and temperature produce extremely high shear energy waves and turbulence in the cavitation zone (Soria and Villamiel, 2010).

Ultrasound is a frequency-dependent technology in which locally produced alternating positive and negative pressures cause material expansion or compression, leading to severe cell damage. In addition, ultrasound causes hydrolysis of water inside the oscillating bubbles, resulting in the formation of H^+ and OH^- free radicals that can be captured in some chemical reactions (Majid et al., 2015). By disrupting glycosidic bonds, such radicals accelerate the disintegration of lignocellulosic materials during ultrasonic treatment. Under the influence of ultrasound technology at optimal operating conditions, the overall phenomenon ultimately facilitates or improves biochemical reactions of LCB (Haldar and Purkait, 2021).

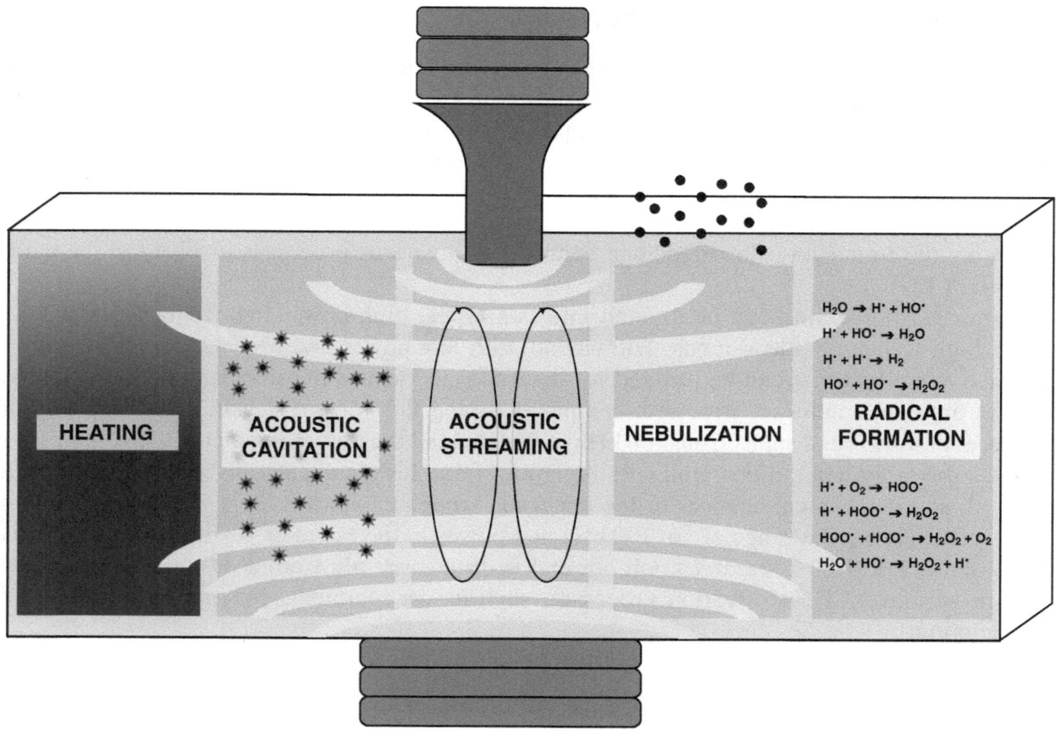

FIGURE 18.2 Schematic representation of the main physicochemical phenomena observed when a liquid medium is directly or indirectly sonicated (Flores et al., 2021). Publisher: Elsevier, Open Access, https://creativecommons.org/licenses/by-nc-nd/4.0/. LCB, lignocellulosic biomass.

18.4 FACTORS INFLUENCING THE ULTRASONIC PRETREATMENT OF LCB

There are several parameters that influence the ultrasound-assisted pretreatment process of LCB. These parameters include the type of ultrasonic processor, frequency, temperature, solvents, period of sonication treatment, sonication intensity, and amplitude. Depending on the type of biomass, these operating process factors need to be optimized to achieve a higher pretreatment efficiency.

18.4.1 Ultrasonic Processor

An ultrasonic processor is primarily composed of three major components: a reactor, a transducer, and an ultrasonic frequency generator (Bussemaker and Zhang, 2013). The design of the transducer and its arrangement, the reactor geometry, and the type of reactors (e.g., tube, flat plate, probe, or bath) influence the efficiency of the ultrasonic processor in biomass pretreatment. The bath type is an indirect irradiation-based sonochemical reactor, whereas the probe type is a direct immersion reactor that produces higher cavitational intensity than the bath type (Subhedar and Gogate, 2013). However, both the probe- and bath-type ultrasonicators are extensively utilized for the pretreatment of LCB in laboratories, despite their low energy efficiency and inhomogeneous acoustic field distribution (Luo et al., 2014).

For the pretreatment of LCB at a large scale, reactors fitted with multiple transducers would be necessary. For direct-immersion-type reactors, it is recommended that a flow loop is formed utilizing many transducers submerged in the loop, which should be in contact with the flowing solution. On the other hand, flow cells with transducers connected to the reactor's bottom or walls can be fabricated for indirect contact-type reactors (Subhedar and Gogate, 2013). Nevertheless, an appropriately designed ultrasonic reactor for an industrial- or laboratory-scale operation should provide a homogenous energy field for LCB, with a proper energy level for reactant activation (Luo et al., 2014).

18.4.2 Frequency

The ultrasonic frequency significantly affects the cavitation process because of its influence on the critical size of the cavitation bubble (Karimi et al., 2014). The change in the duration time of the acoustic cycle, which is long at low frequencies (e.g., 50 s for 20 kHz) and short at comparatively high frequencies (e.g., 2 s for 500 kHz), affects the behavior of bubble cavitation (Luo et al., 2014). The cavitational collapses are very violent because the bubbles have a longer time to grow at lower frequencies (Bussemaker and Zhang, 2013). In contrast, cavitation is less violent at high frequencies. However, there are more cavitational events at high frequency and, therefore, more chances for free radicals to form (Karimi et al., 2014).

The frequency of ultrasound used for a specific application is selected based on the nature of the process. For operations requiring severe physical effects such as polymer breakdown and cell breakage, an ultrasonic frequency range of 10–100 kHz is used. In addition, this frequency range can also be utilized for non-microbial and non-enzymatic ultrasound treatment of LCB. Nevertheless, in comparison to less-recalcitrant starchy biomass, the treatment of recalcitrant biomass including LCB needs more energetic ultrasound (Rehman et al., 2012).

18.4.3 Processing Temperature

Ultrasonic energy is eventually dissipated as heat within the suspension. As a result, when the operating conditions are not controlled, the temperature of the suspension normally increases. For instance, when biomass is pretreated with an alkaline or acidic medium, a higher temperature enables a faster reaction rate. Thus, ultrasound could be utilized in combination with chemical methods to reduce the pretreatment time as well as achieve a high pretreatment efficiency. However, when microbial action or enzymatic catalysis is involved in a process, the controlling of temperature is generally required to avoid biocatalyst deactivation (Rehman et al., 2012).

A high temperature of solvent causes disintegration of strong solute–matrix interactions such as dipole attractions, van der Waals forces, and hydrogen bonding among the active sites on the biomass matrix and solute molecules, resulting in rapid diffusion rates. On the other hand, better cavitation is produced at low temperatures when the system generator's ultrasonic power remains constant. This might be because when the temperature rises, the solvent's vapor pressure rises as well, filling the cavitation bubbles with more solvent vapor, making the collapse less violent, and minimizing the effects of ultrasonication. As a result, in order to reap the benefits, an optimal temperature must be determined for ultrasonic pretreatment of biomass (Subhedar and Gogate, 2013).

18.4.4 Solvent

The ultrasonic pretreatment of LCB is affected by the properties of the liquid in which it is suspended. Generally, the aqueous solution is used to mix LCB for pretreatment under ultrasound; however, ionic liquid (IL) or organic solvents are also utilized (Rehman et al., 2012). The chemical and physical characteristics of the solvents have a significant impact on acoustic propagation and cavitation. The chemical characteristics include the chemical activities related to radical formation as well as the composition in entirely miscible binary/ternary organic mixtures, surfactant solutions, or electrolyte solutions. On the other hand, the physical characteristics include the surface tension, volatility, and viscosity of single- and multi-component solvents (Luo et al., 2014).

Hydrophobic solutes have an impact on the effects of ultrasound in an aqueous medium. The solutes congregate at the interface of the bubble where they can either lower the highest temperature reached during the collapse of the bubbles or play a role as radical scavengers in the hot zone around the bubble, suppressing sonoluminescence. However, the solvent's volatility influences the ultrasonic effects in organic solvents. This is because the volatility of organic solvents is determined by vapor pressure, which controls the ease of evaporation into the cavity. The cavitational collapse intensity diminishes as volatility increases, which lowers the reaction rates and maximum temperatures. On the other hand, ultrasonic effects can still be noticed in ILs, although they have a low vapor pressure. The heated zone surrounding the interface of the bubbles during cavitational oscillations is the reason behind the ultrasonic effects in the IL medium (Bussemaker and Zhang, 2013).

18.4.5 Other Parameters

Several other factors such as treatment time, the intensity of ultrasound, and amplitude affect the ultrasound pretreatment of LCB. The length of sonication has one of the most significant influences on pretreatment efficiency. Increasing the ultrasonic treatment time increases the overall biomass delignification and subsequent sugar release; however, increasing sonication time beyond a certain point has little effect on both delignification and the release of sugar. Besides, the treatment duration is determined based on the type of biomass to be pretreated. For instance, a more-recalcitrant LCB needs a longer ultrasonication treatment than a less-recalcitrant starchy biomass (Rehman et al., 2012).

The sonication intensity is directly proportional to the amplitude of vibration of the ultrasonic source. Hence, increasing the amplitude of vibration increases the sonochemical effects. In addition, high sonication amplitudes can cause the ultrasonic transducer to degrade, resulting in excessive cavitation and poor ultrasound transmission through a liquid medium. However, for a liquid medium having high viscosity, the amplitude should be increased. This is necessary because the resistance of the liquid medium to movement rises as the viscosity increases. Thus, high intensity is usually needed to set the ultrasound device in order to achieve the mechanical vibrations required to cause cavitation (Subhedar and Gogate, 2013).

18.5 IMPACT OF ULTRASONIC PRETREATMENT ON BIOMASS CHARACTERISTICS

During ultrasonic pretreatment, LCB goes through several changes such as changes in its cellulose, hemicellulose and lignin content, cellulose crystallinity, particle size, and surface area. For instance, the delignification of LCB after ultrasonic pretreatment is because of high temperature and pressure during cavitation as well as the oxidative stress caused by the generation of radicals during ultrasonication. On the other hand, the dissolution or removal of hemicellulose after pretreatment is due to its amorphous structure. In contrast, the disintegration of cellulose is more difficult due to hydrogen bonding in cellulose microfibrils. As a result, a reduction in hemicellulose and lignin while an increase in cellulose content is more common after pretreatment (Bundhoo and Mohee, 2018).

The recalcitrant cellulose of LCB exhibits strong hydrophobicity in the reaction media due to a vast number of intermolecular and intramolecular hydrogen bonds. Thus, to reduce crystallinity, the recalcitrant characteristic of cellulose necessitates disintegration with the assistance of high activation energy (Haldar and Purkait, 2021). In this regard, ultrasonic energy can be utilized to treat LCB to reduce the crystallinity of the cellulose fraction. When the cellulose fraction is exposed to ultrasonic waves, intermolecular hydrogen bonds are weakened or potentially delinked. This reduces the organization of cellulose sheets and makes the hydrolytic process easier by improving the accessibility, hydrophilicity, and reactivity of this fraction. However, if the supplied ultrasound is increased further, despite the cellulose's reduced crystallinity, its amorphous areas begin to enhance the polymerization degree, resulting in a more-stable structure (Flores et al., 2021).

The particle size of LCB influences the effectiveness of the hydrolysis process, which affects its reaction rate and mass transfer, as well as the fermentation process. In addition, particle size has an impact on the kinetics diffusion and pretreatment efficiency, sugar yield, and delignification efficiency (Vasic et al., 2021). Ultrasonication assists in disrupting the cellulosic structure, cell lysis, and floc disintegration through thermal, shear, and oxidative stresses during cavitation, and these ultimately lead to an increase in the surface area and a decrease in the particle size of biomass (Bundhoo and Mohee, 2018). The increase in the surface area improves the accessibility of enzymes during enzymatic saccharification, which ultimately leads to an increase in the total reducing sugar yield.

18.6 ULTRASOUND-ASSISTED PRETREATMENT OF LCB

Ultrasonic irradiation has gained popularity as a green pretreatment approach for valorizing LCB for value-added products over the years. When traditional pretreatment procedures are combined with ultrasonication, a synergistic effect causes the kinetics of the process or yield to be improved, resulting in a better yield of desirable products such as sugars. It is necessary to understand the peculiar mechanistic pathway of each conventional pretreatment together with the chemical and physical effects of ultrasonication and cavitation capture to scale up ultrasound-assisted bioalcohol processes successfully (Ranjan et al., 2016). Pretreatment in an aqueous medium, alkaline pretreatment, acid pretreatment, IL pretreatment, deep eutectic solvent (DES) pretreatment, organosolv pretreatment, ammonia pretreatment, pretreatment using metallic salts, and other routes such as $KMnO_4$ treatment and ozonolysis are the most frequently used chemical pretreatments that are combined with ultrasound pretreatments. The pretreatment of different LCB using various ultrasound-assisted pretreatment processes is provided in Table 18.1.

18.6.1 Pretreatment in Aqueous Medium

The sonication process is influenced by the properties of the liquid used to suspend the lignocellulosic particles, which is usually an aqueous solution. Organic solvents and ILs, on the other hand, are also utilized in biomass pretreatment. The following aspects should be considered while choosing a sonication medium: cost-effectiveness and non-hazardousness. The energy barrier for cavitation in water is relatively low, and water sonication produces hydronium ions (H_3O^+) and hydroxyl radicals ($OH\cdot$), which function as oxidants to speed up the depolymerization process of floating biomass (Garcia et al., 2012).

He et al. (2017) used ultrasound to pretreat eucalyptus wood in aqueous solutions of soda, acetic acid, or water, and the findings revealed that ultrasound pretreatment could impact the physicochemical structure of the biomass. The samples' crystallinity improved by 35.5% due to the ultrasound pretreatment, which also removed cellulose, hemicellulose, and lignin and enhanced their exposure to treatment solutions. Nikolic et al. (2010) used *Saccharomyces cerevisiae* var. *ellipsoideus* yeast and pretreated maize meal with ultrasonication in a water medium for simultaneous saccharification and fermentation. Pejin et al. (2012) also studied ultrasound pretreatment in water to improve the sugar yield in triticale crops for biofuel production. In the sonication processing of sunflower seed husk, distilled water was utilized by Sul'man et al. (2011), and they observed that cellulose decomposition was nearly 16% of its absolutely dry mass, whereas lignin decomposition reached 11.4% of its dry mass. As we comprehend, the energy threshold for cavitation in water is relatively low. As a result, compared to studies on other powerful media for ultrasonication, limited research has been reported on water as a medium for ultrasonication of LCB.

TABLE 18.1
Various Ultrasound-Assisted Pretreatment of Lignocellulosic Biomass

Biomass	Ultrasound Conditions	Experimental Medium	Findings	References
Eucalyptus wood	300 W and 28 kHz (0.5–1.5-h, 50°C)	Distilled water, soda solution, and acetic acid	Crystallinity improved to 35.5%	He et al. (2017)
Corn meal	600 W and 40 kHz (5 min, 60°C)	Water	Glucose yield increased by 6.82% and ethanol yield increased by 11.15%	Nikolic et al. (2010)
Triticale crops	125 W and 40 kHz (5 min, 60°C).	Water	Glucose and maltose concentration increased by 15.71% and 52.57% and ethanol concentration increased by 10.89% (84.56%)	Pejin et al. (2012)
Sunflower husks	460 W and 30 kHz (15 min)	Water	Cellulose degradation 16% of dry substance and lignin 11.4 % of dry substance	Sul'man et al. (2011)
Hardy sugarcane, giant reed, narrow-leaved cattail, and pink morning glory	400 W and 24 kHz (30 min)	0.5% (w/v) NaOH	Pink morning glory showed the highest sugar release and a fermentation efficacy of 85%	Muthuvelu et al. (2019)
Rice straw	300 W and 22 kHz (20 min, 50°C)	NaOH solution of 0.4%, 0.6%, 0.8%, and 1.0%, w/v	Sugar release increased 3.5 times	Wu et al. (2017)
Grass clipping	60–650 W and 25 kHz (30 min, 60°C)	NaOH and Ca $(OH)_2$ solution 0.75%	Sugar release increased 3.5 times with $CaOH_2$	Yang and Wang (2019)
Sugarcane bagasse	400 W and 24 kHz (70.15°C, 47.42 min)	NaOH 2.89% (w/v)	Sugar yield reached 92.11%	Velmurugan and Muthukumar (2012)
Sugarcane bagasse	300 W (30°C–40°C, 5–30 min)	0.3 M H_2SO_4	Mass dissolution increased by 4%	Nadar et al. (2020)
Hardy sugarcane, giant reed, narrow-leaved cattail, and pink morning glory	400 W and 24 kHz (60°C and 60 min)	1.0% (w/w) H_2SO_4	Hardy sugarcane showed the highest sugar release of 309.75±3.98 mg/g and narrow-leaved cattail gave the highest percentage of ethanol of 81%	Paramasivan et al. (2021)
Cassava bagasse	400 W (50°C, 60 min)	Metaphosphoric acid 10%	Sugar release increased to about 70% in 30 min.	Polachini et al. (2021)
Deenanath grass and Napier grass	100 W, 353 K	H_2SO_4, H_3PO_4, HNO_3 and HCl (0.5%, 1%, 1.5%, 2% (w/v))	Delignification increased by around 80% for both grass	Mohapatra et al. (2017)
Sugarcane bagasse	100 W and 20 kHz (60 min)	[Bmim]Cl (5 g) and PEG-8000 (300 mg)	Sugar yield increased (16.5%)	Sharma et al. (2019)
Bagasse	35 W and 24 kHz (60 min, 25°C)	15 mL, 5 g ChOAc).	Less inhibitory action by ChOAc. Cellulose 80% and hemicellulose 72% recovery	Ninomiya et al. (2015)
Rice straw	840 W and 40 kHz (30°C–70°C, 30–240 min)	AMIMCl, HCl	Reducing sugar, cellulose formation, and delignification increased by about 28.96%, 35.23%, and 19.33%, respectively	Zhang et al. (2021b)

(*Continued*)

TABLE 18.1 (*Continued*)
Various Ultrasound-Assisted Pretreatment of Lignocellulosic Biomass

Biomass	Ultrasound Conditions	Experimental Medium	Findings	References
Cogongrass	(50 and 100 W, 20 kHz) for 15 and 30 min, 120°C	[EMIM]OAc and rhamnolipid solution	Enzymatic digestibility showed a 1.2-fold increase, and ethanol yield increased 1.6-fold	Goshadrou (2021)
OPEFB	210 W (40°C, 3 h)	Choline chloride: lactic acid	Reducing sugar yield increased	Lee (2021)
Sugarcane bagasse	(20, 28, 40 kHz) (60, 120, 180, 240, and 300 W) (30, 60, 90, 120, and 150 min) and 25°C	Ternary DES (choline chloride: glycerol: $FeCl_3.6H_2O$)	Cellulose yield increased to 47 % and lignin removal reached 82.71%	Ji et al. (2021)
Parthenium hysterophorus	40 W and 20 kHz (20 min)	Choline chloride/ sorbitol (molar ratio 1:5)	Reducing sugar yield increased	Nargotra et al. (2020)
WMR biomass	28 kHz	Chorine chloride + lactic acid)	The lower heating value and associated exergy values showed possibility for bioenergy channeling	Fakayode et al. (2021a)
Bamboo	100 W and 35 kHz (20°C, 5 min)	Methylbenzene and ethanol	Thermal stability of the fractional yield increased	Li et al. (2012)
Olive tree prune residues	40 kHz and 200 W (15–120 min, 25°C)	Ethanol–water (60:40 v/v)	Sugar yield increased by 16%	Garcia et al. (2012)
OPEFB	53 kHz (25, 50°C)	Ethanol concentration (10%–50%, v/v)	Under ambient sonication conditions, 41.3 mg sugar recovered	Lee and Ng (2019)
Sugarcane bagasse	400 W and 24 kHz (15, 30 and 45 min and temperature (40°C, 60°C and 80°C)	Aqueous ammonia solution (5%–15%, v/v),	Cellulose recovery and delignification observed were 95.78% and 58.14%, respectively	Ramadoss and Muthukumar (2014)
Corncob	60–650 W and 20 kHz (60°C, 12 min)	Aqueous ammonia	Cellulose recovery, delignification, and sugar recovery yield reached 83.8%, 84.7%, and 77.6%, respectively	Du et al. (2018)
Corncob, sorghum stalk, and corn stover	90 W and 59 kHz (40°C–70°C, 1–4 h)	Aqueous ammonia	Sugar yield of corn cob, sorghum stalk, and corn stover increased (80.6%, 56.9% 66.2%, respectively)	Xu et al. (2017)
Sugarcane bagasse	400 W and 24 kHz (100°C, 60 min	H_2O_2 and metal salts (ammonium molybdate, cobalt chloride, and titanium dioxide)	94.98% holocellulose and 78.72% delignification	Ramadoss and Muthukumar (2016a)
Moso bamboo	300 W and 40 kHz (1.5 h, 50°C)	5% (w/w) $ZnCl_2$	Water absorption decreased by 10.15%, crystallinity increased by 4.4%, and water contact angle increased 25.2%	Zhang et al. (2021a)
Sugarcane bagasse	400 W and 24 kHz (60°C, 75 min)	H_2O_2 and NaCl, KCl, $MgCl_2$, $CaCl_2$, $CuCl_2$, and $FeCl_3$ $AlCl_3$	94.94% cellulose recovery and 79.42% delignification	Ramadoss and Muthukumar (2016b)

(*Continued*)

TABLE 18.1 (*Continued*)
Various Ultrasound-Assisted Pretreatment of Lignocellulosic Biomass

Biomass	Ultrasound Conditions	Experimental Medium	Findings	References
Corn cob and spelt husk	40 kHz and 30 min, room temp	Ozone concentration of 2%	90% high-purity lignin was recovered. Cellulose crystallinity reduced to 19%	Ibrahim et al. (2021)
Moj, areca nut husk, and bon bogori	30 kHz (35°C, 60–180 min)	(0.25 0.5 0.75) lime	68% (bon bogori), 65% (Areca nut husk), and 64% (moj) lignin removal. Ethanol release and yield of ethanol about 0.35	Sasmal et al. (2012)
Coffee waste	310 W and, 47 kHz (20 min, 25°C)	4% $KMnO_4$	98% cellulose and 46% lignin removal	Ravindran et al. (2017)
Rice straw	22 kHz (25°C, 2 s and an off-time of 4 s)	$FeSO_4{\cdot}7H_2O$ solution and H_2O_2	Reducing sugar release increased four times	Xiong et al. (2017)
Corncob	37 kHz and 340 W (35°C,12 h)	Fe ions, H_2O_2, and TiO_2	The highest release of lignin concentration level (1.03 g/L) and dissolution level (80.25%)	Kawee-Ai et al. (2016)

TRS, total reducing sugars.

18.6.2 Alkaline Pretreatment

Alkaline pretreatment is gaining popularity as a pretreatment method since it is very economic, requires less energy, and is effective on a wide range of lignocellulosic materials. It appears to be a successful way for disrupting the ester connections between lignin, hemicellulose, and cellulose (Gaspar et al., 2007). Standard bases employed in alkaline hydrolysis include NaOH (sodium hydroxide), NH_4OH (ammonium hydroxide), KOH (potassium hydroxide), and $Ca(OH)_2$ (calcium hydroxide). To overcome the problem of long processing time, the integration of ultrasound in the pretreatment procedures affects macropolymers during cavitation and increases the release of fermentable sugar. In general, alkaline pretreatment with ultrasound assistance efficiently deconstructs the lignin structure, increasing porosity and surface area to liberate more cellulose during hydrolysis.

Muthuvelu et al. (2019) studied the potential of bioethanol extraction from *Saccharum arundinaceum*, *Ipomoea carnea*, *Arundo donax*, and *Typha angustifolia*. The bubbling effect of ultrasound-assisted alkaline pretreatment was demonstrated by lower lignin content, larger surface area, crystalline nature changes, and active biomass groups. Yang and Wang (2019) used grass clippings that had been pretreated with ultrasound $Ca(OH)_2$, which exhibited the best results in terms of reducing sugar at low temperatures and enhancing enzymatic hydrolysis. Upon ultrasound-alkaline pretreatment, hemicellulose and lignin solubilization increased, as did the crystallinity values of cellulose. Velmurugan and Muthukumar (2012) studied the use of ultrasound to assist in the alkaline pretreatment of sugarcane bagasse to produce fermentable sugars. One of the most appealing advantages of ultrasound-assisted alkaline pretreatment is the significant reduction in the pretreatment duration and temperature while enhancing productivity. The dissolution rate of hemicellulose and lignin, as well as the crystallinity index of cellulose, increased after pretreatment with an ultrasound-assisted alkaline solution.

18.6.3 Acid Pretreatment

Ultrasound-assisted acid hydrolysis for pretreating LCB is based on the theory that ultrasonic irradiation promotes biomass solubility in acid media. When a low solid loading is used, ultrasonic irradiation will have a more significant impact on performance than temperature as there will be more contact for acid diffusion into the biomass and a greater likelihood of acoustic cavitation (Nadar et al., 2020). Acids used in pretreatment include diluted H_2SO_4 (sulfuric acid), HNO_3 (nitric acid), H_3PO_4 (phosphoric acid), and HCl (hydrochloric acid), with H_2SO_4 being the most popular.

Paramasivan et al. (2021) assessed the bioethanol generation capacity of different agricultural residues such as *Typha angustifolia*, *Saccharum arundinaceum*, *Ipomoea carnea*, and *Arundo donax*. For delignification of these four agro residues, the researchers used ultrasonication in a dilute acid medium. The combined outcome of dilute acid and the acoustic cavitation effect of ultrasound waves resulted in the exposure of more cellulose for the saccharification process and reduced lignin content. Nadar et al. (2020) used sugarcane bagasse in sulfuric acid medium, and as expected, the enhancement in bulk dissolution due to ultrasonication was shown to rise as the reaction time increased. The application of ultrasound was found to eliminate the initiation period for the treatment. Polachini et al. (2021) concluded that the sugar release was 52%–70% higher in ultrasonically aided phosphoric acid trials after only 30 min of processing cassava bagasse. Mohapatra et al. (2017) observed effective lignin removal and cellulose yield when treating two *Pennisetum* grass species in different acid media such as sulfuric acid, phosphoric acid, hydrochloric acid, and nitric acid, out of which ultrasonic-aided HCl treatment exhibited better outcomes. Characterization techniques such as scanning electron microscopy and autofluorescence microscopy revealed that this combined treatment was more successful in making conformational changes.

Ultrasound technology-assisted acid pretreatment represents an alternative way to break down the recalcitrant biomass structure and to release fermentable sugars. It could also explain a drop in chemical consumption and a reduction in reaction time for subsequent bioethanol production processes. Meanwhile, acid hydrolysis is corrosive by nature; therefore, measures should be taken to mitigate this issue during the pretreatment of LCB.

18.6.4 Ionic Liquid Pretreatment

Among the different methods to treat LCB, IL pretreatment is very effective and environment-friendly. ILs are viscous media that can liquefy biomass at reduced temperatures; they are also chemically inert and cost-effective. The main advantages of this method are that this is found to be easy to process, regenerate, and recycle and is also found to be non-toxic to enzymatic and microbial fermentation (Han et al., 2020). However, the main disadvantages of ILs are their high cost and the fact that, due to their high viscosity, combined ultrasound IL can affect the core effects of ultrasonication, such as acoustic cavitation and microbubble disintegration (Bizzi et al., 2021).

Sharma et al. (2019) used 1-butyl-3-methyl-imidazolium chloride ([Bmim]Cl) to pretreat sugarcane bagasse in the presence of a surfactant (PEG-8000). The sugar yield increased by 16.5% after enzymatic saccharification. Ninomiya et al. (2015) reported an effective combination of biocompatible cholinium IL and ultrasonication in treating bagasse. The action of choline acetate (ChOAc) was comparable with the widely accepted ([Emim]OAc), imidazolium IL. In addition, ChOAc had a lower inhibitory impact on cellulase than [Emim]OAc. Zhang et al. (2021b) found that using two ILs, [Amim]Cl and [Bmim]Cl, and an HCl catalyst improved the reducing sugar production from rice straw at room temperature. Goshadrou (2021) investigated the use of ultrasound pretreatment assisted with a surfactant rhamnolipid in [Emim]OAc medium pretreatment to increase the bioethanol production from harmful weed cogongrass. Compared with IL pretreatment, a novel step-by-step approach with ultrasonic assistance gave the greatest bioethanol production and was about 1.6 factor greater than the reference.

18.6.5 Deep Eutectic Solvent Pretreatment

The presence of a proton donor and acceptor in DESs aids the cleavage of intramolecular hydrogen bonds in biomass structures, leading to its high dissolution (Chen and Mu, 2019). For instance, Lee (2021) studied the impact of ultrasound in conjunction with different DESs for processing empty fruit bunches of oil palm (OPEFB). Choline chloride:lactic acid (ChCl:LA), choline chloride:glycerol (ChCl:Glycerol), and choline chloride:urea (ChCl:U) were the medium used under the action of ultrasonication for 15 min. ChCl:LA gave predominant results with the highest sugar yield (36.7%). Under these conditions, ChCl:LA processed OPEFB showed noteworthy structural and morphological changes as well as the lowest crystallinity and lignin concentration, which might be due to its inherent properties of reduced viscosity and surface tension, and the addition of ultrasonication to the DES pretreatment process pointedly increased the production of reducing sugars.

Ji et al. (2021) investigated sugarcane bagasse pretreatment with ultrasound-assisted ethanol administered with DES. Ultrasound assistance was offered for a medium with three DESs (choline chloride, glycerol, and $FeCl_3 6H_2O$) and ethanol, and the final bioethanol conversion rate was evaluated. The combination resulted in a saccharification rate of 90%. When *P. hysterophorus* was saccharified, Nargotra et al. (2020) found that pretreatment in choline chloride/sorbitol produced the highest sugar yield when ultrasonic irradiation was used. Using chorine chloride and lactic acid, Fakayode et al. (2021a) investigated ultrasound-induced DES pretreatment of watermelon rind (WMR) biomass. WMR biomass that has been processed has a high heating value and exergy value, indicating that it has a considerable biofuel production capability. Despite all of the advantages of DES, it should be emphasized that recycling it after treatment is costly, and separating valuable chemicals may be challenging.

18.6.6 Organosolv Pretreatment

Organic solvent pretreatment has gained increasing attention as a pretreatment method due to its inherent advantages. This pretreatment method can fractionate LCB into high-purity components, including hydrolyzed cellulose, lignin, and hemicelluloses. Generally used organic solvents are methanol, ethanol, glycerol, ethylene glycol, acetic acid, and formic acid. Currently, ultrasonication has been added to the organosolv pretreatment process, allowing for a more productive and cheap pretreatment approach to synthesize different value-added products (Zhang et al., 2016). However, organic solvents may include toxic solvents, and they demand extreme conditions such as high temperatures and pressure, which can alter bioactive molecules (Sun et al., 2012).

Yang et al. (2020) used ultrasonic radiation and microwave radiation together in an ethylene glycol medium for the treatment of sawdust from pine. They found that fractionation of pine sawdust into components drastically improved with the help of ultrasonic radiation and microwave heating when compared to oil bath heating. Li et al. (2012) used ultrasound-assisted ethanol pretreatment for bamboo. It was discovered that as the ultrasonic irradiation time was extended, the fraction's lignin concentration increased. To assess the influence of ultrasonic pretreatments (0, 15, 30, 60, and 120 min) on black liquor components, Garcia et al. (2012) applied ultrasonic pretreatments to olive tree pruning wastes in an ethanol–water media. The findings validated sonication as a viable tool for biorefinery process intensification. Lee and Ng (2019) pretreated OPEFB with ethanol solvent using ultrasound. The sonication power, treatment time, and ethanol content had a substantial impact on pretreatment performance. They concluded that this technique gave higher reducing sugars from OPEFB and took less treatment duration than traditional thermal treatment.

18.6.7 Ammonia Pretreatment

One of the best methods for fermentable sugar production is ammonia-aided pretreatment. This procedure is feasible because ammonia is extremely volatile, and a 95% retrieval rate is possible. Besides, ammonia-assisted ultrasound pretreatment is a safe and effective approach with no detrimental effects on the surface of lignocellulosic materials. Also, ultrasound and ammonia have a stimulatory effect in eliminating lignin and cellulose recovery. With all of its merits, this pretreatment method has one drawback, that is, the high cost of ammonia.

Ramadoss and Muthukumar (2014) used a sonication-aided ammonia pretreatment method to process sugarcane bagasse for the generation of fermentable sugar. FTIR and XRD results indicated the successful removal of lignin and the reduction in crystallinity of cellulose, a necessary condition required for the generation of high sugar yield. Du et al. (2018) used ultrasound-assisted aqueous ammonia solution to pretreat corncob. According to elemental studies, the pretreatment preferentially eliminated lignin and hemicelluloses while retaining cellulose. SEM examination revealed the substantial expansion of surface cavity width, which is advantageous to the synthesis of fermentable sugar. Xu et al. (2017) coupled ammonia solution with ultrasonication to improve cellulose exposure and sugar yield in biomass corncobs, corn stover, and sorghum stalks. According to XRD data, the crystalline index increased after pretreatment, indicating the successful removal of hemicellulose and lignin. Wang et al. (2019) reported an efficient strategy for expediting waste-activated sludge breakdown and promoting anaerobic fermentative short-chain fatty acid synthesis using free ammonia, which has been shown to have a bactericidal effect on various bacteria. The findings indicated that free ammonia aided sludge solubilization by ultrasound pretreatment.

18.6.8 Pretreatment Using Metallic Salt

The introduction of metallic salts as aqueous media has recently received interest as one of the most novel pretreatment techniques as it is less corrosive than acid alternatives (Loow et al., 2015). These salts can speed up the hydrolysis process and increase the amount of cellulose and hemicellulose recovery. When inorganic salts are used with other pretreatment techniques, the kinetics of the process improves significantly (Arumugam et al., 2021). Ultrasound-assisted treatments in the presence of metallic salts such as simple metal ion chlorides, such as iron, magnesium, zinc, aluminum, and sodium, as well as unique and complex salts, can enhance the potential to produce cellulosic bioethanol and value-added compounds from LCBs.

Ramadoss and Muthukumar (2016a) observed an increase in reaction velocity while treating sugarcane bagasse with different metal salts. The ultrasonic aided titanium dioxide (TiO_2) pretreatment technique gave the best holocellulose recovery and delignification among the various metal salts utilized. The theoretically anticipated bioethanol production (92%) was achieved after 72 h. TiO_2 combined with hydrogen peroxide and ultrasonication was the optimal combination for biomass disintegration in their study. Zhang et al. (2021a) explored the effect of comparatively unutilized zinc chloride with ultrasonication for pretreatment of Moso bamboo and observed an increase in the crystallinity index by 4.4% and the water contact angle by 25.2%, and enhanced surface wettability. Besides, the pretreatment process also enhanced Moso bamboo's breakdown and reduced the temperature necessary for it, making it a safe and environmentally friendly green technology. Ramadoss and Muthukumar (2016b) examined the efficacy of ferric salt in association with ultrasonication for pretreatment of sugarcane bagasse, reporting a maximum cellulose yield of 95% and superior lignin removal efficiency. After 72 h of fermentation, the hydrolysate was fermented with *Saccharomyces cerevisiae*, yielding roughly 93% potential ethanol yield.

18.6.9 Other Ultrasound-Assisted Pretreatments

Recent studies have reported successful incorporation of Fenton reagent, ozone, permanganate, lime, etc., with ultrasonication as possible modified pretreatment strategies. Ultrasound-assisted lime pretreatment can lessen the processing time. In a previous study, different agricultural biomass materials such as moj (*Albizia lucida*), bon bogori (*Ziziphus rugosa*), and Areca nut husk (*Areca catechu*) were pretreated using ultrasound in combination with lime. The total lignin content in the lignocellulosic fiber was removed 68% (bon bogori), 65% (Areca nut husk), and 64% (moj) (Sasmal et al., 2012). Furthermore, Ravindran et al. (2017) developed a unique pathway for breaking down spent coffee waste that relies on alkaline $KMnO_4$'s increased oxidizing capability combined with sonication effects. The group was able to increase the sugar yield to about 1.7 times when compared to untreated biomass. Therefore, it was concluded as an excellent novel strategy that could be used as a bioethanol feedstock pretreatment method in the future.

Another alternative is the Fenton process, in which reaction between ferrous ion and hydrogen peroxide results in the formation of hydroxyl radical (OH), which oxidizes almost all organic compounds into carbon dioxide (CO_2) and water under environmentally friendly conditions such as normal temperature and pressure. Xiong et al. (2017) pretreated rice straw with Fenton's reagent and ultrasound to boost its enzymatic digestibility for saccharification. The treated rice straw showed the highest specific surface area and enzymatic activity compared with the other pretreatments. The reducing sugar yield was about four times compared with the raw sample and 1.5 times with Fenton's reagent alone. As a result, the Fenton process with sonication assistance is a safe and effective way to treat rice straw. Similarly, Kawee-Ai et al. (2016) found that pretreatment of corncobs in a similar pathway resulted in a lignin content of 1.03 g/L and an increased dissolution level of about 80%.

18.7 ULTRASOUND-ASSISTED ENZYMATIC HYDROLYSIS OF PRETREATED BIOMASS

Ultrasound technology can be applied both for enhancing enzymatic activities and for deactivating enzymatic activities depending on the intensity of applied sound wave and frequencies as well as the process conditions (Dalagnol et al., 2017; De Carvallo Silvello et al., 2021; Islam et al., 2014). The cavitation effect enhances enzyme activity when the ultrasonication stays for a short time at a lower intensity by changing the protein conformation that changes the enzyme functionality (De Carvallo Silvello et al., 2020; Wang et al., 2018). The ultrasonic process enhances the reducing sugar yield, which significantly increases the bioethanol yield during the pretreatment of LCB (De Carvalho Silvello et al., 2019; Subhedar et al., 2018). The enhancement of the reducing sugar yield can be driven by improving the reaction between the enzymes and biomass as well as by improving the delignification process, which is triggered by the energy released from microbubbles during the course of the ultrasonication process (De Carvalho Silvello et al., 2019; Ong and Wu, 2020). Furthermore, the sonication process increases the rate of mass transfer and porosity of the substrate, which facilitates the interaction between the cellulase enzyme and the pretreated biomass by increasing the available surface area for enzymatic reactions (Sulaiman et al., 2013). Enhancement of enzyme activity and an increase in the hydrolysis rate were observed during the cellulase enzyme-assisted ultrasound process of model compounds on cellulose and carboxymethyl cellulose (De Carvalho Silvello et al., 2020; Sulaiman et al., 2013).

Studies on coconut coir, pistachio shells, and groundnut shells showed that the rate of hydrolysis is significantly improved after sonoenzymolysis, a term used to describe enzymatic hydrolysis assisted by ultrasonication (Subhedar et al., 2018). Improved glucose yield, about 90.12%, was obtained from the enzymatic hydrolysis of pretreated palm oil frond by combined ultrasound-assisted NaOH and DES pretreatment (Ong et al., 2021). Ultrasound-assisted acid catalyst and enzymatic hydrolysis of *Gracilaria verrucosa* significantly reduced the formation of furfural and furfural derivative compounds and increased the reducing sugar concentration to 76.26% (Park and Jeong, 2021). Furthermore, they observed that the combination of an acid catalyst and ultrasonication process significantly increased the rate of hydrolysis (reduced the hydrolysis time).

The rate and yield of the delignification process of orange peel biomass were significantly affected after the alkali-assisted ultrasonication process at ambient temperature (Utekar et al., 2021). They observed a lignin removal of 86% and a reducing sugar concentration of 2.069 g/L at optimal conditions, which significantly reduced the amount of enzyme required for hydrolysis (Utekar et al., 2021). Enhancement of reducing sugars, about 0.26 g/g of biomass, was observed after ultrasound exposure of sugarcane bagasse at 50°C for 240 min with 10% enzyme concentration (Lunelli et al., 2014). The yield of total reducing sugar was nearly doubled (100% increment) due to the effect of ultrasound treatment (Lunelli et al., 2014).

On the other hand, about 50% increment of total reducing sugars was observed during ultrasound-assisted enzymatic hydrolysis of xylan at optimal conditions, exposure time of 10 min at 50°C, and ultrasound power of 200 W (Sun et al., 2015). Singh et al. (2015a) observed that the yield of total reducing sugars was increased by 20% due to ultrasonication, while the reaction rate increased by six-fold. Delignification and hydrolysis of LCB with the synergistic effect of ultrasonication and dilute acid pretreatments followed by enzymatic hydrolysis resulted in a maximum reducing sugar yield of 309.75 mg/g in *Saccharum arundinaceum,* while 215.79% in *Ipomoea carnea*, 299.74% in *Typha angustifolia*, and 164.97% in *Arundo donax*, respectively (Paramasivan et al., 2021). The operating conditions and the yields of ultrasound-assisted enzymatic hydrolysis of various LCBs are given in Table 18.2.

Many studies suggested that ultrasonication processes are green and efficient and speed up the rate of hydrolysis and delignification of LCB. The combination of ultrasonication and enzymatic hydrolysis (sonoenzymolysis) of LCB is more promising to enhance the yields of total reducing sugars and increase the rate of the delignification process. However, profound work on optimizing the operating parameters such as temperature, residence time, ultrasound frequency and intensity,

TABLE 18.2
Ultrasound-Assisted Enzymatic Hydrolysis of Different LCBs

Biomass	Pretreatment Process	Optimal Conditions	TRS Yield	References
Orange peel	Ultrasound-assisted delignification	6 h, 60 W, 70% duty cycle, substrate loading (2%, w/v), enzyme loading 1% v/v, 50°C, 300 rpm	1.3 g/L	Utekar et al. (2021)
Oil palm frond	Ultrasonication + NaOH-DES	72 h, 50°C	90.12%	Ong et al. (2021)
Watermelon rind	Ultrasonication + DES	180 W, 40 KHz, 40 min, 120°C	83.03%	Fakayode et al. (2021a)
Gracilaria verrucosa	Ultrasonication + acid catalyst	100 Mm H_2SO_4, 60% ultrasonic amplitude, 60 min	70.26%	Park and Jeong (2021)
Groundnut shells	Ultrasound-assisted enzymatic hydrolysis	0.08% w/v, enzyme loading, substrate loading, 3% w/v, 60 W, 70% duty cycle, 6.5 h	21.3 g/L	Subhedar et al. (2018)
Coconut coir	Ultrasound-assisted enzymatic hydrolysis	0.08% w/v, enzyme loading, substrate loading, 3% w/v, 60 W, 70% duty cycle, 6.5 h	23.9 g/L	Subhedar et al. (2018)
Pistachio shells	Ultrasound-assisted enzymatic hydrolysis	0.08% w/v, enzyme loading, substrate loading, 3% w/v, 60 W, 70% duty cycle, 6.5 h	18.4 g/L	Subhedar et al. (2018)
Sunn hemp	Ultrasound-assisted NaOH	1% NaOH, 10 min, 121°C	7.25 mg/mL	Manasa et al. (2018)
Food industry waste (potatoes)	Ultrasound-assisted HCl hydrolysis	2.1% HCl, 10.7 min,	65.8 mg/L	Suresh et al. (2020)
Food industry waste (potatoes)	Ultrasound-assisted enzyme hydrolysis	54.1 g/L α-amylase, 25.3 U/mL, 10.2 min	54.1 mg/L	Suresh et al. (2020)
Wheat waste biomass	Ultrasound-assisted NaOH	Substrate loading 10%, 2% NaOH (100°C, 30 min), 50°C, US (20 KHz, 2 W/mL, 60 min)	84.5%	Saratale et al. (2020)
Rice straw	Ultrasound-assisted NaOH	48-h enzymatic hydrolysis, 1% NaOH, 50°C, 22 KHz, 300 W	2.91 g/L	Wu et al. (2017)

TRS, total reducing sugars.

solvents, and enzyme load is necessary to gain insights into the feasibility and the environmental impact as well as upscaling aspects of the process for commercial production. Therefore, optimization and the scale aspects of ultrasound-assisted enzymatic hydrolysis of LCB need to be thoroughly investigated in order to harness the full potentials of this promising technology.

18.8 FERMENTATION PROCESS INTENSIFICATION USING ULTRASONICATION

Nowadays, the fermentation technology is becoming a crucial process for the production of various biomolecules that have high market value and demand. Moreover, the process attracts much attention due to its operation under mild environments and its eco-friendly nature. However, due to inadequate mass transfer rates, lower fermentation yield is a main problem associated with the fermentation technology. Many techniques such as genetic engineering and the addition of promoters have been applied to increase the productivity and yields of the fermentation process. However, low product concentration remains a major problem that increases the costs of downstream processing, and product recovery steps pose as an obstacle to using the technology. The application of the ultrasonication technology can overcome all of these problems and help intensify the fermentation process by increasing the rate of mass transfer and changing microbial morphology and enzyme conformation. Ultrasonication intensifies the fermentation process via cavitation effects, which can enhance the rate of mass transfer. Although the exact mechanism of mass transfer during ultrasonication is not fully understood, it is widely assumed that the rate of nutrient transfer within the cell and the surrounding medium is accelerated due to the increased pore size of the cell membrane, which likely increases the permeability of the cell (Pawar and Rathod, 2020; Raskar et al., 2014). Recent studies on the low-intensity ultrasound technology suggested that it accelerates the rate and yield of the fermentation process by improving the performance and growth of microorganisms (Huang et al., 2017, 2019; Ojha et al., 2017; Wang et al., 2021). On the contrary, enzyme deactivation and cell disruption are more commonly carried out using the high-intensity ultrasonication process.

The increase in the pore size of cell membranes is due to microbubbles that are created by the cavitation process (Bekeredjian et al., 2007). The process allows the migration of macromolecules into and/or out of the cells that are originally non-permeable substances, which increases the transmembrane movements (Jomdecha and Prateepasen, 2006). It also permits access to more active sites of the enzymes originally bounded by the cell membranes, leading to an overall increase in the rate and yield of the bioconversion process (Mahajan et al., 2012). The ultrasonic effect can also be introduced by power dissipation mechanisms. Various parameters such as the geometry of the reactor, pressure, viscosity, and temperature of the liquid medium significantly affect power dissipation (Pawar and Rathod, 2020). *Saccharomyces cerevisiae* was best grown and produced maximum bioethanol at 160 W ultrasound power (Subhedar and Gogate, 2015), while *Streptomyces peucetius* MTCC 4332 significantly increased the production of daunorubicin when the ultrasound power was increased to 160 W from 40 W (Raskar et al., 2014). This shows that ultrasound power is one of the most influential parameters that need to be thoroughly investigated and optimized for process intensification of the fermentation process for enhancing bioethanol production. The use of ultrasound in intensification of the fermentation process in bioethanol production from different LCBs is provided in Table 18.3.

The exposure time of the ultrasonication process also greatly influences the fermentation process by changing the morphology of the cell, enzyme conformations, and the rheological property of the fermentation medium (Pawar and Rathod, 2020). Therefore, optimal exposure time is necessary to maximize the rate and yield of the fermentation process. Studies on potato waste treatments by coupling of the ultrasonication process with enzymatic hydrolysis and acid pretreatments significantly enhanced the bioethanol yield, 25.4 and 65.5 g/L, respectively (Suresh et al., 2020). The amount of bioethanol produced was increased by 1.8 fold after ultrasound-assisted fermentation of waste

TABLE 18.3
Application of Ultrasonication in Fermentation Process Intensification for Bioethanol Production

Process Mode	Substrate Used	Process Condition	Findings	References
Ultrasound-assisted simultaneous saccharification and fermentation	Oil palm fronds	37 kHz, 200 W	Maximum ethanol concentration (18.15 g/L) and yield (57.02%)	Ofori-Boateng and Lee (2014)
Ultrasonication-DES + enzymatic hydrolysis + fermentation	Watermelon rind	180 W, 40 KHz, 40 min, 120°C	0.458 g/g bioethanol yield	Fakayode et al. (2021b)
Ultrasound-assisted hydrolysis and fermentation	*Parthenium hysterophorus*	35 kHz, 35 W, 10% duty cycle	Two-fold increase in ethanol production (ethanol titer of 12.14 g/L in 10 h)	Singh et al. (2015c)
Ultrasound-assisted HCl hydrolysis	Food industry waste (potatoes)	2.1% HCl, 10.7 min, 19.2 g/L *S. cerevisiae*	65.5 g/L bioethanol yield	Suresh et al. (2020)
Ultrasound-assisted simultaneous saccharification and fermentation	*P. hysterophorus* biomass	35 kHz, 35 W, 30°C	Four-fold increase in ethanol production (yield of 0.4 g ethanol/g pretreated biomass)	Singh et al. (2015b)
Ultrasound-assisted α-amylase hydrolysis	Food industry waste (potatoes)	54.1 g/L α-amylase, 25.3 U/mL, 10.2 min, 19.2 g/L *S. cerevisiae*	25.4 g/L bioethanol yield	Suresh et al. (2020)
Low-intensity pulsed ultrasound	LCB	1.5 MHz, 20% duty cycle, 80 mW/cm^2 intensity, 5 min exposure and 12 exposures per day	31 ± 10% increase in ethanol production by *S. cerevisiae*	Shaheen et al. (2013)

LCB, lignocellulosic biomass.

newspaper at optimal conditions (Subhedar and Gogate, 2015). Furthermore, low-intensity ultrasound pretreatment (0.0003 W/cm and 43 kHz) activated yeast *S. cerevisiae* activity and reduced the fermentation time by 50%–64% (Pawar and Rathod, 2020).

The continuous ultrasonication process of LCB was used for intensification of the fermentation process. The enzymatic hydrolysis of waste papers such as office papers, newspapers, and carton papers was significantly improved by a continuous ultrasonication process that likely enhanced the yield of bioethanol (Li et al., 2005). The rate of hydrolysis was higher in small-volume reactors, while lower hydrolysis kinetics was observed in large-volume reactors. In order to overcome such problems and improve the cellulose hydrolysis of waste papers, an external loop airlift bubble column was designed for dispersion at higher solid loading (Li et al., 2004, 2006). The combination of ultrasonication and enzymatic hydrolysis of cellulose increased the glucose yield from 37% to 50% while decreasing the hydrolysis time to 24 h (Adewuyi and Deshmane, 2015).

The application of ultrasonication on algal biomass enhances the availability of fermentable sugars in the fermentation broth and improves the bioethanol yield by disintegrating the algal biomass (Jeon et al., 2013; Naveena et al., 2015). An ethanol yield of 5.6 g/L was obtained after applying an ultrasonic frequency of 40 kHz for 15 min of exposure time (Jeon et al., 2013). Moreover, the ultrasonication process provides more carbohydrates (more algal biomass) for the microbes by increasing the hydrophilicity and the electrostatic repulsion of algal biomass in the fermentation media (Jeon et al., 2013; Naveena et al., 2015).

The results from the low-intensity ultrasonication process are seen as promising for fermentation process intensification. However, in-depth investigation into and optimization of process parameters such as ultrasound power and intensity, exposure time, mode of operations (pulsed or continuous), the nature of the medium (viscosity, solubility), ultrasonic amplitude, and frequency are necessary for enhancing the bioethanol yield and intensifying the fermentation process.

18.9 CHALLENGES TO EMPLOYING ULTRASONICATION IN BIOETHANOL FUEL PRODUCTION

The application of the ultrasound technology in bioethanol production from LCB is an emerging field. Ultrasound-assisted enzymatic hydrolysis, ultrasound-assisted chemical hydrolysis, and ultrasound-assisted fermentation processes are promising for process intensification of ethanol production. Each process significantly increases the depolymerization and fractionations of LCB via enhancing hydrolysis and fermentation kinetics by increasing the mass transfer rate. The lab-scale results prove that the ultrasonication process assisted by enzymatic and chemical pretreatments enhanced the rates and yields of bioethanol while reducing the exposure time. The choice of ultrasonic parameters greatly influences the intensification of the bioethanol production process. The mode of operation (pulsed or continuous), the nature of the medium (viscosity, solubility), and the operating temperature affect the distribution of the ultrasonic wave/energy within the medium, which ultimately affects the kinetics and yields of the product. A prolonged exposure of the ultrasonic power probe in the fermenter can cause cell distress and reduce the growth rate, negatively impacting the bioethanol yield. The location/position of the ultrasonic probe in the reactor greatly influences the process. Optimization of the process conditions such as exposure time, ultrasonic frequency, ultrasonic power, enzyme concentration, substrate loading, and duty cycles needs to be further investigated, especially for LCB treatments with particular emphasis on process intensification for bioethanol production as well as to harness the full potential of ultrasonic technology.

Scaling up issues and energy efficiency-related problems are also major challenges facing ultrasonic technology for process intensification of bioethanol production. However, some studies show that ultrasonic technology is better than conventional pretreatment technologies in terms of energy requirement. For instance, ethanol production from cassava chips by ultrasonication process of 20 kHz requires 11 KJ, while 22 KJ is needed in the conventional process (Nitayavardhana et al., 2010). A similar study conducted on the overall economic analysis of applying the ultrasonic process and jet cooking suggested that ultrasound technology was more advantageous in energy efficiency (Montalbo-Lomboy et al., 2011). Therefore, a novel design of an ultrasonic reactor that considers all the aspects raised above is required for process intensification of pretreatment, hydrolysis, and fermentation process for bioethanol production from LCB. This helps design a green, eco-friendly, and sustainable technology as well as helps us utilize both the full potentials of ultrasonic technology and the abundance of LCB for biofuel production.

18.10 CONCLUSION AND PROSPECTS

The structural complexity of LCB hindered exploiting the vast applications of its valorized products. The production of bioethanol from LCB is highly influenced by the cost of pretreatments, hydrolysis, and separation processes. The coupling of two or more pretreatment technologies, either simultaneously or stepwise, has a great advantage in overcoming the biomass's recalcitrant nature and improving the product yields. Ultrasound-assisted pretreatment, hydrolysis, and fermentation technology show promising results for process intensifications of bioethanol production. The technology provides many advantages over conventional processes, such as easy operations and sophistication, reduced production cost, lower enzyme concentration requirements, reduced exposure time requirement, and product enhancement capabilities. Ultrasonication improves the productivity of bioethanol production by increasing the mass transfer kinetics and changing the cell morphology. Therefore, it has great potentials to reduce the cost and enhance the rate and yield of bioethanol production via the cavitation effect. However, to exploit the full potentials of the technology, it needs to address many process-related factors that need to be optimized completely. Comparative analysis with other pretreatments and scale-up aspects of ultrasonic technology are not investigated. Moreover, the introduction of biorefinery, multiple feedstocks, and multiple product scenarios are also not studied yet. Therefore, investigating the depolymerization and fractionations of LCB via ultrasound-assisted enzymatic hydrolysis, ultrasound-assisted pretreatment, and ultrasound-assisted fermentation process, as well as optimizing the processes conditions, opens the path for exploiting the full potentials of the technology as well as use of the renewable resources.

ACKNOWLEDGMENTS

The authors would like to acknowledge the funding from Technological University Dublin under the Researcher Award-2020.

DES, deep eutectic solvent; WMR, watermelon rind; OPEFB, empty fruit bunches of oil palm.

REFERENCES

Adewuyi, Y. G. and V. G. Deshmane. 2015. Intensification of enzymatic hydrolysis of cellulose using high-frequency ultrasound: An investigation of the effects of process parameters on glucose yield. *Energy & Fuels* 29:4998–5006.

Aditiya, H. B., T. M. I. Mahlia, W. T. Chong, H. Nur and A. H. Sebayang. 2016. Second generation bioethanol production: A critical review. *Renewable and Sustainable Energy Reviews* 66:631–653.

Arumugam, A., V. V. Malolan and V. Ponnusami. 2021. Contemporary pretreatment strategies for bioethanol production from corncobs: A comprehensive review. *Waste and Biomass Valorization* 12:577–612.

Bekeredjian, R., C. Bohris and A. Hansen, et al. 2007. Impact of microbubbles on shock wave-mediated dna uptake in cells *in vitro*. *Ultrasound in Medicine & Biology* 33:743–750.

Bizzi, C. A., D. Santos, G. D. Iop, E. M. M. Flores. 2021. Biomass processing assisted by ultrasound. In *Biomass Valorization: Sustainable Methods for the Production of Chemicals,* Eds. D. Ravelli and C. Samori, pp. 315–341. Hoboken, NJ: Wiley.

Bundhoo, Z. M. A. and R. Mohee. 2018. Ultrasound-assisted biological conversion of biomass and waste materials to biofuels: A review. *Ultrasonics Sonochemistry* 40:298–313.

Bussemaker, M. J. and D. Zhang. 2013. Effect of ultrasound on lignocellulosic biomass as a pretreatment for biorefinery and biofuel applications. *Industrial and Engineering Chemistry Research* 52:3563–3580.

Chen, Y. and T. Mu. 2019. Application of deep eutectic solvents in biomass pretreatment and conversion. *Green Energy and Environment* 4:95–115.

Dalagnol, L. M. G.,V. C. C. Silveira, H. B. da Silva, V. Manfroi and R. C. Rodrigues. 2017. Improvement of pectinase, xylanase and cellulase activities by ultrasound: Effects on enzymes and substrates, kinetics and thermodynamic parameters. *Process Biochemistry* 61:80–87.

De Carvalho Silvello, M. A., J. Martinez and R. Goldbeck. 2019. Increase of reducing sugars release by enzymatic hydrolysis of sugarcane bagasse intensified by ultrasonic treatment. *Biomass and Bioenergy* 122:481–489.

De Carvalho Silvello, M. A., J. Martinez and R. Goldbeck. 2020. Low-frequency ultrasound with short application time improves cellulase activity and reducing sugars release. *Applied Biochemistry and Biotechnology* 191:1042–1055.

De Carvallo Silvello, M. A., A. F. Camargo and T. Scapini, et al. 2021. Enzymatic hydrolysis intensification of lignocellulolytic enzymes through ultrasonic treatment. *BioEnergy Research* 15:875–888.

Du, R., R. Su, W. Qi and Z. He. 2018. Enhanced enzymatic hydrolysis of corncob by ultrasound-assisted soaking in aqueous ammonia pretreatment. *3 Biotech* 8:166.

Fakayode, O. A., N. D. K. Akpabli-Tsigbe and H. Wahia, et al. 2021b. Integrated bioprocess for bio-ethanol production from watermelon rind biomass: Ultrasound-assisted deep eutectic solvent pretreatment, enzymatic hydrolysis and fermentation. *Renewable Energy* 180:258–270.

Fakayode, O. A., Z. Wang and H. Wahia, et al. 2021a. Higher heating value, exergy, pyrolysis kinetics and thermodynamic analysis of ultrasound-assisted deep eutectic solvent pretreated watermelon rind biomass. *Bioresource Technology* 332:125040.

Flores, E. M. M., G. Cravotto, C. A. Bizzi, D. Santos and G. D. Iop. 2021. Ultrasound-assisted biomass valorization to industrial interesting products: State-of-the-art, perspectives and challenges. *Ultrasonics Sonochemistry* 72:105455.

Garcia, A., M. G. Alriols and J. Labidi. 2012. Evaluation of the effect of ultrasound on organosolv black liquor from olive tree pruning residues. *Bioresource Technology* 108:155–161.

Gaspar, M., G. Kalman and K. Reczey. 2007. Corn fiber as a raw material for hemicellulose and ethanol production. *Process Biochemistry* 42:1135–1139.

Goshadrou, A. 2021. A novel sequential ultrasound-rhamnolipid assisted [EMIM]OAc pretreatment for enhanced valorization of invasive Cogongrass to bioethanol. *Fuel* 290:119997.

Haldar, D. and M. K. Purkait. 2021. A review on the environment-friendly emerging techniques for pretreatment of lignocellulosic biomass: Mechanistic insight and advancements. *Chemosphere* 264:128523.

Han, S., C. Park, G. Kwon, N. Kim and J. Kim. 2020. Ionic liquid pretreatment of lignocellulosic biomass. *Journal of Forest and Environmental Science* 36:69–77.

Hassan, S. S., G. A. Williams and A. K. Jaiswal. 2018. Emerging technologies for the pretreatment of lignocellulosic biomass. *Bioresource Technology* 262:310–318.

He, Z., Z. Wang and Z. Zhao, et al. 2017. Influence of ultrasound pretreatment on wood physiochemical structure. *Ultrasonics Sonochemistry* 34:136–141.

Huang, G., S. Chen and C. Dai, et al. 2017. Effects of ultrasound on microbial growth and enzyme activity. *Ultrasonics Sonochemistry* 37:144–149.

Huang, G., S. Chen and Y. Tang, et al. 2019. Stimulation of low intensity ultrasound on fermentation of skim milk medium for yield of yoghurt peptides by *Lactobacillus paracasei*. *Ultrasonics Sonochemistry* 51:315–324.

Ibrahim, H. H., P. E. Bilsborrow and A. N. Phan. 2021. Intensification of pre-treatment and fractionation of agricultural residues. *Chemical Engineering and Processing – Process Intensification* 159:108231.

Islam, M. N., M. Zhang and B. Adhikari. 2014. The inactivation of enzymes by ultrasound – A review of potential mechanisms. *Food Reviews International* 30:1–21.

Jeon, B.H., J. A. Choi and H. C. Kim, et al. 2013. Ultrasonic disintegration of microalgal biomass and consequent improvement of bioaccessibility/bioavailability in microbial fermentation. *Biotechnology for Biofuels* 6:37.

Ji, Q., X. Yu and A. E. A. Yagoub, et al. 2021. Enhancement of lignin removal and enzymolysis of sugarcane bagasse by ultrasound-assisted ethanol synergized deep eutectic solvent pretreatment. *Renewable Energy* 172:304–316.

Jomdecha, C. and A. Prateepasen. 2006. The research of low-ultrasonic energy affects to yeast growthin fermentation process. In *12th A-PCNDT 2006 – Asia-Pacific Conference on NDT, 5th.* Auckland, New Zealand.

Karimi, M., B. Jenkins and P. Stroeve. 2014. Ultrasound irradiation in the production of ethanol from biomass. *Renewable and Sustainable Energy Reviews* 40:400–421.

Kawee-Ai, A., A. Srisuwun and N. Tantiwa, et al. 2016. Eco-friendly processing in enzymatic xylooligosaccharides production from corncob: Influence of pretreatment with sonocatalytic-synergistic Fenton reaction and its antioxidant potentials. *Ultrasonics Sonochemistry* 31:184–192.

Lee, K. M. 2021. Combination of ultrasonication and deep eutectic solvent in pretreatment of lignocellulosic biomass for enhanced enzymatic saccharification. *Cellulose* 28:1513–1526.

Lee, K. M. and K. N. Ng. 2019. Effect of ultrasonication in organosolv pretreatment for enhancement of fermentable sugars recovery from palm oil empty fruit bunches. *Progress in Energy and Environment* 11:15–23.

Li, C., K. Seki and T. Matsunaga, et al. 2004. Enzymatic hydrolysis of waste paper in an external loop airlift bubble column with continuous ultrasonic irradiation. *Journal of Chemical Engineering of Japan* 37:1041–1049.

Li, C., M. Yoshimoto and H. Ogata, et al. 2005. Effects of ultrasonic intensity and reactor scale on kinetics of enzymatic saccharification of various waste papers in continuously irradiated stirred tanks. *Ultrasonics Sonochemistry* 12:373–384.

Li, C., T. Matsunaga and K. Seki, et al. 2006. Effects of sparging gas properties and substrate size on enzymatic hydrolysis of waste paper in an ultrasonic external loop airlift. *Chemical Engineering & Technology* 29:1090–1096.

Li, M. F., S. N. Sun, F. Xu and R. C. Sun. 2012. Ultrasound-enhanced extraction of lignin from bamboo (*Neosinocalamus affinis*): Characterization of the ethanol-soluble fractions. *Ultrasonics Sonochemistry* 19:243–249.

Limayem, A. and S. C. Ricke. 2012. Lignocellulosic biomass for bioethanol production: Current perspectives, potential issues and future prospects. *Progress in Energy and Combustion Science* 38:449–467.

Loow, Y. L., T. Y. Wu and K. A. Tan, et al. 2015. Recent advances in the application of inorganic salt pretreatment for transforming lignocellulosic biomass into reducing sugars. *Journal of Agricultural and Food Chemistry* 63:8349–8363.

Lunelli, F. C., P. Sfalcin and M. Souza, et al. 2014. Ultrasound-assisted enzymatic hydrolysis of sugarcane bagasse for the production of fermentable sugars. *Biosystems Engineering* 124:24–28.

Luo, J., Z. Fang and R. L. Smith. 2014. Ultrasound-enhanced conversion of biomass to biofuels. *Progress in Energy and Combustion Science* 41:56–93.

Mahajan, P. M., K. M. Desai and S. S. Lele. 2012. Production of cell membrane-bound α- and β-glucosidase by *Lactobacillus acidophilus*. *Food and Bioprocess Technology* 5:706–718.

Majid, I., G. A. Nayik and V. Nanda. 2015. Ultrasonication and food technology: A review. *Cogent Food & Agriculture* 1:1071022.

Manasa, P., S. Paramjeet and K. Narasimhulu. 2018. Ultrasound-assisted alkaline pretreatment to intensify enzymatic saccharification of *Crotalaria juncea* using a statistical method. *Biomass Conversion and Biorefinery* 8:659–668.

Mohapatra, S., S. J. Dandapat and H. Thatoi, H. 2017. Physicochemical characterization, modelling and optimization of ultrasono-assisted acid pretreatment of two *Pennisetum* sp. using Taguchi and artificial neural networking for enhanced delignification. *Journal of Environmental Management* 187:537–549.

Montalbo-Lomboy, M., S. K. Khanal, J. H. van Leeuwen, D. R. Raman and D. Grewell. 2011. Simultaneous saccharification and fermentation and economic evaluation of ultrasonic and jet cooking pretreatment of corn slurry. *Biotechnology Progress* 27:1561–1569.

Muthuvelu, K. S., R. Rajarathinam and L. P. Kanagaraj, et al. 2019. Evaluation and characterization of novel sources of sustainable lignocellulosic residues for bioethanol production using ultrasound-assisted alkaline pre-treatment. *Waste Management* 87:368–374.

Nadar, D., K. Naicker and D. Lokhat. 2020. Ultrasonically-assisted dissolution of sugarcane bagasse during dilute acid pretreatment: Experiments and kinetic modeling. *Energies* 13:5627.

Nargotra, P., V. Sharma, S. Sharma, N. Kapoor and B. K. Bajaj. 2020. Development of consolidated bioprocess for biofuel-ethanol production from ultrasound-assisted deep eutectic solvent pretreated *Parthenium hysterophorus* biomass. *Biomass Conversion and Biorefinery* 12:5767–5782.

Naveena, B., P. Armshaw and J. T. Pembroke. 2015. Ultrasonic intensification as a tool for enhanced microbial biofuel yields. *Biotechnology for Biofuels* 8:140.

Nikolic, S., L. Mojovic, M. Rakin, D. Pejin and J. Pejin. 2010. Ultrasound-assisted production of bioethanol by simultaneous saccharification and fermentation of corn meal. *Food Chemistry* 122:216–222.

Ninomiya, K., A. Kohori and M. Tatsumi, et al. 2015. Ionic liquid/ultrasound pretreatment and *in situ* enzymatic saccharification of bagasse using biocompatible cholinium ionic liquid. *Bioresource Technology* 176:169–174.

Nitayavardhana, S., P. Shrestha and M. L. Rasmussen, et al. 2010. Ultrasound improved ethanol fermentation from cassava chips in cassava-based ethanol plants. *Bioresource Technology* 101:2741–2747.

Ofori-Boateng, C. and K. T. Lee. 2014. Ultrasonic-assisted simultaneous saccharification and fermentation of pretreated oil palm fronds for sustainable bioethanol production. *Fuel* 119:285–291.

Ojha, K. S., T. J. Mason, C. P. O'Donnell, J. P. Kerry and B. K. Tiwari. 2017. Ultrasound technology for food fermentation applications. *Ultrasonics Sonochemistry* 34:410–417.

Ong, V. Z. and T. Y. Wu. 2020. An application of ultrasonication in lignocellulosic biomass valorisation into bio-energy and bio-based products. *Renewable and Sustainable Energy Reviews* 132:109924.

Ong, V. Z., T. Y. Wu, K. K. L. Chu, W. Y. Sun and K. P. Y. Shak. 2021. A combined pretreatment with ultrasound-assisted alkaline solution and aqueous deep eutectic solvent for enhancing delignification and enzymatic hydrolysis from oil palm fronds. *Industrial Crops and Products* 160:112974.

Paramasivan, S., S. Sankar and R. S. Velavan, et al. 2021. Assessing the potential of lignocellulosic energy crops as an alternative resource for bioethanol production using ultrasound assisted dilute acid pretreatment. *Materials Today: Proceedings* 45:3279–3285.

Park, M. R. and G. T. Jeong. 2021. Production of reducing sugar in *Gracilaria verrucosa* using physio-chemical pretreatment and subsequent enzymatic hydrolysis. *Algal Research* 60:102531.

Pawar, S. V. and V. K. Rathod. 2020. Role of ultrasound in assisted fermentation technologies for process enhancements. *Preparative Biochemistry & Biotechnology* 50:627–634.

Pejin, D. J., L. V. Mojovic and J. D. Pejin, et al. 2012. Increase in bioethanol production yield from triticale by simultaneous saccharification and fermentation with application of ultrasound. *Journal of Chemical Technology and Biotechnology* 87:170–176.

Polachini, T. C., I. Hernando, A. Mulet, J. Telis-Romero and J. A. Carcel. 2021. Ultrasound-assisted acid hydrolysis of cassava (*Manihot esculenta*) bagasse: Kinetics, acoustic field and structural effects. *Ultrasonics Sonochemistry* 70:105318.

Ramadoss, G. and K. Muthukumar. 2014. Ultrasound assisted ammonia pretreatment of sugarcane bagasse for fermentable sugar production. *Biochemical Engineering Journal* 83:33–41.

Ramadoss, G. and K. Muthukumar. 2016a. Mechanistic study on ultrasound assisted pretreatment of sugarcane bagasse using metal salt with hydrogen peroxide for bioethanol production. *Ultrasonics Sonochemistry* 28:207–217.

Ramadoss, G. and K. Muthukumar. 2016b. Ultrasound assisted metal chloride treatment of sugarcane bagasse for bioethanol production. *Renewable Energy* 99:1092–1102.

Ranjan, A., S. Singh, R. S. Malani and V. S. Moholkar. 2016. Ultrasound-assisted bioalcohol synthesis: Review and analysis. *RSC Advances* 6:65541–65562.

Raskar, H. D., D. N. Avhad and V. K. Rathod. 2014. Ultrasound assisted production of daunorubicin: Process intensification approach. *Chemical Engineering and Processing: Process Intensification* 77:7–12.

Ravindran, R., S. Jaiswal, N. Abu-Ghannam and A. K. Jaiswal. 2017. Evaluation of ultrasound assisted potassium permanganate pre-treatment of spent coffee waste. *Bioresource Technology* 224:680–687.

Saratale, G. D., R. G. Saratale and S. Varjani, et al. 2020. Development of ultrasound aided chemical pretreatment methods to enrich saccharification of wheat waste biomass for polyhydroxybutyrate production and its characterization. *Industrial Crops and Products* 150:112425.

Sasmal, S., V. V. Goud and K. Mohanty. 2012. Ultrasound assisted lime pretreatment of lignocellulosic biomass toward bioethanol production. *Energy and Fuels* 26:3777–3784.

Satari, B. and A. K. Jaiswal. 2021. Green fractionation of 2G and 3G feedstocks for ethanol production: Advances, incentives and barriers. *Current Opinion in Food Science* 37:1–9.

Shaheen, M., M. Choi and W. Ang, et al. 2013. Application of low-intensity pulsed ultrasound to increase bioethanol production. *Renewable Energy* 57:462–468.

Sharma, V., P. Nargotra, and B. K. Bajaj. 2019. Ultrasound and surfactant assisted ionic liquid pretreatment of sugarcane bagasse for enhancing saccharification using enzymes from an ionic liquid tolerant *Aspergillus assiutensis* VS34. *Bioresource Technology* 285:121319.

Singh, S., M. Agarwal, A. Bhatt, A. Goyal and V. S. Moholkar. 2015a. Ultrasound enhanced enzymatic hydrolysis of *Parthenium hysterophorus*: A mechanistic investigation. *Bioresource Technology* 192:636–645.

Singh, S., M. Agarwal, S. Sarma, A. Goyal and V. S. Moholkar. 2015b. Mechanistic insight into ultrasound induced enhancement of simultaneous saccharification and fermentation of *Parthenium hysterophorus* for ethanol production. *Ultrasonics Sonochemistry* 26:249–256.

Singh, S., S. Sarma, M. Agarwal, A. Goyal and V. S. Moholkar. 2015c. Ultrasound enhanced ethanol production from *Parthenium hysterophorus*: A mechanistic investigation. *Bioresource Technology* 188:287–294.

Soria, A. C. and M. Villamiel. 2010. Effect of ultrasound on the technological properties and bioactivity of food: A review. *Trends in Food Science & Technology* 21:323–331.

Subhedar, P. B. and P. R. Gogate. 2013. Intensification of enzymatic hydrolysis of lignocellulose using ultrasound for efficient bioethanol production: A review. *Industrial and Engineering Chemistry Research* 52:11816–11828.

Subhedar, P. B. and P. R. Gogate. 2015. Ultrasound-assisted bioethanol production from waste newspaper. *Ultrasonics Sonochemistry* 27:37–45.

Subhedar, P. B., P. Ray and P. R. Gogate. 2018. Intensification of delignification and subsequent hydrolysis for the fermentable sugar production from lignocellulosic biomass using ultrasonic irradiation. *Ultrasonics Sonochemistry* 40:140–150.

Sulaiman, A. Z., A. Ajit and Y. Chisti. 2013. Ultrasound mediated enzymatic hydrolysis of cellulose and carboxymethyl cellulose. *Biotechnology Progress* 29:1448–1457.

Sul'man, E. M., M. G. Sul'man and E. A. Prutenskaya. 2011. Effect of ultrasonic pretreatment on the composition of lignocellulosic material in biotechnological processes. *Catalysis in Industry* 3:28–33.

Sun, H., X. Ge, Y. Lv and A. Wang. 2012. Application of accelerated solvent extraction in the analysis of organic contaminants, bioactive and nutritional compounds in food and feed. *Journal of Chromatography A* 1237:1–23.

Sun, J., Z. Zhang, X. Xiao and X. Jin. 2015. Production of xylooligosaccharides from corncobs using ultrasound-assisted enzymatic hydrolysis. *Food Science and Biotechnology* 24:2077–2081.

Suresh, T., N. Sivarajasekar and K. Balasubramani, et al. 2020. Process intensification and comparison of bioethanol production from food industry waste (potatoes) by ultrasonic assisted acid hydrolysis and enzymatic hydrolysis: Statistical modelling and optimization. *Biomass and Bioenergy* 142:105752.

Rehman, M. S., I. Kim, Y. Chisti and J. I. Han. 2012. Use of ultrasound in the production of bioethanol from lignocellulosic biomass. *Energy Education Science and Technology Part A: Energy Science and Research* 30:359–378.

Utekar, P. G., M. M. Kininge and P. R. Gogate. 2021. Intensification of delignification and enzymatic hydrolysis of orange peel waste using ultrasound for enhanced fermentable sugar production. *Chemical Engineering and Processing – Process Intensification* 168:108556.

Vasic, K., Z. Knez and M. Leitgeb. 2021. Bioethanol production by enzymatic hydrolysis from different lignocellulosic sources. *Molecules* 26:753.

Velmurugan, R. and K. Muthukumar. 2012. Ultrasound-assisted alkaline pretreatment of sugarcane bagasse for fermentable sugar production: Optimization through response surface methodology. *Bioresource Technology* 112:293–299.

Vohra, M., J. Manwar, R. Manmode, S. Padgilwar and S. Patil. 2014. Bioethanol production: Feedstock and current technologies. *Journal of Environmental Chemical Engineering* 2:573–584.

Wang, D., L. Yan and X. Ma, et al. 2018. Ultrasound promotes enzymatic reactions by acting on different targets: Enzymes, substrates and enzymatic reaction systems. *International Journal of Biological Macromolecules* 119:453–461.

Wang, D., Y. Huang and Q. Xu, et al. 2019. Free ammonia aids ultrasound pretreatment to enhance short-chain fatty acids production from waste activated sludge. *Bioresource Technology* 275:163–171.

Wang, H., Y. Tao and Y. Li, et al. 2021. Application of ultrasonication at different microbial growth stages during apple juice fermentation by *Lactobacillus plantarum*: Investigation on the metabolic response. *Ultrasonics Sonochemistry* 73:105486.

Wu, H., X. Dai and S. L. Zhou, et al. 2017. Ultrasound-assisted alkaline pretreatment for enhancing the enzymatic hydrolysis of rice straw by using the heat energy dissipated from ultrasonication. *Bioresource Technology* 241:70–74.

Xiong, Z. Y., Y. H. Qin and J. Y. Ma, et al. 2017. Pretreatment of rice straw by ultrasound-assisted Fenton process. *Bioresource Technology* 227:408–411.

Xu, Q. Q., M. J. Zhao and Z. Z. Yu, et al. 2017. Enhancing enzymatic hydrolysis of corn cob, corn stover and sorghum stalk by dilute aqueous ammonia combined with ultrasonic pretreatment. *Industrial Crops and Products* 109:220–226.

Yang, G. and J. Wang. 2019. Ultrasound combined with dilute acid pretreatment of grass for improvement of fermentative hydrogen production. *Bioresource Technology* 275:10–18.

Yang, X., C. Cui and A. Zheng, et al. 2020. Ultrasonic and microwave assisted organosolv pretreatment of pine wood for producing pyrolytic sugars and phenols. *Industrial Crops and Products* 157:112921.

Yusaf, T. and R. A. Al-Juboori. 2014. Alternative methods of microorganism disruption for agricultural applications. *Applied Energy* 114:909–923. https://doi.org/10.1016/j.apenergy.2013.08.085

Zabed, H., J. N. Sahu and A. Suely, et al. 2017. Bioethanol production from renewable sources: Current perspectives and technological progress. *Renewable and Sustainable Energy Reviews* 71:475–501.

Zhang, K., Z. Pei and D. Wang. 2016. Organic solvent pretreatment of lignocellulosic biomass for biofuels and biochemicals: A review. *Bioresource Technology* 199:21–33.

Zhang, T., J. Gao, L. He, Z. He and S. Yi. 2021a. Effects of combined zinc chloride-ultrasound pretreatment and thermal modification on the physicochemical properties of Moso bamboo. *Journal of the Taiwan Institute of Chemical Engineers* 126:223–230.

Zhang, W., J. Liu and Y. Wang, et al. 2021b. Effect of ultrasound on ionic liquid-hydrochloric acid pretreatment with rice straw. *Biomass Conversion and Biorefinery* 11:1749–1757.

Zhuang, X., W. Wang and Q. Yu, et al. 2016. Liquid hot water pretreatment of lignocellulosic biomass for bioethanol production accompanying with high valuable products. *Bioresource Technology* 199:68–75.

Index

FSC
www.fsc.org
MIX
Papier | Fördert gute Waldnutzung
FSC® C083411

Kolibri 360 GmbH
Ferdinand-Jühlke-Straße 7
99095 Erfurt, Deutschland
produktsicherheit@kolibri360.de